Riverine Systems

Abhijit Mukherjee
Editor

Riverine Systems

Understanding the Hydrological, Hydrosocial and Hydro-heritage Dynamics

Springer

Editor
Abhijit Mukherjee
Indian Institute of Technology Kharagpur
Kharagpur, West Bengal, India

Jointly published with Capital Publishing Company, New Delhi, India

The print edition is not for sale in SAARC countries. Customers from SAARC countries — Afghanistan, Bangladesh, Bhutan, India, the Maldives, Nepal, Pakistan and Sri Lanka — please order the book from: Capital Publishing Company, 7/28, Mahaveer Street, Ansari Road, Daryaganj, New Delhi 110 002, India.

ISBN 978-3-030-87066-9 ISBN 978-3-030-87067-6 (eBook)
https://doi.org/10.1007/978-3-030-87067-6

This Springer imprint is published by the registered company Springer Nature Switzerland AG
The registered company address is: Gewerbestrasse 11, 6330 Cham, Switzerland

Preface

The rivers are considered as the cradle of human civilization. Be it the Nile, the Indus, the Amazon, the Yellow or the Tigris and Euphrates, known history of human have always traced back to the big rivers and their flood plains. For millennia, the early dwellers of fluvial landscape didn't systematically record much knowledge on the science of the rivers, their social values and historical perspectives. However, the ancient legends and epics do provide us some footprints what a river meant to the early "civilized" lives. It is only in the last few centuries, the science of river has flourished as hydrology, eventually evolving from an observational science to fluvial engineering to an all-encompassing environmental science, integrating ecology, biodiversity, water quality and pollution and management of natural hazards e.g. floods, and droughts.

The river basins are not only important for their geographical extent or disposition but also because of their universal processes, diverse ecology, intrinsic historical heritage, different social setups and fundamental importance as a water management unit. This book explores a number of methodological frameworks that delineates the connectivity between these various facets of a river. Most of these are scientific for problem-solving by data collection, analysis, and prediction. Key data are still relatively scarce for the freshwater environment, however, there are possibilities of less quantifiable parameters such as attitudes, preferences, policies, and laws. Rivers are not just symbol of harmony but almost every region across the world has transboundary river related conflicts within two or more neighbouring states or countries. The rifts mainly occur over the access and rights to control the water resources and being largely driven by geopolitical interests. This is actually a historical trend, as wide range of water conflicts have happened over millennia since river water is often considered synonymous with the development of any region.

Further, rivers and groundwater act as linked ecosystems in various physiographic and climatic domains. During interaction between shallow groundwater and rivers, the waters undergo various processes leading to transportation, transformation, degradation and sorption including other chemical and microbial activities. The direction of flow for groundwater and river water while mixing depends on the hydraulic head. On a basin scale river water-groundwater interaction, whether the stream will act as loosing (inflow of groundwater in the aquifer as streamflow capture) or gaining (outflow of groundwater to the adjacent stream as baseflow) is dependent on the natural hydraulic conditions and physio-climatic settings of that particular river reach.

In this book, I attempt to integrate the knowledge-base that exist from various studies on rivers across Asia, and specifically South Asia, extending from the extensively and intricately studied regions of Ganges (Ganga) river of India to less documented river systems of the Middle East. Authored by leading experts, the studies compiled in this book, range from high-resolution, field-scale studies to basin-scale gross estimates, thereby attempting to bridge the gap of scale-of-observation. I have arranged the chapters following logical, thematic thrust areas, such that the readers can easily find out their subject of interest. The twenty-three chapters included in this edited compilation provides a holistic approach to study river basins by considering the utility of multi-faceted state-of-art knowledge of *hydrology*, *hydrosocial* and *hydro-heritage*, to provide a practical understanding of the inherent issues of river ecology, their conservations, chronicles and governance.

Thus, the present book include twelve chapters on *hydrology*, looking at their geography, fluvial processes, environmental attributes and planning through sustainable management of freshwater environments by planners, agriculturalists, industrialists, conservationists, and engineers. The four chapters on *hydrosocial* exposition helps to reorient human approach of studying rivers as an object, to understand water river as being integrally embedded in its social context to adjust to human agencies, power asymmetries, and fostering socially-just solutions. The choice of comprehensive river management system becomes the focal point in the hydrosocial framework that embraces scientific interventions along with socio-economic justice to provide comprehensive knowledge and wisdom. And finally, the seven chapters on *hydro-heritage* explore the link between the river flows and the cultural landscape of a region. Over the ages, the fluvial systems have not only catered the cultural development of human civilizations, but also to its deep ecosystem of community functioning and religious aspirations. The origin and destination of rivers are seen as a flow, which carries these deeper significances, and the imageries are likened to the perennial cycles of life, reincarnation and immortality of the soul. Based on recoded and anecdotal memoirs, along with satellite-based modern age techniques, the authors have tried to unearth the grand linkage between lost fluvial landscapes and its people.

Hope, the book provides the first step towards integration of ideas, history and knowledge of this invaluable resource for this immensely populous and diplomatically important area of the world, such that we would be able to effectively understand, manage and preserve the rivers for our future generations.

Kharagpur, India
May 2021

Abhijit Mukherjee

Contents

Part I
Hydrology

Chapter 1
Environmental Flows and Their Assessment

Sharad K. Jain, P. K. Mishra, and Anupma Sharma

INTRODUCTION

Healthy rivers, floodplains and wetlands provide different services such as freshwater, food, wood and fuel and support primary production, nutrient cycling and soil formations. In addition, natural eco-systems regulate climate, water, extremes such as floods and droughts, and health. They also give cultural, social, spiritual and recreational benefits. However, in recent times, upstream withdrawals, hydro-projects and land use land cover changes have significantly altered the magnitude, timing and duration of river flow regime. Water resources development has provided water–food–energy security to the society. It has almost eliminated famines and considerably reduced malnutrition. But at the same time, these developments have caused significant harm to the natural aquatic ecosystems, rivers, wetlands, floodplains and estuaries (Jain, 2020).

These modifications in river flows are done through building dams and weirs to meet water needs for agriculture, flood control, urban water supply, and navigation (Postel and Richter, 2003; Kumar et al., 2007; Jain and Kumar, 2014; Jain, 2015). These interventions result in significant flow alternation in terms of reduced flow. Rivers are the main carriers of flora, fauna, minerals, etc. In fact, people living near river banks directly depend on the river to meet their livelihood. Hundreds of religious and cultural events are organized regularly on river banks. All these benefits or ecosystems services are provided by naturally flowing rivers "free" and these benefits are for all and for all the times. But, when the flow of rivers is reduced by storage and diversion and polluted water is dumped in the rivers, many of these benefits are sacrificed.

S. K. Jain (✉)
Civil Engineering Department, Indian Institute of Technology Roorkee, Roorkee, India

P. K. Mishra · A. Sharma
National Institute of Hydrology, Roorkee, India

A. Mukherjee (ed.), *Riverine Systems*, https://doi.org/10.1007/978-3-030-87067-6_1

The concept of minimum flows in a river ecosystem evolved in 1970s. In 1992, United Nations Conference on Environment and Development (UNCED) established that to sustain the well-being of humans, it is necessary to ensure the health and integrity of the entire river ecosystem. As per Brisbane Declaration (2007), "environmental flows (EFs) are the quantity, timing, duration, frequency and quality of flows required to sustain freshwater, estuarine and near-shore ecosystems and the human livelihoods and well-being that depend on them" (Arthington, 2012). EFs have critical role in maintaining river health, economic development and poverty alleviation. EFs may be seen as a way to make a delicate equilibrium between the use of river flow to meet societal needs and economic development and delivering ecosystem services (Jain, 2015).

For sustainable management of rivers, EF requirements are necessary, thereby a portion of flows of certain magnitudes, timings, frequencies and durations are ensured in the stream. This small portion of flow sustains the biotic habitats and ecosystem processes in a river. This flow is known as "environmental flow". Environmental flow assessments (EFA) are the methods that have been formulated for determining the minimum flows in a river ecosystem. In the context of EFs, normally four terms are in use: instream flows, ecological reserve, ecological flows and environmental flows. The term instream flows are interpreted by some people to exclude floodplains which are important for lateral connectivity of riverine ecosystem with terrestrial ecosystems. Further, the term 'ecological flows' is perceived as the flow which is required to meet the ecological functions of the flora and fauna present in the water body. Some authors prefer to use the term 'flows' rather than 'flow' because the word 2 flow refers to a single value of discharge whereas 'flows' refers to a complete flow regime with temporal variations. EF requirement of a river basically depends on three important aspects: (i) the characteristics and the sensitivity of the aquatic life to river water, (ii) stages of development of the river, and (iii) social needs. Right magnitude of EFs for a river reach can be established by utilizing streamflow data, river cross-sections and habitat data of the river for the particular reach, stream morphology, sensitivity of aquatic life to flow variability, and basin's stakeholders' preferences (Brown and King, 2003; Acreman and Dunbar, 2004). In monsoonic climate such as in India, a high percentage of annual flows (about 70 to 85%) occurs during monsoon (June to October). In regions with a tropical monsoon hydrology, storage and flow diversion structures are constructed. This helps in augmentation of additional water resources, and meet the sectoral demands as per the availability. Himalayan rivers being snowfed are characterised by perennial flows and steep gradients offering abundant scope for hydropower development. A river reach is deprived of its natural flows due to storage and diversion at control structures. Certainly, there will be reaches in the river where the existing ecosystem prior to the construction of the projects is significantly affected due to the altered flows.

IMPORTANCE OF ENVIRONMENTAL FLOWS IN ECOSYSTEM

An ecosystem is a natural unit consisting of both biotic (living) and abiotic (non-living) factors. Plants, animals and all micro-organisms are attributed as biotic factors in an area living along with non-living physical (abiotic) factors of the environment. A river reach may be considered as aquatic ecosystem and its catchment as terrestrial ecosystem. These two ecosystems support distinct ecologies. There is a growing need to safeguard these ecosystems while abstracting waters for human need (Postel and Richter, 2003). Integrated water resources management (IWRM) optimizes the natural resources without compromising the ecosystems. This is done by meeting social and economic requirement in short term and protection of natural resources in longer term. We are required to optimize the water allocation to different services and sectors. Ecosystems although provide varieties of valuable services to people, they are the frequently neglected from water allocation decision-making.

In India, the livelihoods of many people especially from rural areas significantly depends directly on river ecosystem services. If there is no flow in the river even for some time all will be severely affected. River acts as a medium to transport water and energy. It also transports nutrients, sediments, and aquatic biota from upstream to downstream. A river also ensures not only the longitudinal connectivity, but also lateral connectivity to wetlands. At certain time and certain frequency, flood water spread out from the main channel and inundate the adjoining floodplains. This helps in establishing a hydraulic connection between the river and the floodplains, allowing ecosystem cross-subsidy of food, nutrients and carbon. Hydraulic connectivity between a river and its floodplains is one of the principal driving mechanisms for the interactions, productivity, and diversity of the biota in river floodplain.

It has been found that maintaining ecosystems has proven to be less costly in providing different services. Therefore, it is required to recognize the actual benefits of ecosystem services, and planning should be made for safeguarding the livelihoods for the future. A failure in doing so will jeopardize other efforts in restoring the river ecosystem.

METHODS FOR ENVIRONMENTAL FLOWS ASSESSMENT

Several methods/frameworks have been proposed for assessment of environmental flows, ranging from simple use of hydrological data to advanced procedures considering changes in river discharge with geomorphological and ecological response.

Some methods consider socio-economic aspects also. The hydrologic index methods and hydraulic rating methods are based entirely on hydrological analysis of data such as discharge, velocity, depth of flow, wetted perimeter, and wetted area at different cross-sections of the river reach. Some recent studies have combined different methods within a broader methodological framework designed to determine water allocations for ecosystem protection.

EFs assessment initiatives started in the USA in late 1940s. Australia and South Africa are few other countries where different EFA methodologies and their applications were carried out (Tharme, 2003). Initially, the main objective of EFA was to assess the minimum flow for the instream habitat available so as to match the habitat of few species especially fish (Pusey, 1998). Fish occupy many levels of trophic chain in freshwater and their survival and growth depends on availability and diversity of fluvial habitats. Fishes respond to effects of events that impact habitats. Since they display a consistent ecological response, composition and status of fish community is widely used to evaluate and ecologically validate specific EF. Of course, it is advisable to take care of the entire ecosystem while setting EF. Recent EFA methodologies increasingly take a holistic approach while assessing the EF.

In response to growing requirements, several methodologies for EFs assessments have been developed for different rivers. These can be categorized into four major methods: (i) hydrological, (ii) hydraulic rating, (iii) habitat simulation, and (iv) holistic. Hydrological methods use provide rapid estimates based on limited data; these are easy and fast in calculation. However, advanced methods are data intensive and tedious. These methods utilize data on streamflow, river morphology and river ecology. Note that these methods are rigorous and time consuming. The holistic methods integrate hydrological, hydraulic and habitat simulation models, and adopt ecosystem-based approach.

EFA methodologies have been classified in several ways by different organizations (Table 1). Jain and Acreman (2017) presented a matrix of EFA methods based on several factors (Table 2). EFA approaches are either perspective or interactive. Perspective EFAs recommend a single EF. Interactive EFAs establish relationship between streamflow and other attributes of the river ecosystem. Once a relationship is developed, the same is used to factor the social and economic implications with respect to flow scenarios. Unlike perspective approach, interactive methodologies facilitate trade-offs among various water allocation scenarios.

Tharme (2003) presented a review of the EF methods. A review of EF assessments for projects especially for India shows that hydrology-based methods were mostly used. This is because the indices used are based on streamflow characteristics, and can be easily calculated for both natural and controlled flows. Later, these were compared to determine the change in flow conditions due to operation of the project (MoEF, 2006; Jha et al., 2008; Report by AHEC, 2011; MoWR, 2012; Gopal, 2013). River ecosystem is highly sensitive to degree of stream flow alteration. This helps in assessing the acceptable alteration and associated impact due to project interventions in a river. Many studies undertaken in India have employed the hydrological methods. WWF-India (WWF, 2012, 2013) applied BBM. Different EFA methods are briefly given in Table 3.

Table 1. EFA methodologies

Organization	*Methods/ Approaches*	*Sub-category*	*Examples*
IUCN (Dyson et al., 2003)	Methods	Look-up tables	Q95 index; Tennant Method
		Desktop Analysis	Richter method; Wetted Perimeter Method
		Functional Analysis	BBM; EPAM; Benchmarking Methodology
		Habitat Modelling	PHABSIM
	Approaches		Employ Expert Team; Involve Stakeholders
	Frameworks		IFIM; DRIFT; ELOHA
IWMI (Tharme, 2003)	Hydrological Index Methods		Tennant Method, Desktop Method
	Hydraulic rating Method		Wetted Perimeter Method
	Habitat Simulation Methodologies		IFIM
	Holistic Methodologies		Holistic Approach; IFIM; DRIFT; BBM; EPAM; SPAM; Habitat Analysis Method; ELOHA

Note: BBM: Building Block Methodology; EPAM: Expert Panel Assessment Method; PHABSIM: Physical Habitat Simulation Modelling; IFIM: Instream Flow Incremental Methodology; DRIFT: Downstream Response to Imposed Flow Transformation; ELOHA: Ecological Limits of Hydrological Alteration; Scientific Panel Assessment Method.

Table 2. Matrix of environmental flow assessment methods (after Jain and Acreman, 2017)

Discipline	*Index to whole hydrograph*	*Simple to complex*	*Species to whole ecosystem*	*Spatial scale*	*Examples*
Hydrology + Hydraulics + Biology	Whole hydrograph + ecological data	Complex	Whole ecosystem	Region	ELOHA
					BBM
					DRIFT
					PHABSIM
Hydrology + Hydraulics	Whole hydrograph	Moderately complex	Keystone Species	River reach	Hydraulic geometry
Hydrology	Hydrological Indices	Simple	Proxy data	Single Site	IHA/ RVA/ FDC

CASE STUDIES

The EF in Upper Ganga Basin

Upper Ganga Basin (UGB) consists of two prominent streams, viz. Bhagirathi and Alaknanda. River Bhagirathi is one of the two important streams of Ganga river situated in the district of Uttarkashi, Uttarakhand, (India). Bhagirathi is a Himalayan

Table 3. Summary of EFA methods

S N	*Methods*	*Key points*
I.	Hydrological index method	Also known as desk-top methods; Simplest and widely used; EF is calculated as a percentage of avg. annual flow or as a percentile from the flow duration curve. Assessment is made on a seasonally or monthly; Suitable as preliminary estimates during project inception or in low or noncontroversial situations.
i.	Tennant or Montana method (Tennant (1976))	Three factors are considered viz. wetted width, flow velocity and depth suitable for fishes. In this method flows have been recommended for two six-month blocks viz. October-March and April-September for different flow categories.
ii.	Look-up Tables	Based mainly on hydrological data; Do not take into account the site specific conditions; Helpful in noncontroversial projects; Different countries have different approaches; In France, it has been recommended that the residual flows in the bypass river section must be a minimum of 1/40 of the mean flow for existing projects and 1/10 of the mean flow for new projects. However, in U.K. flow which is equalled or exceeded for 95% of the time (Q95) is taken into consideration.
II.	Desktop method	
i.	Based on hydrological data	
	• Range of Variability Approach (Richter et al., 1997)	Uses different 'indicators of hydrological alteration' (IHA); Identifies components of natural flow regime by flow magnitude, timing (monthly), and frequency of events, duration (moving avg., minima and maxima) and rate of change; Uses observed or modelled daily flow data along with 32 indices; This method does not recommend any EF values.
	• Desktop Reserve Model Hughes and Münster (2000)	In this method CV/BFI, i.e. coefficient of variation of flows divided by the base flow index is calculated; Curves are plotted for the mean annual runoff (MAR) required to provide high and low flows of the environmental flow regime.
	• Flow Duration Curve Based Approach	FDC is the plot of streamflow vs percentage time equalled or exceeded; This approach seeks to introduce some seasonality into the modified flow regime; FDC analysis does not explicitly consider inter-annual variation of discharge.
	• EMC based FDC Approach Smakhtin and Anputhas (2006)	FDC based approach computes EFR to keep river in the desired environmental management classes (EMC); FDCs are represented by flows corresponding to arbitrary fixed 17 percentage points; Six EMCs are used in this approach and EWR are suggested corresponding to these.

(continued)

Table 3. (continued)

S N	*Methods*	*Key points*
	• Sustainability Boundary Approach	Assess the allowable natural flow regime for alteration without significantly affecting the river ecosystem; This is done through stakeholders consultations; This approach restricts the hydrological alterations within range based on percentage around natural flow regime.
ii.	Based on hydrological and ecological data	Uses ecological data along with the hydrological data; addresses the flow and ecology. The nature of the river in question is taken into account. Few disadvantages are:
		Biotic indices sensitive only to flow. They are sensitive to attributes such as habitat structure, water quality, etc. Therefore, it becomes very difficult to derive these indices. (Armitage and Petts, 1992).
		Inadequate hydrological and biological data
		Time series of ecological data are dependent. This violates the assumptions of independency required in classical statistical techniques.
III.	Hydraulic rating method	Relationships are developed by relating flow regime change with response of species and their communities; uses indices of habitat health corresponding to the given discharge.
IV.	Habitat simulation	Computes EF by the use of hydraulic, hydrologic data and biological response data. This method relates discharge with available habitat situations and their suitability for the identified biota. EF is estimate by using the habitat-discharge relations or habitat time and exceedance series.
		PHABSIM, by Bovee, 1986 is a popular habitat simulation methodology, use hydraulic, habitat and discharge relationships, and analyses both the magnitude and suitability of the river habitat for the biota under consideration.
V.	Holistic approach	These are frameworks that integrate hydrological, hydraulic and habitat simulation models. An approach based on ecosystem is adopted for EFA. Various approaches adopted are:
		IFIM by Bovee et al., 1998
		DRIFT by King et al., 2003
		BBM by King and Tharme (1994)
		EPAM
		HAM by Burgess and Vanderbyl, 1996
		ELOHA by Poff et al. 2009.

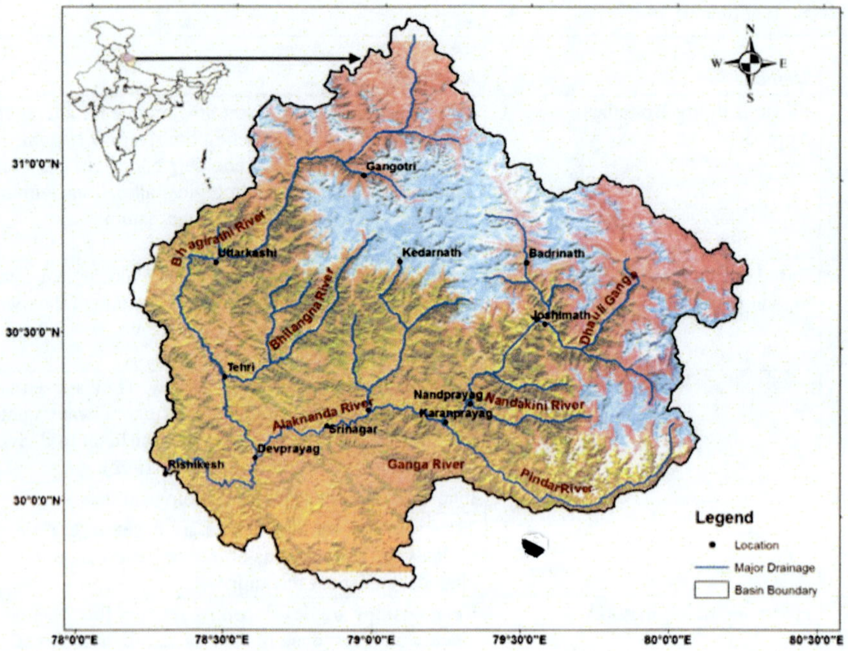

Figure 1. Upper Ganga Basin upto Rishikesh.

river arising at Gaumukh (3892 m) from the Gangotri glacier and Western face of Chaukamba peaks. Gaumukh is the source of Ganga river. Bhagirathi basin till Devprayag has a drainage area of 8847 km^2. The Alaknanda basin, with an area of 12587 km^2, is located in the eastern part of the Garhwal Himalaya. The basin rises at the confluence and foot of Satopanth and Bhagirathi Kharak glacier. River Alaknanda runs a total 224 km distance before confluence with Bhagirathi at Devprayag. The mean-annual rainfall reduces from 200 cm to 25 cm as altitude rises 1000 m to 4000 m in the Alaknanda basin. Alaknanda basin receives snowfall during winter at places above 2000 m altitude. Bhagirathi river flows for 217 km to reach Devprayag (elevation 475 m) where it merges with the Alaknanda River. From this confluence onwards, the combined river is known as Ganga. The UGB experiences considerable climatic variations. Most of the annual flow takes place during June-September, when both rainfall and snow melt are high. The index map showing the Bhagirathi, Alaknanda, and Ganga river systems of the UGB is shown in Fig. 1.

Streamflow Variability

Central Water Commission (CWC) has established gauging sites in the Upper Ganga basin to measure river flows. Location and corresponding catchment areas at six sites

Table 4. Basic statistics for different gauging sites for the discharge data

Gauge site	*River*	*Latitude*	*Longitude*	*Catchment area (km^2)*	*Mean annual flow (cumec)*
Uttarkashi	Bhagirathi	30°43'32.8"	78°26'22.4"	4555	45789
Tehri	Bhagirathi	30°19'51"	78°29'18"	7208	87189
Joshimath	Alaknanda	30°33'21.0"	79°31'55.1"	4508	65093
Rudraprayag_G5	Alaknanda	30°17'15"	78°59'04"	10675	131160
Karnaprayag	Pinder	30°15'20"	79°13'16"	2294	30639
Devprayag_Z9	Ganga	30°08'29"	78°35'49"	19600	235215

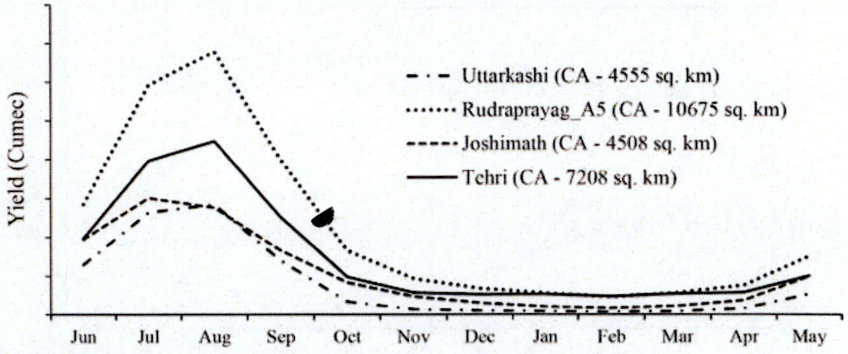

Figure 2. Variation of discharge at selected sites in Upper Ganga Basin.

are given in Table 4. Variation of discharge at selected sites in the UGB is shown in Fig. 2 (since the flow data is confidential, only the variation is shown).

Environmental Flow (EF) Assessment Using Modified Empirical Method

The depths of water in the river for obtaining environmental flow have been obtained in the stretch of Gangotri to Rishikesh. Based on information about species present in the river, it has been observed that at Joshimath G&D site, only phytoplanktons are present. Hence, a depth of 0.30 m may be recommended for e-flow estimation at such places. For the other G&D sites, on the basis of keystone species (fishes), the desirable value of river depth has been recommended as given in Table 5. At each G&D site, the river widths were taken from the remote sensing data and riverbed slopes were calculated from the DEM. Based on the required depth, width and slope, the corresponding discharge at each G&D site were calculated by the Manning's equation as follows:

$$Q = VA = \left(\frac{1}{n}\right) AR^{2/3} S^{1/2} \tag{1}$$

Table 5. Minimum flow requirements at various G&D sites to meet needs of keystone species for lean period (November-May)

	Width (m)	*Depth (m)*	*Area (m^2)*	*Velocity (m/s)*	*Discharge (cumec)*
G&D Sites	*W*	*D*	*A*	*V*	*Q*
Joshimath	12.0	0.30	3.87	0.61	2.36
Alaknanda at Rudraprayag	12.0	0.80	11.52	1.20	13.82
Uttarkashi	13.0	0.80	12.32	1.14	14.06

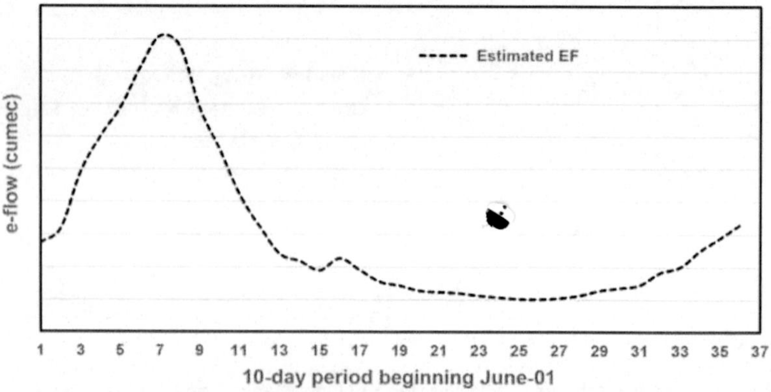

Figure 3. Estimated EF using Modified Empirical method for a representative site.

where Q = Flow rate (m^3/s), V = Velocity (m/s), A = Flow area (m^2), n = Manning's Roughness Coefficient (which was 0.035 to 0.032 in the present case), R = Hydraulic radius (m), and S = Channel bottom slope. Note that these values need to be refined by actual measurements of river cross-section, slope and Manning's roughness. Due to the presence of boulders in the Himalayan rivers, flow area is less than the cross-section area as the water flows in multiple paths between the boulders.

To develop the EF hydrograph, the discharge so computed is assumed to be the EF for 10-daily period corresponding minimum 10-daily flow for the spawning and lean-flow seasons. Flows for the other 10-daily periods were obtained by increasing this flow in the proportion of 10-daily flow for the target period in an average year and the flow of minimum 10-daily flow for the spawning and lean-flow seasons. This way, environmental flows for the Rudraprayag has been estimated. In place of the data for an average year, data for the 90% dependable year may also be used. The environmental flow hydrograph developed in this way will mimic the natural flow variation. The estimated environmental flow time series (10-daily) for a representative site is presented in Fig. 3 (due to classified nature of the data, only the variation is shown).

Gazette Notification for EF Assessment in Upper Ganga Basin

An Inter-Ministerial Group (IMG) was constituted by the Govt. of India to suggest EFR that could be prescribed for different stretches of Bhagirathi, Alaknanda and other tributaries in the Upper Ganga basin. This topic was further studied by various organizations such as National Institute of Hydrology, Central Water Commission, the Consortium of IITs, Wild Life Institute of India, etc. In Oct. 2018, a Gazette notification was published by the Govt. of India (The Gazette of India Extraordinary, Part II—Section 3—Sub-section (ii), No. 4009, dated Oct. 10, 2018, New Delhi) to notify the minimum environmental flows that are to be maintained at specified locations downstream of the projects constructed to divert river flows for various needs.

1. **Upper Ganga River Basin stretch** includes stretches from the originating glaciers, flowing through various tributaries, confluences, and finally up to Haridwar:

Season	*Months*	*Percentage of monthly average flow observed during each of preceding 10-daily period*
Dry	November to March	20
Lean	October, April and May	25
High flow season	June to September	30*

* 30% of monthly flow of high flow season.

2. **Stretch of main stem of River Ganga** includes stretches from Haridwar (Uttrakhand) to Unnao (Uttar Pradesh)

Location of Barrage	*Minimum flow releases (in cumecs) immediately downstream of barrages non-monsoon (October to May)*	*Minimum flow releases (in cumecs) immediately downstream of barrages monsoon (June to September)*
Bhimgoda (Haridwar)	36	57
Bijnor	24	48
Narora	24	48
Kanpur	24	48

The Gazette notification also says that these environmental flows are subject to certain conditions. For example, the compliance of minimum environmental flow is applicable to all existing, under-construction and future projects; and the existing projects, which currently do not meet these environmental flows norms shall comply and ensure that the desired norms are complied within a period of 3 years from the date of issue of the above order. It may be noted that for the Upper Ganga River stretch up to Haridwar, the EF are in terms of percentages monthly average flow observed during each of preceding 10-daily period. For the main stem of Ganga river from Haridwar to Unnao, the EF are in terms of absolute numbers for monsoon and non-monsoon seasons.

The EF in Yamuna Basin

The Yamuna river originates from the Yamunotri glacier near Banderpoonch peaks in the Mussoorie range of the lower Himalayas at an elevation of about 6387 m above msl in the Uttarakhand state. The river traverses through Uttarakhand and Himachal Pradesh in the upper stretch of 200 km drawing water from several major streams. Yamuna enters the plains at Dak-Pathar in Uttarakhand. The Hathnikund barrage in Yamuna Nagar district of Haryana state diverts the river water into Western Yamuna Canal (WYC) and Eastern Yamuna Canal (EYC) for irrigation. Yamuna river enters Delhi near Palla village, about 224 km downstream of Hathnikund. The river water is used for irrigation, drinking and industries, mass bathing, cattle bathing, washing clothes, etc.

Figure 4 shows the location of Yamuna basin within the Ganga basin. The study area, the river reach between Hathnikund and Okhla barrage, is illustrated in the inset map. Hathnikund barrage diverts a part of waters of Yamuna for irrigation in states of Haryana and Uttar Pradesh through the Western Yamuna Canal and the Eastern Yamuna Canal, as well as water supply to Delhi. To meet the water demand of the districts adjacent to Yamuna in the states of Haryana and Uttar Pradesh, about 10 cumecs of water is released downstream of Hathnikund Barrage. However, this amount seems to be inadequate to meet ecosystem needs in this reach.

A plot of 18 years (2001-2018) average monthly inflows and releases from Hathnikund is shown in Fig. 5. During the monsoon season about 57% of inflows, while during the non-monsoon season more than 91% of inflows at Hathnikund barrage, are diverted to WYC and EYC commands. Thus, about 43% and 9% of inflows at Hathnikund barrage are released into the river during monsoon and non-monsoon seasons, respectively. A portion of the released water evaporates or percolates into the ground as it flows downstream, on account of which the river remains dry in several stretches between Hathnikund and Delhi during the non-monsoon period.

Yamuna supports the livelihood of millions of people in its basin states of Uttarakhand, Haryana, the National Capital Territory of Delhi and Uttar Pradesh. To fulfil the growing needs of drinking water, irrigation and industry, several structures have been built which store and divert large volumes of water from the river as it runs its course through the riparian states. During the lean season, Yamuna river carries small flow in its stretch from Hathnikund to Okhla barrage, which adversely affects the quantity and quality of river water. Large volumes of treated/untreated sewage and industrial effluent are discharged into the river from various towns in Haryana and Uttar Pradesh and through several drains in Delhi segment. Encroachment and dumping of municipal and construction waste on the flood plains

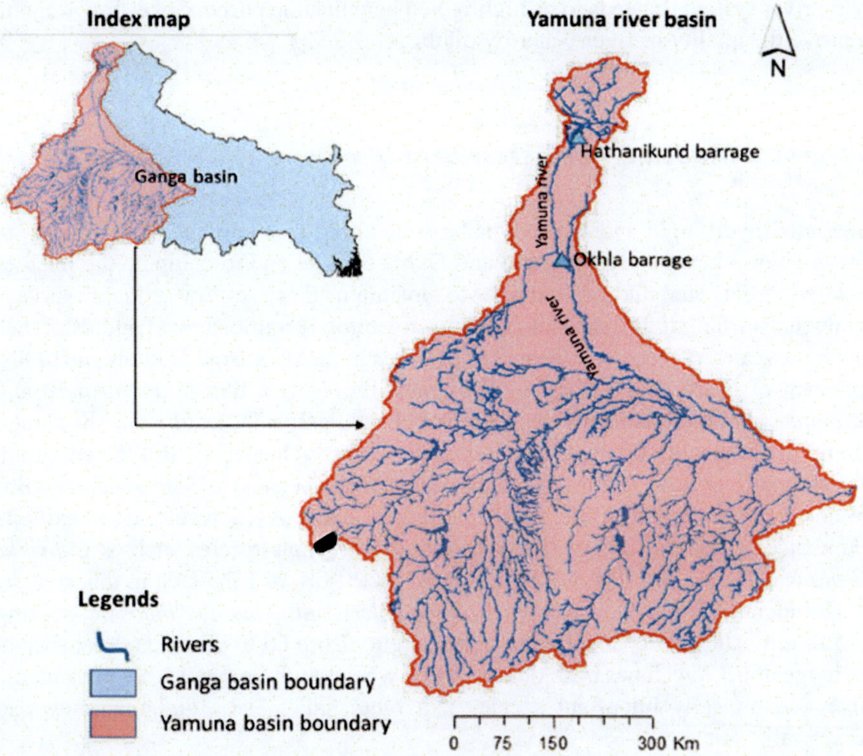

Figure 4. The river stretch between Hathnikund and Okhla Barrage.

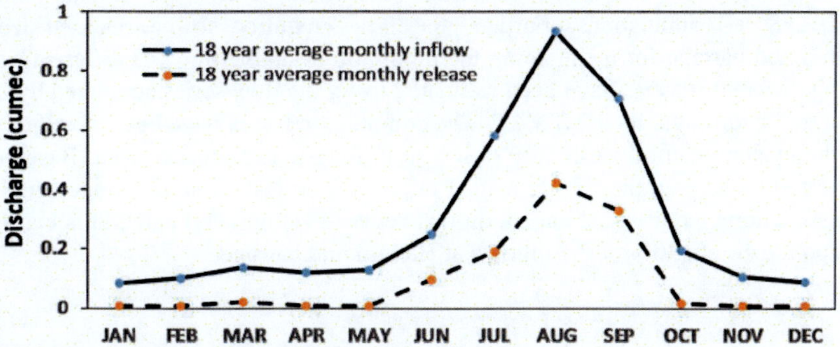

Figure 5. Plot of average monthly inflows and releases from Hathnikund.

has impeded the flow and impaired its natural ability to rejuvenate itself. This has not only affected the well-being of river, but also the health of citizens exposed to the hazards of polluted water in the river and possible seepages in groundwater. Environmental flows are essential for Yamuna to regain the characteristics of a healthy living river system. It needs to be highlighted here that the purpose of e-flow is not to improve the quality of river water by dilution.

Integrated Modeling for Assessment of e-flows

Integrated hydrologic and hydrodynamic modelling was adopted in this study to assess e-flows between Hathnikund and Okhla barrage and to compute the releases required from Hathnikund barrage to maintain these e-flows. Field surveys conducted in the study reach of Yamuna river for identification of indicator fish species revealed presence of 40 species of fishes in the river from Hathnikund to just upstream of Delhi. Good population of reophilic species like *Bangana dero* and *Raiamas bola* was found between Hathnikund and Panipat indicating that the habitat condition and quality of water is still conducive for such species. But the condition of Yamuna river downstream of Panipat deteriorates because of large-scale modifications in river bed and increasing pollutant loads. As a result, river exhibits abundance of non-native invasive and pollution tolerant species such as *Cyprinus carpio* and *Tilapia nilotica*, which constitute over 90% of fish catch in this stretch.

The identified indicator species, *Bangana dero* and *Raiamas bola* are thriving well in run habitat of channel with depth ranging from 60 to 90 cm and velocity in the range of 0.1 m/s. It has been observed that wherever these species are present, the other economically important species and other small barbs and minnows also thrive well.

For converting the habitat suitability depth values of indicator fish species (identified during the field survey) into the flow values, depth versus discharge curves have been developed covering the whole hydrologic regime using HEC-RAS 1D simulations. Further, the flows required to be released from Hathnikund barrage for maintaining the minimum desirable flows at different sites during different seasons have been estimated using the flow series simulated by the calibrated hydrologic model SWAT. The computed releases have been modified as per the existing natural variability observed in long-term historical data. Based on integrated modelling, the recommended releases from Hathnikund barrage (Fig. 6) for maintaining required habitat conditions between Hathnikund and Okhla barrage are approximately 30% of the inflows at Hathnikund barrage.

IMPLEMENTATION OF ENVIRONMENTAL FLOWS

Viewed from quality and environment perspective of water, many rivers in India are in poor condition. The science of EFA in India is evolving as the required data, knowledge base and capacity is now expanding. There is a growing concern and desire for improving the ecological conditions in large rivers, such as Ganga and Yamuna. This can be done through multi-prong strategies so as to address all the

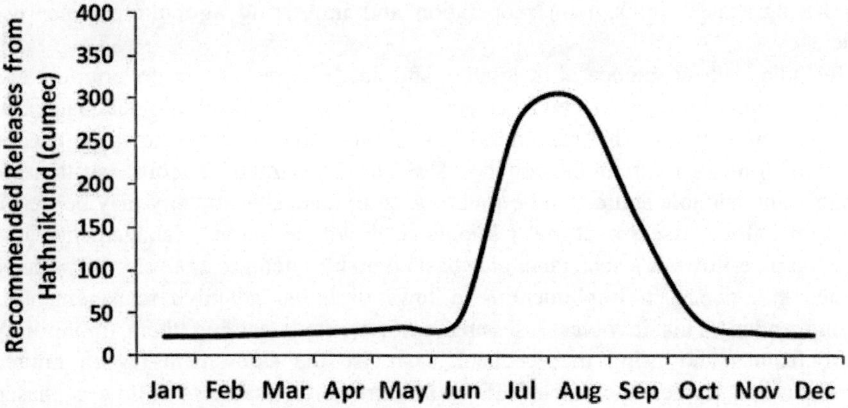

Figure 6. Recommended releases from Hathnikund barrage.

related issues. A few strategies would be to develop a nationwide programme to design and strengthen the data observation networks for hydro-biological variables, to understand habitat requirements of keystone species in different regions, to develop hydrology-ecology relations, and to assess ecosystem services. It is also important that the decision makers and political leaders are sensitized on the importance and need for developing this science. Evidently, there are many gaps in data and knowledge base which still remain to be filled and it would be necessary to initiate efforts to fill these gaps. Based on the results and experiences gained from large number of studies already attempted so far, it is needed to make concerted efforts to take the EF science in India to higher level. Already many research schemes that aim to develop EF regimes by following scientific processes have been and are being initiated. We believe that many such steps are needed for the sake of sustainable management of water resources in India. To that end, it will be helpful to carefully record what assets and ecosystem services we have in India at present and how these have changed with time. It would also be necessary to understand the reasons behind the change (water diversions, land-use, land-cover changes, pollution of water bodies, sand mining from river beds, climate change, and so on) and assess magnitude and significance of changes. A clear and better understanding of flow-environment relationship will definitely contribute to better assessment of EF. Cost and benefits and trade-offs of EFs for different sectors can be presented to all stakeholders so that the most preferred solution can be arrived at and the corresponding EF can be set.

Implementation of e-flows is challenging due to many reasons. Usually, different regulatory bodies look after different components of environment, e.g. rivers, fishery, wetlands, wildlife, forests, etc. Importance of these components changes during the course of a river. Yamuna river is an over-allocated water system and finding water for e-flows in an over-allocated system is very challenging because supplies to some existing uses will have to be reduced and this typically results in economic loss and resistance. In a few systems, water for environmental needs was found by purchasing water rights, dam reoperation and improving agriculture water use efficiency.

In India, e-flow science is in infancy and needs support from government and scientific community. We need to create and strengthen hydro-biological databases and develop indicators that respond to flow. Studies to estimate E-flows are needed to cover all major rivers in the country. The e-flow programmes require sustainable funding and suitable sources to be found. A comprehensive strategy may be developed for e-flows assessment under Integrated Water Resources Management. Final decision on e-flows is a societal choice backed up by scientific analysis. As we have limited experience in implementing e-flows in India, adaptive management is recommended. This involves assessing e-flows, implementing them, monitoring the outcomes and using the feedback to revise the assessments. With limited experience in implementation of EF in India, it is advisable to follow a phased approach; many successful programmes began with simple approaches in selected locations, and then evolved as more data were collected, expertise was gained and tools collected.

SUMMARY

Many ill impacts of river ecology can be addressed by careful planning and allocation of water. Based on the experience gained from environmental studies undertaken, following actions are recommended for future EFA research and applications:

- Collect data pertaining to hydrological, ecological and socio-economic variables and store in shareable databases,
- Create an environmental flow knowledge base by compiling case studies, research papers, and reports,
- Assess hydraulic habitat requirements for existing species (particularly keystone species) for different rivers/regions of India,
- Develop hydrology-ecology relationships for different rivers and their reaches,
- Create inventory of pollution in the rivers due to municipal, agricultural, and industrial sources and initiate sustainable programmes to reduce pollution,
- Carry out assessment of ecosystem services,
- Develop new methods or refine/customize existing EFA methods for Indian rivers,
- Implement EFs in adaptive management mode and monitor the status of rivers for periodic review and feedback,
- Promote multidisciplinary research on EFA through allocation of funds,
- Enhance interaction between scientists, public, bureaucrats and policy makers.

References

Acreman, M. and Dunbar, M.J. (2004). Defining environmental river flow requirements – A review. *Hydrology and Earth System Sciences*, 8(5): 861-876.

Armitage, P. and Petts, G.E. (1992). Biotic score and prediction to assess the effects of water abstraction on river macroinvertebrates for conservation purposes. *Aquat. Conserv.*, 2: 1-17.

Arthington, A.H. (2012). Environmental Flows: Saving Rivers in the Third Millennium. University of California Press, Berkeley. 424 pages.

Bovee, K.D. (1986). Development and evaluation of Habitat Suitability Criteria for use in Instream Flow Incremental Methodology. *U.S. Fish and Wildlife Service Biological Report*, 86(7), U.S. Fish and Wildlife Service.

Bovee, K.D., Lamb, B.L., Bartholow, J.M., Stalnaker, C.D., Taylor, J. and Henriksen, J. (1998). Stream habitat analysis using the instream flow incremental methodology. Biological Resource Division. *Information and Technical Report*, 4, U.S. Geological Survey, Fort Collins, CO, USA.

Brown, C. and King, J. 2003. Environmental flow assessment: Concepts and methods. *Water Resources and Environment*, Technical Note C.1., World Bank, Washington D.C.

Burgess, G.K. and Vanderbyl, T.L. 1996. Habitat analysis method for determining environmental flow requirements. *In:* Proceedings of Water and the Environment, the 23rd Hydrology and Water Resources Symposium. Barton, ACT, Australian Institution of Engineers, pp. 203-206.

Gopal, B. (2013). Methodologies for the assessment of environmental flows. *In:* Gopal, B. (ed.), Environmental Flows: An Introduction for Water Resources Managers. National Institute of Ecology. New Delhi, pp. 129-182.

Hughes, D.A. and Münster, F. (2000). Hydrological information and techniques to support the determination of the water quantity component of the ecological reserve. Water Research Commission Report TT 137/00, Pretoria, South Africa. 91 pp.

Jain, S.K. (2015). Assessment of environmental flow requirements for hydropower projects in India. *Current Science*, 108(10): 1815-1825.

Jain, S.K. (2020). Assessment and implementation of environmental flow. *Current Science*, 118(11): 1639-1640.

Jain, Sharad K. and Acreman, M.C. (2017). Environmental flows. Chapter 134. *In:* V.P. Singh (ed.), Handbook of Applied Hydrology, McGraw-Hill, New York.

Jain, S.K. and Kumar, P. (2014). Environmental flows in India: Towards sustainable water management. *Hydrological Sciences Journal*, 59(3-4): 751-769.

Jha, R., Sharma. K.D. and Singh, V.P. (2008). Critical appraisal of methods for the assessment of environmental flows and their application in two river systems of India. *KSCE Journal of Civil Engineering*, 12(3): 213-219.

King, J., Brown, C. and Sabet, H. (2003). A scenario-based holistic approach to environmental flow assessments for rivers. *River Research and Applications*, 19(5-6): 619-639.

King, J.M. and Tharme, R.E. 1994. Assessment of the Instream Flow Incremental Methodology (IFIM) and initial development of alternative instream flow methodologies for South Africa. Water Research Commission, Report No. 295/1/94. Pretoria, SA. 590 pp.

Kumar, P., Chaubey, U.C. and Mishra S.K. (2007). Environmental Flows for Hydropower Projects – A Case Study. *In:* Proceedings, International Conference on Small Hydropower - Hydro Sri Lanka, 22-24 October 2007.

MoEF (2006). Environmental Impact Assessment Notification – 2006, Ministry of Environment and Forests (MoEF). Government of India.

MoWR (2012). National Water Policy. Ministry of Water Resources. Government of India, New Delhi. 9 pp.

Poff et al. (2009). The ecological limits of hydrologic alteration (ELOHA): A new framework for developing regional environmental flow standards. *Freshwater Biology*, doi:https://doi.org/10.1111/j.1365-2427.2009.02204.x

Postel, S. and Richter, B. (2003). *Rivers for Life: Managing Water for People and Nature*. Washington, DC: Island.

Pusey, B.J. (1998). Methods addressing the flow requirements of fish. *In:* A.H. Arthington and J.M. Zalucki (Eds.), Comparative Evaluation of Environmental Flow Assessment Techniques: Review of Methods. pp. 66-105. LWRRDC Occasional Paper No. 27/98. LWRRDC: Canberra.

Report by AHEC (2011). Assessment of cumulative impact of hydropower projects in Alaknanda and Bhagirathi basins, Chapter 6: Water Quality, Biodiversity and River Ecology, pp. 1-62.

Report by Consortium of 7 IIT's (2013). Ganga river basin environment management plan: Interim report, pp. 1-133.

Richter, B.D., Baumgartner, J.V., Wigington, R. and Braun, D.P. (1997). How much water does a river need? *Freshwater Biology*, 37(1): 231-249.

Smakhtin, V.U. and Anputhas, M. (2006). An assessment of environmental flow requirements of Indian river basins. Colombo, Sri Lanka: International Water Management Institute. 42 p. (IWMI Research Report 107).

Tennant, D.L. (1976). Instream flow regimes for fish, wildlife, recreation and related environment resources. *Fisheries*, 1(4): 6-10.

Tharme, R.E. (2003). A global perspective on environmental flow assessment: Emerging trends in the development and application of environmental flow methodologies for rivers. *River Research and Applications*, 19: 397-442.

WWF (2012). Summary Report. Assessment of Environment Flows for Upper Ganga Basin. WWF, New Delhi.

WWF (2013). Environmental Flows for Kumbha, Triveni Sangham Allahabad. WWF, New Delhi.

Chapter 2
Story of the Ganga River: Its Pollution and Rejuvenation

Monika Simon and Himanshu Joshi

INTRODUCTION

Water is indispensable for the basic subsistence of human beings. No wonder, most of the civilisations have come upon the banks of rivers or in the river valleys as elsewhere in the world (Chaturvedi, 2019). India is a blessed country in terms of having numerous rivers in this regard (Hudda, 2011). Unfortunately, in 2017, the Ganga River, the National Legacy, and the life support of millions of people was classified as the world's highly polluted river (Mariya et al., 2019). Ganga, with over 2,525 km long main-stem along with her tributaries has constantly provided material, spiritual and cultural sustenance to millions of people living in and around its basin. The riverine water resources provide irrigation, drinking water, economical transportation, electricity, recreation and religious fulfilment, support to the aquatic ecosystem as well as livelihoods for many stakeholders. The myths and anecdotes about the river and its connection with the people and nature date back to ancient times (Kaushal et al., 2019).

India has rich water resources and roughly 45,000 km of riverine systems criss-crossing the country's length and breadth. It has 12 major river basins, 46 medium river basins and 14 minor river basins. The Ganga River Basin (22°30′ to 31°30′ N, 73°30′ to 89°00′ E) is the largest of these and approximately 43% of India's population lives in it, which extends over 860,000 km^2, making up 26% of the country's landmass; supporting 30 cities, 70 towns and thousands of villages in the areas of the fertile river basin (NRCD and MoEF, 2009; Trivedi, 2010; Nandi et al., 2016; Pathak et al., 2018; FAO, 2020). The Ganga stem covers the states of Uttarakhand, Himachal Pradesh, Haryana, Delhi, Uttar Pradesh, Bihar, Jharkhand, Rajasthan, Madhya Pradesh, Chhattisgarh and West Bengal (NRCD and MoEF,

M. Simon (✉) · H. Joshi
Department of Hydrology, Indian Institute of Technology Roorkee, Haridwar, India
e-mail: msimon@hy.iitr.ac.in

A. Mukherjee (ed.), *Riverine Systems*, https://doi.org/10.1007/978-3-030-87067-6_2

2009). It is one of the world's most agriculturally productive and populous regions. The Ganga encounters about 15 tributaries during its journey from Gomukh to the Bay of Bengal, which contribute to 60% of its total water, making it the third largest river in the world in terms of the volume of water released (Zhang et al., 2019). In 2008, Ganga River was professed as the 'National River' of India. There are more than 29 cities, 97 towns and thousands of villages along the stem of Ganga (Bhutiani et al., 2016).

Natural water resources have been exposed to an exponentially growing population, rising standards of living and accelerated industrialisation and urbanisation (Sharmila and Arockiarani, 2016). The Ganga is facing various human-induced risks across its basin. A wide range of pollutants falling in different categories like organic, inorganic, hazardous etc. emanating primarily from Municipal, Industrial and Agricultural sectors have severely affected its water quality, ecosystem and health of living beings dependent on it. Growing demands on the river water for irrigation, domestic, industrial use and hydroelectric projects exacerbate the challenges of ecosystem preservation. Due to the complexity of the issues facing Ganga, it has logically become the source of an intense discussion on benefits of river, ownership and even rights with an unclear line between use and misuse (Sanghi and Kaushal, 2014).

Considering that the river Ganga has a very special place in the collective consciousness of this country and recognising that the country's rivers were in a severe state of deterioration, an inception towards their rejuvenation was initiated with the introduction of the Ganga Action Plan-I (GAP-I) in 1985 with the support of several indigenous and international agencies (Hamner et al., 2006; IITs, 2013; Sharmila and Arockiarani, 2016). GAP-I was promoted to lessen the quantity of direct disposal of wastewater into the river, although the river water quality remained a challenging issue and was not even fit for bathing (Tare et al., 2003). GAP-II was consequently started in 1993 with the objective of managing the pollution load of the three main tributaries of the Ganga river (e.g. Yamuna, Gomti and Damodar) and the 25 towns that were not considered in GAP-I (Das and Tamminga, 2012). GAP-II had also been unable to decrease the pollution load, mainly due to a lack of strategic planning and less stakeholder engagement (Ching and Mukherjee, 2015). The National Ganga River Basin Authority (NGRBA) was established in 2009 for restoration and conservation, but even after nine years of the launch of NGRBA, the improvement in the river water quality was apparently less than desired and there still remained a paucity in the availability of systematic data on the status of various pollutants in the Ganga river. In 2015, the Integrated Ganga Conservation Mission was launched by the Union Government as a "Namami Gange" flagship scheme in order to integrate previous and current river rejuvenation initiatives with a budgetary provision, but the strategic goal could not be obtained due to long delays in execution (CAG, 2017).

Government of India has of late taken several steps and has made substantial investment for the conservation and improvement of the Ganga. Under the direction of various ministries, several actions have been undertaken viz. source control of sewage, industrial wastewater and solid waste; river catchment/basin management (controlled groundwater extraction, environmental flow management); along with other restoration measures such as riverfront development on the main stem and the tributaries, afforestation and biodiversity conservation, protection of flood plain zone and spreading community awareness.

STATUS AND IMPACT OF POLLUTION

The deterioration in the water quality of river Ganga has been of great concern for quite some time, as it has lost its fitness to be used for a variety of the designated uses in most of its stretches. Both point and non-point sources, including untreated and partially treated municipal and industrial wastewater, agricultural and urban runoff, leaching from dumping sites, open defecation in its basin, navigation, its use for recreational and religious purposes and last but not the least, severe depletion in its flow is adversely influencing its quality (NMCG, 2017; Dutta et al., 2020). The following sections discuss the nature of pollutants and various sources which are responsible for pollution.

Municipal Sewage and Open Defecation

The untreated sewage from urban and peri-urban areas is a significant cause of Ganga water quality deterioration. It contains constituents such as nutrients (N, P, K), toxic chemicals (heavy metal, pesticides), pathogens, organic and inorganic matter (FAO, 1992; Vega et al., 1998). The literature has revealed that Ganga water quality in the upper stretch is not even suitable for bathing and other nonportable purposes. The middle and lower stretch water are not even beneficial for livelihood activities (Mishra and Mohapatra, 2009; Tare, 2010; NMCG-NEERI, 2017). In the Ganga basin, around 12,000 Million Liters per day (MLD) of sewage is generated while the treatment capacity is only about 5,000 MLD (Sharmila and Arockiarani, 2016). The contribution of disposal of wastewater from municipal sewage and industries are 70-80% and 15-20%, respectively (Das, 2011; CPCB, 2013a; NMCG, 2017).

Inadequate and sub-optimal functioning of sewage treatment plants (STPs)

In India, a 50% difference between the sewage production and treatment has persisted for the last 4-5 years in spite of continuous augmentation in the capacity of treatment plants (CPCB, 2016a). It mainly occurs due to a rapid population growth, improper adoption of technology, and dearth of in-situ remediation of sewage carrying drains. In 2013, CPCB audited 51 out of 64 STPs along the main starches of the Ganga (CPCB, 2013b), and it was reported that only 60% of the installed volume of STP was operational while 30% of the STPs were non-functional. Thus, the actual treatment is although lesser, and raw sewage is flowing into the river. Such release of sewage results in an increase in the routinely analysed parameters like Biological Oxygen Demand (BOD) and coliform count, which is beyond the prescribed critical limit for outdoor bathing. A concern has also been raised that the conventional secondary treatment provided in the STPs is unable to bring the coliform count below the permissible limit, and the ensuing disinfection of the treated effluent has more issues than benefits. The water quality representing the Total and Faecal coliform (TC and FC) count, BOD and nutrients (nitrate and phosphate) has also been found poor in the downstream of Haridwar due to release of sewage (Dwivedi et al., 2018; Dutta et al., 2020) even after implementation of various federal schemes (Fig. 1). The dissolved oxygen (DO) trend has also depleted despite massive efforts and investment by the Indian government. However, clean water through the Yamuna at Allahabad somewhat improves water quality (CPCB, 2016a). Many STPs are observed to be "(a) overflowing during the wet period, (b) ineffective for the treatment of hazardous waste, (c) non-operational by recurrent electrical shortages, (d) often non-operational due to huge running cost". In the light of present treatment capacity (4994 MLD) and wastewater generation (11314 MLD) scenario in the riparian states have been represented in Fig. 2A, the difference in treatment capacity in these states is relatively huge which is about 6321 MLD (Dutta, 2019). It may also be noted that though we are at present generally concerned about the performance of our STPs in terms of reduction in the conventional water quality parameters, yet the current trends point towards a strong need to consider their efficiency in handling of the emerging pollutants also (Mathew and Kanmani, 2020).

Open defecation and unsatisfactory sanitation

Despite the continuous campaign towards control of open defecation, the reality remains that a huge portion of population still uses the flood plains as areas for defecation. Thus, resulting faeces finally reach the river water which leads to an increase in pathogenic bacterial and organic contamination during flood conditions (Chaturvedi, 2019). Prevalent open defecation considered to be a significant public health issue (Spears et al., 2013; Sahoo et al., 2015; O'Reilly et al., 2017). Approximately 524 million people defecate openly in India. Besides, unsatisfactory sanitation measures, social and religious activities also accelerate pathogenic pollution (Srinivas et al., 2020). As per the data of National Family Health Survey (NFHS, 2015), the status of sanitation facilities is unsatisfactory, particularly in the states of Uttar Pradesh, Bihar and West Bengal (Fig. 2B). Recently, 4465 villages situated alongside the Ganga river have been made open defecation free (ODF) under Swachh Bharat Mission (SBM) (SwachhindiaNDTV, 2019). Total 10,83,688 domestic toilets have been built by Ministry of Drinking Water and Sanitation (MoDWS). NMCG has funded Rs. 829 Cr to MoDWS for this work (PIB, 2018). 83 out of the 97 Ganga towns have already achieved ODF (open defecation free)

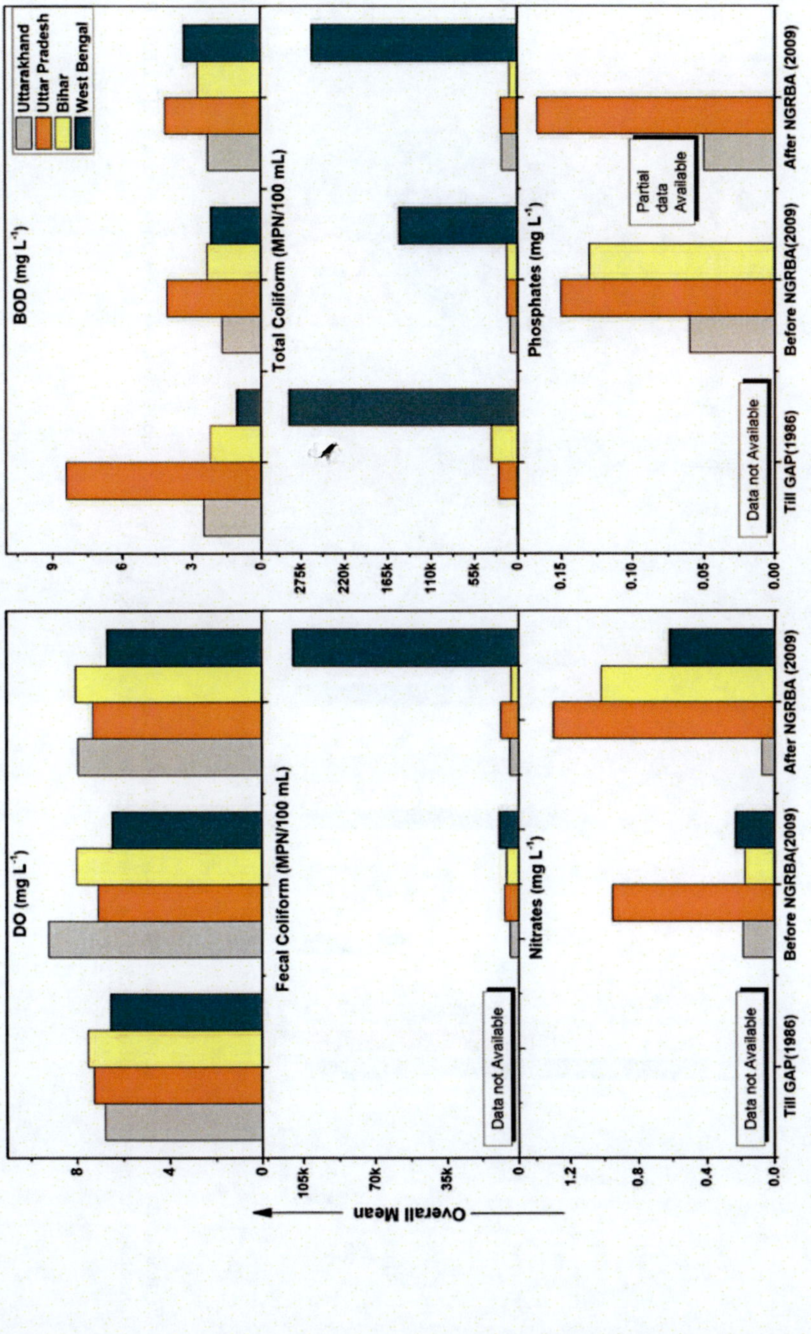

Figure 1. Water quality variation in the states of Ganga basin with the various federal schemes of Ganga (Dwivedi et al., 2018).

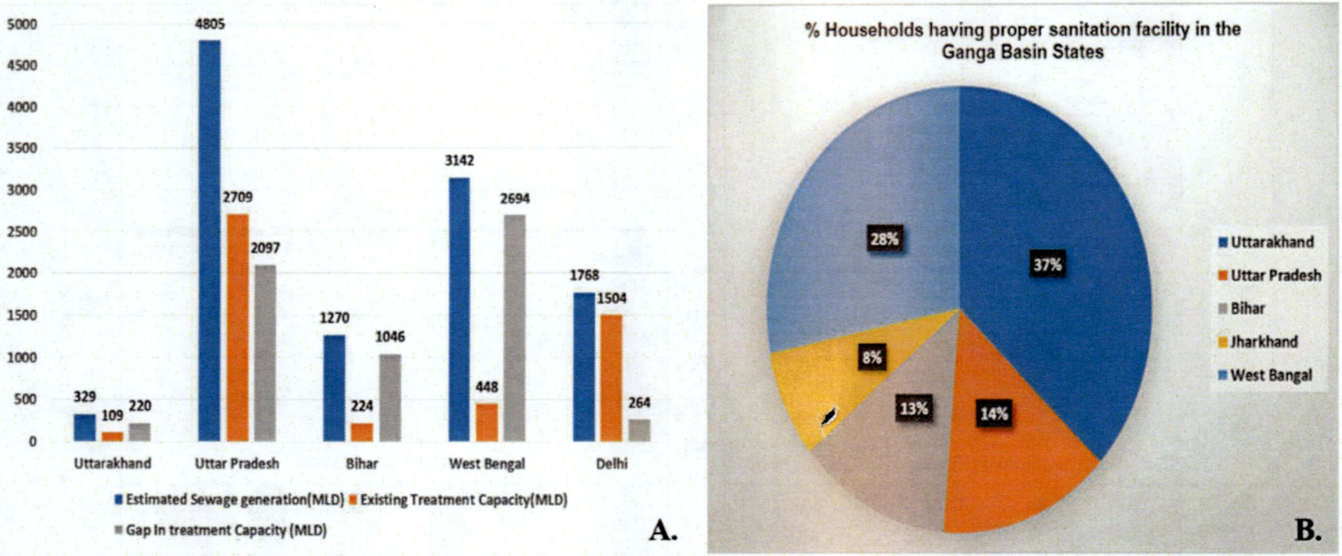

Figure 2. A. Scenario of sewage generation, existing treatment capacity and gap in the states along main stem of Ganga (Dutta et al., 2020). B. Status of households having proper sanitation facility in states under Ganga Basin (NFHS, 2015).

status, of which 44 have been certified ODF through third-party verification (SwachhindiaNDTV, 2019). Still, the type of twin-pit toilets employed and the way in which their construction has been reportedly implemented at many places raises doubts due to a strong possibility of contamination of groundwater and subsequently the river through river-aquifer interaction. The management of excreta is hence still a challenge. The total count of faecal coliform ranging from 2,500 to 2,40,000 per 100 ml depicts high pathogenic pollution along the Ganga basin of five states (Kaur, 2018). The consumption of water apparently polluted by these pathogens may cause a gastrointestinal contagion followed by restricted growth, undernourishment and diarrhoea. It is a known fact that approximately 0.2 million people die each year due to lack of proper access to safe drinking water (CWMI, 2018). Moreover, 2,000 children aged less than five die every year due to diarrhoea, as reported by WHO.

Industrial Wastewater Discharge

Discharge of industrial effluents to water bodies is one of the main cause of environmental poisoning that threatens the aquatic ecosystem and impairs the quality of water (Sinha and Paul, 2012). As per CPCB state and industry-wise waste consumption, wastewater generation and treatment have been illustrated in Fig. 3A. and B. (CPCB, 2013b). An increase in the industrialization and urbanization along the river Ganga has resulted in degraded water quality (MoEF, 2016). The effluents include the primary pollutants such as volatile, biodegradable, persistent organic compounds, hazardous metals, nutrients, suspended solids and microbial pathogens etc. (Paul, 2017). In the upper and middle stretch, industries present in the basin of Ramganga, Kali rivers and Kanpur city catchments have been reported as the main contributors to industrial pollution (MoEF, 2016; Pandey, 2019). The main industrial sectors responsible for Ganga's pollution reportedly include sugar, distillery, pulp and paper, tannery, textiles, thermal power plants, electro-processing industries and wood and jute mills, which are contributing 15-20% of the total effluent (NMCG, 2017; Dwivedi et al., 2018). These industries are responsible for the huge quantity of industrial wastewater also containing toxic compounds and heavy metals etc. (Tripathi et al., 2016; Mariya et al., 2019).

According to NMCG 1109, Grossly Polluting Industries (GPI) have been reported on the main stretches of the Ganga which contributes to about 669 MLD of industrial wastewater (NMCG, 2017). GPI from UP only are reportedly discharging over 100 kg/d of BOD load into the river Ganga (CPCB, 2016b; Chaudhary and Walker, 2019). Contribution of industrial wastewater has been reported at different times around 15-20% of the total quantity of wastewater discharged in the Ganga (Das, 2011; CPCB, 2013a; NMCG, 2017). Although their volume-wise contribution appears to be relatively low, still they are a serious cause of negative impact on river water due to their toxic and non-biodegradable nature (CPCB, 2013b; Pathak et al., 2018).

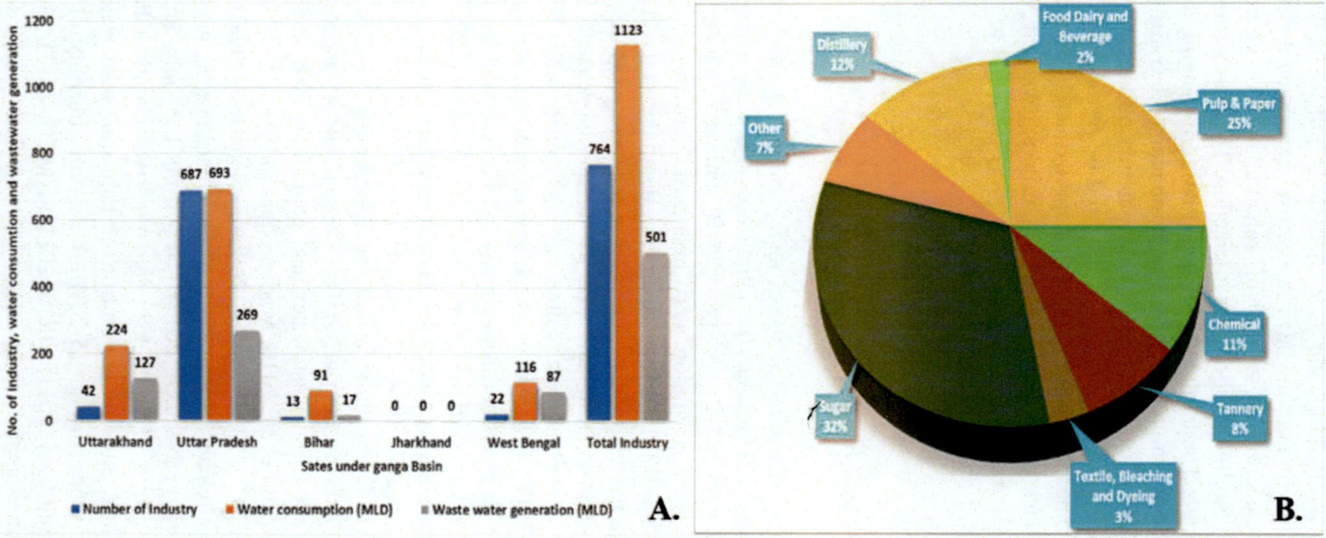

Figure 3. A. State wise status of industrial unit, water consumption and wastewater generation (CPCB, 2013b). B. Industry-wise wastewater generation in Uttar Pradesh (CPCB, 2013b).

A large volume of untreated water has been discharged into the river after the development of several industrial clusters along the bank of the Ganga in Kanpur, Prayagraj, Varanasi, and Patna (Chaturvedi, 2019). Besides, these point sources, its tributaries such as Yamuna, Kali, Ramganga, Damodar, Gomti, Ghaghra, and Son also have been responsible for further increasing the pollutant load in the river (Misra, 2010; Mishra et al., 2015). The Yamuna is Ganga's most polluted tributary. There are almost 359 industries in the Yamuna basin, that dump their wastewater directly or indirectly in the river Ganga (Misra, 2010). Approximately 400 tanneries reportedly contribute 8% of the total industrial effluent discharge along the river catchment of Unnao, causing severe chromium pollution and health hazard to the human population (CPCB, 2016a) (Homa et al., 2016). It is responsible for health problems such as cancer, diarrhoea, stomach, intestinal bleeding, cramps, liver, and kidney damage (Paul, 2017; Srinivas et al., 2020). The pulp and paper industries generate the largest amount of BOD load (65%) and account for 57% of wastewater flow and 44% of wastewater dumping in the river Ganga (CPCB, 2016b). A coal-based power plant on Pandu River burns 600 MT of coal every year generating 210 MT of fly ash, which leads to the addition of Pb and Cu into the river (Chaturvedi, 2019).

Agricultural Runoff and Soil Erosion

The agricultural runoff containing varying concentrations of pesticides, fertilisers and heavy metals is a primary non-point source that increases the pollutant load to water bodies at a large scale and their concentration found beyond the standard norms of WHO (Kumari et al., 2001; Guzzella et al., 2005; Singh et al., 2012; Mutiyar and Mittal, 2013; Srinivas et al., 2020). Extensive and improper utilization of fertilizers degrades the surface water by increasing the nitrate load (Agrawal et al., 2010). In Ganga Basin, the annual consumption of pesticides is about 60000 MT (Mohapatra et al., 1995; Ghose et al., 2009; Kumar et al., 2013). The runoff from arable land leads to an increase in the nutrient levels of nitrogen and phosphorous with the value of 70 and 0.05-1.1 mg/l, respectively (CPCB, 2016a; NMCG, 2017) (NMCG, 2017) and this further causes eutrophication (Zhang et al., 2017). The agricultural runoff resulting to eutrophication which in further causes a severe threat to riverine ecosystem and human lives. It leads to an imbalance in the river ecosystem that includes algal blooming, habitat destruction, and an decrement in self-purifying efficiency (Diaz and Rosenberg, 2008; Conley et al., 2009). In addition to the general agricultural activities in the Ganga Basin, the production of vegetables and fruits in the dry bed regions during summer also leads to pesticides in the riverine water. The maximum pesticide level has been observed in Ganga water of Uttar Pradesh before NGRBA, accompanied by West Bengal, Bihar and Uttarakhand, whereas Bihar was the highly polluted state with regard to pesticide content in Ganga water since NGRBA. A lack of data showing the pesticide contamination is reported in Bihar and West Bengal of Ganga Basin (Dwivedi et al., 2018). Organochlorine and organophosphorus pesticides are being widely utilized in India. However, some of them are still in practice even after the ban on highly hazardous pesticides such DDT, aldrin and hexachlorocyclohexane (Abhilash and Singh, 2009; Vijgen et al., 2011). These are potentially carcinogenic and mutagens and cause significant attention due to their harmful health impacts (Ejaz et al., 2004; Kumar et al., 2013). Thirteen banned organochlorine pesticides have been reported in the surface water of Ganga lower stem, causing critical ecological concerns (Sah et al., 2020). In another study, Varanasi's urban stretch has been identified as affected by high concentrations of heavy metals viz. Iron, followed by manganese, zinc, chromium, copper, nickel, lead, and cadmium, resulting from anthropogenic activities (Pandey and Singh, 2017). Non-judicious use of agrochemicals has reportedly led to massive damage of soil fertility, river ecosystem, destruction of conventional knowledge base, groundwater quality, and nutritive value of staple food (CPCB, 2016a).

Sediment load is also a crucial factor that causes pollution to Ganga Basin. It is prominent in the middle, and lower stretch, primarily covering the highly erodible alluvial plains (NRCD and MoEF, 2009; Sanghi and Kaushal, 2014) due to which floodplains and banks are more susceptible to erosion by run-off. Moreover, the massive deforestation to increase urbanisation and industrialization along the river banks has caused extreme flooded conditions (primarily in Bihar), also causing soil erosion on a large scale (Srinivas et al., 2020). Intense deforestation in the river catchment zone for the extension of agricultural lands, industries and residential areas causes river deterioration due to enhanced sediment loads. The anthropogenic activities along the Ganga have reportedly transformed the physicochemical characteristics of water and sediments and it also transformed the habitat distribution of several species (WII-GACMC, 2017).

Religious Activities and Inefficient Solid Waste Management

Religious Activities

Ganga Basin holds more than 300 million people, having an average density (persons/km^2) of 520, including the countries India, Bangladesh, and Nepal (Gopal, 2000; Das and Tamminga, 2012). The Ganga river basin is enriched with culture, heritage, and religious values from ancient times (Welcomme, 1985; Panigrahi and Pattnaik, 2019). Several ritualistic practices unfortunately contaminate the sacred Ganga water, leading to degradation in its limnological properties (Rani et al., 2014; Dwivedi et al., 2018; Chaudhary and Walker, 2019; Panigrahi and Pattnaik, 2019).

Several authors have evaluated the impacts of ritualistic activities occurred during the festive seasons of Kumbh, Maha Kumbh etc. and reported high deterioration in the Ganga water quality (Khanna et al., 2012; Srivastava et al., 2013; Tyagi et al., 2013; Sanghi and Kaushal, 2014; Dwivedi et al., 2018; Chaudhary and Walker,

2019; Panigrahi and Pattnaik, 2019). A high value of TSS, TDS, SPC, BOD, COD, and faecal counts were reported earlier during the period of Ardh Kumbh at Haridwar (Kulshrestha and Sharma, 2006); Tyagi et al., 2013; (Chauhan and Bhardwaj, 2018; Dwivedi et al., 2020).

The heavy metal/metalloid contamination, including As, Cd, Cr, Hg and Pb, is known to get released during the ritual of Idol immersion, which causes serious health problems like Cancer (WHO, 2003). The addition of gypsum, sulfur, phosphorus, and magnesium from the plaster of Paris which is used for making idols also increases the pollutant load of river. (Das et al., 2012; Kaur et al., 2013; Bhattacharya et al., 2014). A significant increase in temperature, pH, EC, BOD, COD, total alkalinity, Cl^-, Ca and Mg, and PO_4^{3-} has also been reported during the festival of Durga Pooja (Sarkar, 2013). High Pb and Cr content due to the addition of Sindhur (vermilion) in water bodies is also reported, and concentrations are found beyond the acceptable limit set by BIS, WHO, ICMR, and ISO 10500:1991 (Panigrahi and Pattnaik, 2019). Direct addition of offerings like milk adds to the organic load in the river. Further, heavy metal pollution due to these religious activities reportedly causes bioaccumulation and biomagnification issues across the food chain (Goswami et al., 2012; Dubey and Dubey, 2016).

Ineffective solid waste disposal and management

Solid waste management, treatment, and disposal in river basins are significant factors, which influence the quality of the receiving water bodies. A lack or inadequacy of proper management practices as well as processing facilities in especially the urban areas located along the Ganga Basin results in a significant increase in the pollutant loads. As per older estimates, Class I and II cities were generating around 14,000 TPD of municipal solid waste (CPCB, 2013b) along the main stretches of the River Ganga. According to the Ministry of Housing and Urban Affairs (MoH&UA), the solid waste generation in 97 towns along River Ganga is 11,428 TPD. Among these, in the states West Bengal, Uttar Pradesh, Bihar, Uttarakhand and Jharkhand, per day generation of solid waste (TPD) is 6001, 3282, 1771, and 347 respectively. Several SWM projects are being implemented to overcome the gap between solid waste generation and existing treatment capacity (UrbanUpdate, 2018). Plastic waste has also gained considerable attention as an emerging water pollutant. The Plastic Waste (management and handling) Rules 2011 was framed for combating the use of plastic bags below 40 microns (CPCB, 2016b). But this rustled in massive use of plastic bags of more than 40 microns. During anaerobic degradation of solid waste, a greenhouse gas methane (up to 50%) is produced from the solid processing sites (Alam and Ahmade, 2013).

Besides, uncollected solid waste may also inhibit stormwater runoff from reaching the receiving water bodies, which is likely to result in flash flooding. Solid waste management is an essential part of flood risk prevention. Floodwaters are also known to cause debris to flow to the river, which increases pollutant load by leaching out the hazardous compounds and also leads to groundwater pollution

(Lamond et al., 2012). Solid waste processing and disposal area also pose potential health threat to humans in neighbouring areas and groundwater pollution by leaching toxins (Srinivas et al., 2020).

Further, more than 100 MT of flowers, garlands, and earthen lamps are being reportedly thrown annually in the Ganga river. At Kashi Vishwanatha temple in Varanasi, floral waste per day is about 2 MT, whereas this limit increases to five times during the month of Shrawan (Dwivedi et al., 2018). An earlier study revealed that an average of 32,000 dead bodies is cremated (2009–2011) only at Harishchandra and Manikarnika Ghats in Varanasi and their half burnt ash is disposed into the river (Tripathi and Tripathi, 2014). Thus, all these activities continue contaminating water quality of Ganga.

Flow Obstructions and Water Abstractions

The ability of a river to self-purify and recharge itself is substantially damaged due to surface and groundwater over-extraction for irrigation, industrial and domestic utilisation. Mishandling of the Ganga water resources and inefficient utilisation has also resulted in increased levels of pollution in the river. Several hydropower projects have induced large stretches of the river to run dry and have resulted in many irretrievable environmental and social costs due to their cumulative impact (GAP, 2020). The barrages and dams have also impacted the ecological flow pattern and altered the habitat of riverine fauna, including Gangetic dolphins, wetland birds, several fish species, and freshwater turtles.

As river Ganga opens up to Gangetic plains (Bhimgoda barrage, Haridwar), an estimated volume of water (295 m^3/s) is allowed to divert into Upper Ganga Canal for irrigation purposes (Acharya et al., 2016). Additionally, during monsoon season, about 74 km from Haridwar, a barrage diverts the water into Madhya Ganga Canal at Bijnor. Further, a diversion of water into the lower Ganga Canal occurs downstream of the Bijnor barrage. About 492 major projects lead to the abstraction of water from the river for irrigation purpose. Among these, maximum projects have been implemented in Uttar Pradesh by forming 74,000 km canal systems (Shah and Rajan, 2019). It has been reported that around 30 diversion projects have been made along the main stem of Ganga and her tributaries which diverts 40-60% Ganga's yearly flow for the irrigation through the canal

This has resulted to a nominal quantity of water to flow into the river during the dry season, which is quite insufficient to provide adequate dilution to the added pollutant loads. Further, due to the excess extraction of groundwater in the river catchment, which has a significant impact on the quality of the river water, the river flow has also been seriously impacted in the recent past. If the dry period diversions from Ganga's major parts are blocked, there would be a 25% hike in base flows when the river reaches to Bihar (Khan et al., 2014).

OVERVIEW OF EFFORTS UNDERTAKEN FOR THE REJUVENATION FOR RIVER GANGA

Ganga rejuvenation has been attempted to restore the wholesomeness of the Ganga river in terms of ensuring pollution control (Nirmal Dhara) and river flow maintenance (Aviral Dhara). A holistic river basin approach has been adopted to rejuvenate the Ganga River, including its tributaries, under one framework. Pollution reduction is though the main objective, but not the only component; with the others being flow maintenance, rural sanitation, conservation of biodiversity, afforestation, environmental monitoring, ghats and crematoria development, enhancing public awareness, riverfront development in order to reinstate the magnificence of the Ganga river. For these activities, 254 projects with total cost of Rs. 24,672 cr have been approved under Namami Gange Programme (NGM) (PIB, 2018).

Source Control of Pollution

The principal source control includes treatment and disposal of domestic sewage, industrial pollution control and solid waste management as discussed below.

Channelization, treatment, utilization and disposal of treated domestic sewage

The Government of India has implemented several ambitious programmes to overcome the increasing problem of pollution in the Ganga water from GAP-I, and GAP-II to the Namami Gange project to restore the Ganga river (Fig. 4). The GAP-I with financial support of more than Rs. 462.04 cr focused to achieve the expansion of new and rehabilitation of existing STPs in urban settings, development of sewer networks in disjointed areas and installation of electric crematoria on the banks of the river (Birol and Das, 2010). GAP-I was promoted to lessen the quantity of direct disposal of wastewater into the river, although the river water quality remained a challenging issue and was not even fit for bathing (Tare et al., 2003).

GAP-II was consequently started in 1993 with the objective of managing the pollution load of the three major tributaries of the Ganga River (e.g. Yamuna, Gomti and Damodar) and the 25 towns that were not considered in GAP-II (Das and Tamminga, 2012). GAP-II had also been unable to decrease the pollution load, mainly due to a dearth of strategic planning and less stakeholder engagement (Ching and Mukherjee, 2015). The NGRBA was established in 2009 for restoration and conservation, but after nine years of the launch of NGRBA, the river water quality did not seem to improve and there is still a paucity of systematic data on the status of various pollutants in the Ganga river. Integrated Ganga Conservation Mission was launched in 2015 by the Union Government as a "Namami Gange" flagship scheme

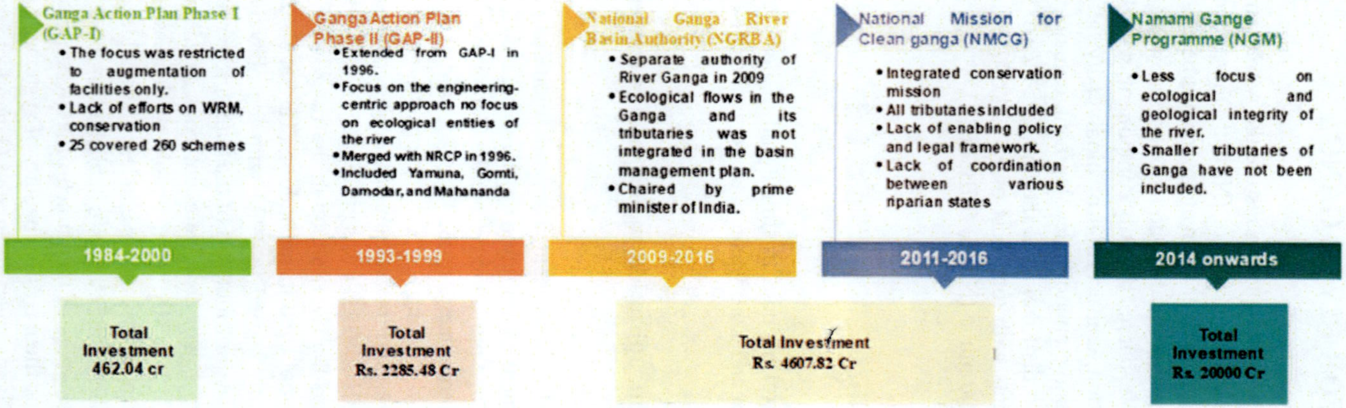

Figure 4. Various river rejuvenation schemes to clean river Ganga (Dutta et al., 2020; NMCG, 2020a).

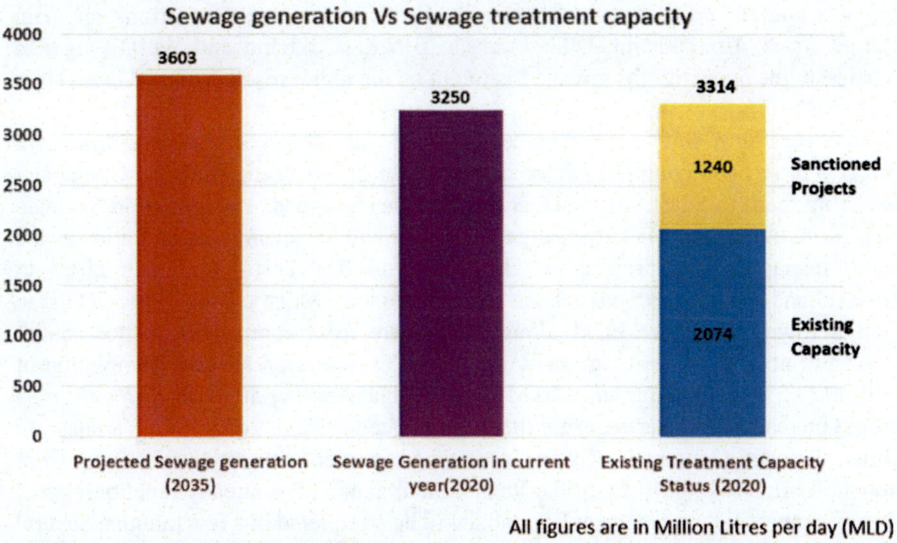

Figure 5. Status of existing infrastructure and interventions in 97 towns along Ganga main stem (Dutta et al., 2020; NMCG, 2020a).

in order to integrate previous and current river rejuvenation initiatives and their tributaries having a financial support of Rs. 20,000 Cr for the year 2015 to 2020, yet the strategic goal could not be obtained due to long delays in execution (CAG, 2017).

Nevertheless, in order to restore the glory of Ganga, the government has been constantly putting efforts. Since sewage pollution plays a prime role in the deterioration of the Ganga River, enormous investments and initiatives have been made to reduce this pollution. Under NGP, out of 254 projects 131 were approved for construction of new STPs with total capacity of 3076 MLD, refurbishment of existing STPs of total volume 887 MLD and laying/refurbishment of 4942 km sewer network for control of pollution load in river Ganga and Yamuna (TOI, 2017; UrbanUpdate, 2018). Recently, 97 towns have been identified along the major reaches of the river Ganga, contributing around 3250 MLD (current figure for the year 2020) and expected to generate 3603 MLD of sewage (projected for the year 2035). The current capacity for sewage treatment in these towns is 2074 MLD and the remaining 1176 MLD untreated sewage released into the Ganga. NMCG has approved several projects for construction of an extra 1240 MLD of STP which will enhance the capacity to 3314 MLD, and these projects are at various stages of execution (Fig. 5) (Dutta et al., 2020; NMCG, 2020a). All the sewage treatment infrastructure requirements in main towns that contribute around 64% of the current sewage generation of these 97 cities have been fully addressed (MoWRRD & GR, 2018; PIB, 2018). However, present treatment capacity (4994) and wastewater generation (11314) in the riparian states are very high and the difference in treatment volume (6321 MLD) in these states is also huge (Fig. 2A) (Dutta, 2019). The projects undertaken so far would address all the necessary interventions for main Ganga states; Uttarakhand, Uttar Pradesh, Bihar, Jharkhand and West Bengal in relation to the necessity for sewage treatment on the major reaches of the Ganga river until 2035.

In a paradigm shift in sewage sector in the past few years, several landmark projects have been completed intercepting major drains falling into the Ganga and diverting them to STPs. 80 major drains have been tapped, 120 years old Sisamau nala discharging 140 MLD in Ganga at Kanpur and kasawan Nala in Haridwar are some examples. All projects in Haridwar and Rishikesh—the main cities in Uttarakhand on Ganga, have been commissioned. Most of the other STPs in Ganga towns in Uttarakhand along Ganga are also completed. Almost entire Prayagraj now has a sewerage network and STPs. Varanasi saw the completion of 140 MLD STP at Dinapur and 120 MLD at Goitha. Another 50 MLD STP at Ramna would be ready this year to ensure that no untreated sewage flows from Varanasi. In Bihar, Namami Gange projects are increasing treatment capacity by 10 times from about 60 MLD to 650 MLD. In Jharkhand, Sahibganj STP is already functioning and the only other STP on Ganga at Rajmahal will be completed in a few months. Several projects in West Bengal too are making progress (Chakraborty, 2020).

Industrial Pollution Control

Due to accelerated Industrial development, the quantity of effluent discharged to the river has increased, resulting in different forms of deterioration, affecting the aquatic ecosystem and quality of life. There are several polluting industries, viz. pulp, distilleries, tanneries, sugar, fertilizers, textiles, automobiles, synthetic rubber and electro-processing industries along the Ganga tributaries which lead to the pollution load in the Ganga (CPCB, 2016b). Central Pollution Control Board (2019) has identified 1,072 Grossly Polluting Industries (GPIs) from Haridwar to Kanpur which discharge 219.18 MLD of effluents into river Ganga (TOI, 2017; NGM, 2020). Connection of online continuous effluent monitoring stations (OCEMS) to the CPCB server in 885 out of 1072 GPIs has been completed (NGM, 2020).

As far as industrial pollution abatement is concerned, it was noted that only 45% of the GPI units had effluent treatment plants (ETPs) and 18% did not work efficiently or comply with quality standards. These industries dispose wastewater into the Ganges River to the extent of 2667.16 MLD (NMCG, 2017; Srinivas et al., 2020). Nearly 11% (almost 275) units keep operating by violating country's pollution control norms. Approximately 50% of these non-complying GPIs are located in Uttar Pradesh and are liable for contaminating rivers such as Ganga (Pandey, 2019). Considerable improvement has been observed in few towns along Ganga stretch in Kanpur, where a significant amount of industrial wastewater is disposed into the Ganga. In 2019, under the Namami Gange project, the Sisamau drain in Kanpur, which used to release nearly 183.29 MLD of wastewater into the river, was stopped (Dutta et al., 2020). As per the latest report of NMCG, the improvement in compliance of industries was observed from 51.5% (in 2016-18) to 87.45% and construction of three common effluent treatment plant (CETP) having capacity 31.18 MLD is under process (NMCG, 2020a). The existing and future rejuvenation actions have been suggested to emphasise on source control using effective measures like wider use of CETPs, serious considerations in effective enforcement, regulation and even future socio-economic changes in long-term industrial expansion while utilising advanced eco-friendly technologies (Walker et al., 2015; Hoffman et al., 2017; Chaudhary and Walker, 2019). In view of the necessity of having an effective monitoring and evaluation framework, the NMCG plans to carry out Lidar mapping of 2525 km of the Ganga river, which will aid in precisely defining pollution and provide a more attentive method for long-term rejuvenation plans (Del Bello, 2018).

Proper enforcement, however, is necessary to restrict the known pollution point sources to control industrial pollution. In order to protect the future of the sacred Ganga, state and central government organisations are required to enforce the 'polluter pays' concept and also get harsher on violating industries. Enforcement of regulation of GPIs is conducted through periodic and surprise audits for compliance confirmation against specified pollution standards and process amendment, wherever mandated by third party technical organization. The first round of inspection of GPIs by the technical organization was conducted in 2017 and the second round of audit carried out in 2018. Out of 1072 GPIs, 961 were inspected, out of which 636 industries were observed adhering to the norms, 110 found noncomplying and 215 were shut.

Solid Waste Management

Solid Waste Management (SWM) is a significant part of Namami Gange Programme and Ganga river would not be fully clean unless the issue of SWM is addressed promptly. It is a domain of the province and the sole responsibility of the State Governments/Urban Local Bodies. This activity falls within the directive of the Ministry of Housing and Urban Affairs (MoHUA) at the Central level. The Namami Gange programme specifies that NMCG will fund STP projects whereas the other work will be carried out by ministries in accordance with their mandates. Thus, under the Swachh Bharat Mission, 97 towns along the main stem of Ganga have been identified and SWM projects have been implemented under the directives of MoHUA (UrbanUpdate, 2018). The total solid waste generation from these towns is 11428 tonnes per day (TPD). West Bengal produces 6001 TPD, the highest of the five main Ganga states, accompanied by Uttar Pradesh, where 3282 TPD solid wastes are produced every day while solid waste generation in Uttarakhand, Jharkhand and Bihar is 347 TPD, 27 TPD and 3282 TPD, respectively. The SWM programmes are being taken up on a high priority to fill the gap between the production of solid waste and the existing treatment capacity. Nearly 667 detailed project reports (DPRs) have now been sanctioned, while 413 are under review for approval in the five states of the Ganga basin. The SWM projects being carried out by MoHUA are in addition to several other initiatives undertaken by MNCG, such as river surface cleaning by trash skimmers in few Ganga towns and ghat cleaning in

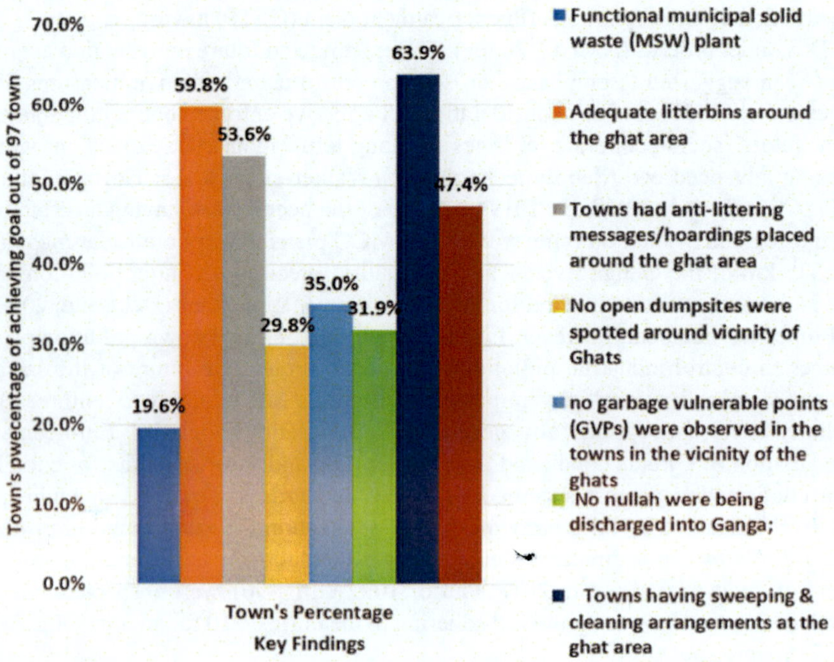

Figure 6. Results of Solid Waste Management Initiatives in the 97 main towns along the river Ganga (MoHUA, 2020).

Varanasi. Similar projects have been sanctioned by NMCG in Bithoor, Kanpur, Allahabad, Haridwar, and Mathura-Vrindavan, following the success of the ghat cleaning project in Varanasi (UrbanUpdate, 2018; NGM, 2020).

In addition, MoHUA has encouraged these states to install bar screens on drains falling into the river Ganga to avert the flowing solid waste into the river and to segregate waste at source and to ensure that process of recycling waste are carried out in productive and environmentally benign means (UrbanUpdate, 2018). Solid waste from 243 ghats is being managed under NMCG and river surface cleaning has been done through trash skimmer at 11 locations. All states under the Ganga Basin have been directed to prevent solid waste dumping in drains with special attention on 1 km stretch from the river and ensure no garbage is dumped in the flood plains (NMCG, 2020a).

As a result of effective SWM projects implementation by MoHUA under the flagship of Namami Ganga programme, a substantial improvement has been observed in 97 main towns in the states of Ganga. The implemented project has reflected the cleanliness status of the surveyed towns. The key results of this projects are (Fig. 6) that 19 (19.6%) towns had functional municipal solid waste (MSW) plant within the town; 58 (59.8%) towns had adequate litter bins around the ghat area; 52 (53.6%) towns had anti-littering messages/hoardings placed around the ghat area; no open dumpsites were spotted around the vicinity of ghats in 29 (29.8%) towns; no garbage vulnerable points (GVPs) were observed in 34 (35%) towns in the vicinity of the ghats; no nullah was being discharged into Ganga in 31 (31.9%) towns; 62 (63.9%) towns had sweeping and cleaning arrangements at the ghat area; across 46 (47.4%) towns, no solid waste was found floating on river surface (MoHUA, 2020).

River Catchment/Basin Management

River basin management is a holistic approach and it has been adopted to restore the wholesomeness of the Ganga ecosystem and revive its ecological health, taking due account of the problem of competing water use in the river catchment. It includes environmental flow Management, periodic quality assessment and controlled groundwater extraction. These measures have been discussed in the subsequent heads.

Maintenances of environmental flows (e-flow)

The maintenance of uninterrupted and continuous river water flow (Aviral Dhara) in the natural course of the river is the backbone for river ecosystems and facilitates river self-rejuvenation. Sufficient e-flow in the river is needed to sustain aquatic ecosystems, transport of sediments, recharge of groundwater through aquifers and improve the quality of river water (MoEF, 2016). The ability of a river to self-purify and recharge itself is substantially damaged due to surface and groundwater over-extraction for irrigation, industrial and domestic utilisation. Mishandling of the Ganga water resources and inefficient utilisation has also resulted in increased levels of pollution in the river. Several hydropower projects have induced large stretches of the river to run dry and have resulted in many irretrievable environmental and social costs due to their cumulative impact (GAP, 2020).

A consortium of 7 IITs (Indian Institute of Technology) advised that the minimum e-flow need to maintain river integrity are less than one-third of virgin flow in the wet period while more than half of that in the lean (dry) period flows, the river flows being minimum in winter (IITs, 2013). In 2018, the Central Government of India mandated the minimum e-flows for the river Ganga that have to be maintained at different river locations even after the river flow is diverted by projects and structures for purposes such as hydropower, irrigation domestic and industrial utilization, etc. It was a significant move to maintain the continuous/uninterrupted (Aviral) flow of the river (PIB, 2018). It has been recommended recently to maintain an average flow of 51% of the river water in the natural stream bed of the Ganga throughout the year, to adequately support the rich biodiversity of the Ganga, to dilute hazardous waste and to allow users to access their inherent water rights. Besides, other measures such as restoration of all water bodies on the banks and in the vicinity of river Ganga; new water bodies to store rainwater, interlinking of various rivers and adoption of water-saving irrigation practice have been directed by concerned central agencies (GAP, 2020). According to the Central Water Commission (2020), 4 out of the 11 hydropower plants on the upper stretches of the river Ganga river still need to comply with ecological flow standards.

Moreover, high priority has been given to the Inter-Linking of Rivers (ILR) schemes to maintain the ecological flow of the river and to have a balancing act between water surplus and deficit basins. "It has been conducted in an advisory manner by the Government of India and four priority links under Peninsular River Component have been identified for preparation of DPR namely Ken-Betwa Link project (KBLP), Damanganga-Pinjal link project, Par-Tapi-Narmada link project and Mahanadi-Godavari link project". A large number of people or farmers are expected to be benefitted from the Ken-Betwa Interlinking Project, leading to additional irrigation advantages of around 8.98 lakh hectares yearly (MoWRRD & GR, 2018).

Water quality monitoring

Numerous water quality monitoring projects have been taken up by the Ministry of Water Resources, River Development and Ganga Rejuvenation and significant results have been achieved by the implementation of some of these projects (PIB, 2018). As per the latest report of NMCG 44, Real-Time Water Quality Monitoring Stations (RTWQMS) are operational and additional 44 RTWQMS stations being set up under Namami Gange programme (besides other manual monitoring stations). Improvement in water quality trends have been observed as compared to 2017 and monitoring at 138 sites (20 km interval) (MoWRRD & GR, 2018) of Ganga depicts that DO levels has improved at 39 sites, BOD improved at 42 sites and coliform count improved at 47 sites (PIB, 2018).

Ganga river bio-monitoring at different locations (Haridwar to West Bengal) is being also conducted on regular basis to investigate benthic macroinvertebrates which shows the biological health of the river. It was noticed that Ganga water maintains a diverse community composition and river reaches exhibit low to medium levels of pollution with relation to Biological Water Quality Criteria (BWQC), which is an indicative standard accepted by CPCB. At some of the locations of the Ganga stretch, the biological water quality has improved significantly (PIB, 2018).

Controlled groundwater extraction

Groundwater (GW) recharge is impeded by the construction and development along the banks of rivers. It is also affected by the disruption of the natural flow of water. The self-revival capacity of the river flow is affected by non-scientific and severe sand mining in select stretches, including in environmentally sensitive regions. An appropriate regulatory framework is a must to ensure that river banks are developed only in an environmentally sustainable way to support the coexistence of river flow and economic activities (MoEF, 2016). India is now the biggest user of groundwater for agriculture in the world. In some parts of Ganga, improper GW development initiatives have resulted to decline in GW levels, drying up of shallow wells, decreases in viability of wells, increased impairment GW quality and decreases in base flow in many small rivers etc. (PIB, 2017).

In 2012, with Central Ground Water Board (CGWB) as the executing body, a promising National Aquifer Mapping and Management Program (NAQUIM) was initiated with an aim to delineate and characterize the aquifers to develop plans for groundwater management. The total targeted area for NAQUIM is 12.91 lakh km^2 till March 2020 against 24 lakh km^2 area identified for mapping in the country (Downtoearth, 2019; PIB, 2019). Aquifer mapping and management plans prepared by the CGWB are finally provided to concerned state agencies in order to improve the groundwater situation by implementing the management plans (MoWRRD & GR, 2018; PIB, 2018). Aquifer mapping and management work performed under NAQUIM have been used to pursue successful projects, that will reduce wet season water dependence. For example, in the Tapi Mega Recharge (Managed Aquifer Recharge) plans for Chhatarpur basin, wells (18 Nos.) were drilled in water stressed villages in Tikamgarh district of Bundelkhand region for State PHED. In Latur, Maharashtra, 25 wells were drilled for drought mitigation in the Year 2016 and handed over to the state agency for water supply. Multi-layered aquifers were identified in intensely irrigated water scares areas of Haryana (MoWRRD & GR, 2018; PIB, 2018). For the tapping of arsenic-free water 72 innovative water wells have been created in Ballia and Gazipur districts (U.P), and 62 wells in Hugli districts, West Bengal. Additionally, two mobile apps "Jal Sanchayan" and "Mera Bhujal" commenced to enable and improve citizen's understanding to take up water conservation, groundwater recharge and quality measures (MoWRRD & GR, 2018).

Recently a comprehensive guideline has been released by Government of India under the purview of Jal Shakti Ministry (CGWA) to have sustainable management of groundwater resources. It has been prepared to control groundwater extraction and conserve the scarce groundwater resources in the country. This will regulate the groundwater abstraction from agricultural and commercial (industrial use, mining projects and infrastructure development projects sectors) (CGWA, 2020).

Other Restoration Measures

In addition to the major rejuvenation approaches discussed above, several other efforts have been initiated in order to revive the Ganga river in a holistic manner. Some of these efforts are discussed as follows.

Variety of Riverfront Development Projects on Ganga and its Tributaries

The ghats and river fronts-based interventions enable a better connection between citizens and pave the way for river-centred urban planning processes. The government has been proactive in taking up these projects of riverfront development and crematoria works. In selected cities, works for 151 ghats and 54 crematoria have been completed in 2019. In West Bengal (2011 to 2014), 24 ghats were also completed (MoWRRD & GR, 2018; PIB, 2018). Under the scheme of Namami Gange flagship programme, approximately 28 projects related to riverfront development and 33 entry level projects were reported to be launched for the infrastructural set-up of 182 ghats and 118 crematoria (NGM, 2020).

Afforestation and Biodiversity Conservation (BC)

Afforestation and BC interventions have been taken up with the objective to promote community driven sustainable land and ecosystem management of the riverscape, while improving and maintaining the forest/vegetation cover in the buffer zone along the course of river Ganga and its tributaries, and protection and conservation of the representative biodiversity of the Ganga riverscape (WWF et al., 2019). Wildlife Institute of India (WII), Central Inland Fisheries Research Institute (CIFRI) and Centre for Environment Education (CEE) have been already asked to take up the afforestation projects along the Ganga. They have been executed as per the DPR prepared by Forest Research Institute (FRI) for a duration of 5 years (2016-2021) at a worth of Rs. 2300 cr. Afforestation work has already been initiated in 7 districts of Uttarakhand for medicinal plants (NGM, 2020). Additionally, numerous integrated projects for medicinal plantation are being executed in 10 districts, 180 gram panchayats, 60 clusters of Uttar Pradesh and 800 Ha corridor of medicinal plantation along Ganga by National Medicinal Plant Board (NMPB) (NMCG, 2020a). According to NMCG, Rs.114 cr. have already been invested on afforestation.

In view of the biodiversity conservation of river Ganga, several interventions have already been commenced such as biodiversity conservation and Ganga rejuvenation, fish and fishery conservation in Ganga river, Ganges river dolphin conservation education programme etc. Five bio-diversity centres at various locations, viz. Dehradun, Narora, Allahabad, Varanasi and Barrackpore has been established for the protection of identified priority species (NGM, 2020). Furthermore, six projects related to the conservation and rehabilitation of river Ganga biodiversity with respect to ghariyal, fisheries, water birds, otter, and dolphin have been carried out. Among these, two rehabilitation projects have already been completed. For afforestation schemes in the Ganga basin, a funding of Rs. 190.3 cr has been provided to the state forest departments of Uttarakhand, Uttar Pradesh, Bihar, Jharkhand and West-Bengal for the year 2018-19 (PIB, 2018).

Protection of Flood Plains

Flood plains are an integral part of overall river ecology and their protection is important for the health of the river. The domain of flood management and flood zone protection along with erosion control comes under the control of the States. The state governments engage other concerned agencies to undertake projects related to river management, flood and erosion abatement, drainage expansion, flood proofing works and refurbishment of damaged flood management projects. These schemes are planned, investigated and executed as a primary consideration within the state by state governments with their own resource base. The central government provides technical direction and promotional monetary support in essential areas to supplement the work of the states (NMCG, 2020a).

"Flood Management Programme (FMP)" and "River Management Activities & Works related to Border Areas (RMBA)" schemes have now been integrated into the single scheme "Flood Management and Border Areas Programme (FMBAP)" for the 3 year duration from 2017-18 to 2019-20. Under this scheme monetary support worth of Rs. 5435.74 cr has been provided upto the year 2018 to the various states as grant-in-aid under FMBAP (PIB, 2018).

In-situ Bioremediation

Municipal sewage and industrial effluent control initiatives have been undertaken in towns along the major streaches of Ganga under the Namami Gange programme to target the main river and the tributaries, through the construction of STPs and ETPs/CETPs. These plants have a lengthy period of completion, and during this interim period, the effluents keep flowing into the water bodies, hence these pollution loads need to be addressed employing various available technological innovations. The NMCG is in the process of evaluating and promoting in-situ bioremediation technology.

In-situ bioremediation is the process of treating polluted wastewater at the site itself using microbial/phytoremediation technologies with no major structural modification of the site. In-situ bioremediation technologies can be implemented (with display of results) in a small time (few months only), are simple to operate and consumes less energy, unlike traditional treatment technologies. Primarily Microbial Bioremediation (Bacterial Inoculants, Nualgi, Ecobio Block, RENEU etc.), Phytoremediation (Constructed Wetland System and Root Zone Treatment, Green Bridge, Artificial Island and Phytorid technology etc) have been used for the treatment of these effluent carrying drains (CPCB, 2020).

The NMCG is proactively working towards the treatment of smaller Ganga tributaries and drains. After successful implementation of the in-situ bioremediation technology at Bakarganj drain in Patna, NMCG further executed two more pilot projects one each in Patna and Allahabad. Further, it was planned to scale it up by identifying 54 drains across the four states namely Uttar Pradesh, West Bengal,

Bihar and Jharkhand where in-situ bioremediation technologies could be adopted to avert wastewater flowing directly into the Ganga (Mohan, 2017). So far, projects worth Rs.100.8 cr have been sanctioned for the in-situ bioremediation of 52 drains carrying 1369 MLD of wastewater along the main stem of Ganga, which are directly or indirectly impacting the river health. Out of these drains, 29 drains carrying 903 MLD of sewage have been identified in Uttar Pradesh, followed by 20 drains having 442 MLD of sewage in West Bengal, and 3 drains carrying 24 MLD sewage in Bihar (NMCG, 2020b). Recently, National Environmental Engineering Research Institute (NEERI) has successfully implemented an integrated in-situ bioremediation technology called RENEU (Restoration of Nallah with Ecological Units) for the treatment of 21 MLD of Sewage in drains at various locations in Prayagraj (NEERI, 2020). This technology has been found to be encouraging and is able to remove BOD up to 80-95%.

Public Awareness

Public awareness programmes play a vital role in the dissemination of information on river health and to take up the river conservation and rejuvenation initiatives at the local level. Awareness initiatives need to include stakeholders from diverse professions and levels that are from top of local representative bodies to municipal mayors, city/town planners, urban local officials and association of inhabitants. The main goal is to educate and train the community to make them more aware of their important responsibilities in restoring the deteriorated health scenario of rivers. The awareness programmes would not be so effective if carried out at a later stage of development, Thus, education concerning rivers has to therefore start at the level of primary education. Further, in any rejuvenation mission, governmental organisations, non-profit organisations, activists and society members must be welcomed.

For the awareness among various stakeholders in the Ganga Basin, several initiatives such as Ganga Manthan, Great Ganga Run, Ganga Utsav, Ganga Amantran, Ganga Quest, Ganga Cleanathon have been frequently organised by the Government of India through various stakeholders (NMCG, 2020a). These programmes have been implemented through numerous innovative awareness activities such as rallies, campaigns, exhibitions, shram daan, cleanliness drive, competitions, plantation drives and development and dispersal of resource materials. For large scale promotions, various media such as TV/radio, print media advertisements, advertorials, featured articles and advertorials are being used (NGM, 2020). In order to establish a successful pitch for public outreach and community involvement in this programme, a series of events like workshops, seminars and conferences and numerous IEC (Information, Education and Communication) activities has been regularly organised.

LESSONS FROM FEW GLOBAL RIVER REJUVENATION ATTEMPTS

Planning and management of river water at the basin level have been carried out for many decades on a global scale moving towards a more scientific and engineering-based approach during its gradual evolution over time in such a manner that the limited water resources of a river can be efficiently managed to achieve all possible requirements for the socio-economic development of a country.

In developed countries, the management of rivers and river basins has been fairly more holistic and exhaustive than in developing countries, (Newson, 1992; Shah et al., 2001). European Union, USA and Australia have been adopting well-structured advanced policies since the last few decades towards more holistic watershed management as governed by various social, environmental, political and economic factors (Newson, 1992; Heinz et al., 2007; Molle, 2009; Boon and Raven, 2012; Euler and Heldt, 2018). Appropriate policy execution at the national, local and regional level is also extremely essential for the sustainable management of the river basins which assures a fair dispersion of the gains of technological and other innovations in hydropower and water conservation etc. in the river catchment (Dourojeanni, 2001). Most of the river basin management committees across the world do have well-framed policies towards achieving the sustainable goals of basin management (Srinivas et al., 2020).

The Nile river basin has a legislative framework that includes educating the public about downstream and upstream rights, understanding the relationship between water quality and quantity, decreasing sedimentation, considering the impact of climate change on future development plans and focusing on reinstating biodiversity in the river basin (Hoag and Mohamoda, 2003; Misgan, 2013). Mississippi river basin management mainly encourages the perennial farming systems, reduction of nutrients, the management of chloride, the augmentation of fish research, the establishment of nitrate standards, the development of fertiliser management and the execution of tile drainage in agricultural watersheds (Hooper, 2012; Fang et al., 2017; MDEQ, 2020).

Thames river basin management plan focuses primarily on the innovative sewage treatment works including construction of the Thames tideway tunnel, artificial aquifer recharge, construction and management of wetlands, empowering communities through community modelling strategies and public awareness on eco-friendly agricultural practices (Environment Agency, 2015). Volga river basin authorities focus on management of flood plain, reviving ecosystem health, improving monitoring of biological parameters, advancement of wastewater treatment infrastructure and management of river channels (Schletterer et al., 2018). Amazon river basin management is designing a policy framework for integrated water and land management, conservation of biodiversity, evaluating the susceptibility of biodiversity and local communities to climate change, and improving mechanisms to recognise and address the root causes of river water pollution and community involvement in integrated water management (FAO, 2015; Souza-Filho et al., 2016). Similarly, Rio de la Plata river basin policy focuses on inland navigation, hydropower viability, exploring land-water interactions, introducing new management tools to riverine states and cross-border management (FAO, 2016; Villar et al., 2018).

CHALLENGES CONCERNING TO REJUVENATION STRATEGIES

Despite serious attempts at river rejuvenation, the level of alteration in river basins across the globe has often indicated that it is physically or financially unfeasible to restore rivers to the pre-development situation in most of the cases. Many rejuvenation projects have also been unsuccessful to address issues at the accurate spatial scale and even sometimes failed to properly understand basin level processes. It is also difficult to ensure that the rejuvenated rivers are perfectly suitable for the future world, because of the huge uncertainty about the future of river basins. Among many other aspects, uncertainties remain about climate change, land use and urbanization. Recently, a serious shortfall was reported by CAG of India in "Namami Gange" programme, when it was observed that even after years of NGRBA notification, the NMCG has no river basin management plan. The Ganga rejuvenation attempts were also queried by the National Green Tribunal, the Supreme Court of India and the state governments (SANDRP, 2018). Key challenges experienced during implementation of Ganga rejuvenation plans have been discussed further.

- Even after huge investment and efforts on implementing control measures for sewage pollution, Ganga rejuvenation has still not been able to reach the desired level due to several factors such as sewage infrastructure interventions limited in government schemes to Class I and Class II; Operation and maintenance issues and inadequate and sub-optimal functioning of STPs; Delay and poor quality DPRs; Inadequate citizen involvement and Ineffective data generation and analysis; Lack of Inter-Ministerial and Centre-State coordination and inadequately addressed non-point source pollution. It is also observed that efforts for expansion of sewage treatment infrastructure in order to achieve cleaner Ganga goals are not in line with rapid urbanization rate. Another issue with sewage treatment is the lack of sewerage networks in the many cities, they are not able to receive the designed volume of wastewater. There is also a discrepancy in sewage generation design consideration. "The estimation of sewage generation assumes that 80% of the water supplied is returned as wastewater. Some recent data compiled by CPCB shows that actual measured discharge of wastewater generation is 123% higher than the estimated discharge of wastewater (Kaur, 2018)."
- Ganga river basin have been made Open Defecation Free (ODF) now under rural sanitation mission, however, as per the FSSM (2017), household sewage management in rural areas will still pose a major challenge because most of the villages have twin-pit technology which is not recommended in low-lying areas surveyed by CSE (Kaur, 2018). Therefore, while the suppression of human waste

will be largely achieved, its safe disposal still remains an enormous challenging issue.

- Numerous comprehensive plans with fund allocation have been prepared in the past to regulate industrial pollution, so far it is found that water quality improvement due to Ganga rejuvenation initiatives are still transient as there is no major scheme to regulate the GPIs. There is an inefficient control of industrial wastewater discharge, even after the Namani Gange Programme, which shows week execution of pollution regulations and enforcement by central/state agencies (WWF et al., 2019). In addition, enforcement framework of small-scale industries is still weaker as compared to large and medium scale industries.
- The riverfront projects, when taken up in large stretches specially, though environmentally aesthetic, but are viewed by ecohydrologists as undesirable initiatives that are truly not able to reinstate the river to its original state in the manner executed. River engineers need to understand where to concretise the river banks and river fronts so that ecological well being can also be protected. Rivers naturally create their own water fronts and planners should invest in ecological health not creating ecosystems that are liable for their destruction (WWF et al., 2019). It is needed to sincerely integrate future ecological and sustainability goals, and cannot simply be a transient rightist step (Dutta, 2019).
- Maintenance of ecological flow is another critical issue, as the Ganga, except during monsoons, is unable to maintain the minimum required flow and, thus, it is not only a concern about the polluted Ganga but also about the existence of Ganga. It was also noticed that the water velocity tends to decrease and siltation increases due to reduction in the flow, and settling down of sediments at the river bottom leads to decrease in the flow cross-section area (WWF et al., 2019). In addition, dams are the primary cause of flow impediment in the upstream inducing large stretches of the river to run dry and result in many irretrievable environmental impacts; whereas on the other side, barrages divert huge amount of water for downstream irrigation and other purposes (Dutta et al., 2020; Trivedi, 2020).
- Mishandling and inefficient utilisation of the Ganga Basin water resources as a whole have also resulted to increased levels of pollution in the river. Appropriate policies for use of groundwater and treated/recycled effluents in agriculture are still not in place or lacking proper implementation.
- Seamless coordination within the existing institutional set-up for Ganga rejuvenation comprising various government organizations is still a major constraint in an efficient policy formulation and implementation. For instance, the Water Resources Ministry has MOUs with 10 other ministries for improved execution (Kaur, 2018). This is apparently leading to redundancy and ambiguity of tasks at both the federal and state levels and discourages centralised assessment of this limited resource. It does lead to several governance glitches such as lack of coordination, co-operation and accountability etc. among the various concerned agencies which ultimately affects the goal of Ganga river rejuvenation.

INTERVENTIONS RECOMMENDED

In order to strengthen the rejuvenation programme following interventions are recommended in order to restore pristine nature of the river Ganga.

- Regular assessment of all the implemented programmes should be done and programmes need to be decentralised as much as possible. In order to achieve the goals of the Ganga rejuvenation, all the stakeholders residing in the basin must be encouraged to participate. Unless it has a bottom-up approach, the programme cannot receive proper feedback at different stages from all relevant quarters, which is essential for its success.
- There is a need of good governance and seamless coordination among the various engaged agencies and the policy framework should be stringent while at the same time, sufficiently flexible to introduce changes with time. Pollution control schemes need to be more transparent, accountable and inclusive at every stage.
- In the light of non-compliance of wastewater treatment and discharge guidelines, more stringent regulation and enforcement are needed. For existing ETPs and STPs, there should be a requirement for strict third-party compliance audit to make the system more transparent and efficient. In order to protect the future of the sacred Ganga, state and central government organisations require to enforce the 'polluter pays' concept and also get harsher on violating industries. Additionally, enforcement framework for small-scale industries is also needed to be reviewed especially with respect to their connectivity to the Common Effluent Treatment Plants (CETPs) and the genuine problems faced by them. The effluent discharge standards need to be based on the concept of "Allowable Loads" rather than the independent parameter values as at present. Besides, for the design and monitoring of STPs, emerging contaminants needs to be duly considered as this has become a rising concern now.
- It seems that environmental flow standards have primarily been formulated with implicit reference to hydrological approaches primarily, and other crucial associated factors have been overlooked, like transported sediments, instream biota (ecosystem features), pollutants and other demands of water users. Thus, there is need to revisit the e-flow standards in a holistic manner.
- Agriculture sector should receive a lot of attention, considering its role in contributing non-point pollution on the one hand, as well as presenting an ever-increasing demand for surface and ground water. Rational employment of conjunctive use guidelines must be enforced, along with strictly adopting sustainable irrigation practises for achieving proper conservation of water. It should be ensured that the implementation of projects approved by NMCG is completed within stipulated time, while the monitoring and evaluation frameworks need to be more stringent.
- Although ODF problem has been completely eradicated but the problem of ensuing pollution has not been completely addressed because the field scale construction of toilets with lack of concern for safety protocols is raising concerns

regarding severe risk to the groundwater pollution and its adverse influence on the river quality also. These schemes need to be critically evaluated.

- Rejuvenation programmes should focus on the strengthening of capacity building by designing and providing novel learning and training tools for the scientific and efficient implementation of schemes. Moreover, collaborative frameworks should be developed for various sectors and at different stages between educational institutions, employer institutions (public, private, non-governmental organisations) and corporate sector, and appropriate incentives should be provided to encourage and retain staff.
- Regular and comprehensive monitoring of the river system should be ensured with further inclusion of the emerging pollutants, and all related information should be available on a common platform along with the information of all the ongoing or completed Ganga rejuvenation initiatives. The information should be regularly updated and arrangements should be made for a hassle-free access by stakeholders, taxpayers and research community.

References

Abhilash, P.C., and Singh, N. (2009). Pesticide use and application: An Indian scenario. *J. Hazard Mater,* 165: 1-12. https://doi.org/10.1016/j.jhazmat.2008.10.061

Acharya, S., Pandey, A., Mishra, S.K. and Chaube, U.C. (2016) GIS based graphical user interface for irrigation Management. Water Sci Technol Water Supply, 16: 1536-1551. https://doi.org/10.2166/ws.2016.081

Agrawal, A., Pandey, R.S., Sharma, B. (2010). Water Pollution with Special Reference to Pesticide Contamination in India. *J Water Resour Prot,* 02: 432-448. https://doi.org/10.4236/jwarp.2010.25050

Alam, P., and Ahmade, K. (2013). Impact of Solid Waste on Health and the Environment. *Int J Sustain Dev . . .,* 2: 165-168.

Bhattacharya, S., Bera, A., Dutta, A. and Ghosh, U.C. (2014). Effects of Idol Immersion on the Water Quality Parameters of Indian Water Bodies: Environmental Health Perspectives. *Int Lett Chem Phys Astron,* 39: 234-263. https://doi.org/10.18052/www.scipress.com/ilcpa.39.234.

Bhutiani, R., Khanna, D.R., Kulkarni, D.B. and Ruhela, M. (2016). Assessment of Ganga river ecosystem at Haridwar, Uttarakhand, India with reference to water quality indices. *Appl Water Sci,* 6: 107-113. https://doi.org/10.1007/s13201-014-0206-6.

Birol, E. and Das, S. (2010). Estimating the value of improved wastewater treatment: The case of River Ganga, India. *J Environ Manage.,* 91: 2163-2171. https://doi.org/10.1016/j.jenvman.2010.05.008

Boon, P.J. and Raven, P.J. (2012). River Conservation and Management. River Conserv Manag, 1-412. https://doi.org/10.1002/9781119961819.

CAG (2017). Rejuvenation of River Ganga (Namami Gange).

CGWA (2020). Notification-Regulation and control of Ground Water management and development.

Chakraborty, P. (2020). Ganga Rejuvenation Enhancing Urban Renewal Conditions. BW Businessworld 1-3.

Chaturvedi, A.K. (2019). River Water Pollution—A New Threat to India: A Case Study of River Ganga.

Chaudhary, M. and Walker, T.R. (2019). River Ganga pollution: Causes and failed management plans (correspondence on Dwivedi et al., 2018. Ganga water pollution: A potential health threat to inhabitants of Ganga basin. Environment International, 117: 327-338). *Environmental International* 126: 202-206. https://doi.org/10.1016/j.envint.2019.02.033.

Chauhan, P. and Bhardwaj, N. (2018). Assessment of Ganga water contamination at Haridwar: Studies on Some Physico-Chemical and Microbiological Characteristics. 12: 65-73. https://doi.org/10.9790/2402-1212016573

Ching, L. and Mukherjee, M. (2015). Managing the socio-ecology of very large rivers: Collective choice rules in IWRM narratives. *Glob Environ Chang*, 34: 172-184. https://doi.org/10.1016/j.gloenvcha.2015.06.012

Conley, D.J., Paerl, H.W., Howarth, R.W., et al. (2009). Ecology - Controlling eutrophication: Nitrogen and phosphorus. *Science*, (80-) 323: 1014-1015. https://doi.org/10.1126/science.1167755.

CPCB (2013a). Performance Evaluation of Sewage Treatment Plant Under NRCD

CPCB (2016a). Restoration/rejuvenation of River Ganga. 2016.

CPCB (2013b). Pollution Assessment: River Ganga.

CPCB (2016b). CPCB Bulletin Vol. I. CPCB Bull Vol. I, 1-26

CPCB (2020). In-situ Bioremediation Techniques for Wastewater Treatment.

CWMI (2018). Composite water management index India.

Das, K.K., Panigrahi, T. and Panda, R.B. (2012). Idol Immersion Activities Cause Heavy Metal Contamination in River. *Int J Mod Eng Res*, 2: 4540-4542.

Das, P. and Tamminga, K.R. (2012). The ganges and the GAP: An assessment of efforts to clean a sacred river. *Sustainability*, 4: 1647-1668. https://doi.org/10.3390/su4081647

Das, S. (2011). Cleaning of the Ganga. *J Geol Soc India*, 78: 124-130. https://doi.org/10.1007/s12594-011-0073-9

Del Bello, L. (2018). Indian scientists race to map Ganges river in 3D. *Nature*, 560: 149-150. https://doi.org/10.1038/d41586-018-05872-w

Diaz, R.J., Rosenberg, R. (2008). Spreading dead zones and consequences for marine ecosystems. *Science*, (80-) 321: 926-929. https://doi.org/10.1126/science.1156401

Dourojeanni, A. (2001). Water management at the river basin level: Challenges in Latin America.

Downtoearth (2019). Aquifer mapping programme critical to raise groundwater levels. 1-25.

Dubey, R.S. and Dubey, A.R. (2016). Comparative Effects of Idols Immersion on the Quality of Flowing Holy Ganga Water and Stagnant Water of Ganga Sarovar: A Case Study at Varanasi. 1-9.

Dutta, V. (2019). 10 critical steps for Ganga revival. 1-5.

Dutta, V., Dubey, D. and Kumar, S. (2020). Cleaning the River Ganga: Impact of lockdown on water quality and future implications on river rejuvenation strategies. *Sci Total Environ* 743: 140756. https://doi.org/10.1016/j.scitotenv.2020.140756.

Dwivedi, S., Chauhan, P.S., Mishra, S., et al. (2020). Self-cleansing properties of Ganga during Maha-Kumbh. Env Monit Assess 15.

Dwivedi, S., Mishra, S. and Tripathi, R.D. (2018). Ganga water pollution: A potential health threat to inhabitants of Ganga basin. *Environ Int J*, 117: 327-338. https://doi.org/10.1016/j.envint.2018.05.015

Ejaz, S., Akram, W., Lim, C.W. et al. (2004) Endocrine disrupting pesticides: A leading cause of cancer among rural people in Pakistan. *Exp Oncol*, 26: 98-105.

Environment Agency (2015). Water for life and livelihoods Part 1: Humber river basin district river basin management plan. 107.

Euler, J. and Heldt, S. (2018). From information to participation and self-organization: Visions for European river basin management. *Sci Total Environ*, 621: 905-914. https://doi.org/10.1016/j.scitotenv.2017.11.072

Fang, K. Sivakumar, B. and Woldemeskel, F.M. (2017) Complex networks, community structure, and catchment classification in a large-scale river basin. *J Hydrol*, 545: 478-493. https://doi.org/10.1016/j.jhydrol.2016.11.056

FAO (2020) AQUASTAT _ Land & Water _ Food and Agriculture Organization of the United Nations _ Land & Water _ Food and Agriculture Organization of the United Nations. In: FAO.

FAO (1992). Wastewater characteristics and effluent quality parameters.

FAO (2015). AQUASTAT Transboundary River Basin Overview – Amazon.

FAO (2016). Transboundary River Basin Overview - La Plata.

GAP (2020). Restoration of Flow. In: Ganga Action Parivar. https://www.gangaaction.org/restoration-of-flow/

Ghose, N.C., Saha, D. and Gupta, A. (2009). Synthetic Detergents (Surfactants) and Organochlorine Pesticide Signatures in Surface Water and Groundwater of Greater Kolkata, *India. J Water Resour Prot*, 01: 290-298. https://doi.org/10.4236/jwarp.2009.14036

Gopal, B. (2000). River conservation in the Indian subcontinent. Glob Perspect River *Conserv Sci*, 233-261.

Goswami, K., Gachhui, R. and Goswami, I. (2012). The Idol Immersion in Ganges Cause Heavy Metal Contamination. 84: 54-56.

Guzzella, L., Roscioli, C. and Viganò, L., et al. (2005). Evaluation of the concentration of HCH, DDT, HCB, PCB and PAH in the sediments along the lower stretch of Hugli estuary, West Bengal, northeast India. *Environ Int*, 31: 523-534. https://doi.org/10.1016/j.envint.2004.10.014

Hamner, S., Tripathi, A. and Mishra, R.K., et al. (2006). The role of water use patterns and sewage pollution in incidence of water-borne/enteric diseases along the Ganges River in Varanasi, India. Int J Environ Health Res, 16: 113-132. https://doi.org/10.1080/09603120500538226

Heinz, I., Pulido-Velazquez, M., Lund, J.R. and Andreu, J. (2007). Hydro-economic modeling in river basin management: Implications and applications for the European water framework directive. *Water Resour Manag*, 21: 1103-1125. https://doi.org/10.1007/s11269-006-9101-8

Hoag, H.J. and Mohamoda, D.Y. (2003). Nile Basin Cooperation: A Review of the Literature.

Hoffman, E., Lyons, J. and Boxall, J., et al. (2017). Spatiotemporal assessment (quarter century) of pulp mill metal(loid) contaminated sediment to inform remediation decisions. *Environ Monit Assess*, 189:. https://doi.org/10.1007/s10661-017-5952-0

Homa, D., Haile, E. and Washe, A.P. (2016). Determination of Spatial Chromium Contamination of the Environment around Industrial Zones. *Int J Anal Chem*, 2016:. https://doi.org/10.1155/2016/7214932

Hooper, B.P. (2012). Advancing integrated river basin management in the Mississippi basin.

Hudda, S. (2011). River Pollution: Causes and Actions "For a better tomorrow, act today".

IITs (2013). Ganga River Basin environment management plan: interim report.

Kaur, B. (2018). Namami Gange: Five reasons why Ganga will not be clean by 2020. Down To Earth 1-16.

Kaur, B.J., George, M.P. and Mishra, S. (2013). Water quality assessment of river Yamuna in Delhi stretch during Idol immersion. *Int J Environ Sci*, 3: 2122-130. https://doi.org/10.6088/ijes.2013030600028

Kaushal, N., Babu, S. and Mishra, A., et al (2019). Improving River Flows- Towards a Healthy Ganga. *Front Environ Sci*, 7.

Khan, M.R., Voss, CI., Yu, W. and Michael, H.A. (2014). Water Resources Management in the Ganges Basin: A Comparison of Three Strategies for Conjunctive Use of Groundwater and Surface Water. *Water Resour Manag*, 28: 1235-1250. https://doi.org/10.1007/s11269-014-0537-y

Khanna, D.R., Bhutiani, R. and Tyagi, B., et al (2012). Assessment of water quality of River Ganges during Kumbh mela 2010. 2-7.

Kulshrestha, H. and Sharma, S. (2006). Impact of mass bathing during Ardhkumbh on water quality status of river Ganga. *J Environ Biol*, 27: 437-440.

Kumar, B., Verma, V.K. and Naskar, A.K., et al (2013). Human health risk from hexachlorocyclohexane and dichlorodiphenyltrichloroethane pesticides, through consumption of vegetables: estimation of daily intake and hazard quotients. *J Xenobiotics*, 3: 6. https://doi.org/10.4081/xeno.2013.e6

Kumari, A., Sinha, R.K. and Gopal, K. (2001). Concentration of organochlorine pesticide residues in Ganga water in Bihar, *India. Environ Ecol*, 19: 351-356.

Lamond, J., Bhattacharya, N. and Bloch, R. (2012). The role of solid waste management as a response to urban flood risk in developing countries, a case study analysis. *WIT Trans Ecol Environ*, 159: 193-204. https://doi.org/10.2495/FRIAR120161

Mariya, A., Kumar, C., Masood, M. and Kumar, N. (2019). The pristine nature of river Ganges: its qualitative deterioration and suggestive restoration strategies. *Environ Monit Assess*, 191:. https://doi.org/10.1007/s10661-019-7625-7

Mathew, R.A. and Kanmani, S. (2020). A review on emerging contaminants in Indian waters and their treatment technologies. *Nat Environ Pollut Technol*, 19:549–562. https://doi.org/10.46488/NEPT.2020.V19I02.010

MDEQ (2020). Mississippi's Basin Management Approach. https://www.mdeq.ms.gov/water/surface-water/watershed-management/basin-management-approach/

Misgan, S. (2013). The Nile Basin States: The need for genuine cooperation.

Mishra, N.K. and Mohapatra, S.C. (2009). Effect of Gangetic Pollution on Water Borne Diseases in Varanasi: A Case Study. *Indian J Prev Soc Med*, 40: 39-42.

Mishra, S., Kumar, A., Yadav, S. and Singhal, M.K. (2015). Assessment of heavy metal contamination in Kali river, Uttar Pradesh, *India. J Appl Nat Sci*, 7: 1016-1020. https://doi.org/10.31018/jans.v7i2.724

Misra, A.K. (2010). A River about to Die: Yamuna. *J Water Resour Prot*, 02: 489-500. https://doi.org/10.4236/jwarp.2010.25056

MoEF (2016). Swachh Bharat and Ganga Rejuvenation.

Mohan, V. (2017). Centre turns to 'sewage- eating' microbes to treat Ganga water at 54 new sites. Dev. News 1-3

Mohapatra, S.P., Gajbhiye, V.T., Agnihotri, N.P. and Raina, M. (1995). Insecticide pollution of Indian rivers. *Environmentalist*, 15: 41-44. https://doi.org/10.1007/BF01888888

MoHUA (2020). Assessment of 97 Ganga Towns.

Molle, F. (2009). River-basin planning and management: The social life of a concept. *Geoforum* 40: 484–494. https://doi.org/10.1016/j.geoforum.2009.03.004

MoWRRD & GR (2018). Achievements of four years (2014-15 to 2017-18).

Mutiyar, P.K. and Mittal, A.K. (2013). Status of organochlorine pesticides in Ganga river basin: Anthropogenic or glacial? *Drink Water Eng Sci*, 6: 69–80. https://doi.org/10.5194/dwes-6-69-2013.

Nandi, I., Tewari, A. and Shah, K. (2016). Evolving human dimensions and the need for continuous health assessment of Indian rivers. *Curr Sci*, 111: 263-271. https://doi.org/10.18520/cs/v111/i2/263-271

NEERI (2020). Sustainable Treatment Options for Sewage, In-situ Drain and Lake / River Rejuvenation in Indian context.

Newson, M. (1992). Land, water and development. River basin systems and their sustainable management. Land, water Dev River basin Syst their Sustain Manag 505378. https://doi.org/10.1016/0022-1694(93)90292-h

NFHS (2015). National Family Health Survey-4 (NFHS-4). New Delhi: Ministry of Health and Family Welfare, Government of India

NGM (2020). Namami Gange Programme: The key achievements under Namami Gange programme [Online]. Available: https://nmcg.nic.in/hi/NamamiGanga.aspx. (Accessed 5 Nov. 2020)

NMCG-NEERI (2017). Assessment of Water Quality and Sediment to understand the Special Properties of River Ganga.

NMCG (2017). Reference Note.

NMCG (2020a). Leading River Rejuvenation A case of Namami Gange.

NMCG (2020b). Treatment of sewage carrying drains joining Ribver ganga. https://nmcg.nic.in/csr/biodrains.aspx

NRCD, MoEF (2009). STATUS PAPER ON RIVER GANGA State of Environment and Water Quality.

O'Reilly, K., Dhanju, R. and Goel, A. (2017). Exploring " The Remote" and "The Rural": Open Defecation and Latrine Use in Uttarakhand, India. 93: 193-205. https://doi.org/10.1016/j.worlddev.2016.12.022

Pandey, J. and Singh, R. (2017). Heavy metals in sediments of Ganga River: up- and downstream urban influences. *Appl Water Sci*, 7: 1669-1678. https://doi.org/10.1007/s13201-015-0334-7

Pandey, K. (2019). Grossly polluting industries more than doubled in 8 years: SOE in Figures. Down to Earth.

Panigrahi, A.K. and Pattnaik, S. (2019). A Review on Pollution Status of River Bhagirathi-Hooghly in the Stretch of West Bengal, India. 9: 5

Pathak, D., Whitehead, P.G., Futter, M.N. and Sinha, R. (2018). Water quality assessment and catchment-scale nutrient flux modeling in the Ramganga River Basin in north India: An application of INCA model. Sci Total Environ 631-632: 201-215. https://doi.org/10.1016/j.scitotenv.2018.03.022

Paul, D. (2017). Research on heavy metal pollution of river Ganga: A review. *Ann Agrar Sci*, 15: 278-286. https://doi.org/10.1016/j.aasci.2017.04.001

PIB (2018). Achievements of ministry of water resources-river development and ganga rejuvenation during 2018. Press Inf. Bur. 1-15

PIB (2017). Ministry of Water Resources, River Development and Ganga Rejuvenation During the Last Three Years. 1-5

PIB (2019). Management of Ground Water. *J. Am. Water Works Assoc.*, 60: 640-644.

Rani, N., Vajpayee, P. and Bhatti, S., et al (2014). Quantification of Salmonella Typhi in water and sediments by molecular-beacon based qPCR. *Ecotoxicol Environ* Saf 108: 58-64. https://doi.org/10.1016/j.ecoenv.2014.06.033

Sah, R., Baroth, A. and Hussain, S.A. (2020). First account of spatio-temporal analysis, historical trends, source apportionment and ecological risk assessment of banned organochlorine pesticides along the Ganga River. *Environ Pollut*, 263: 114229. https://doi.org/10.1016/j.envpol.2020.114229

Sahoo, K.C., Hulland, K.R.S. and Caruso, B.A., et al (2015). Sanitation-related psychosocial stress: A grounded theory study of women across the life-course in Odisha, India. *Soc Sci Med* 139:80–89. https://doi.org/10.1016/j.socscimed.2015.06.031

SANDRP (2018). Is there hope from National Mission for Clean Ganga ? Listen to official agencies. 1-6.

Sanghi, R. and Kaushal, N. (2014). Introduction to Our National River Ganga via cmaps.

Sarkar, R. (2013). Study on the Impact of Idol Immersion on Water Quality of River Ganga At Ranighat, Chandernagore (W.B.). 3: 24-29.

Schletterer, M., Shaporenko, S.I., and Kuzovlev, V.V., et al (2018). The Volga: Management issues in the largest river basin in Europe. *River Res Appl*, 35: 510-529.

Shah, T., Makin, I. and Sakthivadivel, R. (2001). Limits to leapfrogging: Issues in transposing successful river basin management institutions in the developing world. Irrig River Basin Manag Options Gov Institutions 31-49. https://doi.org/10.1079/9780851996721.0031

Shah, T. and Rajan, A. (2019). Cleaning the Ganga. Econ Polit Wkly 39: 57-66.

Sharmila, S. and Arockiarani, I. (2016). A pollution model of the river ganges through inter criteria analysis. *Int J Ocean Oceanogr*, 10: 81-91.

Singh, L., Choudhary, S. and Singh, P. (2012). Pesticide concentration in water and sediment of River Ganga at selected sites in middle Ganga plain. *Int J Environ Sci*, 3: 260-274. https://doi.org/10.6088/ijes.2012030131026

Sinha, S.N. and Paul, D. (2012). Detoxification of Heavy Metals by Biosurfactants. *Bull Environ Sci Res*, 1: 1-3. https://doi.org/10.6084/m9.figshare.1352038

Souza-Filho PWM, de Souza, E.B. and Silva Júnior, R.O., et al (2016). Four decades of land-cover, land-use and hydroclimatology changes in the Itacaiúnas River watershed, southeastern Amazon. *J Environ Manage*, 167: 175-184. https://doi.org/10.1016/j.jenvman.2015.11.039

Spears, D., Ghosh, A. and Cumming, O. (2013). Open Defecation and Childhood Stunting in India: An Ecological Analysis of New Data from 112 Districts. PLoS One 8:1–9. https://doi.org/10.1371/journal.pone.0073784

Srinivas, R., Singh, A.P. and Shankar, D. (2020). Understanding the threats and challenges concerning Ganges River basin for effective policy recommendations towards sustainable development. Springer Netherlands.

Srivastava, P., Burande, A. and Sharma, N. (2013). Fuzzy Environmental Model for Evaluating Water Quality of Sangam Zone during Maha Kumbh 2013. *Appl Comput Intell Soft Comput* 2013: 1-7. https://doi.org/10.1155/2013/265924

SwachhindiaNDTV (2019). All 97 Ganga Towns Will Achieve ODF Status By March 2019. SwachhindiaNDTV

Tare, V. (2010). River Ganga at a Glance: Identification of Issues and Priority Actions for Restoration.

Tare, V., Bose, P. and Gupta, S.K. (2003). Suggestions for a modified approach towards implementation and assessment of Ganga action plan and other similar river action plans in India. *Water Qual Res J* Canada, 38: 607-626. https://doi.org/10.2166/wqrj.2003.039

TOI (2017). NGT asks NMCG to give detail of industrial clusters near Ganga. 18-19.

Tripathi, A., Tripathi, D.K., Chauhan, D.K. and Kumar, N. (2016). Chromium (VI)-induced phytotoxicity in river catchment agriculture: evidence from physiological, biochemical and anatomical alterations in *Cucumis sativus* (L.) used as model species. 7540:. https://doi.org/10.1080/02757540.2015.1115841

Tripathi, B.D. and Tripathi, S. (2014). Issues and Challenges of River Ganga. In: Our National River Ganga: Lifeline of Millions. pp 211-220

Trivedi, A. (2020). River Rejuvenation: An Innovative and Logistic Approach. In: Recent Trends in Agricultural Sciences & Technology. pp 195-207

Trivedi, R.C. (2010). Water quality of the Ganga River - An overview. *Aquat Ecosyst Heal Manag*, 13: 347–351. https://doi.org/10.1080/14634988.2010.528740

Tyagi, V.K., Bhatia, A. and Gaur, R.Z., et al (2013). Impairment in water quality of Ganges River and consequential health risks on account of mass ritualistic bathing. *Desalin Water Treat*, 51: 2121-2129. https://doi.org/10.1080/19443994.2013.734677

UrbanUpdate (2018). Solid waste management projects in 97 towns along Ganga. UrbanUpdate 3-5.

Vega, M., Pardo, R., Barrado, E. and Debán, L. (1998). Assessment of seasonal and polluting effects on the quality of river water by exploratory data analysis. *Water Res*, 32: 3581-3592. https://doi.org/10.1016/S0043-1354(98)00138-9

Vijgen, J., Abhilash, P.C. and Li, Y.F., et al (2011). Hexachlorocyclohexane (HCH) as new Stockholm Convention POPs-a global perspective on the management of Lindane and its waste isomers. *Environ Sci Pollut Res*, 18: 152-162. https://doi.org/10.1007/s11356-010-0417-9

Villar, P.C., Ribeiro, W.C. and Sant'Anna, F.M. (2018). Transboundary governance in the La Plata River basin: status and prospects. *Water Int*, 43: 978-995. https://doi.org/10.1080/02508060.2018.1490879

Walker, T.R., Willis, R. and Gray, T., et al (2015). Ecological Risk Assessment of Sediments in Sydney Harbour, Nova Scotia, Canada. *Soil Sediment Contam*, 24: 471-493. https://doi.org/10.1080/15320383.2015.982244

Welcomme, R.L. (1985). River Fisheries.

WHO (2003). Characteristics and quality assessment of surface water and groundwater resources of Akwa Town, Southeast, Nigeria. *J Niger Assoc Hydrol Geol,* 14:71-77.

WII-GACMC (2017). Aquatic fauna of the ganga river-Status and Conservation.
WWF, INTACH, Toxic link, SANDRP (2019). Rejuvenating Ganga—A Citizen's Report.
Zhang, S.Y., Tsementzi, D. and Hatt, J.K., et al (2019). Intensive allochthonous inputs along the Ganges River and their effect on microbial community composition and dynamics.
Zhang, W., Jin, X., Liu, D., et al (2017). Temporal and spatial variation of nitrogen and phosphorus and eutrophication assessment for a typical arid river—Fuyang River in northern China. *J Environ Sci* (China), 55: 41-48. https://doi.org/10.1016/j.jes.2016.07.004

Chapter 3
Ganges Groundwater Interaction at Varanasi

Prerona Das, Abhijit Mukherjee, Atrayee Haldar, Ashok Shaw, and Mrinal Kanti Layek

INTRODUCTION

Surface water and groundwater act as linked ecosystems in various physiographic and climatic domains (Sophocleous, 2002). Although both show differences in physical and chemical properties, the hyporheic zone which is an area of interaction between shallow groundwater and surface water undergoes various processes leading to transportation, transformation, degradation and sorption of substances, including other chemical and microbial activities (Briody et al., 2016). This intermediate zone is characterized by porous and permeable sediment along with continual saturation and low flow rate thus resembling aquifer system (Kalbus et al., 2006). The direction of flow for groundwater and surface water while mixing depends on the hydraulic head. Whether the stream will act as loosing (inflow of groundwater in the aquifer (streamflow capture)) or gaining (outflow of groundwater to the adjacent stream (baseflow)) (Mukherjee et al., 2018) is dependent on the natural hydraulic conditions and physio-climatic settings of that particular river reach. On a basin scale river water-groundwater interaction depends on: (a) the stream channel position with respect to the basin morphology; (b) hydraulic conductivity of the sediment deposit both at the river bed along with the associated alluvial zone; (c) seasonal change in river water level with respect to groundwater and; (d) other fluvial anthropogenic and geological processes (Winter and Harvey, 1998; Yu et al., 2013; Menció et al., 2014; Martinez et al. 2015). During monsoon due to heavy runoff, river stage remains much higher thus acting as loosing stream whereas the opposite is witnessed in dry season. Thus contamination of one has drastic effects on the other (Sophocleous, 2002; Winter and Harvey, 1998). Growing demand of fresh water and increasing uncertainty in sustainable water supply poses new challenges to researchers in

P. Das · A. Mukherjee (✉) · A. Haldar · A. Shaw · M. K. Layek
Department of Geology and Geophysics, Indian Institute of Technology Kharagpur, Kharagpur, India

A. Mukherjee (ed.), *Riverine Systems*, https://doi.org/10.1007/978-3-030-87067-6_3

preservation and management of groundwater for a riparian ecosystem. Allocation of water resource and evaluation of potential impact for increasing water usage requires a clear understanding of the groundwater and surface water pathways including flux quantification (Sophocleous, 2002; Kalbus et al., 2006; Banks et al., 2011; Martinez et al., 2015). This quantification of mass flow between them eventually leads to protection of water resources from unwanted contamination.

Emerging trends of groundwater depletion in South Asian countries have suggested eminent fresh water availability crisis (Rodell et al., 2018). Such crisis have been more pronounced in river Ganges basin which is the 6th largest river system hosting approximately 10% of the global population. This river acts as a sewerage disposal in northern India and is presently believed to be one of the most polluted mega rivers in South Asia jeopardizing the health of the population thriving on its alluvial plains. Mukherjee et al. (2018) have reported a decadal summer drying of Ganges river water due to excessive and unsupervised pumping. Planning Commission 2014 has reported an estimate of 4 million groundwater extraction points in Uttar Pradesh. This emerging crisis suggests an urgent need to develop a proper understanding of the conjunctive hydrologic system, especially in highly urbanized areas.

STUDY AREA

Gangetic River System

River Ganges, flowing along the northern part of India is a perennial river initiating from the Himalayas and concluding its course at the Bay of Bengal forming the largest delta system with a catchment area of approximately 250,000 km^2. The river is fed by meltwater from more than 4000 glaciers from Himalayas thus determining its headwater contribution (Maurya et al., 2011) the signature of which is being carried along to its lower stages. River Ganges emerges from the Gharwal Himalayas at the Gangotri glacier and enters its middle course at Haridwar. It encounters a lot of tributaries on its pathway and finally merges in the Bay of Bengal after covering 2200 km along the north Indian provinces (Shukla et al., 2012). The river nearly traces the foreland bulge of the Himalayan orogenic system and at places flow along the sub-parallel lineament formed due to the tectonism. The northern part of the river is mostly covered with alluvium which is deposited by the river as well as its tributaries from the Himalayas. The southern part has lesser alluvial deposits underlain by the Deccan volcanic traps and Vindhyan Kaimur sandstones. River Ganges carries signature of highly evaporated glacial melt from the Himalayas mixed with the groundwater from the encountered aquifers on its way.

The central Gangetic basin on the other hand is encountered by the highly meandering middle course of the river and its tributaries at places. The sediment that it carries includes detritus of argillaceous metamorphics and mafic volcanics from Himalayas (gray micaceous sub greywacke in nature) and felsics from the

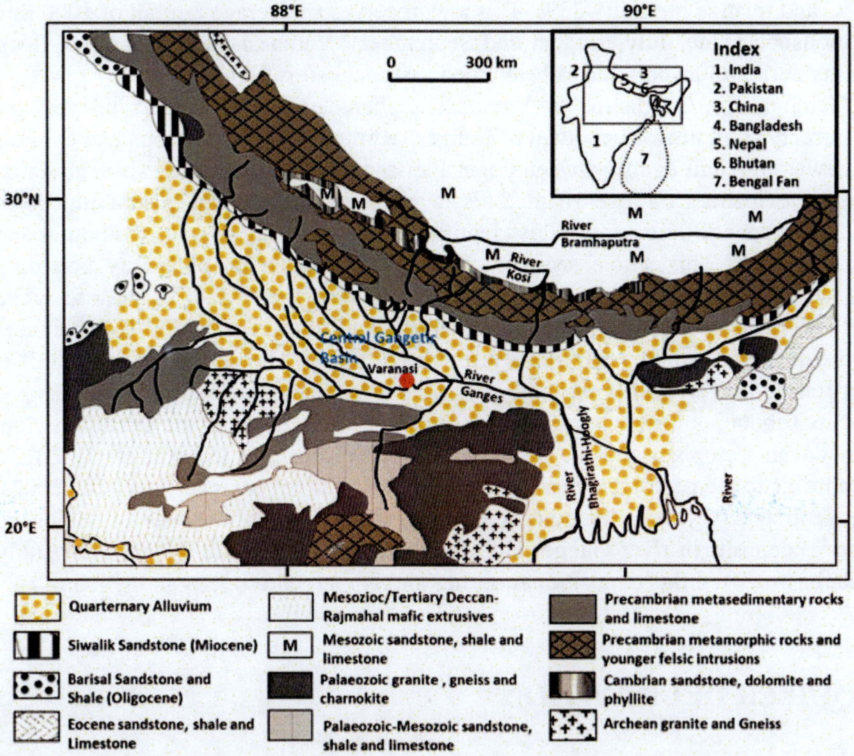

Figure 1. Geological map of central Gangetic plain along with the river Ganges and its tributaries. Location of study area is marked in red colour (modified after Mukherjee et al., 2009).

cratonic Peninsular India (brown subarkosic in nature) (Mukherjee et al., 2012). The river acts as a major hydrological boundary and regional discharge zone between the Himalayan foreland basin aquifers and the Indian peninsular cratonic aquifers (Shukla et al., 2012). Figure 1 shows the pathway of Ganges and its tributaries in order to delineate the sedimentary sources in central Gangetic plain.

Varanasi

Varanasi is a highly populated, culturally rich, historical city residing on the bank of river Ganges with a population of 1.4 million (2011 census) relying on groundwater and the river for their regular supply of water for domestic usage, irrigation and potability. At Varanasi the river flows in the direction from south to north. The city, on the convex side of river Ganges, is enclosed with river Varuna in the north and Assi in the south, both meeting river Ganges. Varanasi encounters a complex and diverse climate annually with summer (March, April, May) temperature rising upto

45°C and relative humidity 33%. This area receives an average rainfall of 1000 mm in monsoon (June, July, August and September). Varanasi faces occasional flood events during monsoon due to heavy bed load deposit in Ganges.

Geologically, the district of Varanasi is characterized into older and younger quaternary alluvium as a resultant of Ganges sediment deposit. The top layer consists of newer alluvium aging between upper Pleistocene to recent which also show rich cultural evidences up to 8-10 m. This layer is underlain by older alluvium from middle to upper Pleistocene and is identified by the presence of calcrete nodules and gravelly sand. This whole sediment package sits on an unconformity boundary below which lies the upper Vindhyan Kaimur sandstone which is grey to white and arkosic in nature with a capping of laterite and bauxite at places. The thickness of the quaternary alluvium increases from east to west. The alluvial geomorphological setting in Varanasi is primarily dependent on sediments deposited by Ganges and its tributaries upstream and partially by the changing climatic conditions, the Himalayan tectonism and the sequential sea level changes (Shukla et al., 2012). According to the geological map of Uttar Pradesh (Surveyor General of India 2002), the sedimentary setting of central Gangetic plain mostly consists of Holocene alluvial deposits in river Ganges and its tributaries. The city of Varansi, our study area, has suffered this depositional setting and is marked properly in the map.

AQUIFER DELINEATION

Figure 2 represents the subsurface lithology of Varanasi as modelled from 110 borehole lithologs in and around Varanasi. The entire study area is capped by surficial clay layer which resulted from multiple annual flood events. At around 30-40 mbgl lies the silty clay to fine sand layer with sudden intercalations of calcrete layers thus defining the shallow aquifer. Starting from 60 mbgl lays a thick layer of medium to coarse sand which confirms the presence of deeper aquifer in our study region. The shallow and the deep aquifer is separated by presence of a semi-continuous clay layer throughout the subsurface of Varanasi.

GROUNDWATER CHEMISTRY

Chemistry of infiltrating water, sediment-water interaction and groundwater residence time are the major factors controlling the groundwater chemistry of a region. In Varanasi the fate of groundwater is primarily controlled by the river Ganges. Piper plots (Fig. 3) of the water samples showed that the water is relatively young and does not show any prominent evolutionary trend. The Ca-HCO_3 type hydrochemical facies represents most of the groundwater samples with the mean TDS value below 500 mg/l, which is a typical signature of alluvial setting. Bivariate mixing diagram of Na normalized Ca versus Na normalized Mg and HCO_3 (Fig. 4) tend to

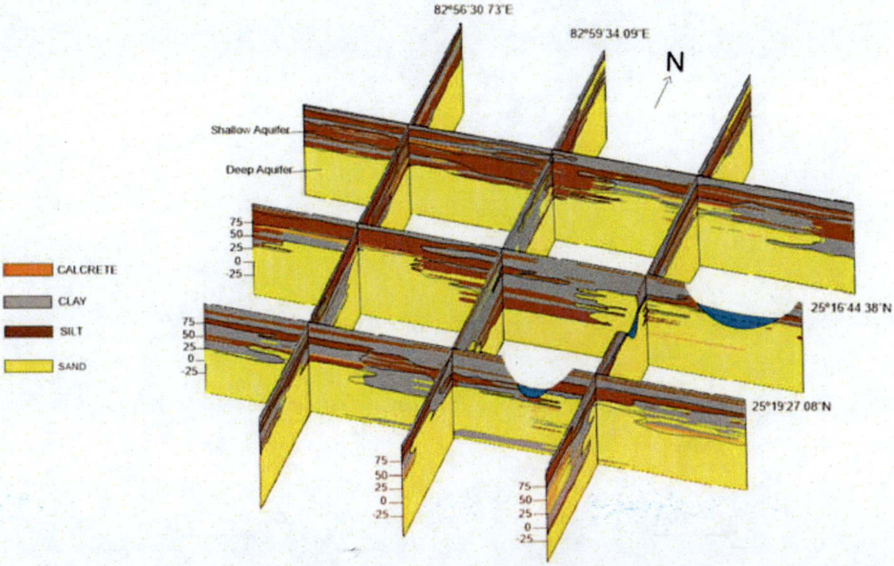

Figure 2. Subsurface lithologic model of Varansi delineating the shallow and deep aquifer acquired from 110 borehole logs in and around the study area. The depths are marked in meters (0 represents MSL) and the position of the river is shown in blue colour. Dotted line in river represents stream stage in dry season.

fall within and close to the global average silicate weathering domain thus suggesting incongruent leaching of argillaceous metamorphics and volcanics from Himalayas, and surrounding precambrian (lesser Himalayas and Indian craton) and mesozoic (Deccan-Rajmahal Trap) units along with other solute transfer mechanism that controls the solute chemistry of the groundwater in study area. The sediment provenance is evident from predominance of Ca and Mg relative to total cationic concentration suggests primary dominance of alkaline earth silicates from Himalayas followed by carbonates from paleozoic-mesozoic and precambrian cratonic units and quarternary alluvium, since silicate weathering is the primary source of bivalent cations in groundwater. The groundwater in the city area is reported to be contaminated with nitrate, faecal coliforms (Mohan et al., 2011) and other microorganic contaminants which are probably a resultant of river water-groundwater interaction.

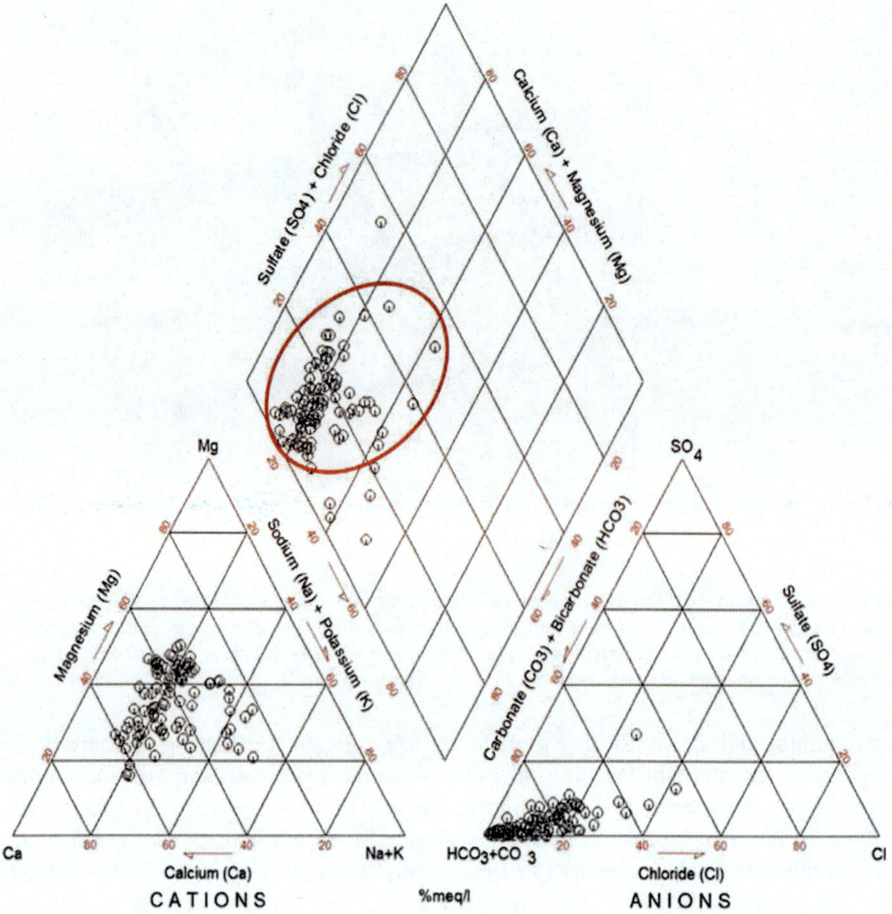

Figure 3. Piper plot represents hydrochemical facies of groundwater samples to be Ca-HCO_3 type (circled).

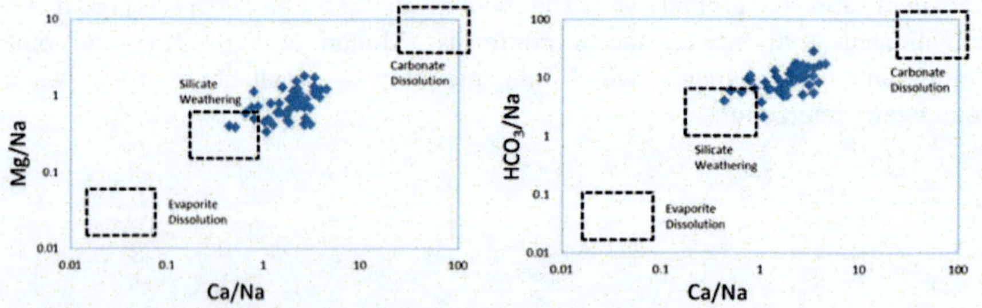

Figure 4. Bivariate mixing of Na normalized Ca versus Na normalized Mg and HCO_3 plots show the groundwater to fall within and close to global silicate weathering zone thus suggesting incongruent leaching of argillaceous metamorphics and volcanics from source rocks.

SURFACE WATER-GROUNDWATER INTERACTION

Water Level Fluctuation

Both river water and spatially distributed groundwater level fluctuation was recorded using piezometric level loggers. Two of these groundwater hydrographs (one on each bank), tapping the deeper aquifer, nearest to the river is compared with the stream fluctuation and is shown in Fig. 5. The temporal gaining and losing phases of river is evident from the seasonal hydrographs. During dry seasons the stream stage being at a lower elevation acts as gaining stream whereas during monsoon the stream stage is at a higher elevation thus recharging the adjacent aquifer. The stream stage increase is directly proportional with precipitation event which is evident from the meteoric data (Fig. 5). The shallow aquifer gets directly affected by the river fluctuations due to proper connectivity but the deeper aquifer gets a secondary influence where the aquifers are disconnected. The fluctuation on the left bank show no significant response to the river stage since the shallow and deeper aquifer on the western side have poor connectivity whereas the one on the eastern bank show notable response of lower intensity as a result of proper aquifer connectivity.

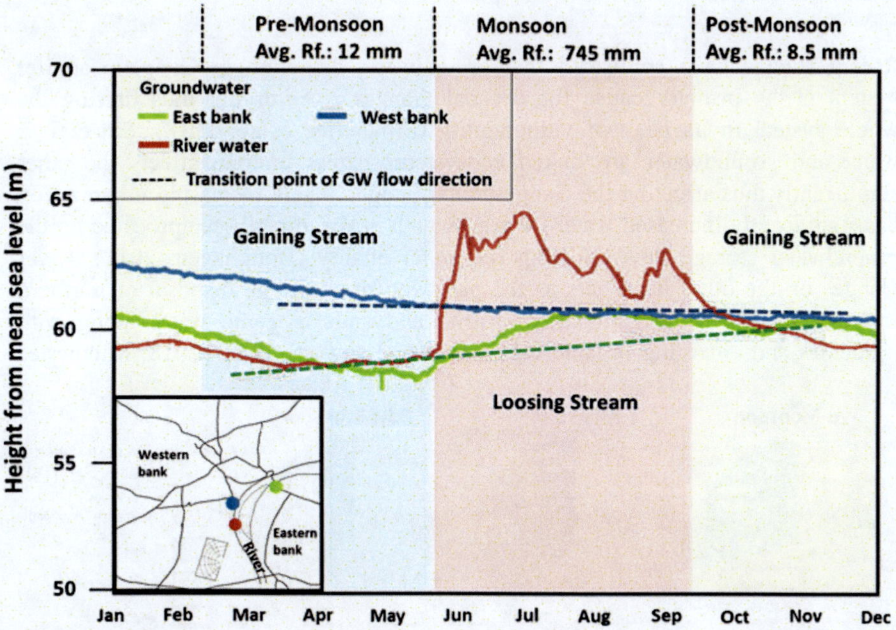

Figure 5. Hydrographs from two piezometric wells, one on each side of the river (West: blue, east: green colour and river: red), demarcates the gaining and loosing phases of the river seasonally. The average rainfall (2017) in each three seasons is sourced from Indian Meteorological Department.

Mixing Model

Seasonal surface water–groundwater interaction have been evaluated through three component mixing models using water isotope $\delta^{18}O$ and electrical conductivity (EC) values of both groundwater and river water. Considering precipitation, glacial melt, upstream discharge and groundwater exfiltration to be the end components (depending on seasonality) a mixing model is being carried out using mass conservation equation, which verified proper interaction throughout the year. The glacial contribution was found to be more than 20% during summer. Precipitation and surface runoff fraction was obviously higher in monsoon relative to the dry seasons. Groundwater contribution is found to be the highest in summer which is almost 30% of the total discharge followed by winter which goes down to approximately 25%. But during monsoon the groundwater contribution falls to a negligible amount which indicates an opposite directional flow from river to the aquifer thus recharging the aquifer. The upstream contribution is highest during monsoon due to heavy water flux from upstream of the river carrying monsoonal runoff from locations upstream of the river. Groundwater contribution is also found to vary proportionally with river cross-sectional area. The mixing model is presented in Fig. 6.

CONCLUSION

Recent studies have confirmed that groundwater depletion in Gangetic aquifer system is the primary cause for decadal Ganges river drying thus driving the whole system to its peak of vulnerability (Mukherjee et al., 2018). Since river water and groundwater are linked ecosystems, stress on one affects the other significantly thus affecting the Ganges water budget. Again increasing urbanization has significantly increased water demand which is also directly reciprocated on the groundwater storage thus affecting the water quality (Bookhagen, 2012). River Ganges on the other hand acts as the pathway for sewerage disposal of northern India which carries irrigational, industrial and anthropogenic waste along with pesticides and emerging organic contaminants (Lapworth et al., 2018; Mukherjee

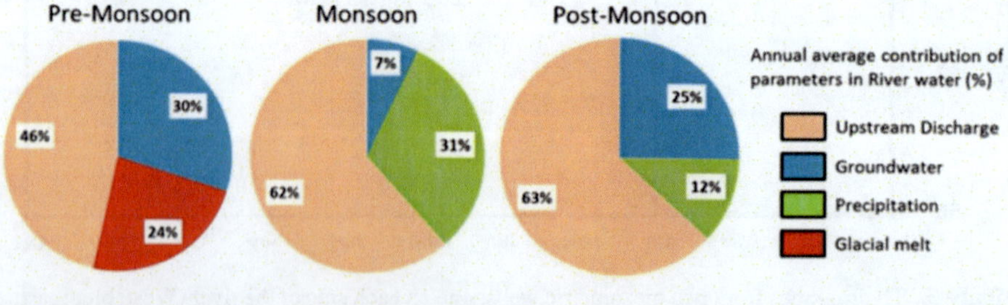

Figure 6. Seasonally variable contribution of various components in river water resulting from mixing model. The groundwater contribution is highest in dry seasons.

et al., 2018) thus affecting the Gangetic aquifer system indirectly. Till date groundwater has been the major source of drinkable fresh water in Varanasi but recently Lapworth et al. (2018) has reported presence of contaminant in deeper groundwater also which is an effect of excessive groundwater abstraction thus making the whole aquifer system vulnerable. Proper treatment of sewage system and regular monitoring of groundwater through deep pumping sites can provide a plausible solution towards protection of long term sustainable water sources. Implementation of proper irrigational framework, groundwater recharge management policies and supervised groundwater abstraction will ensure water and food security for the present as well as future population in Varanasi.

References

Banks, E.W., Simmons, C.T., Love, A.J. and Shand, P. (2011). Assessing spatial and temporal connectivity between surface water and groundwater in a regional catchment: Implications for regional scale water quantity and quality. *Journal of Hydrology,* 404(1-2): 30-49.

Bookhagen, B. (2012). Himalayan groundwater. *Nature Geoscience,* 5(2): 97-98.

Briody, A.C., Cardenas, M.B., Shuai, P., Knappett, P.S. and Bennett, P.C. (2016). Groundwater flow, nutrient, and stable isotope dynamics in the parafluvial-hyporheic zone of the regulated Lower Colorado River (Texas, USA) over the course of a small flood. *Hydrogeology Journal,* 24(4): 923-935.

Kalbus, E., Reinstorf, F. and Schirmer, M. (2006). Measuring methods for groundwater-surface water interactions: A review. *Hydrology and Earth System Sciences Discussions,* 10(6): 873-887.

Lapworth, D. J., Das, P., Shaw, A., Mukherjee, A., Civil, W., Petersen, J.O., Gooddy, D.C., Wakefield, O., Finlayson, A., Krishan, G. and Sengupta, P. (2018). Deep urban groundwater vulnerability in India revealed through the use of emerging organic contaminants and residence time tracers. *Environmental Pollution,* 240: 938-949.

Martinez, J.L., Raiber, M. and Cox, M.E. (2015). Assessment of groundwater–surface water interaction using long-term hydrochemical data and isotope hydrology: Headwaters of the Condamine River, Southeast Queensland, Australia. *Science of the Total Environment,* 536: 499-516.

Maurya, A.S., Shah, M., Deshpande, R.D., Bhardwaj, R.M., Prasad, A. and Gupta, S.K. (2011). Hydrograph separation and precipitation source identification using stable water isotopes and conductivity: River Ganga at Himalayan foothills. *Hydrological Processes,* 25(10): 1521-1530.

Menció, A., Galán, M., Boix, D. and Mas-Pla, J. (2014). Analysis of stream–aquifer relationships: A comparison between mass balance and Darcy's law approaches. *Journal of Hydrology,* 517: 157-172.

Mohan, K., Srivastava, A. and Rai, P. (2011). Ground Water in the City of Varanasi, India: Present status and prospects. *Quaestiones Geographicae,* 30(3): 47-60.

Mukherjee, A., Bhattacharya, P., Shi, F., Fryar, A.E., Mukherjee, A.B., Xie, Z.M. and Bundschuh, J. (2009). Chemical evolution in the high arsenic groundwater of the Huhhot basin (Inner Mongolia, PR China) and its difference from the western Bengal basin (India). *Applied Geochemistry,* 24(10): 1835-1851.

Mukherjee, A., Scanlon, B.R., Fryar, A.E., Saha, D., Ghosh, A., Chowdhuri, S. and Mishra, R. (2012). Solute chemistry and arsenic fate in aquifers between the Himalayan foothills and Indian craton (including central Gangetic plain): Influence of geology and geomorphology. *Geochimicaet Cosmochimica Acta,* 90: 283-302.

Mukherjee, A., Bhanja, S.N. and Wada, Y. (2018). Groundwater depletion causing reduction of baseflow triggering Ganges river summer drying. *Scientific Reports*, 8(1): 12049.

Rodell, M., Famiglietti, J.S., Wiese, D.N., Reager, J.T., Beaudoing, H.K., Landerer, F.W. and Lo, M.H. (2018). Emerging trends in global freshwater availability. *Nature*, 557(7707): 651.

Shukla, U.K., Srivastava, P. and Singh, I.B. (2012). Migration of the Ganga River and development of cliffs in the Varanasi region, India during the late Quaternary: Role of active tectonics. *Geomorphology*, 171: 101-113.

Sophocleous, M. (2002). Interactions between groundwater and surface water: The state of the science. *Hydrogeology Journal*, 10(1): 52-67.

Winter, T.C. and Harvey, J.W. (1998). Ground water and surface water: A single resource. Vol. 1139. Denver, CO: US Geological Survey.

Yu, M.C.L., Cartwright, I., Braden, J.L. and De Bree, S.T. (2013). Examining the spatial and temporal variation of groundwater inflows to a valley-to-floodplain river using 222Rn, geochemistry and river discharge: The ovens river, southeast Australia. *Hydrol. Earth Syst. Sci.*, 17: 4907-4924.

Chapter 4
Microbial Water Quality Assessment in River Ganga (Bihar)

Ashok Kumar Ghosh and Ranjit Kumar

RIVER SYSTEM

Bihar is located in the eastern part of India and is bounded by Himalaya from the north and Chota Nagpur plateau from the south. Both Himalayan and Chota Nagpur plateau are sources of the entire river system flowing in Bihar. It is an entirely land–locked area but it is also very rich in ground water and surface water resources.

Since the ancient times the rivers were integral parts in development of human civilisation. The river Ganga forms the largest river basin in the Indian subcontinent. River Ganga contains its many tributaries like Gandak, Sone, Ghaghra, Koshi etc. Many tributaries joining river Ganga are increasing in load of dissolved particulate matters from both anthropogenic and natural sources. Increase in population exponentially increases pollution load in different rivers due to industrialisation and urbanisation, various municipal and industrial waste disposed in the rivers are causing great deterioration in ecosystem of rivers. The water quality of these rivers reflects many major changes in phytoplankton and zooplankton, atmospheric input, climatic conditions, lithology of river basin, and many anthropogenic activities.

These rivers carry industrial and municipal waste material and also runoff from agriculture land, which contains high load of heavy metal, pesticides and other toxic materials. Industrial and municipal waste water entering in river without treatment constitute constant polluting sources in river. In rainy season surface runoff heavily pollute the river ecosystem and finally causes changes in river basin ecosystem due to non-point source of pollution (Singh et al., 2004).

A. K. Ghosh (✉)
Mahavir Cancer Sansthan and Research Centre, Patna, India

R. Kumar
Central University of Himachal Pradesh, Dharamshala, H.P., India

A. Mukherjee (ed.), *Riverine Systems*, https://doi.org/10.1007/978-3-030-87067-6_4

The Holy Ganga in Bihar

The Ganga is not an ordinary river. It is a symbol of purity and virtue for Indian population. Ganga is also the main river of India from the geographical point of view. This perennial river originating from Himalaya flows through many states before its confluence in the Bay of Bengal (Fig. 1). Ganga water was considered purest and sacred even in the ancient medical journal *Caharaksanhita*. It is believed from medieval times that Ganga water has no germs bacteria or fungi and it can be kept safe for many years. But the condition has changed during past many decades. Sewage and industrial effluents as well as runoff from chemical fertilizer industries and pesticides used in agriculture field get mixed in Ganga, which also includes thousands of animals' carcasses and human crapes are now dumped in river Ganga (Yu et al., 2001).

Water is the most precious natural element required for socio-economic development of any country. Since most of human civilisation develop around the river, rivers were the main source of surface water from ancient times, due to which civilisation grew near rivers in ancient times. India is very rich in river system which is composed of large number of major, medium and small rivers, which covers around 252.8 million hectares of land area. The Ganga basin is the largest river basin of Indian subcontinent. It covers more than 25% of the country's total geographical area and it originates in the state of Uttarakhand. The Ganga basin

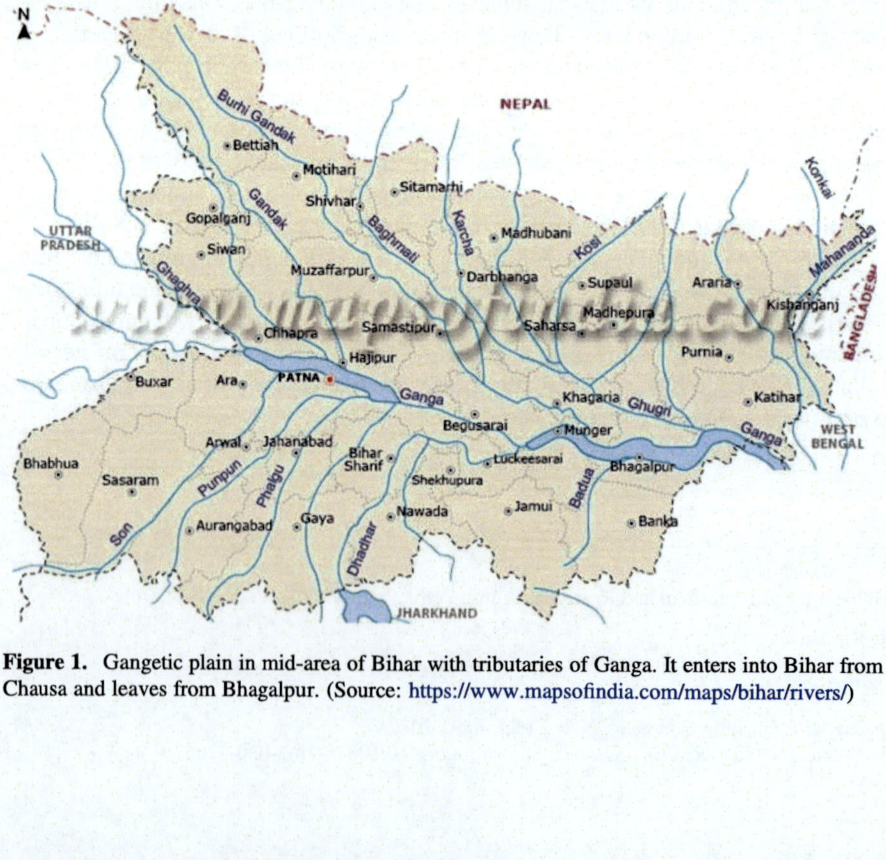

Figure 1. Gangetic plain in mid-area of Bihar with tributaries of Ganga. It enters into Bihar from Chausa and leaves from Bhagalpur. (Source: https://www.mapsofindia.com/maps/bihar/rivers/)

accounts for nearly 50% of class 1 and class 2 towns of the country and the mode of discharge of municipal waste is mainly into the river. Water pollution is alteration in the physical, chemical and biological entities of water, which may lead to adverse effects on living organism and it faces several forced and unforced human activities. Water pollution is very important in terms of public health as well as for aesthetics, preservation and conservation of natural beauty and resources. The current status of river Ganga is highly polluted both from public health point of view as well as preservation of biodiversity.

Sources of Ganga Pollution

The main sources polluting Ganga are:

1. River Gandak, Sone, Punpun, Koshi, which are main tributaries of Ganga, contain significant number of industries around them and adversely affect water quality of Ganga.
2. Many drains carrying domestic and industrial waste water with heavy load of microbial and heavy metal contaminations are discharged into Ganga.
3. Ganga is sign of faith for millions of people taking bath in it whole year which leads to increase in bacteriological contamination.
4. It is also polluted due to non-availability of dumping ponds, proper crematoria and washerman (dhobi) ghats.
5. The agriculture area runoff in the river basin catchment area that contains heavy load of chemical fertilizers and pesticides further aggravates the problem.

COLIFORM BACTERIA IN RIVER WATER

Coliform bacteria consists of very large group of bacterial species which include both faecal and non-faecal coliform bacterial species. Faecal coliforms are bacteria found naturally in gastrointestinal tract of warm-blooded animals. Faecal coliforms are sometimes pathogenic, as many of them cause different disease, though many species are also non-pathogenic. The faecal coliform indicates the contamination of water body through human or animal faecal material. The faecal coliform concentration in any lake or pond increases with addition of sewage waste or sewage manure (Kay et al., 1999).

Faecal coliform is an indicator bacterial organism for contamination from animal or human being. High amount of faecal coliform in river can affect the economy, environmental measures and public health of different communities residing there (Kay et al., 2001). There are long history of illness epidemics and outbreaks have demonstrated that faecal coliform are responsible for many types of diseases causing bacteria and viruses, called pathogens. Pathogens can be ingested with water or food accidentally. Swimmers and other people in water are highly prone to exposure if they have even small cuts in body or abrasions in mucus membrane. Common symptoms of illness include stomach upset, ear infections, diarrhoea and red rashes. However, few pathogens like *Salmonella* and *E. coli* cause serious health hazards on animals or humans. Common problems caused by faecal coliform are as follows:

- Deplete oxygen in the water that is needed by fish and other aquatic animals.
- Affect the natural acidic/alkaline (pH) balance of water.
- Interfere with recreational activities such as boating, fishing, skiing and swimming.
- Create foul odour problems and unpleasant views.
- Affect property values.

Bacterial Indicators Use in Water Quality

Bacteria are bioindicator for monitoring drinking water quality globally. The world's two major international bodies, United States Environmental Protection Agency (USEPA) and European Union (EU), included *E. coli* as a mandatory microbial water quality indicator. The USEPA has given detailed guidelines for estimation of total coliforms through total coliform rule. Guideline for drinking water quality refer to total estimates of overall bacterial numbers in drinking water. This method is known as 'standard plate count method' or 'total heterotrophic plate count' (HPC), and it represents cleanliness guidelines for assessment of drinking water quality. Heterotrophic plate count is not a confirmatory for assessment of health risk. These HPC bacteria are not commonly considered as a compliance measure; rather are regularly monitored in drinking water to understand changes in bacterial numbers over times. With growing acceptance and understanding of total coliforms limitations, there were changes of focus in microbial bioindicator in Europe. In 1998, the European Union introduced enterococci as primary indicator of drinking water quality and removed total coliforms from the list of primary indicator. The World Health Organization has given very extensive regulations for maintenance of drinking water standards. Second edition of WHO has given detailed guidelines for microbial water quality assessment especially for the total coliforms and *E. coli* detection in drinking-water (WHO, 1993). In second edition of volume 2 of WHO guidelines detailed discussion on the merits of alternative indicators such as enterococci and sulfite-reducing *Clostridium* bacteria in inadequacies of total coliforms as faecal pollution indicator are given. According to Australian drinking water Guidelines 1996, and the WHO drinking-water Quality Guidelines, there are detailed emphases on total drinking water risk management for acceptable water quality. The WHO Drinking-water Quality Guidelines in third edition were considered removal of total coliforms as a major primary compliance factor. The New Zealand health ministry also amended their Drinking Water quality Standards in August 2000, for people (NZMoH, 2000). These changes in drinking water quality standards consider only presence of *E. coli* as a primary bacterial indicator of faecal contamination, and it never relies on coloney count of faecal or total coliforms. The change in the water quality indicator is based on the fact that both total and faecal coliforms may be found in natural waters. Presence of total or faecal coliform does not impose any health risk on animals or humans, due to which *E. coli* is considered as microbial indicator for measurement of water quality in New Zealand according to new guidelines there.

Water Quality Indicator Microbes

Indicator Microbes

Pathogens are disease causing microorganisms. Bacteria, viruses and protozoa are common pathogens which causes different waterborne disease in living organism. Different groups of pathogens found in water which leads to very diverse types of diseases in animals and humans. We cannot test every water sample for presence of diverse group of pathogen. Different types of viruses reside in gut of mammals. More than 100 virus species are isolated from human faecal samples and few were isolated from sewage effluent (Payment and Franco 1993). It is very difficult to identify these viruses, and even few species are not isolated properly in pure form. Many times these pathogens are found in drinking water due to contamination of faecal materials, a perfect detection technique is required to alert about the presence of coliform bacteria in water. Microbial water quality indicator is very important which gives detailed idea about safety of drinking water and also accounts for measurement of all diverse categories of microorganisms which may cause health risk. There are many parameters which will be helpful for assessment of the water quality and indicates its purity (NHMRC-ARMCANZ, 1996; WHO, 1996).

The coliforms bacterial indicators give idea about presence of faecal pollution in drinking water and acts like microbial indicators of organism in water samples. Coliform group of bacteria are not considered very hazardous for health risk with few exceptions, but presence of coliform group indicates that water is contaminated with faecal pollution and different pathogens groups might be present. The term 'coliform' was coined for rod-shaped bacillus bacteria in 1880s which was first isolated and analysed from human faeces. All coliform bacteria belong to *Enterobacteriaceae* group which are functionally-related and consists of many genera and species. The *Salmonella* and *Shigella* also belong to the *Enterobacteriaceae* family, but they are not considered in coliforms.

Total Coliform

The total coliform bacteria were rich in population of *E. coli,* which was considered as prime indicator of faecal contamination due to following reasons:

1. Coliform group of bacteria were easily obtained from human faecal samples and faecal contaminated water samples.
2. *E. coli* are major group isolated from human faecal samples, and it was considered that *E. coli* presence are indicator of total coliforms presence.
3. The technology which causes isolation of *E. coli* in early 1900s was not found suitable for total coliform analysis in present times due to diversity of bacterial species and advancement in technology.

Originally total coliform bacteria has four genera and it belongs to family Enterobacteriaceae which would cause lactose fermentation. These groups includes *Escherichia, Enterobacter, Citrobacter* and *Klebsiella* respectively. The *Escherichia coli* (*E. coli*) is main indicator group found in total coliform, it represents the majority of microbes population present in gut. Total coliforms constitute only about 1% of overall bacterial populations of bacteria found in faecal samples out of 109 bacteria species present in per gram of faecal samples (Brenner et al., 1982). Faecal coliform bacteria constitutes very diverse group. It was found in normal conditions of different water and soil environments which was not affected by faecal contamination. *E. coli* presence in any sample is considered as faecal contamination indicator; uncertainty indicates the presence of total or faecal coliforms, which is very effective health indicator for animals and humans. Microbiological techniques for isolation and identification of pathogens are required for understanding disease etiology through different groups of viruses, bacteria and protozoa found in water and are pathogenic in nature. These techniques are still not reliable, sensible, reproducible, specific or economical to replace the use of existing bacterial water quality indicator system.

Escherichia coli (E. coli)

Scientists discovered bacteria in human faecal samples more than 100 years ago, which also contaminate water samples and its presence in water makes it unsafe for drinking. Renowned microbiologist Escherich in 1885 reported that two types of organisms are found in faecal contaminated samples, viz. *Bacterium coli*, now known as *Escherichia coli* (*E. coli*) and was considered that *B. coli* presence causes water contamination. Van Fritsch in 1880 has given concept of indicator on the basis of presence of *Klebsiellae* in water samples contaminated with human faeces (Hendricks, 1978). *B. coli* group of bacteria are hardly distinguished from other groups of total coliform bacteria in faeces and water samples. The development of new method was needed to identify *B. coli* from other coliform bacteria, and more detailed method for analysis would be developed to carry out analysis of *B. coli* through MPN method. Bacteriologists has developed many methods to confirm the presence of *Bacterium coli* in water samples and described it in detail than any other bacteria found in gut. Through advancement in bacteria identification methods it becomes easier to distinguish *Bacterium coli* from the typhoid causing bacteria like *Salmonella typhi*. The *B. coli* causes production of lactose from acid and gas, while

Salmonella typhi does not produce lactic acid. Keene et al., 1994 confirmed that the techniques developed in the early 1900 are still very useful for estimation of total coliform in drinking water.

SAMPLE COLLECTION SITE FOR MICROBIAL ANALYSIS IN GANGA WATER

Since river Ganga enters into Bihar at Chausa and leaves Bihar at Bhagalpur, we have collected water samples from river Ganga at 33 sites from Chausa to Bhagalpur to analyse microbial contamination in different seasons. The details of sampling site from river Ganga are:

1. Chausa karmnasa, Confluence, Buxar
2. Ramrekha ghat, Dist. - Buxar
3. Near Road Bridge, Buxar
4. Jail Ghat, Buxar
5. Near Ara - Chhapra Road Bridge Ara, Bhojpur.
6. Mauzampur water intake point, Barhara block, Bhojpur
7. Near Pipapul, Danapur, Dist. - Patna
8. Kurji, Dist. - Patna
9. Gulabi Ghat, Patna
10. Gandhighat, NIT, Patna
11. Gaighat Dist. Patna.
12. Malsalami, Patna City
13. Trveni Ghat, near Punpun, Fatuha, Patna.
14. Kewala ghat, Fatuha, Patna
15. Kachchi Dargah-Bidupur Road, Patna
16. Bakhtiyarpur-Tajpur Road, Athamalgola, Patna
17. Mokama, Mahadeo Asthan, Patna
18. Mokama, Rajendra Bridge, Simaria Ghat Patna
19. Umanath Ghat Barh, Patna
20. NTPC Water Point, Barh, Patna
21. NTPC Water intake Point, Barh, Patna
22. Barahiya, Nepaly Tola, Maranchi, Lakhisarai
23. Munger, Kasth harini ghat, Munger
24. Crematorium Ghat, Munger
25. Sultanganj, Ajgaibi Nath, Bhagalpur
26. Sultanganj Cremation ghat, Bhagalpur
27. Champa Nagar, Bhagalpur
28. Near Bararighat, Bhagalpur
29. Kuppa ghat, Bhagalpur
30. Near Jahajghat, Kahalgaon
31. Near Cremation Ghat, Kahalgaon, Bhagalpur

32. The Confluece of Sone River, Doriganj, Chapra, Saran
33. Maa Ambika Sthan, Aami, Dist. - Saran

Samples were collected and analysed for Most Probable Number (MPN) method to study microbial contamination in Ganga water in Bihar State Pollution Control Board, Patna Laboratory.

ANALYSIS OF MICROBIAL CONTAMINATION OF GANGA

It was observed that microbial contamination was very high in the Ganga water throughout Bihar. It was also observed that Bihar does not contribute to extra microbial contamination in the Ganga water.

Microbial contamination observation

Rivers from north Bihar were containing high microbial contaminated water as we have observed high level of microbial contamination in terms of total coliform and faecal coliform (Fig. 6). That is why we have observed many peaks at the site of Gaighat, Triveni ghat, Simaria ghat, different ghats of Munger and Bhagalpur. Faecal coliform contamination of Ganga water was observed more near river confluence and bathing area (Fig. 2).

The pH of Ganga water was found alkaline due to presence of microbes. The area where microbial contamination was very high, pH was more than 7.35 in water samples. It clearly indicates that presence of excessive total coliform and faecal coliform make Ganga water more alkaline. In good quality Ganga water dissolved oxygen is above 8 while dissolved oxygen below 8 shows heavy contamination of coliform bacteria (Fig. 3).

Biochemical oxygen demand (BOD) is below 2 in good quality of Ganga water with least microbial contamination (Figs. 7 and 8). While microbial contamination above 2 shows heavy microbial contamination of coliform bacteria in the Ganga water (Fig. 5). Faecal coliform is one of the major contributor in the Ganga water contamination while total coliform indicates source of 50% contamination which includes *E. coli, Schizella, Salmonella* etc. which are highly contagious microbial agent.

Least microbial contamination were found in Ara near road bridge and Mauzampur block of Bhojpur. It was also minimum at Kewla Ghat Fatuha, Kacchi Dargah and Bakhtiyarpur (Figs. 3 and 4).

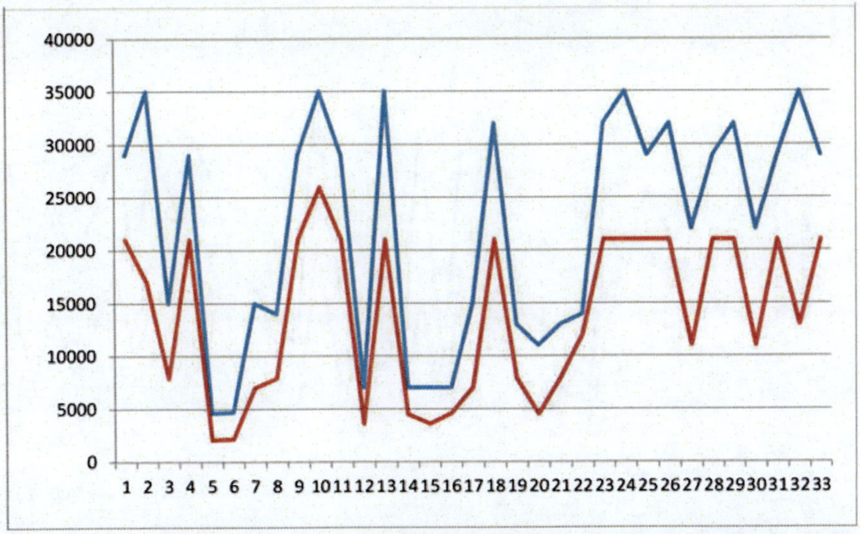

Figure 2. Maximum microbial contamination in Ganga water in Bihar. Maximum level of total coliform (blue line) and faecal coliform (red line) contamination in river Ganga from Chausa to Bhagalpur in whole stretch of river Ganga on 33 sites in Bihar. The minimum contamination was reported from Bhojpur and Ara. While at confluence of other rivers its contamination levels were increased, indicating contribution of its tributary river in increasing pollution of Ganga river. At the exit point's total coliform are also high.

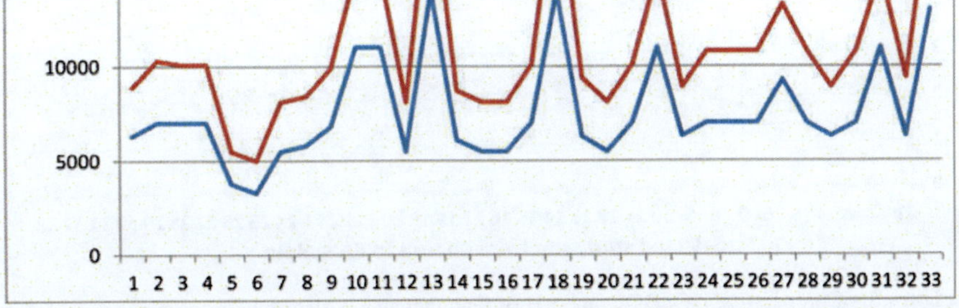

Figure 3. Minimum microbial contamination in Ganga water. Minimum level of total coliform (blue line) and faecal coliform (red line) contamination in River Ganga from Chausa to Bhagalpur in whole stretch of River Ganga on 33 site in Bihar. It was highest around Patna industrial area and near NTPC sub station. It supports that Industrial waste also contributes for heavy microbial contamination in river water.

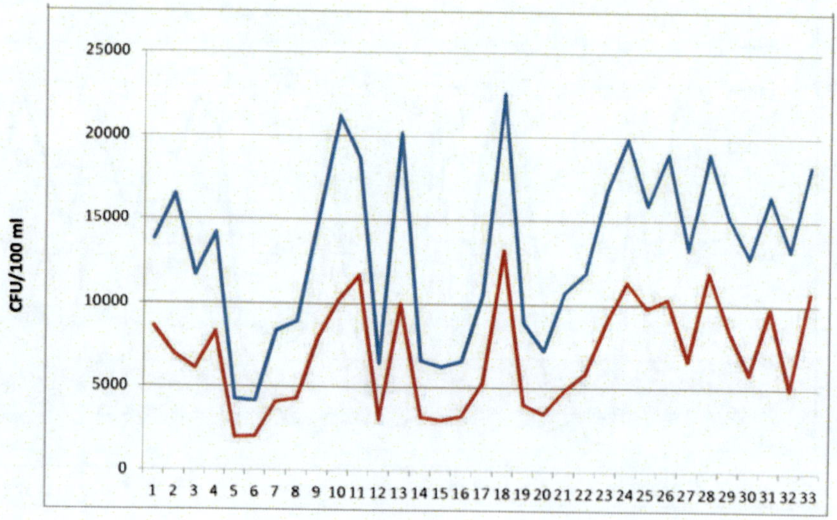

Figure 4. Avarage microbial contamination in Ganga water. Avarage level of total coliform (blue line) and faecal coliform (red line) contamination in River Ganga from Chausa to Bhagalpur in whole stretch of River Ganga on 33 site in Bihar. It was high at the point where Gandak and Koshi joins river ganga. It was also high in Patna Industrial area and after crossing NTPC Power sub station. It clearly indicates that tributary of River Ganga from north Bihar also plays a vital role in its contamination with industrial waste.

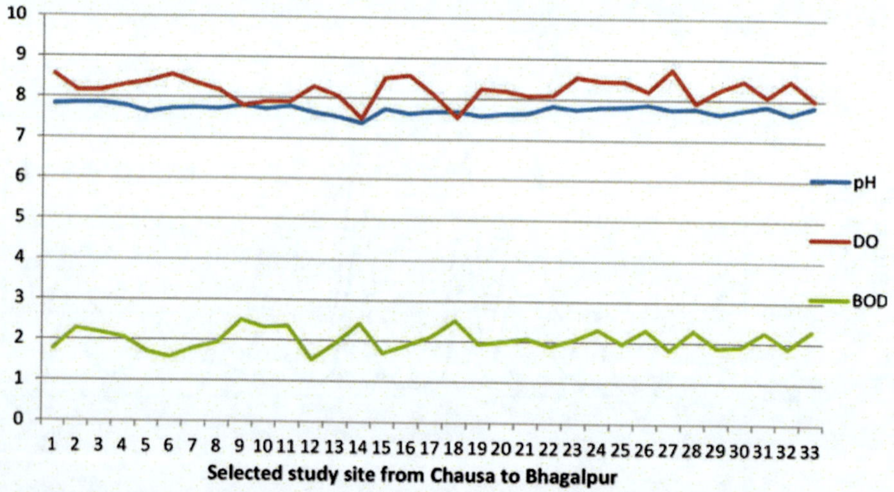

Figure 5. Comparative DO, BOD and pH of Ganga water. Relation between biochemical oxygen demand (BOD), dissolved oxygen (DO), and pH of river Ganga water. It was noted that in area of high microbial contamination dissolved oxygen was least and biochemical oxygen demand was high. While pH was more than 7.4 in high microbial contaminated area.

Figure 6. Bacterial contamination in Ganga water. High level of average total coliform and faecal coliform bacteria throughout Ganga river in Bihar. It shows that Ganga water is highly contaminated with microbial waste in whole stretch.

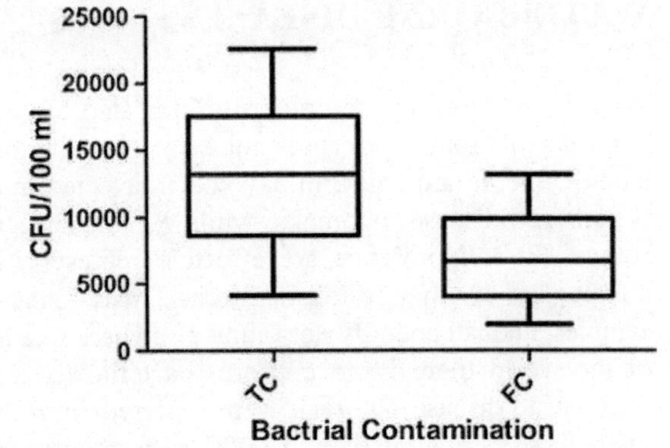

Figure 7. Overall comparison of pH and DO in Ganga water. pH and dissolved oxygen ratio in Ganga river water of Bihar. In area of high microbial contamination dissolved oxygen was decreased and pH value are more alkaline in river water system it varies from 7.25 to 7.90 in whole stretch of Bihar.

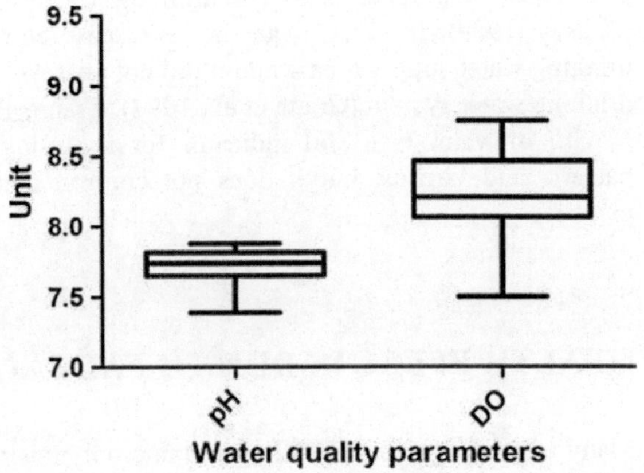

Figure 8. Overall comparison of pH, DO and BOD in Ganga water. pH, biochemical oxygen demand (BOD) and dissolved oxygen (DO) ratio in Ganga river water of Bihar. It shows that when DO was decreased BOD was increased at that point.

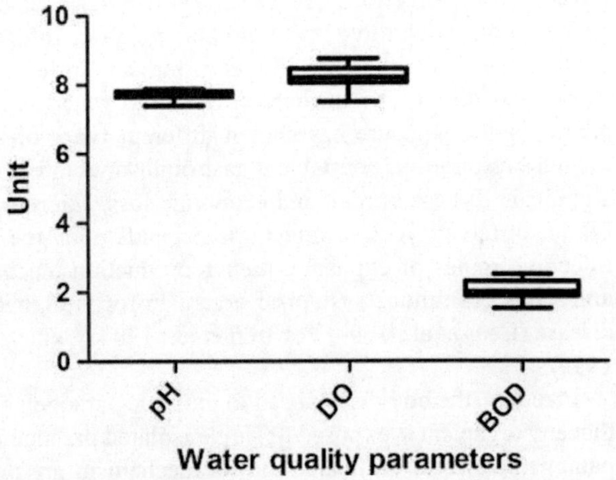

WATERBORNE DISEASES

Waterborne diseases are not easily indicated by presence of total coliforms which confirms that total coliform is not a potent risk indicator of these diseases. In many studies *E. coli* and coliform have been detected in drinking water leading to many waterborne disease outbreaks, while *E. coli* and coliform was not found in other source. *E. coli* presence are effective representative of faecal contamination in comparison to other coliforms, because it is found in very large quantity in faecal samples, and is generally not found elsewhere free in environment. In many regions of the world there is no coliform bacteria detected in water samples but many waterborne disease outbreak were reported from there. In USA, between 1978 to 1986, there were more than 110,000 cases of gastrointestinal disorders reported from 502 reported outbreaks of waterborne disease. USEPA were made the coliform compliance requirements in the drinking water for preventing disease outbreak (Sobsey, 1989). In USA, water borne disease outbreak are found in one-third of drinking water supply area while it did not possess total coliforms contamination in drinking water system (Craun et al., 1997). Craun, 1997 concluded that detection of *E. coli* in water is useful indicator for analysing water contamination through bacteria and viruses, but it does not confirm contamination through protozoan parasites.

HEALTH EFFECTS DUE TO FAECAL POLLUTION

Many nonpathogenic and pathogenic microorganisms may be isolated from household effluents, livestock, sewage, farming activities, domestic animals, industrial processes and wildlife. These pathogenic organisms can cause gastrointestinal infections and infections of respiratory tract, nasal cavity, skin, eyes and ears.

It is very difficult to detect mild and moderate infection due to recreational water contact, which may lead to illness in humans in long term use. Even more severe illness was difficult to attribute only due to water exposure. Faecal polluted recreational water exposure has shown different types of adverse health effects which includes respiratory, dermal and gastrointestinal infections. This leads to increase in significant disease burden and economic loss. Microorganisms which are responsible for different types of infections depends upon the specific pathogen, modes of infection, routes of exposure, their reproduction mechanism, susceptibility of host and status of immune system of organism for viral, bacterial and protozoan caused disease (Haas et al., 1999; Fewtrell et al., 1994; Okhuysen et al., 1999; Teunis et al., 1999).

In reality, the body is exposed to multiple pathogen at single time, it was very rare that any organism is exposed to single isolated pathogen but the cumulative multiple pathogenic effects on health and its mechanism are poorly explored (Esrey et al., 1985). The number and types of pathogens vary widely in sewage effluent. It depends on disease prevalence and presence of different carrier stages in contributing animal and humans as well as seasonal variation in infections rate, due to which wide variety of pathogens are found in different countries across the world with seasonal variation. Both marine and freshwater studies on faecal pollution on water users showed severe health impact. Many bacteria, including *streptococci* and *enterococci*, are also known as faecal index bacteria and important for assessment of water quality. These species of bacteria does not show any adverse health effect in swimmers, but cause significant illness on consumption like other faecally derived pathogens (Prüss, 1998). Pathogen transmission in gastrointestinal system can cause gastroenteritis, which is biologically plausible and is due to contaminated drinking water. The association between diarrhoea and respiratory disorders with these pathogens was reported in many epidemiological studies, which demonstrate a dose dependent response with microbial count (Prüss, 1998).

CONCLUSION

It is concluded from entire study that river water in the Ganga is highly polluted with microbial contamination due to industrial and anthropogenic activity. The pH of water is tending to high alkalinity due to presence of faecal and total coliform bacteria. Dissolved oxygen was decreasing and biochemical oxygen demand was increasing due to presence of high coliform. Coliform contamination was more observed at the site of Ganga where tributaries like Gandak, Punpun, Koshi and Ghaghra join the river. It indicates that rivers from north Bihar has high microbial contamination after crossing Patna Industrial area and NTPC, Barh unit. It indicates that industrial waste is also responsible for contaminating the holy Ganga. In total microbial contamination, faecal coliform almost contributes for 50% of overall microbial contamination. Overall the Ganga is a sign of religious faith but now condition has changed with heavy pollution burden leading to complete filth.

References

Brenner, D.J., David, B.R. and Steigerwalt, A.G. (1982). A typical biogroup of *Escherichia coli* found in clinical specimens and description of *Escherichia hermanii* sp. nov. *Journal of Clinical Microbiology*, 15: 703-713.

Craun, G.F., Berger, P.S. and Calderon, R.L. (1997). Coliform bacteria and waterborne disease outbreaks. *Journal of the American Water Works Association*, 89(3): 96-104.

Esrey, S.A., Feachem, R.G. and Hughes, J.M. (1985). Interventions for the control of diarrhoeal disease among young children: Improving water supplies and excreta disposal facilities. *Bulletin of the World Health Organization*, 63(4): 757-772.

Fewtrell, L., Godfree, A., Jones, F., Kay, D. and Merrett, H. (1994). Pathogenic Microorganisms in Temperate Environmental Waters. United Kingdom, Samara Press, 207 pp.

Haas, C.N., Rose, J.B. and Gerba, C.P. (1999). Quantitative Microbial Risk Assessment. New York, John Wiley & Sons, Inc. 449 pp.

Hendricks, C.W. (1978). Exceptions to the coliform and the faecal coliform tests. *In:* Berg, G. (Ed.), Indicators of Viruses in Water and Food. p. 99. Ann. Arbor. Science, Michigan.

Kay, D., Wyer, M.D., Crowther, J. and Fewtrell, L. (1999). Faecal indicator impacts on recreational waters: Budget studies and diffuse source modelling. *Journal of Applied Microbiology, Symposium Supplement*, 85: 70S-82S.

Kay, D., Fleisher, J., Wyer, M.D. and Salmon, R.L. (2001). Re-analysis of the seabathing data from the UK randomised trials. A Report to DETR. Aberystwyth, University of Wales, Centre for Research into Environment and Health. 17 pp.

Keene, W.E., McAnulty, J.M., Hoesly, F.C., Williams, L.P., Hedber, K., Oxman, G.L., Barrett, T. J., Pfaller, M.A. and Fleming, D.W. (1994). A swimming-associated outbreak of hemorrhagic colitis caused by *Escherichia coli* O157:H7 and *Shigella sonnei*. *New England Journal of Medicine*, 331: 579-584.

NHMRC-ARMCANZ (1996). Australian Drinking Water Guidelines. National Water Quality Management Strategy. National Health and Medical Research Council and Agriculture and Resource Management Council of Australia and New Zealand. Commonwealth of Australia.

NZMoH (2000). Drinking-Water Standards for New Zealand 2000. Ministry of Health, New Zealand.

Okhuysen, P.C., Chappell, C.L., Crabb, J.H., Sterling, C.R. and DuPont, H.L. (1999). Virulence of three distinct *Cryptosporidium parvum* isolates for healthy adults. *Journal of Infectious Diseases*, 180: 1275-1281.

Payment, P. and Franco, E. (1993). *Clostridium perfringens* and somatic coliphages as indicators of the efficiency of drinking water treatment for viruses and protozoan cysts. *Applied and Environmental Microbiology*, 59: 2418-2424.

Prüss, A. (1998). A review of epidemiological studies from exposure to recreational water. *International Journal of Epidemiology*, 27: 1-9.

Singh, K.P., Malik, A., Mohan, D. and Sinha, S. (2004). Multivariate statistical techniques for the evaluation of spatial and temporal variations in water quality of Gomti River (India)—A case study. *Water Research*, 38: 3980-3992.

Sobsey, M.D. (1989). Inactivation of health related microorganisms in water by disinfection processes. *Water Science and Technology*, 21(3): 179-195.

Teunis, P.F.M., Nagelkerke, N.J.D. and Haas, C.N. (1999). Dose response models for infectious gastroenteritis. *Risk Analysis*, 19: 1251-1260.

WHO (1993). Guidelines for Drinking Water Quality. Second Edition, Volume 1: Recommendations. World Health Organization, Geneva

WHO (1996). Guidelines for Drinking Water Quality. Second Edition, Volume 2: Health Criteria and Other Supporting Information. World Health Organization, Geneva.

Yu, Bin, Zhang, Y., Shukla, A., Shukla, S. and Dorris, K. (2001). The removal of heavy metals from aqueous solutions by sawdust adsorption - Removal of lead and comparison of its adsorption with copper. *Journal of Hazardous Materials*, 84(1): 83-94.

Chapter 5
Vulnerability and Resilience Status of River Systems of North-Eastern India: A Special Reference to Brahmaputra

Alok Kumar Thakur, Aparna Das, and Manish Kumar

INTRODUCTION

Rivers are one of the significant components of water cycle on a global scale. These river systems are essential for the geochemical cycling of the elements and rivers do have their geochemistry from their point of origin but on their due course this water chemistry changes due to the various factors. River geochemistry effects due to changes in natural and anthropogenic sources including the process of chemical weathering of silicates and carbonates which convert atmospheric CO_2 into dissolved HCO_3^- (Das et al., 2016).

Along with these weathered products, there has been an issue of a varied range of organic and inorganic contaminants distributed along with the river systems. The presence of geogenic contaminants like arsenic is the most exhaustive contaminant to be remediated in the recent times, a large chunk of it is due to the aquifers containing it in the sedimentary basins present along with the young mountains belts (Saunders et al. 2005). Changes in organic and inorganic water contaminants along with weathering products cause a negative impact in the river system. The natural waters have been continuously noticed, changing their status due to frequent variations in climates in the last few decades. Climate change played an essential role in the changes in natural water during last decades.

A. K. Thakur · M. Kumar (✉)
Discipline of Earth Sciences, Indian Institute of Technology Gandhinagar, Gandhinagar, Gujarat, India
e-mail: manish.kumar@iitgn.ac.in

A. Das
District Agriculture Office, Nalbari, Assam, India

A. Mukherjee (ed.), *Riverine Systems*, https://doi.org/10.1007/978-3-030-87067-6_5

RIVER SYSTEMS IN NORTH-EASTERN INDIA

There are many rivers other than the Brahmaputra basin in north-eastern territory of India like Barak, Gaumti, Myntdu, Kaladan, Manipur and Tizu river systems. North-eastern territory of India has Barak, Gaumti, Myntdu, Kaladan, Manipur, Tizu rivers except Brahmaputra basins. The first three river systems flow into Bangladesh, and other three rivers in the state of Mizoram, Manipur and Nagaland are flowing into Myanmar. Out of these river systems, Barak River is the largest river basin with 41,000 sq. m area excluding the state of Arunachal Pradesh. This river basin has its across the six sisters of northeast. There are many other minor river basins as well in the area which makes up an area of close to 36,000 sq. m. These river systems other than supporting the livelihoods of many, also carry a plentiful amount of biodiversity, the primary two biodiversity hotspots are Indo-Myanmar and Eastern Himalaya. Table 1 shows the growth rate of water demand by the year 2050.

The Brahmaputra river basin

The great Brahmaputra houses millions of people making it the most significant river in Asia. This river basin supports livelihoods in India (*Brahmaputra*), China (*Yarlung-Tsangpo*), Bhutan and Bangladesh (*Jamuna*), thus supporting varied biodiversity as well (Yasuda et al., 2017). The Brahmaputra is the most significant river of India in Asian continent which endorses a varied range of livelihood and biodiversity like other rivers Yarlung-Tsangpo in China and Jamuna in Bhutan and Bangladesh. Brahmaputra river basin has always been characterized by a significant variation in flow, seasonally. A major variability in sediment transport and channel configuration has also been noticed (Goswami, 1985).

Table 1. The probable Brahmaputra water demand (gross and net) by the year 2050 (India)

Sectors	*Area*	*Gross Demand (MCM*)*	*Consumption (%)*	*Net Demand (MCM*)*
Rural	Domestic	2,920	-	-
	Livestock	694	-	1,807 (Total rural)
Urban	Housings	1,533	30	459
Agricultural	Agricultural	31,680	-	-
	Surface	3,520	44	15,500
	Irrigation/ GW	16,900	50	8,500
Industrial	Subtotal	5,147	-	1,060
Total	-	62,394	55	27,630

*MCM – Million cubic meter
Adapted and Improvised (Mahanta et al., 2014).

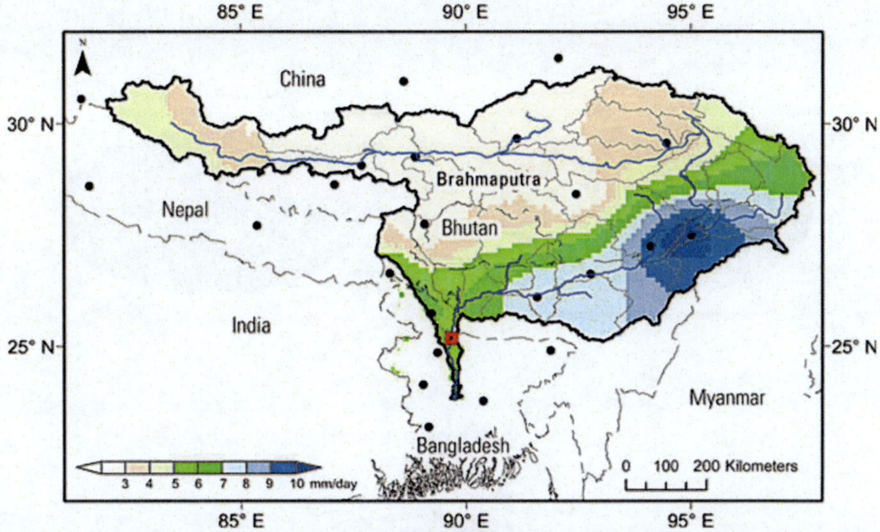

Figure 1. The Brahmaputra River Basin (Pervez and Henebry, 2015).

The remarkable variation in flow, sediment transport and channel geography have been noticed seasonally (Goswami, 1985). Figures 1 and 2 show various tributaries of the major Brahmaputra river system and their other detailed contributions.

VULNERABILITY OF URBAN WATER BODIES

The vulnerability study of water bodies in urban localities are in direct relationship with its resilience capability. This study includes all the aspects of life from ecological benefits to the socio-economic benefits and all the engineering knowledge so that the designed framework can be executed well. These studies try to include policy makers as well as stake holders to walk parallel in order to have a better interaction between the two so that our rivers could be less susceptible to emerging contaminants of concern. The next segment gives a detailed understanding of urban resilience of river systems by the impact of natural and anthropogenic activities. A developing nation like India needs an applicable strict policy for reclaimed water usage for large scale areas by fulfilling agendas of policy makers. Table 2 focusses on rivers of north-eastern India.

There has been an increase in occurrences of heavy metals, virus (pathogenic), ARBs (Antibiotic Resistance Bacteria), PPCPs (Pharmaceutical and Personal Care Products) in the urban water which has disrupted the food-water-energy nexus. Correlation of water contamination elements needs to be interpreted with a detailed framework to have a better insight about climate change and preferable water management practices. These frameworks should be such that, it does not require

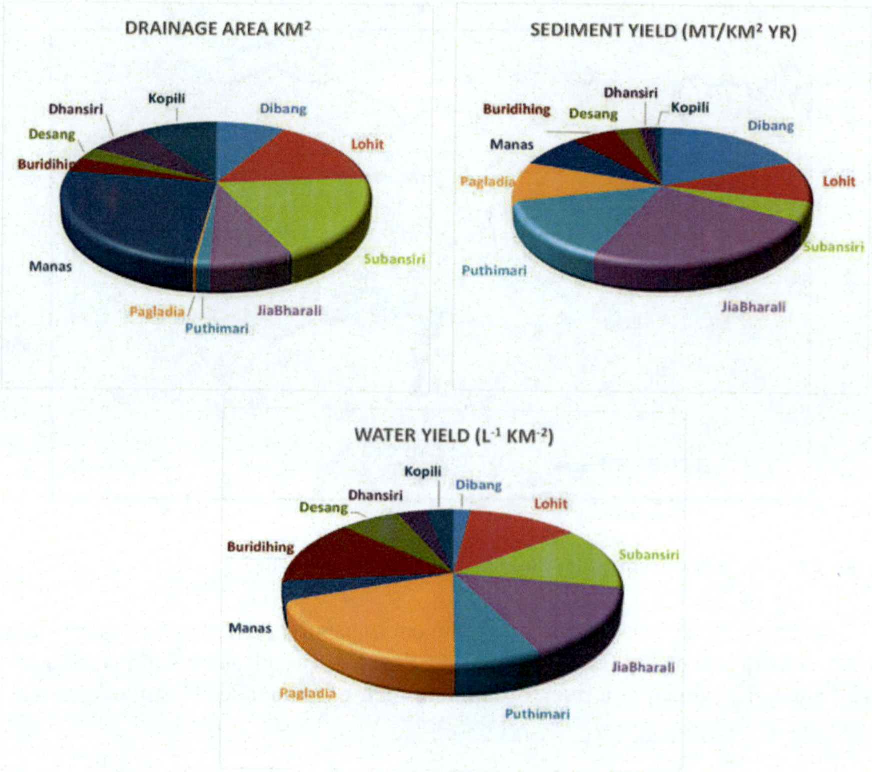

Figure 2. Tributaries of the Brahmaputra and their detailed contribution (adapted from Pahuja & Goswami, 2006).

frequent changes and is valid for at least 2 to 3 decades i.e. should be built including the climate change adaptabilities. Figure 3 shows a logical framework for the assessment of any river system.

RESEARCH STUDIES ON NORTH-EASTERN RIVERS (BRAHMAPUTRA)

The north-eastern rivers of India have caught large attention in the last few decades. There has been studies from a geological process like channel process, sedimentation to the flood management studies and how to prevent them in the future. The recent studies are more focussed towards the contamination, both geogenic and anthropogenic ones. These studies will help gain knowledge of these river basins, which are supporting millions of lives. Figure 4 shows the number of studies on the Brahmaputra and its tributaries over the past few years.

Table 2. Major river system of north–eastern India

Rivers	*Source and States Covered*	*Tributaries*	*Length*	*Flow Rate*	*Annual Precipitation*	*Flora and Fauna*	*Studies Conducted Till Now*
Barak	Manipur Hills and Manipur, Nagaland, Mizoram, Assam[4]	Jiri, Dhaleswari, Singla, Longsai, Sonai, Katakhal[4]	900 km 524 km (India), 31 km (Indo-Ban Border) 345 km (Bangladesh)	7786.08 m^3/sec	4194 mm (Silchar)	Swamps, Mangroves and Grasslands, Siamese Crocodile, Susu Dolphin,	Molluscan diversity[1], Pattern Change Detection[2], Erosional Vulnerability and Spatio-Temporal Variability[3]
Brahmaputra	Kailash Range, Tibet and AP, WB, Assam, Meghalaya, Nagaland and Sikkim		1625 km (China), 918 km (India), 337 km (Bangladesh)[5]	20,000 m^3/sec	1350 mm	Clouded Leopard, Asian Elephants, Malayan Sun Bear, Canopy Trees, Bamboos, Understory Trees.	Geochemistry[6], Arsenic Chemistry[7], Impact of Climate Change[8], Channel Process and Sedimentation[9], Emerging Contaminants[10], Spatial Distribution of Persistent Organic Pollutant[11]
Teesta	TsoLhamo Lake, Sikkim, WB and Bangladesh	Great Rangeet	414 km	1500 m^3/sec (Summer Monsoon)	3000–6000 mm	Intense Fish Fauna, Tropical Deciduous, Temperate broad-leaved, Alpine	Climatic Imprints in Quaternary Valley[12], Assessment of physio-chemical prop., heavy metals and Arsenic[13]

1 - Roy & Gupta (2010), 2 - Das, Dutta and Saraf (2007), 3 - Laskar and Phukon (2012), 4 - Nath et al., 2017, 5 - Singh et al., 2004, 6 - Subramanian et al., 1985, 7 - Singh, 2006, 8 - Nepal & Shrestha, 2015, 9 - Coleman 1969, 10 - Sharma et al., 2019, 11 - Chakraborty et al. 2014, 12 - Meetei et al., 2007, 13 - Saha et al., 2017.

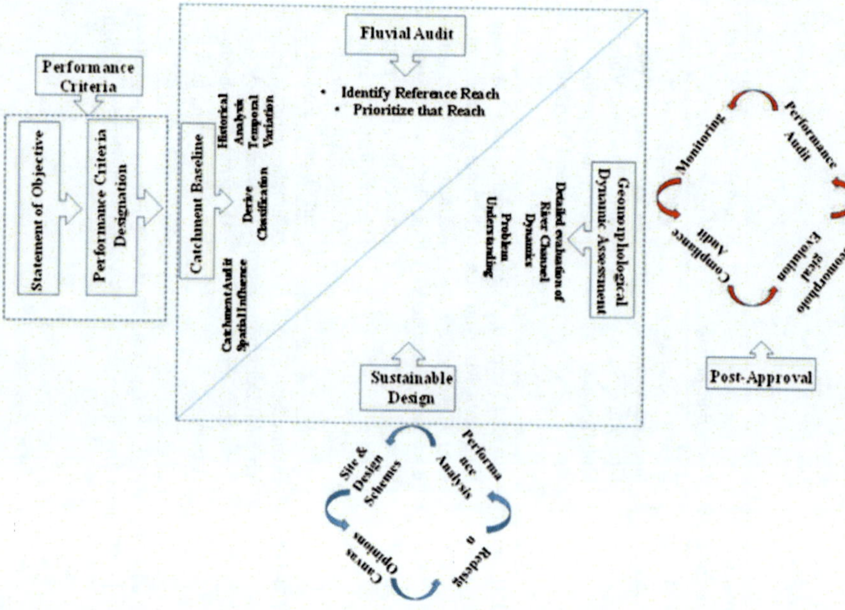

Figure 3. Geomorphological assessment framework for river basins (Adapted and improvised from Thorne, 2002).

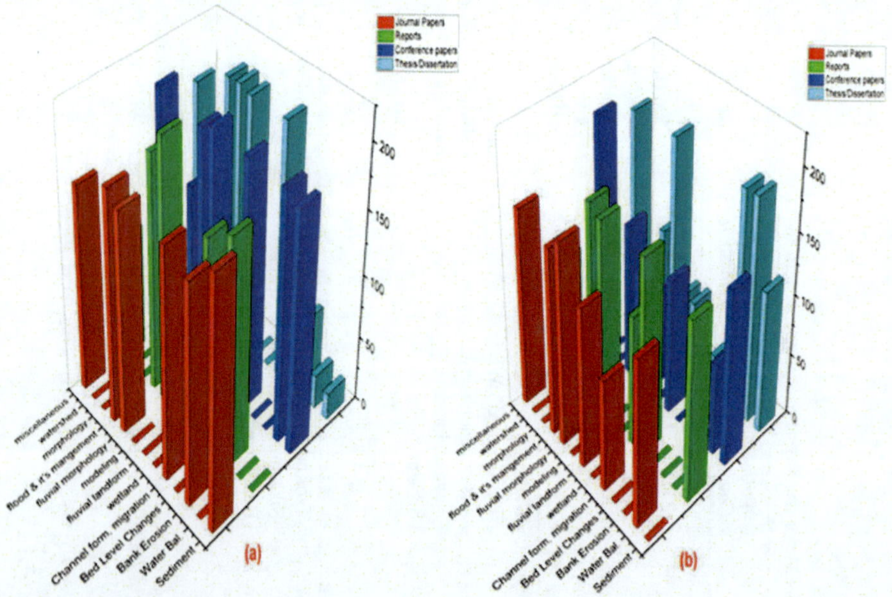

Figure 4. Number of research studies on Brahmaputra river basin: (a) Main stem of Brahmaputra river and (b) Tributaries of Brahmaputra river.

(i) **Chemical Weathering Cyclicity**

In a study by Das et al., 2016, the focus was on evaluating the effect of weathering on CO_2 consumption rate of the river. This in long term helps in evaluating its effect on climate change and was done by observing the weathering analysis along the entire continental stretch of Brahmaputra. This analysis was done alongside the study of CO_2 consumption rate. This way, a relation was established between CO_2 consumption rate, rate of weathering and the variation in the dominant ions. The samples collected were on the scale of both the variation, i.e. temporal and spatial. After the analysis, weathering was found to be a significant contributor, and anthro factors played a minor role. The CO_2 consumption rate of Brahmaputra was more than its counterparts Himalayan rivers. Another study by Samanta, Tripathi and Das observed the weakly variation for a year to have an understanding of pattern in weathering and the contribution of possible solute sources. There was no appreciable change noticed after the data comparison with the data of previous decades. Ion concentration and Ca/Na-Cl varied inversely with the discharge of the river. The main reason for variation in weathering intensification was due to regional conditions than due to climatic conditions.

(ii) **Contamination in the Brahmaputra and Its Nearby River Plains**

Chetia et al., 2011 studied the contamination in the region of Golaghat, Assam. Where more than 67% of the samples were found to be contaminated by arsenic (>10 ppb), 76% by iron (>1 ppm) and 28% by manganese (300 ppb). Gamariguri blocks were worst affected with almost 100% of the samples being contaminated. Two more conclusions were that, the contamination level was found higher in tubewell depth varying between 170 and 190 ft, and secondly a strong co-relation between arsenic and iron, indicating the reductive dissolution of As-Fe bearing materials. Another study by Chakraborti et al. (2013) showed thousand-fold samples than the previous study and also a significant area of 0.5 million km^2. In the entire stretch, a total of 0.5 billion population are prone to As contamination with more than 50% consuming more than the permissible limit of As by WHO (10 ppb). The main inferred statement lies in managing the available water resources more effectively as there is plenty of it.

Emerging Contaminants

A study by Barman and Gokhale (2019) presented the BC in the urban scenarios. This study performed trajectories analysis using HYSPLIT model to examine the correlation of BC pollutants with climate parameters and found out the positive correlation with relative humidity and negative in terms of other climate parameters like temperature, wind speed and solar radiation. Another study by Rajkumar and Sharma, 2013 focussed on the variation of pathogenic bacteria with different seasons. The coliform count was more in monsoon season, *Pseudomonas* spp. was recorded as 16 MPN/ml in the rainy season and around 1.1 MPN/ml in winters.

(iii) **Chemical Speciation of Brahmaputra Sediments for Heavy Metals**

A study by Das et al. noted that the universal nature of copper, lead and zinc are prominent because of anthropogenic and geogenic forms. The destiny of these metals in nature is to a great extent administered by their speciation, which is affected by silt properties. This way, present examination focusses on the dispersion of Pb, Zn and Cu in various compound structures in sediments. Four silt centres were gathered from upstream to downstream of the Brahmaputra waterway. Metal speciation study uncovered that the more significant part of the overwhelming metals was related with the leftover division. Among non-leftover part, Zn and Pb were mainly connected with reducible division while Cu was related with oxidisable portion. The bioavailable portions for Pb are seen as similarly higher. Connection investigation was done among various geochemical divisions of overwhelming metals and silt properties. It was discovered that the vast majority of the non-remaining division of overwhelming metals is fundamentally connected with grain size, pH and fundamental issue. A noteworthy connection was found between metals (Lead, Copper and Zinc) in the oxidisable division and fundamental issue. Hazard appraisal code recommends low to medium hazard for every overwhelming metal.

A study by Saha et al. (2006) had given a detailed outlook on the variation of heavy metals and their respective concentration in the wetland of Sunderban Mangroves. The study was done on the tissues of benthic polychaetes, bivalve molluscs and finfishes. These species were segregated from the north-eastern coast of India. The bioaccumulation analysis study was performed which resulted in setting up of a trend where zinc was at the top followed by manganese, copper, chromium, selenium and mercury. The conclusion being how molluscs can play an essential role in trace metal studies. Another study by Saha et al. (2017) for the Teesta River concludes that the DO and BOD levels are well within the prescribed limit and is thus useful for fishing and agriculture practices. The waters were arsenic-free and other heavy metals were also below the WHO prescribed limit.

(iv) **Water soluble ions in PM_{10}**

In one of the studies by Bhuyan et al. (2016a), a year-long investigation between 2012 and 2013 was done in order to have an insight about the seasonal and spatial variation of water-soluble PM_{10} ions. These ions included sulphate, nitrate, fluoride, chloride, ammonium, calcium, sodium, magnesium and were analysed on ion chromatography. A significant amount of difference was noticed among all the three sampling stations. The urban area showed a concentration varying in the range of 15.1 to 127.1 μg/m^3. In anions, sulphate and chloride were more prominent and in cations, ammonium and potassium were more profound. Seasonal analysis showed the SO_4^{2-} being the most prominent ion all around the year, and also a significant abundance was noticed around winters. Except for winters, all the seasons were inferring towards the atmosphere to be alkaline. Another study of Bhuyan et al. (2016b), highlighted the relationship between PM_{10} aerosols and AOD (Aerosol Optical Depth). The study year was between 2010 and 2014, and all the principal seasons and significant ions were taken into account to study the particulate loadings. There was a strong correlation observed between sulphate, nitrate, ammonium, potassium and the organic C fractions of PM_{10}. This resulted in the variation of incoming lights.

(v) **Redox domination**

A study by Patel et al. attempted to establish a relation between arsenic mobility in an aquifer; this mobility will be seen in variation with the depth of the concerned aquifer. Also, a potential co-relation of arsenic with heavy metals such as zinc and lead is analysed. This study was done utilising pilot-scale staggered groundwater observing framework (MGWS). The samples of groundwater were collected fortnightly for one whole year with the help of multi-level sampler. These samplers were known to collect the samples at depths of 4.6, 9.2 and 13.8 m, and the study area was Tezpur located at Brahmaputra floodplain (BFP). Impacts (geo and anthro) were found to influence the concentrated spring (unconfined). The higher pH along with the oxidizing zone around 4.6 m depth played an important role for arsenic release through desorption. Statistical analysis examinations uncovered that ORP values remained the essential governer of arsenic discharge at each of the 3 profundities. With profundity, more grounded conditions (anoxic) brought about the predominance of hydrolysis (reductive) prompting a simultaneous situation of As (max 4.6 mg L^{-1}) with Zn (max 2514 mg L^{-1}) and Pb (max 740 mg L^{-1}) with impacts of anthro activities of exercises like horticulture and dry testimony from a blocked furnace. Multi-component improvement is a developing concern; however, the master plan has comprehended the idiosyncrasies of individual springs, as speculation can prompt missing a considerable amount of data. In such manner, long haul staggered observing can falicitate in the prescient comprehension of the stratification (vertical) and simultaneous of multi-metals that can in this way be effective for water generation at more secure profundities.

(vi) **Nutrient Dynamics of River System**

The river flood plains are considered as the most endangered ecosystems worldwide due to degraded status of the land and water for non-implementation of stringent rules on ambient water and land-use patterns. The river flood plains are extremely fertile; therefore, they act as a source of income and survivability globally (Arrigoni et al., 2008). Due to seasonal and periodical changes the flood plains undergo, results in the change of mobility of the river leading to ultimate change in land-use pattern.

Nutrient generated at a particular location is utilised as well as displaced in the flowing water to the distant downstream through nutrient spiralling (Ensign and Doyle, 2006). The dissolved inorganic nutrient dynamics is mainly associated with its transportation from catchment to the water column and subsequently to the stream bed (Allen, 1995). The clear point and diffuse sources along with the internal biogeochemical mineralisation processes are the contributing factors of nutrient inputs to the aquatic system. In addition, the hydrologic regime greatly influences the highly dynamic and complex physio-chemical characteristics and ecology of the streams.

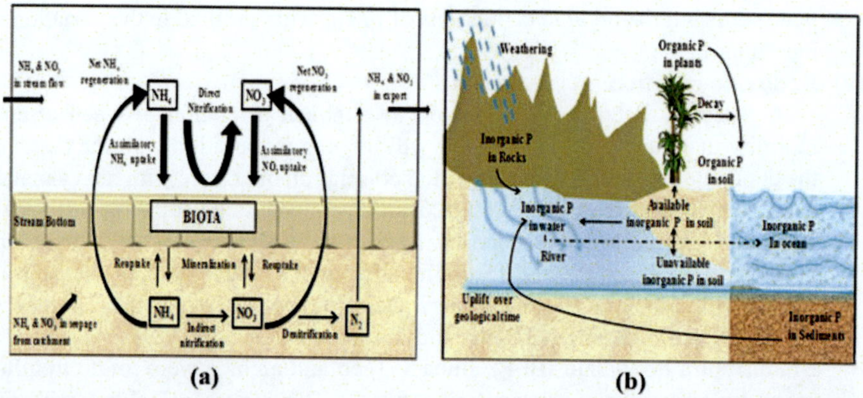

Figure 5. Nutrient dynamics of Brahmaputra river basin: (a) Nitrogen cycle and (b) Phosphorus cycle (PhD Thesis, Das, 2016).

The entanglement of physio-chemical changes during the peak flow period and dry periods are essential parameters to understand the nutrient dynamics of the system. The less amount of discharge along with later half of high temperature conditions during spring and subsequently in summer, may enhance nutrient dynamics associated with column interface of sediment and water. Besides, the benthic mineralisation to trophic chain increases as the water column gets shallower (Capone and Kiene, 1998). Therefore, significant biogeochemical processes might be taking place, in coalescence with water shallowness, thermal conditions and low oxygen concentration. Figure 5 shows the nitrogen and phosphorus cycle of a river system.

Extensive population growth increases problems related to water—qualitatively and quantitatively. This includes the elevated release of nutrient from diffuse and point-sources causing acidification of surface water and eventually eutrophication of inland and coastal waters. (Lindgren et al., 2007; Onderka, 2007; Prasad et al., 2005; Rothwell et al., 2011). Extensive population growth increases problems related to water in terms of quality and quantity, including the elevated release of nutrient from diffuse and point-sources causing acidification of surface water and subsequent eutrophication of inland and coastal waters. (Duda, 1993; Lam et al., 2010). For effective mitigation and remediation of impaired water quality, we need to assess spatial and temporal variations along with point and diffuse sources of pollution to manage the sustainability of watershed. Thus, understanding factors affecting nutrient fluxes in streams is crucial for further evaluation to design nutrient export models at local and regional scales (Rothwell et al., 2011).

Influence of Anthropogenic Factors

Nutrients are released both from natural and anthropogenic sources. The primary sources of nutrients (N and P) to aquatic ecosystems are agricultural and urban activities (Conley et al., 2009; Duan et al., 2011). An agricultural activity like fertiliser application, in combination with channelisation of streams in the agricultural areas (Riseng et al., 2011; Yan et al., 2009; Allan, 2004), has led to increased N

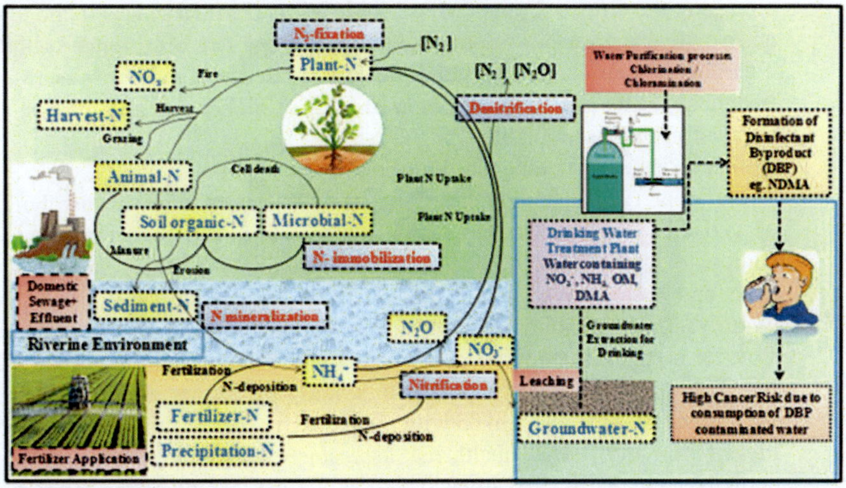

Figure 6. Anthropogenic activities altering nutrient cycling of nitrogen (PhD Thesis, Das, 2016).

and P and this is increasing at an alarming rate. The dwindling number of trees along with the extensive agriculture has lead to increased soil erosion and nutrient transport through river systems (Seitzinger et al., 2010). The other sources include animal wastes (sewage, poultry), which causes atmospheric deposition and groundwater inflow of these contaminants. The effluent released from wastewater (or sewage) treatment plants is discharged into the river, which acts as a source of nutrients. The most unresolved source of nutrients in rivers is weathering. Pollutants that get adsorbed in the sediments after sometime get build upon the open surface waters such as river and lakes. Figure 6 shows the influence of human factors in deteriorating the nutrient dynamics of rivers.

(vii) **Co-occurrence of Contaminants**

1. Co-occurrence of ARBs and PPCPs

   Study by Kumar et al. (2019) shows the co-occurrences of multiple emerging contaminants of concern. These contaminants included the variety of microbes which are devoid of any drug effect and heavy metals, viruses, and a wide range of PPCPs (Pharmaceutical and Personal Care Products). Water contaminants become a critical issue in urban areas in terms of water management and health aspects of the community. There is need for vulnerability research issues for urban areas from origins to remediation of pollutants including faecal pointers. The study was analysing the urban water of Guwahati, India, where co-occurrences of PPCPs, ARBs along with heavy metals and *E. coli* were studied. The study was focussed on identifying a marker for pollution. Samples from lake, tributaries, river and Sewage Treatment Plant (STP) were collected and were analysed for the generally used drugs like carbamazepine, caffeine and others. While analysing the viruses, Pepper Mild Mottle Virus (PMMoV) was found to be an excellent marker for anthropogenic pollution. The occurrence of these emerging contaminants and with little or no presence of modern wastewater treatment plants put the health of entire city at higher risk and making the population more vulnerable.

2. Co-occurrence of Arsenic and Fluoride

Das et al. (2016) studied the effects of river proximity on the distribution of geogenic contamination i.e. As and F. Here, the higher amount of carbonate weathering was increasing the alkalinity of the river at south bank and thereby influencing the levels of the contaminants. Reductive hydrolysis was found to be a significant phenomenon and thus was responsible for higher As levels. Also, the newer sediments were more responsible for having As than the older ones. In a study by Kumar et al. (2017), the occurrence of As is also linked with the reductive dissolution of the hydroxides of iron and further health risk assessment was done in order to have entire Nagaon district investigated. Das et al. (2017a) studied the correlation of all these three contaminants in the upper, middle and lower Brahmaputra flood plain. Cluster and component analysis was used to find the cause of As release, which was later concluded to be Fe-hydroxides. The desorption of As in the presence of Fe^{-} ions at a higher pH suggested a high co-occurrence possibility. These behaviours were mainly noticed in the aquifers of the upper Brahmaputra plain which were isolated. Also, these hydroxides of Fe were acting as a sink of uranium. Also, a further health study showed that kids between the age of 3 to 8 years were at a higher risk than their parents. Das et al. (2017b) also noticed the seasonal variation in these co-occurrences over the Brahmaputra plain.

(viii) **Water Quality in North-eastern Rivers of India**

This study helps us in determining the type of weathering which is behind it on a full basin scale, with the help of which exogenic cycles of different elements can be understood at a continental scale. Studies have already been performed for major river systems around the world like Amazon, Mackenzie, Mekong and Chiang (Sarin et al., 1989). Carbonate weathering was noticed in the region with the significant chunk of bicarbonates of Ca and Mg. Illite comprised a significant portion of the clay noticed in bedload sediments, and only a minor percentage was of kaolinite and chlorite, which shows the rocks being acidic. Similar study was done for the Teestariver as well (Tsering et al., 2019). Attention was paid mainly to the kind of weathering taking place and composition of water in terms of quality. Carbonate weathering was predominated with the presence of HCO_3^- ions of Ca and Mg, but the domination of silicate weathering was more in the downstream, i.e. lowland levels. The parameters affecting the type of weathering were gradient, contact time, vegetation and temperature. After the rock weathering, evaporative dissolution was another significant factor in controlling the chemistry of major ions. Similar study was carried out for Barak river basin, the second-longest river in north-eastern valley (Khangembam and Kshetrimayum, 2019) where the composition of surface water was also noticed as Ca-Mg-HCO_3, while groundwater did not have any fixed composition. The limits of As, Fe and Mn were much above the prescribed limit for drinking.

(ix) **Climate Change Impacts**

A study by Pervez and Henebry (2015) focusses on three responsible factors for the freshwater availability in Brahmaputra river system, which are climate change, land use and change in land cover pattern. The SWAT calibration model was used and daily observed discharge was to calibrate further and validate the model. The results were concluded that there will be an increase in the freshwater availability with increasing seasonal variations and also the frequencies of drought and flood could get much worse. Khan and Ali (2019) studied the changes in the water balance of Teesta influenced by climate change, SWAT (Soil-Water Assessment Tool) was used with the GCM (Global Climate Models) solution for the 2080s. After comparing water balance of climate change and base models, a 48% increase in precipitation during monsoon and 43% decrease in precipitation during dry seasons was noticed. Table 3 gives some of the literature studied on the Brahmaputra river basins.

(x) **Effects of Reservoirs and Dams**

In a study by Wiejaczka et al. (2018), the water analysis was done before and after the reservoir along with the river's longitudinal profile. A decrease in the concentration of Cl^-, K^+, Na^+, Mg^{2+}, NO_3^-, PO_4^{3-}, an increase in the concentration of Ca^{2+}, SO_4^{2-}, NH_4^+ and no change in fluoride concentration was noticed. There was a slight increase in the concentration of heavy metals. Another study (Islam and Islam, 2016) highlights the major environmental issues regarding Tipaimukh dam in Assam region on Barak river. The environmental effects were quite disastrous, including drying up of Surma and Kushiara and thus effecting a significant chunk of Bangladesh population. A study by Bora and Goswami (2017) on the Kolong river in terms of water quality index is analysed. The effects of after dam is such that it has become a contaminated pool with little or no ecological and economic benefits at all. WQI assessment has gone up to

Table 3. Literature Studies predicting future climate change on Brahmaputra River Basin

Impact (>10,000 km^2)
Decrease in upstream water supply by around 20%
Decrease in the availability of water due to the rise in *Evapotranspiration*
Increase in melting of glaciers till 2040, decrease after that
Significant increase in mean peak discharge using GCM (General Circulation Model)
Increase in duration of flood-waves (A2 Case)
Increase in Runoff (Factors: Precipitation and Glacier Melt)
Increase in the peak of the flood

Adapted from (Nepal and Shrestha, 2015)

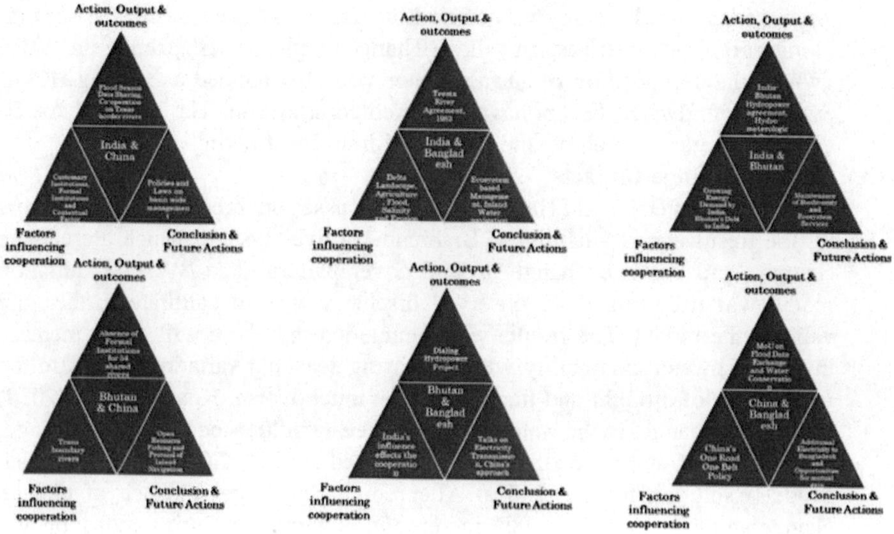

Figure 7. Different cooperation standards on Brahmaputra river basin.

122.47 in monsoon, making it one of the top polluted rivers in India. Many countries sharing the Brahmaputra river, are India, Bangladesh, Bhutan and China. Figure 7 shows the bilateral agreements between different countries sharing the Brahmaputra.

CONCLUSION

Existing approaches and improvement programmes by the policy makers have not yet perceived the change in climate as a critical factor and the impact it can have on the hydrogeological features and socio-economic dynamics of entire north-eastern India. These changes will be even more critical for the state like Assam which is located at the downstream of the major river system. The idea of it remains with the three basic principles of identification of the problem, identification of its cause and remedial measures for mitigation. This can be done with the potential mapping of the area of interest. Dependable appraisal and mapping of powerlessness and hazard and advancing proper adjustment systems are too similarly essential to guarantee relief of the water and atmosphere initiated powerlessness in the area. Quick response framework ought to be started by state governments in the north-eastern area through proper strategies and practices can help attain an ecological balance in the area along with the sustainable development. Other than legitimate logical evaluation of the environmental change impacts, limited working of powerless populace to diminish dangers and adjust to atmosphere driven changes in typical biological systems and human social orders are the urgencies of significant phenomena that is going to revolve around Brahmaputra river.

References

Allan, J.D. (2004). Landscapes and riverscapes: the influence of land use on stream ecosystems. *Annual Review of Ecology, Evolution and Systematics*, 257-284.

Allen, J.D. (1995). Nutrient dynamics. *In:* Stream Ecology: Structure and Functions of Running Waters (pp. 283–303). London: Chapman & Hall.

Arrigoni, A., Findlay, S., Fischer, D. and Tockner, K. (2008). Predicting carbon and nutrient transformations in tidal freshwater wetlands of the Hudson River. *Ecosystems,* 11(5): 790-802.

Barman, N. and Gokhale, S. (2019). Urban black carbon-source apportionment, emissions and long-range transport over the Brahmaputra River Valley. *Science of the Total Environment*, 693: 133577.

Bora, M. and Goswami, D.C. (2017). Water quality assessment in terms of water quality index (WQI): Case study of the Kolong River, Assam, India. *Applied Water Science*, 7(6): 3125-3135.

Bhuyan, P., Barman, N., Bora, J., Daimari, R., Deka, P. and Hoque, R.R. (2016a). Attributes of aerosol bound water soluble ions and carbon, and their relationships with AOD over the Brahmaputra Valley. *Atmospheric Environment*, 142: 194-209.

Bhuyan, P., Barman, N., Begum, S., Gogoi, D., Borah, S., Kumar, M., Sarma, K.P. and Hoque, R.R. (2016b). Spatial and Seasonal Variations of Water Soluble Ions in PM 10 of Mid-Brahmaputra Plain of Assam Valley. *Asian Journal of Water, Environment and Pollution*, 13(2): 69-81.

Capone, D.G. and Kiene, R.P. (1998). Comparison of microbial dynamics in marine and freshwater sediments: Contrasts in anaerobic carbon metabolism. *Limnology and Oceanography*, 33: 725-749.

Chakraborti, D., Rahman, M.M., Das, B., Nayak, B., Pal, A., Sengupta, M.K., Hossain, M.A., Ahamed, S., Sahu, M., Saha, K.C. and Mukherjee, S.C. (2013). Groundwater arsenic contamination in Ganga–Meghna–Brahmaputra plain, its health effects and an approach for mitigation. *Environmental Earth Sciences*, 70(5): 1993-2008.

Chakraborty, P., Sakthivel, S., Kumar, B., Kumar, S., Mishra, M., Verma, V.K. and Gaur, R. (2014). Spatial distribution of persistent organic pollutants in the surface water of River Brahmaputra and River Ganga in India. *Reviews on Environmental Health*, 29(1-2): 45-48.

Chetia, M., Chatterjee, S., Banerjee, S., Nath, M.J., Singh, L., Srivastava, R.B. and Sarma, H.P. (2011). Groundwater arsenic contamination in Brahmaputra river basin: A water quality assessment in Golaghat (Assam), India. *Environmental Monitoring and Assessment*, 173(1-4): 371-385.

Coleman, J.M. (1969). Brahmaputra River: Channel processes and sedimentation. *Sedimentary Geology*, 3(2-3): 129-239.

Conley et al. (2009). Controlling eutrophication: Nitrogen and phosphorus. *Science*, 323: 1014-1015.

Das, J.D., Dutta, T. and Saraf, A.K. (2007). Remote sensing and GIS application in change detection of the Barak river channel, NE India. *Journal of the Indian Society of Remote Sensing*, 35(4): 301-312.

Das, N., Deka, J.P., Shim, J., Patel, A.K., Kumar, A., Sarma, K.P. and Kumar, M., (2016). Effect of river proximity on the arsenic and fluoride distribution in the aquifers of the Brahmaputra Floodplains, Assam, Northeast India.

Das, N., Sarma, K.P., Patel, A.K., Deka, J.P., Das, A., Kumar, A., Shea, P.J. and Kumar, M. (2017a). Seasonal disparity in the co-occurrence of arsenic and fluoride in the aquifers of the Brahmaputra flood plains, Northeast India. *Environmental Earth Sciences,* 76(4): 183.

Das, N., Sarma, K.P., Patel, A.K., Deka, J.P., Das, A., Kumar, A., Shea, P.J. and Kumar, M. (2017b). Seasonal disparity in the co-occurrence of arsenic and fluoride in the aquifers of the Brahmaputra flood plains, Northeast India. *Environ. Earth Sci.*

Das A. (2016) Coupled application of Geochemistry, Isotope and SWAT modeling to understand nutrient dynamics in the Brahmaputra river system. Tejpur University, India

Das, P., Sarma, K.P., Jha, P.K., Ranjan, R., Herbert, R. and Kumar, M. (2016). Understanding the Cyclicity of Chemical Weathering and Associated CO_2 Consumption in the Brahmaputra River Basin (India): The Role of Major Rivers in Climate Change Mitigation Perspective. *Aquatic Geochemistry*, 22(3): 225-251.

Duan et al. (2011). Temperature Control on Soluble Reactive Phosphorus in the Lower Mississippi River? *Estuaries and Coasts*, 34: 78-89.

Duda, A.M. (1993). Addressing non-point sources of water pollution must become an international priority. *Water Sci. Technol.*, 28(3-5): 1-11.

Ensign, S.H. and Doyle, M.W. (2006). Nutrient spiraling in streams and river networks. *Journal of Geophysical Research: Biogeosciences*, 111(G4).

Goswami, D.C. (1985). Brahmaputra River, Assam, India: Physiography, basin denudation and channel aggradation. *Water Resources Research*, 21(7): 959-978.

Islam, M.S. and Islam, M.N. (2016). "Environmentalism of the poor": The Tipaimukh Dam, ecological disasters and environmental resistance beyond borders. *Bandung: Journal of the Global South*, 3(1): 27.

Khan, I. and Ali, M. (2019). Potential Changes to the Water Balance of the Teesta River Basin Due to Climate Change. *American Journal of Water Resources*, 7(3): 95-105.

Khangembam, S. and Kshetrimayum, K.S. (2019). Evaluation of hydrogeochemical controlling factors and water quality index of water resources of the Barak valley of Assam, Northeast India. *Groundwater for Sustainable Development*, 8: 541-553.

Kumar, M., Patel, A.K., Das, A., Das, N. and Goswami, R. (2017). Comparative understanding of arsenic enrichment and mobilization in the aquifers of the river Ganges and Brahmaputra: A provenance, prevalence and health perspective.

Kumar, M., Patel, A.K., Das, A., Kumar, P., Goswami, R., Deka, P. and Das, N. (2017). Hydrogeochemical controls on mobilization of arsenic and associated health risk in Nagaon district of the central Brahmaputra Plain, India. *Environmental Geochemistry and Health*, 39(1): 161-178.

Kumar, M., Chaminda, T., Honda, R. and Furumai, H. (2019). Vulnerability of urban waters to emerging contaminants in India and Sri Lanka: Resilience framework and strategy. *APN Science Bulletin*, 9(1). doi:https://doi.org/10.30852/sb.2019.799.

Lam, Q.D., Schmalz, B. and Fohrer, N. (2010). Modelling point and diffuse source pollution of nitrate in a rural lowland catchment using the SWAT model. *Agr. Water Manage.*, 97: 317-325.

Laskar, A.A. and Phukon, P. (2012). Erosional vulnerability and spatio-temporal variability of the Barak River, NE India. *Current Science*, 80-86.

Lindgren, G.A., Wrede, S., Seibert, J. and Wallin, M. (2007). Nitrogen source apportionment modeling and the effect of land-use class related runoff contributions. *Nord. Hydrol.*, 38(4-5): 317-331.

Mahanta, C., Zaman, A.M., Newaz, S.S., Rahman, S.M.M., Mazumdar, T.K., Choudhury, R., Borah, P.J. and Saikia, L. (2014). Physical assessment of the Brahmaputra River. Ecosystems for Life: A Bangladesh-India Initiative.

Meetei, L.I., Pattanayak, S.K., Bhaskar, A., Pandit, M.K. and Tandon, S.K. (2007). Climatic imprints in Quaternary valley fill deposits of the middle Teesta valley, Sikkim Himalaya. *Quaternary International*, 159(1): 32-46.

Nath, K.D., Borah, S., Yadav, A.K., Bhattacharjya, B.K., Das, P., Deka, P.M., Darngawn, O. and Nath, D.V.J. (2017). Length-weight and Length-length relationship of four native fish species from Barak River, Assam, India. *Journal of Experimental Zoology India*, 20(2): 977-979.

Nepal, S. and Shrestha, A.B. (2015). Impact of climate change on the hydrological regime of the Indus, Ganges and Brahmaputra river basins: A review of the literature. *International Journal of Water Resources Development*, 31(2): 201-218.

Onderka, M. (2007). Correlations between several environmental factors affecting the bloom events of cyanobacteria in Liptovska Mara reservoir (Slovakia) – A simple regression model. *Ecol. Model.*, 209: 412-416.

Pahuja, S. and Goswami, D.C. (2006). A fluvial geomorphology perspective on the knowledge base of the Brahmaputra. *Background Paper*, 3.

Pervez, M.S. and Henebry, G.M. (2015). Assessing the impacts of climate and land use and land cover change on the freshwater availability in the Brahmaputra River basin. *Journal of Hydrology: Regional Studies*, 3: 285-311.

Prasad, V.K., Ortiz, A., Stinner, B., McCartney, D., Parker, J. et al. (2005). Exploring the relationship between hydrologic parameters and nutrient loads using digital elevation model and GIS – A case study from Sugarcreek headwaters. Ohio, USA. *Environ. Monit. Assess.*, 110: 141-169.

Rajkumar, B. and Sharma, G.D. (2013). Seasonal bacteriological analysis of Barak River, Assam, India. *Applied Water Science*, 3(3): 625-630.

Riseng, C.M., Wiley, M.J., Black, R.W. and Munn, M.D. (2011). Impacts of agricultural land use on biological integrity: A causal analysis. *Ecological Applications*, 21(8): 3128-3146.

Rothwell, J.J., Dise, N.B., Taylor, K.G., Allott, T.E.H., Scholefield, P., Davies, H. and Neal, C. (2011). Predicting river water quality across North West England using catchment characteristics. *J. Hydrol.*, 395(3-4): 153-162.

Roy, S. and Gupta, A. (2010). Molluscan diversity in river Barak and its tributaries, Assam, India. *Assam University Journal of Science and Technology*, 5(1): 9-113.

Saha, M., Sarkar, S.K. and Bhattacharya, B. (2006). Interspecific variation in heavy metal body concentrations in biota of Sunderban mangrove wetland, northeast India. *Environment International*, 32(2): 203-207.

Saha, M., Sengupta, S., Sinha, B. and Mishra, D.K. (2017). Assessment of physico-chemical properties, some heavy metals and arsenic of river teesta in Jalpaiguri district, West Bengal, India. *Asian Journal of Research in Chemistry*, 10(3): 399-404.

Sarin, M.M., Krishnaswami, S., Dilli, K., Somayajulu, B.L.K. and Moore, W.S. (1989). Major ion chemistry of the Ganga-Brahmaputra river system: Weathering processes and fluxes to the Bay of Bengal. *Geochimica et cosmochimica acta*, 53(5): 997-1009.

Saunders, J.A., Lee, M.K., Uddin, A., Mohammad, S., Wilkin, R.T., Fayek, M. and Korte, N.E. (2005). Natural arsenic contamination of Holocene alluvial aquifers by linked tectonic, weathering, and microbial processes. *Geochemistry, Geophysics, Geosystems*, 6(4).

Seitzinger, S.P., Mayorga, E., Bouwman, A.F., Kroeze, C., Beusen, A.H.W., Billen, G. and Harrison, J.A. (2010). Global river nutrient export: A scenario analysis of past and future trends. *Global Biogeochemical Cycles*, 24(4).

Sharma, B.M., Bečanová, J., Scheringer, M., Sharma, A., Bharat, G.K., Whitehead, P.G., Klánová, J. and Nizzetto, L. (2019). Health and ecological risk assessment of emerging contaminants (pharmaceuticals, personal care products, and artificial sweeteners) in surface and groundwater (drinking water) in the Ganges River Basin, India. *Science of the Total Environment*, 646: 1459-1467.

Saunders, J.A., Lee, M.K., Uddin, A., Mohammad, S., Wilkin, R.T., Fayek, M. and Korte, N.E., 2005. Natural arsenic contamination of Holocene alluvial aquifers by linked tectonic, weathering, and microbial processes. Geochemistry, Geophysics, Geosystems, 6(4).

Singh, A.K. (2006). Chemistry of arsenic in groundwater of Ganges–Brahmaputra river basin. *Current Science*, 599-606.

Singh, V., Sharma, N. and Ojha, C.S.P. eds. (2004). *The Brahmaputra Water Resources* (Vol. 47). Springer Science & Business Media.

Subramanian, V., Richey, J.E. and Abbas, N. (1985). Geochemistry of river basins of India, Pt II: Preliminary studies on the particulate C and N in the Ganges-Brahmaputra river system. *Transport of Carbon and Minerals in Major World Rivers*, Pt 3: 513-518.

Thorne, C.R. (2002). Geomorphic analysis of large alluvial rivers. *Geomorphology*, 44(3-4): 203-219.

Tsering, T., Wahed, M.S.A., Iftekhar, S. and Sillanpää, M. (2019). Major ion chemistry of the Teesta River in Sikkim Himalaya, India: Chemical weathering and assessment of water quality. *Journal of Hydrology: Regional Studies*, 24: 100612.

Wiejaczka, Ł., Prokop, P., Kozłowski, R. and Sarkar, S. (2018). Reservoir's Impact on the Water Chemistry of the Teesta River Mountain Course (Darjeeling Himalaya). *Ecological Chemistry and Engineering S*, 25(1): 73-88.

Yan, H., Liu, J., Huang, H.Q., Tao, B. and Cao, M. (2009). Assessing the consequence of land use change on agricultural productivity in China. *Global and Planetary Change*, 67(1): 13-19.

Yasuda, Y., Aich, D., Hill, D., Huntjens, P. and Swain, A. (2017). Transboundary water cooperation over the Brahmaputra River. Legal political economy analysis of current and future potential cooperation. Hague, the Netherlands: The Hague Institute for Global Justice.

Chapter 6
Flood Hazard of the Brahmaputra River in Assam: Current Mitigation Approaches, Challenges and Sustainable Solution Options

Lalit Saikia

INTRODUCTION

The Brahmaputra is a trans-boundary river originating from the Angsi glacier near Manasarovar lake in southern Tibet; flowing through China, India and Bangladesh into Bay of Bengal. The Brahmaputra basin in India is shared by six states: Arunachal Pradesh (41.9%), Assam (36.3%), Nagaland (5.5%), Meghalaya (6.1%), Sikkim (3.7%) and West Bengal (6.5%) (Ojha and Singh, 2004). The basin is enclosed by the Himalayas on the north, the Patkai range of hills on the east, the Mikir hills and Shillong plateau on the south and the ridge separating it from Ganga basin on the west. The Brahmaputra valley is 80 km in width, where the river itself occupies about 1.5 to 25 km during its course in Assam. Spatial variability in morphodynamics of the river is strongly linked to central uplift, the slopes, and depressions (Lahiri and Sinha, 2012).

About 26 tributaries on the north bank and 13 on the south bank join the river in Assam (Fig. 1). The main river Brahmaputra along with the well-knit network of its tributaries controls the geomorphic regime of Brahmaputra valley. The river is unique due to its peculiar drainage pattern, diverse geological setting, high sediment load and critical bank erosion problem (Mahanta and Saikia, 2015). The ferocity or devastation capacity of the Brahmaputra is unparalleled. The Brahmaputra and its tributaries and sub-tributaries cause major problems during the monsoon months every year by flood and bank erosion.

Amongst all flood impacted states of India, the recurrence of flood continues to be the burning problem for Assam state. Flood prone areas of the country as a whole stands at about 10.2% of the total area of the country, but flood prone area of Assam is 39.58% of the area which implies that the flood prone area of Assam is four times

L. Saikia (✉)
Department of Earth Science, University of Science and Technology, Meghalaya, Ri-Bhoi, India

A. Mukherjee (ed.), *Riverine Systems*, https://doi.org/10.1007/978-3-030-87067-6_6

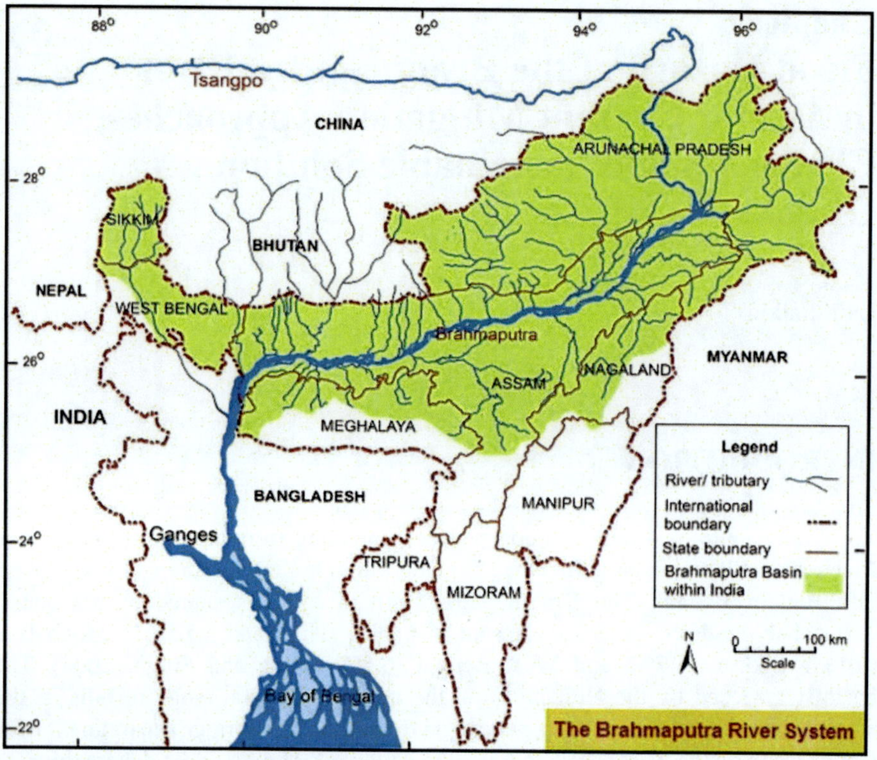

Figure 1. Brahmaputra river system within India.

Table 1. Flood affected areas of Assam and India during high flood years

Year of high flood in Assam	*Total area affected in India (Mha)*	*Area affected in Assam (Mha)*	*% share of Assam in India (flood affected area)*
1954	7.49	3.15	42
1959	5.77	1.04	18
1962	6.12	1.62	26
1966	4.74	1.78	38
1972	4.10	1.10	26.8
1987	8.89	1.53	17.2
1988	16.29	3.82	23.5
1993	4.63	1.25	36.9

Source: NEC, 1993

the national mark of the flood prone area. Records show that average annual area affected by flood in Assam is 9.31 lakh hectares. The Brahmaputra valley, surrounded by hilly terrain, is worst affected due to floods. Damage in Assam during high flood years were much higher in comparison to the fraction of total area affected in India (Table 1).

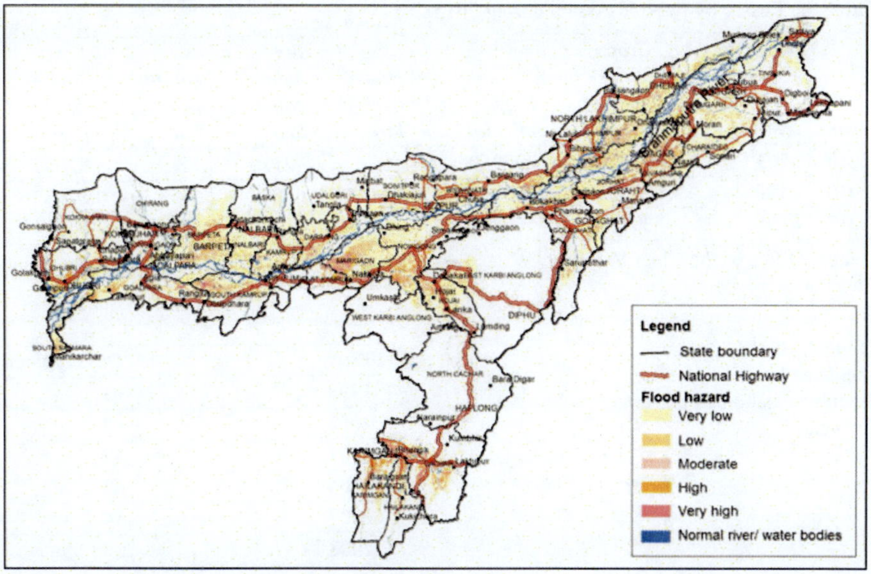

Figure 2. Flood hazard zones of Assam.

Table 2. Flood hazard classification (ASDMA, 2019)

Flood hazard classification	*Number of times/year (area subjected to flood inundation during 1998-2007)*
Very Low	1-2 times
Low	3-4 times
Moderate	5-6 times
High	7-8 times
Very High	9-10 times (almost every year)

Flood hazard map (created by NRSC-Hyderabad, collected and modified from ASDMA 2019) for 10 years (1998-2007) showed flood hazard zones of Assam with five classifications, viz. very low, low, moderate, high and very high (Fig. 2, Table 2).

Being an agrarian state, economy of Assam basically emerges from agriculture. Around 75% population of the state depends directly or indirectly on agricultural activities. But three or four devastating floods annually wipes out major chunk of the fruit of people's labour. Thus, flood has virtually destroyed the economy, more particularly the agro-economy of the state. Average annual loss due to flood in Assam is about Rs 200 crore but in 1998, the loss suffered was about Rs 500 crore and during the year 2004 it was about Rs 771 crore. Statement of tangible losses by flood in Assam during 2015-2017 are given in Table 3.

Floods cause havoc to wildlife also. Kaziranga National Park located in flood plains of Brahmaputra River is an example where flood annually submerges

Table 3. Losses by flood in Assam during 2015-2017

Year	*No. of Villages/ Localities affected*	*Crop area affected (Hect)*	*Population affected*	*Human lives lost*	*Big animal/ cattle loss*	*Houses damaged fully*	*Houses damaged partially*
2015	4763	329303	4203609	64	212	1519	82095
2016	4471	282060	4008602	64	5580	5953	27652
2017	6151	397910	5602090	160	111	3991	74354

Source: Office of the Chief Engineer, Water Resources Department, Govt. of Assam, Chandmari, Guwahati

Table 4. Number of days inundated by flood in Kaziranga National Park in a few years

Year	*Days of flood*
1998	91
2002	42
2003	17
2004	63
2007	14
2008	4

Source: Office of Director, Kaziranga National Park, Bokakhat.

maximum area affecting large number of wild animals including famous one-horned rhino. Total 2479 number of rhinos died during 1986-2015 and 26% deaths were in monsoon season. Large number of rhinos died in monsoon months of 1988, 1991, 1998, 2012 and those years witnessed high flood. Plight of different wild animals in flood situation due to loss of habitat and scarcity of food is easily understood from number of days inundated in Kaziranga in flood years (Table 4).

CAUSES OF FLOODS IN BRAHMAPUTRA

Main factors for extensive floods in Brahmaputra are adverse physiography of the region, heavy rainfall, landslides, excessive sedimentation in valley, frequent earthquakes, reduction of forest cover and human encroachment in the riverine area.

The Brahmaputra is flowing through a tectonically complex and seismically active zone (V). Long-range interactions of mega tectonic units like north-south convergence of the Indian-Eurasian plate along the Himalayas and east-west convergence in the Indo-Burma mountain and under thrusting of the Indian plate below the Eurasian plate (Tiwari et al., 2004) make the region seismically active.

Slope of Brahmaputra river is very steep when it crosses the Himalayas (Fig. 3). Average gradient of the Brahmaputra river is 1.52 m/km. In the course of the river, ~100 km reach known as *Tsangpo* gorge where the river abruptly bends southward is a locus of extremely rapid and focused erosion (Finlayson et al., 2002; Finnegan et al., 2008; Stewart et al., 2008; Larsen and Montgomery, 2012). The geologically

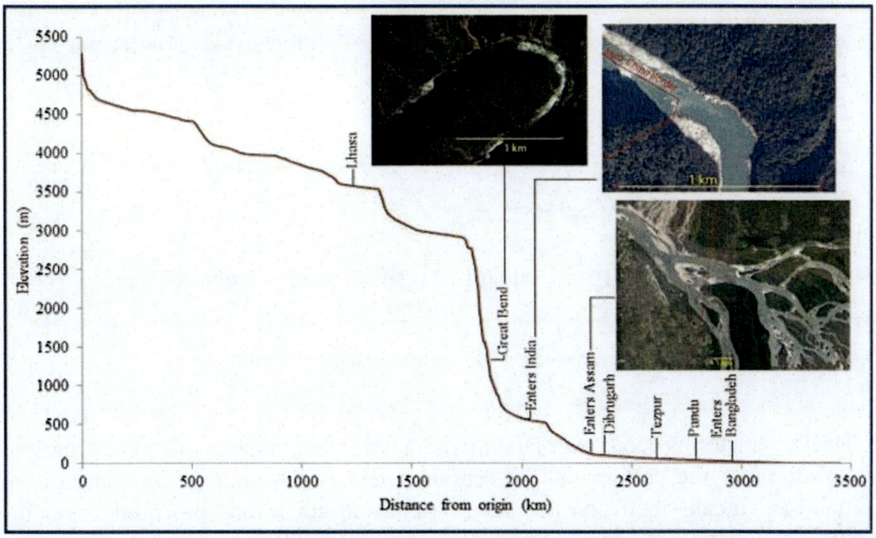

Figure 3. Slope of Brahmaputra River (Saikia, 2017; Mahanta and Saikia, 2017).

youngest lithology of the Himalayas under the influence of intense rainfall and seismicity, load the main river and tributaries with large amount of sediment. The north bank tributaries like *Subansiri, Ranganadi, Jia Bharali, Noa Nadi, Moridhal, Jiadhal* and *Pagladiya* have very steep slopes and shallow braided channels for a considerable distance from the foot hills and in some cases right upto the outfall. Steep slopes of the Brahmaputra and tributaries in the mountainous reaches led to high amount of sediment generation and transportation. The sediment transporting capacity of a flow is highly sensitive to the velocity. Theoretical considerations have shown that the rate of sediment transport is proportional to the fourth to sixth power of velocity. The 10% decrease in flow velocity would result in the reduction of 30 to 40% of its original transport capacity. Sudden decrease in slope of Brahmaputra results in a large amount of sediment deposition developing a prominent braided pattern near *Pasighat* in Arunachal Pradesh, where slope is abruptly decreased. The slope is further decreased during course of river in Assam plains.

Excessive riverbed aggradation due to siltation is considered one of the main short term causes of flood of the Brahmaputra. Due to 1950 earthquake (epicenter in China at about 50 km from north-east border of India, magnitude 8.6 in the Richter scale) and resulting extensive landslides, the river beds of Brahmaputra and most of the tributaries have been raised considerably by heavy deposit of silts, leaving very little scope for sufficient drainage during the peak monsoon period. The Brahmaputra silted up by about 3 m at Dibrugarh within 10 years, whereas the annual average silting rate at Dibrugarh was only 3 cm prior to 1950 (WRD, 2004). Abrupt change in water levels (yearly observed lowest water levels) of Brahmaputra at Dibrugarh after 1950 is discernible (Fig. 4). Since then, floods have been occurring regularly in Brahmaputra as a periodic and anticipated phenomenon.

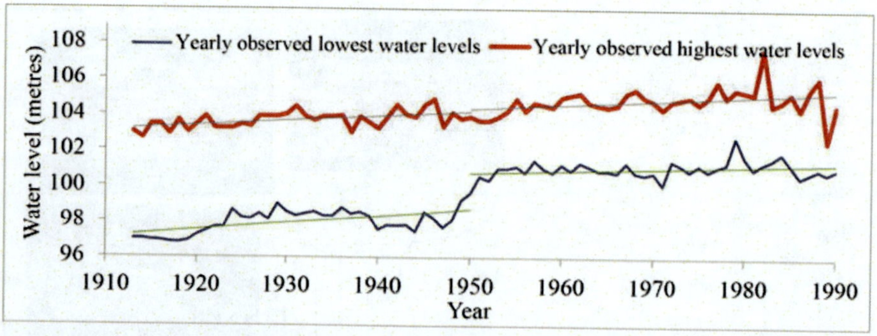

Figure 4. Water levels of Brahmaputra at Dibrugarh during 1913-1990.

Heavy rainstorms occur over Brahmaputra basin due to meteorological situations like shifting of the eastern end of seasonal monsoon trough to the foot-hills of Himalayas, 'break' monsoon and active monsoon conditions prevailing over the region. Prolonged and heavy rainfall ranging from 2,480 mm in the valley to 6,350 mm or more in hilly areas is largely concentrated during monsoon months. Synchronization of heavy rainfall over a smaller time window across a large part of watershed covering both river and tributaries is major reason of high floods in the region. Breaking of artificial lake upstream of river dams and further release of excess water from the reservoir flowing over dams are the immediate causes of a few devastating floods in recent times.

The width of Brahmaputra valley between the foot hills is very narrow, i.e., only 80-90 km where the river itself is 6-10 km. Forest covers a width of only few kilometers, mostly along foot hills, the tea gardens in some districts occupy much of the higher areas. The remaining portion of land in the narrow valley is occupied by villages for settlement, cultivation etc. Population of the state has been increasing faster and this narrow strip of land has to sustain a heavy pressure of increasing population. Large scale deforestation and loss of forest cover in north-east India are already reported due to shifting cultivation in hilly areas (FSI, 2015). Total area under shifting cultivation in the region is about 8,500 sq. km (MoSPI, 2014). Deforestation in the upstream region, soil erosion due to tilling and faulty land use practice are significant long term factors of flood problem. Severity of flood hazard in the state has been further aggravated by the acuteness of bank erosion with permanent loss of land on both banks of river Brahmaputra and its tributaries.

The Brahmaputra valley has been facing a heavy instability of landmass due to river bank erosion, believed to be accelerated after the 1950 earthquake. The stretch falling within Assam (India), has already lost about 7.4% of its total land due to bank erosion and channel migration (Baruah and Goswami, 2013). From a study using remote sensing and GIS tools, total area of erosion in Brahmaputra river was found to be 1557 km^2 during 1973-2014 (Saikia et al., 2019).

EXISTING FLOOD MANAGEMENT PRACTICES

Flood management measures started in Assam after the declaration of National Policy in 1954. Accordingly, a huge network of flood embankments along Brahmaputra and its tributaries was erected across the state. Extensive riverbank protection or anti-erosion measures were initiated only when the existing embankments became more and more vulnerable to breaches from erosion. Different protection works/ technologies taken up in different erosion affected reaches of Brahmaputra since 1970 are construction of:

- Boulder revetment with launching apron
- Deflector/bull head along with up-stream and down-stream protection work.
- Spurs/land spurs.

RCC porcupines have been used as pro-siltation devices. Sediment inducing devices perform satisfactorily in river Brahmaputra by inducing siltation in the channels. High sediment load and comparatively lower velocity of the river provides favourable conditions for the sediment inducing devices. Originally, porcupines were made of timber or bamboo, but these have a limited lifespan. It has been observed that the porcupine performs very satisfactorily as silt inducing device in rivers having the silt factor from 0.7 to 1.50 (grain size 0.158 mm to 0.725 mm). The silt factor of Brahmaputra ranges from 0.9 to 1.0, and the silt inducing performance of porcupines is excellent.

Flood control works executed by the Water Resources Department (WRD), Government of Assam upto December 2018 are as follows:

1. Total length of embankment	4486.44 km
2. Brahmaputra embankment	1031.807 km
3. Brahmaputra tributaries embankment	2697.657 km
4. Barak embankment	251.08 km
5. Barak tributaries embankment	505.896 km
6. Drainage channels	881.966 km
7. Major sluices	100 Nos
8. Minor sluices	545 Nos
9. Town protection works	983 Nos

Courtesy: Office of the Chief Engineer, WRD, Government of Assam, Chandmari, Guwahati, Assam

Implementation of the above structures have afforded reasonable benefit to about 16.173 lakh hectares of land. WRD has adopted use of geo-synthetics materials, imported Amphibian Mini Dredger and hydraulic driving method for bank protection, dredging of river bed and bamboo palisading works. WRD has started use of geo-textile technology in the form of geo-bags and geo-tubular mattress in selected areas. Advantages of geo-textiles are:

- Usage of sand filled geo-textiles in different forms, size, shape is found as perfect replacement for boulder.
- Tubular mattress filled with sand serve dual purposes of preventing bank erosion and bank sloughing due to seepage of water as the seepage water is filtered out by the sand filled mattress.
- Can be manufactured at factories according to site requirements.
- Geo-tubular mattress coupled with bio-engineering techniques provides solution for erosion control.
- Sand for filling and installing the geo-textile bags/mattresses/tubes, is abundantly available at site. Unskilled labours required for filling are also locally available.

CHALLENGES

It has been observed that flood management measures implemented so far could provide reasonable protection to the area from low and medium floods but during high floods there have been large-scale breaches. Structural measures executed so far have remained inadequate to provide satisfactory solution as in most cases these have treated the problem section of a particular portion in isolation without looking into the environmental consequences or downstream impacts. Most of the embankments in Assam were constructed during the first five year plan (1951-1956) and second five year plan (1956-61). The embankments were apparently constructed and planned on inadequate hydrological and hydro-meteorological data, sub-surface information and were designed on specifications short of what is actually required (WRD, 2005). Again, as these rivers were being confined to relatively narrow channels, the water rises quicker and higher and flows with increasing speed and force. In case of a breach, the flood water submerges the surroundings often with devastating impact. Table 5 gives number of breach/cut in embankments of Brahmaputra and its tributaries during 2009-2018.

Table 5. Number of breach/cut during 2009-2018

Year	*Number of breach/cut in entire Assam*	*Number of breach/cut in Brahmaputra basin*	*Number of breach/cut in Brahmaputra river*
2009	41	41	1
2010	60	48	3
2011	22	15	1
2012	74	73	6
2013	45	41	1
2014	28	17	2
2015	27	26	0
2016	37	23	4
2017	26	22	6
2018	32	07	0

Maximum breaches occurred in the tributaries and the causes were overtopping of flood water with a few cases of public cut. On the contrary, embankment breaches in the main Brahmaputra are due to heavy thrust of water, slumping of slope and bank erosion. Irrespective of debate on effectiveness and apparently flawed execution, damages in embankments almost every year since 1962, indicated that the embankment system with the present condition cannot provide assured protection from medium and high flood. Hence, examination, continuous maintenance and improvement of embankments at right time (i.e. not during flood time but after flood in post- or pre-monsoon months) is need of the hour to protect embankments from damage as well as to ensure secured flood protection. High cost and the situation at failure of the Geo-tubes are still putting a question mark on feasibility of geo-textile technique, as long as it is executed and demonstrated as a fool-proof technology for Brahmaputra. Fate of damaged RCC porcupines and the environmental consequences in river ecosystem are other concerns.

In 2016, Government of Assam decided to start dredging of Brahmaputra with objectives of erosion control, sediment management, flood control and water transport development (NDTV, 2016). Although, dredging is practiced in different rivers of the world for navigation, remediation and flood protection; there are evidences of negative impacts from dredging, e.g., increasing risk downstream, ecological risks affecting habitat, disrupting riverine ecosystem and reduced connectivity with the floodplain (Barbe et al., 2000; Gob et al., 2005; Freedman and Stauffer, 2013). Alluvial channels which have been artificially deepened by dredging silt-up more frequently as they tend to return to their pre-dredged state (Environment Agency, 2013). Regular repetitions of dredging is needed and hence, dredging cannot be considered as a sustainable option for large alluvial river Brahmaputra.

Brahmaputra valley is located at transitional zones between different climatic regions and different distinct geomorphologies, e.g., cold dry climate of the Tibetan plateau, warm tropical humid climate of Assam-Bangladesh plains. The valley is vulnerable to climate change impacts due to its location in the eastern Himalayan periphery, fragile geo-environmental setting and economic underdevelopment, in spite of abundant natural resources. Eastern Himalayas are experiencing warming of 0.1 to 0.4°C/decade; the highest rates of warming are in winter (ICIMOD, 2009). Simulation model indicates widespread warming of NE India by 1.8 to 2.1°C during 2030s. Rainfall of higher intensity is likely to occur during monsoon (INNCA, 2010). Thus, climate change is adding additional risk to the flood hazard in Assam. There is a great challenge before the scientists regarding the correct assessment of the causes of disastrous floods and their possible solutions adding climate change dimension as an additional layer of risk assessment and management.

WAY FORWARD

It has been realized that floods in Brahmaputra valley can only be managed, they cannot be controlled. Due to financial and geographic constraints, no structural measure can be adopted for total (or absolute) protection against all conceivable magnitudes of floods. No method can be termed as the best and any method can be adopted according to the circumstances. A combination of various measures is possibly the best way of flood management; for instance, the construction of embankments along with the reservoir or a combination of structural and nonstructural measures (Fig. 5).

Structural measures provide protection to areas hit by direct impact of flood, where public safety is the greatest concern. Increased emphasis should be given, however, on non-structural measures as well, which provide protection while reducing adverse environmental impacts to sensitive areas. Many new technologies are holding some promises, but they should be administered properly with sufficient research and development, and pre-planning.

Embankments can be strengthened to super levee as developed in Japan (Takahasi and Uitto, 2004). Super levee is a high standard river embankment with a broad width (Stalenberg and Kikumori, 2008) having mild slope of 1:30 (Arakawa-Karyu River Office and MLIT, 2006) and is resistant to overflow and seepage. The mild slope of the super levee prevents sliding of the top layer. The great width of the super levee reduces seepage (Arakawa-Karyu River Office and MLIT, 2006). Super levee concept will overcome the drawbacks of existing embankment system in Assam.

Afforestation and conservation of water bodies like ponds, wetlands can help in flood minimization. There are more than 3,500 wetlands covering an area of around 1,01,232 ha in Brahmaputra floodplains. Wetlands can soak up flood water like sponges and have potential in flood cushioning. Grass vegetation and porous concrete surface are highly encouraged in urban areas to improve infiltration along with provision for natural drainage to reduce flash floods.

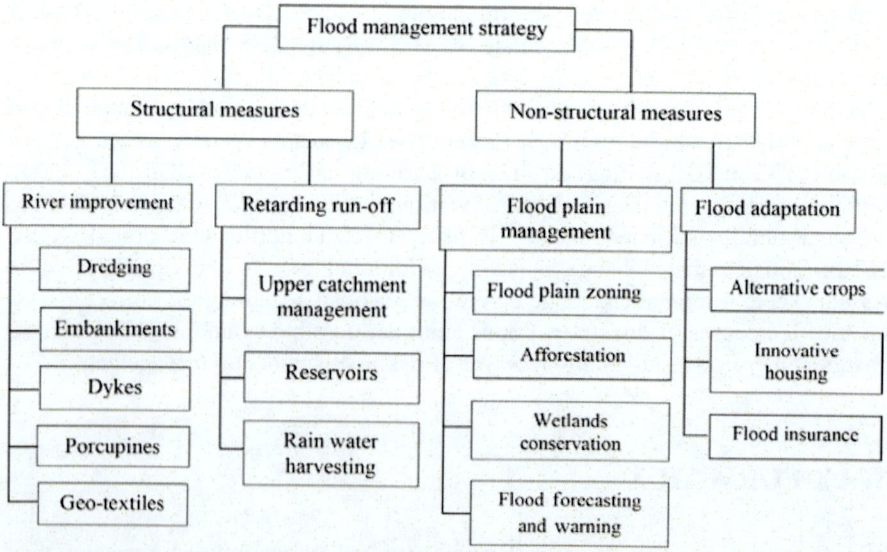

Figure 5. Flood management strategy for Brahmaputra River.

Thousands of farmers in Assam live and cultivate in vulnerable areas in the flood plains. Capacity building should be an integral component of flood management to empower people to live within the dictate of their ecological situations. Flood plain zoning aims at identification of locations and determination of damage from floods of different magnitudes/frequencies and development of those areas to reduce the damage to minimum. Flood proofing measures are essentially a combination of structural change and emergency action which can help greatly in the mitigation of distress and provide immediate relief in the flood inundated areas. Flood forecasting provides advance warning about when the river is going to rise and to what extent.

Private adaptation by individual farmers encompasses steps like switching on to crops and production methods that are best suited for the climate they live in. To compensate crop damage in flood, cultivation of some flood resistant varieties should be recommended with necessary technological intervention. Advancement of sowing season to March-April and direct seeding can ensure early crop establishment escaping floods and higher plant population. Agronomic practices like basal fertilizer application to tolerate submergence are to be developed and popularized. Transplanting two months old seedlings of suitable variety after flood water recedes in the first week of September is a promising alternative (Bhowmick et al., 2000). The rejuvenation of fertility of flood plains by the flood derived sediments can be best utilized by knowing the renewed nutrient status of different croplands and thereby opting for crops which are best suited for a particular land.

Social changes in terms of innovative dwelling house, suitable occupation and raised platforms are also need of the present. Some tribes of flood prone areas are traditionally using 'chang ghor' (a typical house on stilt practiced by the *Mishing* tribe) and boat. These can be encouraged with more innovative ideas to live with the floods. Flood insurance facility, drinking water and sanitation support, provision of other basic amenities including medical facilities during and after flood events need attention. More shelters in high lands of protected areas are to be constructed to reduce death of wild animals in floods.

Due to unique topography and abundant surface water resources, north-east India is endowed with a hydropower potential of 66,000 megawatts, which represents about 40% of the national potential (World Bank, 2007). Hydropower sector emerges as one of the best opportunities for development in the region provided that the projects are developed in a manner appropriate for the region's geographical, social and environmental contexts including their potential for flood cushioning. The dams can reduce flood dimension in Assam if storage facilities with proper engineering structure to withstand in earthquakes were an integral part of the hydropower projects in Arunachal Pradesh.

It is essential to study the whole river to identify effected and erosion prone areas based on severity of river bank erosion problem and probability of occurrence. Site specific protection measures considering causes and mechanisms of erosion and impact assessment studies for protection of other vulnerable reaches can be suggested for erosion prevention.

In northeast India, climate change issue is still at a backstage with no clear responsibilities assigned to any of the departments. Policy interventions and institutional mechanism with participatory processes, capacity building, inter-state and international cooperation are essential to cope with climate change impacts on vast water resources and agriculture sector of Brahmaputra floodplains. Finally, a holistic approach for the entire region and honesty in real sense while addressing flood issue is utmost important.

ACKNOWLEDGEMENT Draft of the manuscript was prepared in IIT Guwahati during PhD programme of the author in Department of Civil Engineering. The author acknowledges his thesis supervisor Prof. C. Mahanta, Central Library and Computer Centre of IIT Guwahati for support. He is thankful to Office of the Chief Engineer, Water Resources Department, Government of Assam, Chandmari and Office of Director, Kaziranga National Park, Bokakhat for providing data related to flood damage in Assam.

References

Arakawa-Karyu River Office and MLIT (2006). The Arakawa: River of the Metropolis; A comprehensive guide to the lower Arakawa, Arakawa-Karyu River Office. Ministry of Land, Infrastructure and Transport, Japan.

ASDMA (2019). Flood hazard atlas, Assam State Disaster Management Authority. Retrieved November 09, 2019 from sdmassam.nic.in.

Barbe, D.E., Fagot, K. and McCorquodale, J.A. (2000). Effects on Dredging Due to Diversions from the Lower Mississippi River. *Journal of Waterway Port Coastal and Ocean Engineering*, 12(3): 121-129.

Baruah, U. and Goswami, R.K. (2013). River Bank Erosion Management in Assam. *In:* Kakati H. and Changkakati P.P. (Eds.), *Proceedings of the Assam Water Conference 2013*. Water Resources Department, Government of Assam, Guwahati.

Bhowmick, B.C., Pandey, S., Villano, R.A. and Gogoi, J.K. (2000). Characterizing risk and strategies for managing risk in flood-prone rice cultivation in Assam. IRRI limited proceeding series No. 5, Risk analysis and management in rainfed rice systems. International Rice Research Institute, Philippines, pp. 143-156.

Environment Agency (2013). *Evidence: Impacts of Dredging*. Environment Agency, Bristol, U.K.

Finlayson, D.P., Montgomery, D.R. and Hallet, B. (2002). Spatial Coincidence of Rapid Inferred Erosion with Young Metamorphic Massifs in the Himalayas. *Geology*, 30: 219-222.

Finnegan, N.J., Hallet, B., Montgomery, D.R., Zeitler, P.K., Stone, J.O., Anders, A.M. and Liu, Y. (2008). Coupling of Rock Uplift and River Incision in the Namche Barwa – Gyala Peri Massif, Tibet, China. *Geological Society of America Bulletin*, 120: 142-155.

Freedman, J.A. and Stauffer, J.R. (2013). Gravel Dredging Alters Diversity and Structure of Riverine Fish Assemblages. *Freshwater Biology*, 5(2): 261-274.

FSI (2015). State of Forest Report, 2015. Forest Survey of India, Dehradun, Uttarakhand, India.

Gob, F., Houbrechts, G., Hiver, J.M. and Petit, F. (2005). River Dredging, Channel Dynamics and Bed Load Transport in an Incised Meandering River (the River Semois, Belgium), *River Research and Applications*, 21(7): 791-804.

ICIMOD (2009). The changing Himalayas, impact of climate change on water resources and livelihoods in the greater Himalayas: Preprint for discussion and comments. The International Centre for Integrated Mountain Development, Nepal, 6-7.

INNCA (2010). Indian Network for Climate Change Assessment. Ministry of Environment and Forest, Govt. of India.

Lahiri, S.K. and Sinha, R. (2012). Tectonic Controls on the Morphodynamics of the Brahmaputra River System in the Upper Assam Valley, India. *Geomorphology*, 169-170, 74-85.

Larsen, I.J. and Montgomery, D.R. (2012). Landslide Erosion Coupled to Tectonics and River Incision. *Nature Geoscience*, 5: 468-473.

Mahanta, C. and Saikia, L. (2015). The Brahmaputra and Other Rivers of the North-East. *In:* R.R. Iyer (Ed.), Living Rivers Dying Rivers. Oxford University Press, New Delhi, p. 155.

Mahanta, C. and Saikia, L. (2017). Sediment Dynamics in a Large Alluvial River: Characterization of Materials and Processes and Management Challenges. *In:* N. Sharma (Ed.), River System Analysis and Management. Springer Singapore.

MoSPI (2014). Ministry of Statistics and Programme Implementation Year Book, 2014. Govt. of India.

NDTV (2016). Dredging of Brahmaputra Soon to Stop Floods, Erosion: Sarbananda Sonowal. Accessed from http://www.ndtv.com/india-news/dredging-of-brahmaputra-soon-to-stop-floods-erosion-sarbananda-sonowal-1449306 on 09.11.2016.

NEC (1993). Chapter VIII, Morphological Studies of River Brahmaputra, Northeastern Council, WAPCOS, VIII – 2.

Ojha, C.S.P. and Singh, V.P. (2004). Introduction. *In:* Singh, V., Sharma, N. and Ojha, C.S.P. (Eds.), The Brahmaputra Basin Water Resources. Springer, Netherlands.

Saikia, L. (2017). Sediment properties and processes influencing key geoenvironmental aspects of a large alluvial river the Brahmaputra in Assam. Unpublished PhD. thesis. Department of Civil Engineering, Indian Institute of Technology Guwahati, Assam, India.

Saikia, L., Mahanta, C., Mukherjee, A. and Borah, S.B. (2019). Erosion deposition and land use land cover in Brahmaputra river in Assam, India. *Journal of Earth System Science*, 128: 211, https://doi.org/10.1007/s12040-019-1233-3.

Stalenberg, B. and Y. Kikumori (2008). Urban Flood Control on the Rivers of Tokyo Metropolitan. *In:* Graaf, R.D. and Hooimeijer, F. (Eds.), Urban Water in Japan. Taylor & Francis Group, London, UK.

Stewart, R.J., Hallet, B., Zeitler, P.K., Malloy, M.A., Allen, C.M. and Trippett, D. (2008). Brahmaputra Sediment Flux Dominated by Highly Localized Rapid Erosion from the Eastern Most Himalaya. *Geology*, 36: 711-714.

Takahasi, Y. and J.I. Uitto (2004). Evolution of River Management in Japan: From Focus on Economic Benefits to a Comprehensive View. *Global Environmental Change*, 14.

Tiwari, R.K., Sri Lakshmi, S. and Rao, K.N.N. (2004). Characterization of Earthquake Dynamics in Northeastern India Regions: A Modern Nonlinear Forecasting Approach. *Pure and Applied Geophysics*, 161: 865-880.

WRD (2004). The Task Force for Flood Management/Erosion Control. Water Resources Department, Government of Assam.

WRD (2005). Achievements under water resource sector. Water Resources Department, Government of Assam, 1996-2005.

World Bank (2007). Development and Growth in Northeast India: The Natural Resources, Water, and Environment Nexus-Strategy Report, 36397-IN, 57.

Chapter 7
Integrated Simulation of Surfacewater-Groundwater (SW-GW) Interactions Using SWAT-MODFLOW (Case study: Shiraz Basin, Iran)

Tina Jafari, Saman Javadi, and Anthony S. Kiem

INTRODUCTION

Global water demand has increased dramatically due to rapid population growth in recent decades. For arid and semi-arid countries like Iran, water supply has become a major concern. Management of water resources was previously focused on surface water (SW) or groundwater (GW) individually (Winter et al., 1998). However, SW and GW are components of the hydrological cycle that interact with each other (Deb et al., 2019). The SW bodies like streams, rivers, reservoirs, and wetlands are in direct or indirect interaction with GW (Eini et al., 2019). Therefore, effective water management requires a comprehensive understanding of SW-GW interactions. However, we have limited knowledge about SW-GW interactions because of the lack of available field measurement/data and complexities associated with SW, GW and SW-GW processes. Thus, SW-GW interactions have frequently been ignored. Lack of proper quantification of SW-GW interactions in water resources management has resulted in unrealistic outputs and therefore, unsuitable policies and decisions (e.g. Aliyari et al., 2019; Dehghanipour et al., 2019; Pai, 2015; Tian et al., 2015; Pérez-Martín et al., 2014; Cho et al., 2010). Integrated simulation of the whole water cycle is required for realistically simulating exchange between SW and GW (Guzman et al., 2015).

The GW resources are considered as the main source of water in many parts of the world (e.g. Kahinda et al., 2010; Javadi et al., 2011). Where the interactions between SW and GW are high GW head depletion could lead to reduction in or ceasing of

T. Jafari · A. S. Kiem
Centre for Water, Climate and Land (CWCL), School of Environmental and Life Sciences, Faculty of Science, University of Newcastle, Callaghan, NSW, Australia

S. Javadi (✉)
Department of Water Engineering, College of Aburaihan, University of Tehran, Tehran, Iran
e-mail: javadis@ut.ac.ir

A. Mukherjee (ed.), *Riverine Systems*, https://doi.org/10.1007/978-3-030-87067-6_7

spring discharges and decrease in river runoff (e.g. Esteller and Diaz-Delgado, 2002; Molina-Navarro et al., 2019). In many regions of Iran, GW level is decreasing considerably and this has resulted in declines in SW runoff and increased subsidence (Jafari et al., 2016a).

Fars province is one of the most important areas for agricultural activities in Iran and has 10% of Iran's total water resources allocation. Fars region has been facing declining trends in GW levels and storage due to the over-exploitation of GW for agriculture irrigation. Integrated water resources management is critically important in Fars because its crop production and socio-economic development is highly dependent on water availability.

Applying integrated hydrological models to simulate SW-GW interactions and determine the spatio-temporal distribution of water is imperative to different sectors like agriculture and water resources management. In general, hydrological models are divided into three main categories: (i) Hydrological models which simulate SW such as Soil Water Assessment Tool (SWAT) (Arnold et al., 1998); (ii) Hydrogeological models which consider GW systems such as the modular finite-difference flow model (MODFLOW) (Harbaugh, 2005); and (iii) Physically based integrated surface-subsurface hydrological models (ISSHM) which simultaneously simulate SW and GW flow equations. The SWAT for SW modeling and MODFLOW for GW simulation are among the most commonly used water resource models (Eini et al., 2020; Sheikhipour et al., 2018; Martínez-Santos and Andreu, 2010; Jafari et al., 2016b; Bieger et al., 2017; Amin et al., 2017).

SWAT is a semi-distributed structure SW model that simulates surface runoff, water budget and water quality. A summary of 250 journal articles that successfully applied SWAT is represented by Gassman et al. The SWAT has been shown to have good performance for stream flow simulation in a wide range of watersheds and has been tested in many countries, scenarios and time periods. Despite the wide application, SWAT has limited capabilities in simulating the GW (Neitsch et al., 2011). In SWAT, an aquifer is represented by a lumped module which is divided into shallow and deep GW. In both shallow and deep aquifers, distributed parameters such as hydraulic conductivity (K) and storage coefficients (S) are not considered. This simplification of GW movement results in misleading evaluation of the aquifer head especially when streamflow is strongly dependent on GW discharge.

MODFLOW is a fully distributed three-dimensional model which simulates GW head and water budget of the aquifer that has been successfully used in many regional studies and is the most popular model for GW simulation (Harbaugh et al., 2000; Harbaugh, 2005; Jafari et al., 2016a, 2016b; de Graaf et al., 2017; Deb et al., 2019). Even though MODFLOW is commonly used, it is not capable of simulating land-atmosphere interactions, agricultural processes (e.g. irrigation), Unsaturated Zone Flow (UZF) and comprehensive surface runoff (Lachaal et al., 2012; Surinaidu et al., 2014). Therefore, MODFLOW does not account for differences in GW recharge rates as a result of changes in precipitation and stream infiltration due to land use and irrigation (Guzman et al., 2015).

To address the above limitations, Bailey et al. developed SWAT-MODFLOW, which links SWAT with MODFLOW and simulates SW-GW interactions comprehensively. SWAT-MODFLOW geographically locates and integrates Hydrologic Response Units (HRUs) in SWAT sub-basins with the spatially gridded MODFLOW model. In SWAT-MODFLOW, percolation from SWAT HRU is infiltrated to the MODFLOW cells as recharge, and MODFLOW-recalculated SW-GW interaction fluxes are passed to the stream channels of SWAT. Therefore, the combination model calculates interactions between SW and GW which enables better computation and conception of the spatial-temporal layouts of SW-GW interactions. The SWAT-MODFLOW has been applied to different catchments in the USA, Canada, Denmark (Molina-Navarro et al., 2019), Iran and Japan. The SWAT-MODFLOW has been further improved by Aliyari et al. (2019) by adding the linkage code between MODFLOW pumping cells and SWAT HRUs for irrigation. In this regard, we applied this latest Aliyari et al. (2019) version of SWAT-MODFLOW (with Newtonian Solver) for the Shiraz catchment in southwest Iran (Fig. 1).

AIM AND SCOPE

The aims of this research are to:

- Simulate the interaction between SW and GW using the integrated SWAT-MODFLOW model to obtain insights about water balance and SW-GW exchanges for the Shiraz catchment in southwest Iran.
- Improve the performance of SWAT-MODFLOW by developing a tool to calibrate GW parameters along with SW parameters using two variables: (a) river runoff from SW observation stations and (b) GW level.

CONCEPTUAL FRAMEWORK OF THE STUDY

In order to achieve the aims of this research the following methodology was followed:

1. *Data collection:* The required data were collected from various sources. Hydrological, hydrogeological and meteorological information including DEM, land use, rainfall, river gage, observations and exploitation wells were collected during the simulation period (2003 to 2019).
2. *GW model set-up (using MODFLOW with Newtonian Solver):* In MODFLOW, the GW flow equation for the hydraulic head in each grid cell in the model domain is dependent on aquifer inputs and outputs (e.g. recharge, pumping),

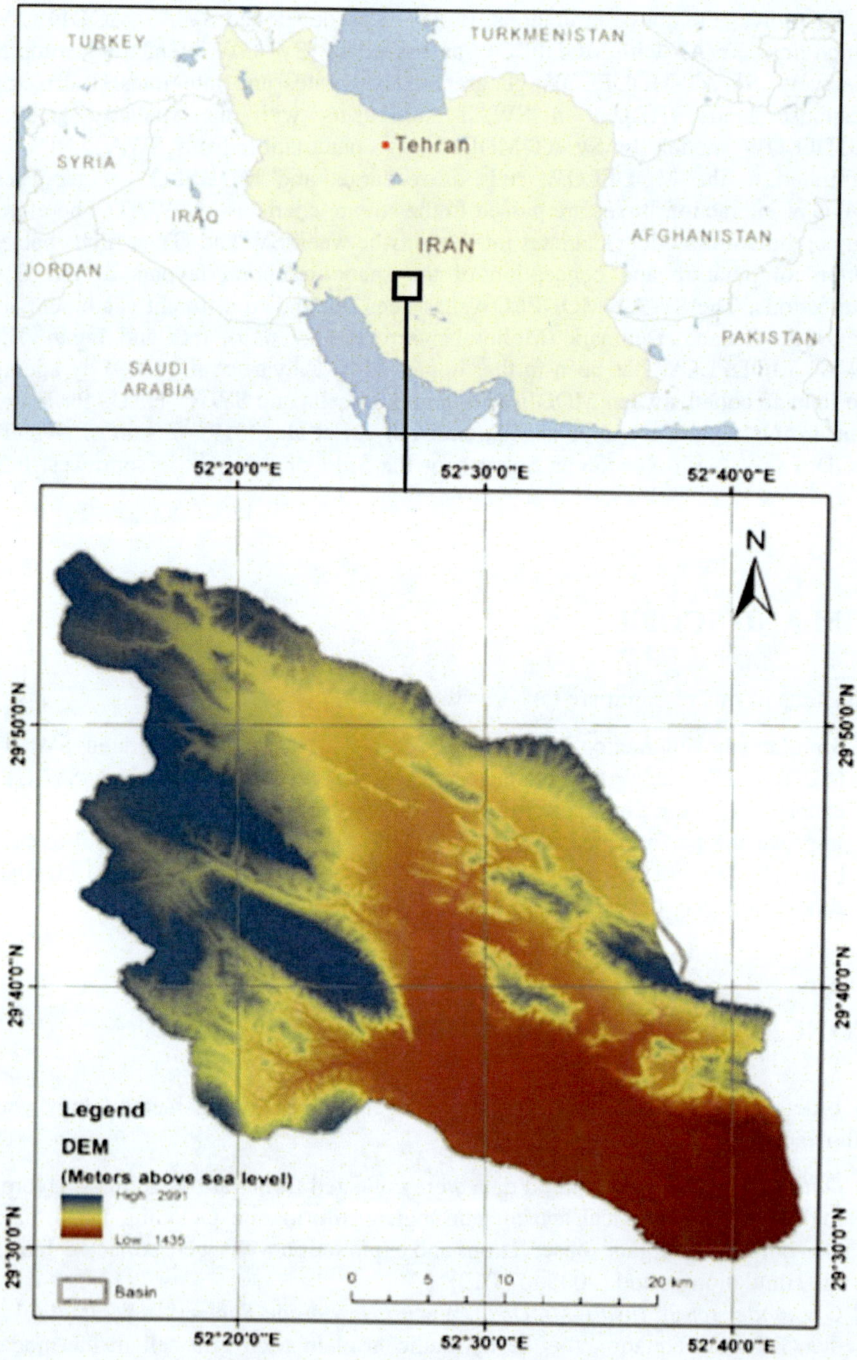

Figure 1. Location and elevation of the Shiraz watershed study area.

boundary conditions, initial conditions, and GW parameters (e.g. hydraulic conductivity, specific yield and riverbed conductance).
3. *Develop SWAT-MODFLOW:* linkage files between HRUs from SWAT and MODFLOW grid cells were created using the joint links codes developed by Bailey.
4. *SWAT-MODFLOW calibration:* In order to calibrate GW parameters along with SW parameters in the SWAT-MODFLOW model, a calibration code was developed (GWCal). The GWCal is linked with the latest version of SWAT-MODFLOW and applied to the Shiraz catchment in southwest Iran.

Figure 2 shows the applied simulation strategy in this research.

THEORY OF SW-GW INTERACTION

As discussed in the introduction, SWAT only simulates shallow GW movements. Figure 3 shows a schematic of the hydrologic cycle and SWAT simulation processes (Neitsch et al., 2011). The rate of percolation below the shallow aquifer (set at a maximum value of 6 m below the ground surface), as represented in Fig. 3, is assumed to be lost out of the system (Neitsch et al., 2011).

In this work, SW-GW interactions are simulated using the updated version of SWAT-MODFLOW with Newtonian Solver (Aliyari et al., 2019). In SWAT-MODFLOW, SWAT simulates components related to SW including crop yield, river processes and soil and unsaturated zone processes, whereas MODFLOW simulates GW head and water budget. In SWAT-MODFLOW, deep percolation from SWAT is passed to the grid cells of MODFLOW as recharge, and MODFLOW-calculated SW-GW interaction fluxes are passed to the stream channels of SWAT. The "River" package from MODFLOW calculates the volumetric flow of exchanged water Q_{leak} $[L^3/T]$ between river and aquifer based on Darcy's equation 1:

$$Q_{\text{leak}} = K_{bed}(L_{str}P_{str})\left(\frac{h_{str} - h_{gw}}{z_{bed}}\right) \tag{1}$$

K_{bed}: the river bed hydraulic conductivity $[L/T]$
L_{str}: river length $[L]$
P_{str}: wetted perimeter of the river [L]
H_{str}: river stage $[L]$
H_{gw}: hydraulic head of GW $[L]$
Z_{bed}: riverbed thickness $[L]$

The value of Q_{leak} will be negative when aquifer recharge the river (gaining stream) and it will be positive when river infiltrate to the GW (losing stream). Data from SWAT is passed to MODFLOW grids, according to the percent area of the

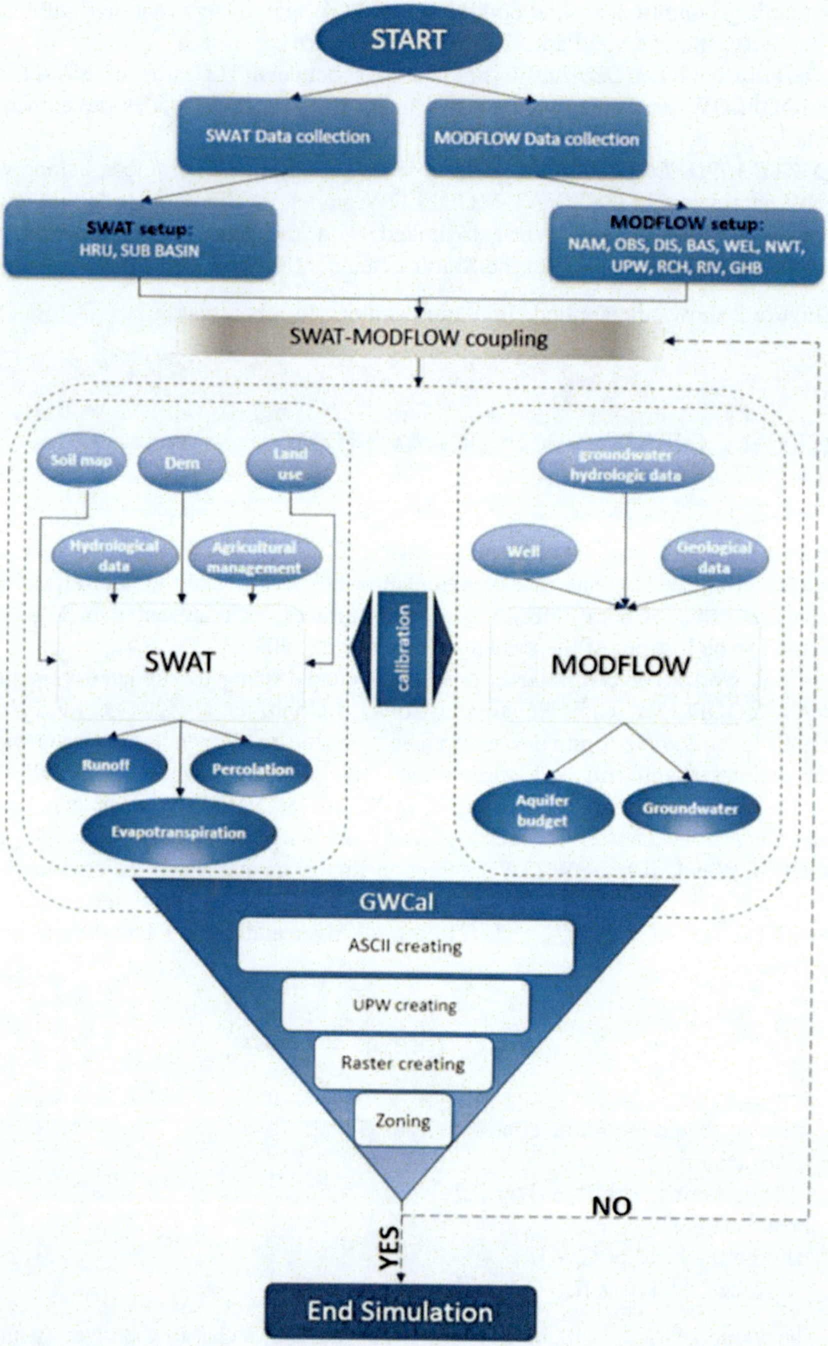

Figure 2. Simulation strategy used.

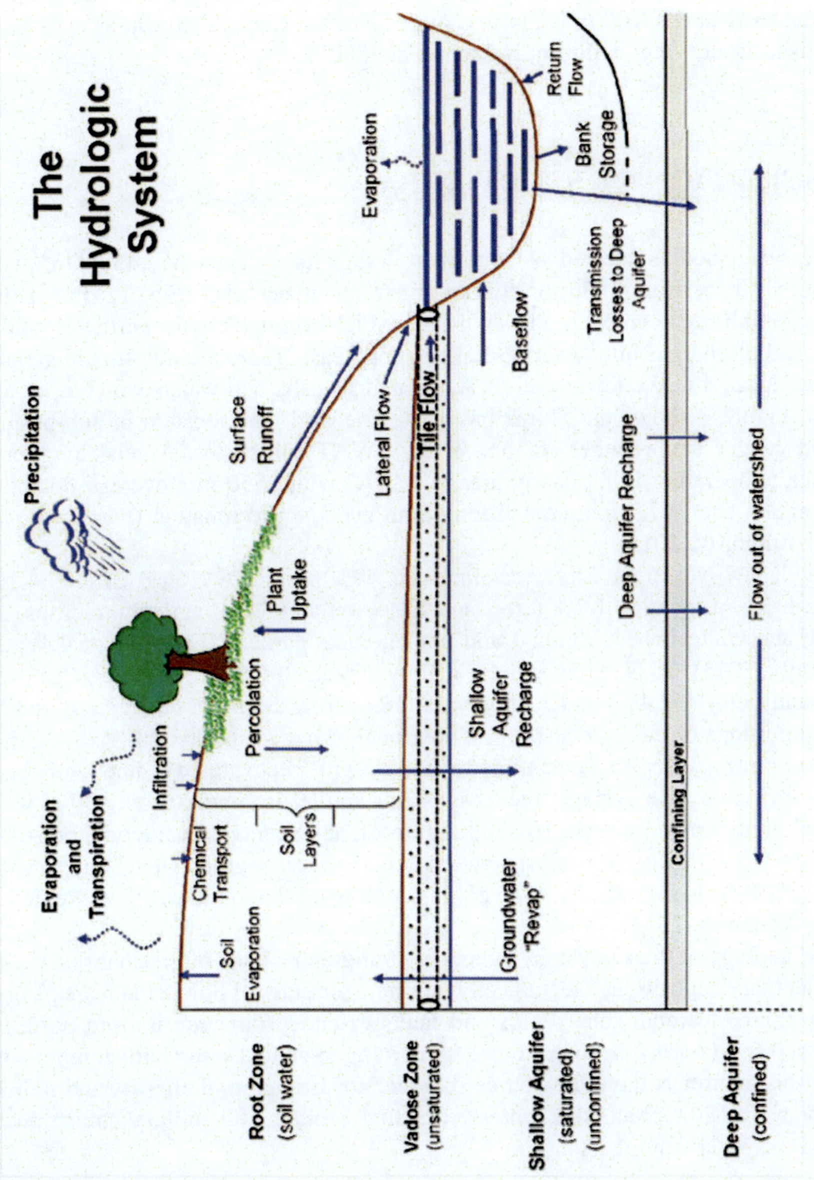

Figure 3. Schematic of the hydrologic cycle and SWAT simulation processes (Neitsch et al., 2011).

DHRU (Disaggregated HRU) contained within the grid cell for use by the recharge package in MODFLOW. This integrated approach in simulating SW and GW provides more realistic impact assessment of climate variability and change (Brown and Funk, 2008; Wheeler and von Braun, 2013), irrigation technology and management (Playan and Mateos, 2006), land use change (Scanlon et al., 2005; Chu et al., 2013), and disturbances (e.g., wildfire; Beeson et al., 2001).

DESCRIPTION OF STUDY AREA

Shiraz watershed is located at southwest Iran with an area of 1452 km^2 with elevation ranges from 1436 to 2991 m above mean sea level (Fig. 1). The main-stream in Shiraz catchment is Khoshk River, which originates in the northwest of the basin and discharges into Mahrloo Lake in southeast. There are two stream gauges that are located at the lower end of Azam and Khoshk sub-watershed (Eini et al., 2019). Figure 4 shows the Shiraz basin river network and location of sub-basins, stream gauges and weather stations in the SWAT sub-basins. Annual long-term average temperature in the study area is 17.9°C with 2550 mm average potential evaporation and 361.5 mm annual long-term average precipitation (Fars Regional Water Authority, 2015).

The land cover in the Shiraz catchment is 79% pasture, 8.9% agriculture, 4.1% orchard, 3.6 residential, 1.7% forest and 1.4% wetland. Main agriculture crops are spring and winter wheat, almond and pomegranate gardens (Fars Regional Water Authority, 2015). Soils in this area are mainly clay loam (49.2%) in the eastern part, fine sandy clay (32.4%) and loamy soils (18.4%) in the north-central and south-central portions of the watershed (Fars Regional Water Authority, 2015).

The Shiraz aquifer is an unconfined shallow aquifer constituted of fine sandy clay and loamy soils. The average thickness of this aquifer is approximately 30 m with transmissivity range between 100 and 200 m^2/d, and average specific yield of 0.05 unit. The GGW depth varies from between 3 to 35 m below ground level (mbgl) with general GW flow direction from the north-west toward the south-east (Fars Regional Water Authority, 2015).

The geology of the study area comprises stratigraphic units of Precambrian rocks to Quaternary deposits, and tectonically, northern altitudes of Shiraz Plain are placed in the Zagros tectonic zone. Folds and faults extended direction is from north to south and from northwest to southeast. According to annual water budget reports in 2019, this aquifer is the main source of water for irrigation, domestic, and public supply with 1802 production bores extracting around 189 million cubic meter (MCM) of water from the aquifer.

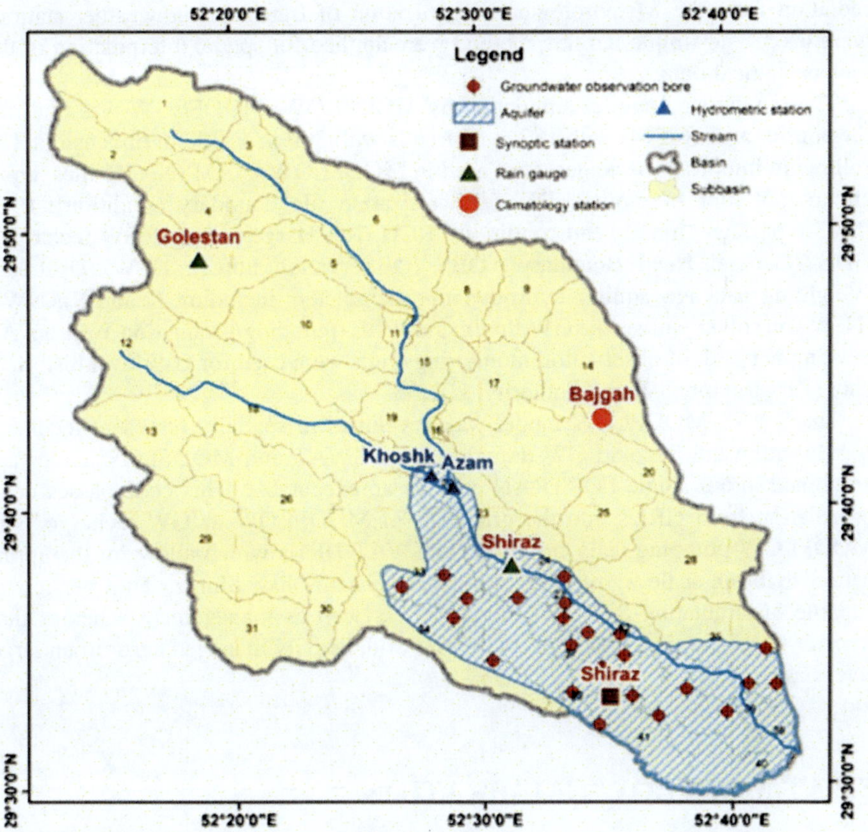

Figure 4. Shiraz catchment river network and location of sub-basins, stream gauges, weather stations, aquifer boundary and observation bores.

SWAT-MODFLOW MODEL DEVELOPMENT

The first step in the development of the SWAT-MODFLOW model is the simulation of SW with SWAT (user interface: ArcSWAT). SWAT was built based on conceptual units of homogeneous land use, management, slope, and soil characteristics that extend below the surface to a soil profile depth called HRUs (Arnold et al., 1998). The SWAT model was built with a total of 32 sub-basins and 188 HRUs with sub-basin areas that ranged from (approximately) 1.2 to 7.1 km^2. Slope, soil data, and land use were the main SWAT inputs. The National Elevation Dataset from the Iranian Ministry of Energy has been used to develop the slope map with grid size 15 $\times$ 15 m. The Food and Agriculture Organization (FAO) soil data set and United State Geological Survey (USGS) global land use with 1 $\times$ 1 km raster layer were used to obtain soil and land use types in the study area, respectively. Precipitation, wind, relative humidity, maximum and minimum air temperatures and solar radiation from the Meteorological Organization of Iran for four weather stations were used. The simple Kriging method were applied for spatial interpolation of the meteorological data.

The second step aquifer simulation by MODFLOW. MODFLOW was built for the aquifer area (Fig. 4) with 183 × 171 cells with 300 m × 300 m dimension. The following important packages were used in MODFLOW: NAM (NAMe files_input and output files of model); DIS (DIScretization file_boundary condition); BAS (BASic package_initial head condition); RCH (ReCHarge package_GW recharge); GHB (General Head Boundary); OBS (OBServation bores); UPW (Upstream Weighting package_aquifer hydrodynamic parameters including K and S); NWT (NeWTon solver_solves the GW flow); and WEL (pumping/production WELls). An existing network of observation monitoring wells were used for GW level historical data (Fars Regional Water Authority, 2015).

The SWAT-MODFLOW model was developed to spatially represent feedback fluxes within the SW and GW domains. Both SWAT and MODFLOW codes are combined into a single FORTRAN code as an executable. This code passes deep percolation from HRUs into the grid cells of MODFLOW as GW recharge. The MODFLOW pumping cells are linked to SWAT HRUs which allow for the application to basin-scale agro-urban watershed systems like Shiraz. This version is capable of simulating SW-GW interactions as well as management schemes that transfer water between the domains in semi-arid, heavily managed agro-urban large river basins.

SWAT-MODFLOW CALIBRATION

To calibrate SW parameters the Sequential Uncertainty Fitting 2 (SUFI-2) approach with SWAT Calibration Uncertainty Procedure (SWAT-CUP) was used. The calibration and validation of the model within SWAT-CUP was conducted considering the hydrologic key parameters based on detailed previous literature sources (Abbaspour, 2015). Each SWAT parameter was set to default lower and upper values as suggested by the SWAT expert group and the best fitted parameter values obtained from SWAT-CUP were incorporated into the database for stream discharge simulations. SWAT model performance was evaluated using *P*-factor, *R*-factor, Nash-Sutcliffe efficiency (NSE), (Nash and Sutcliffe, 1970), coefficient of determination (R^2), and percentage of bias (PBIAS) (Moriasi et al., 2007). Root mean square method (RMSE) was only used to evaluate MODFLOW.

To calibrate GW parameters, five bounds of hydraulic conductivity (range between 0.5 and 35 M/d) and specific yield values (range between 0.01 and 0.25 unit) were generated in the modelling domain and the GW head was manually calibrated using the observation data. In order to make the calibration of mentioned parameters in SWAT-MODFLOW more efficient a calibration code called "GWCal" was developed. GWCal is written in Visual Basic for Applications (VBA) and linked to the SWAT-MODFLOW model to allow calibration of GW parameters in

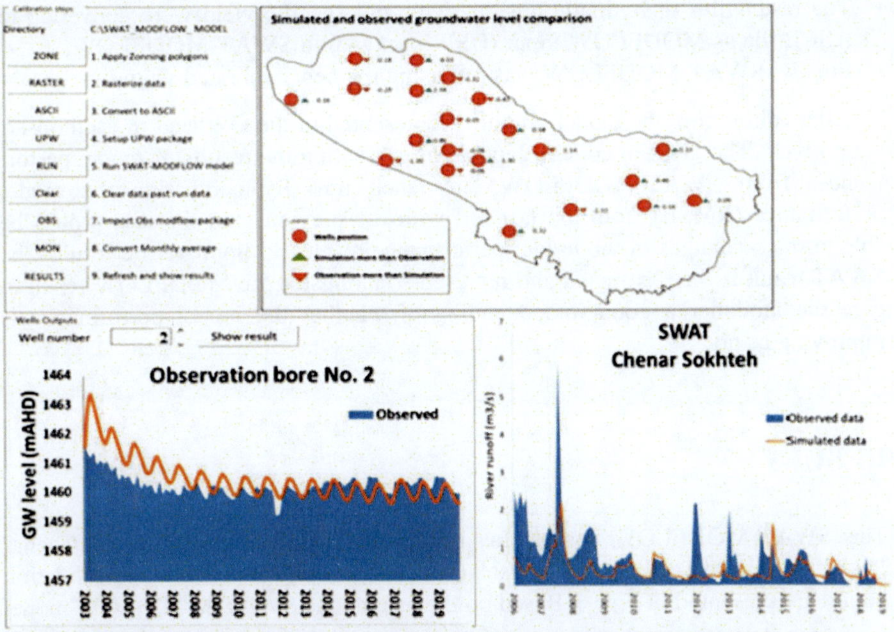

ZONE COMMAND: When the user changes MODFLOW parameters such as K and S in each zone in the main window, GWCal creates corresponding K and S in that particular zone in the text format.
RASTER COMMAND: Create raster data from new S and K in ArcGIS.
ASCII COMMAND: Convert raster data from new raster data to ASCII.
UPW COMMAND: Set up UPW package and create SWAT-MODFLOWinput file in UPW format from ASCII data.
RUN COMMAND: Run SWAT-MODFLOW automatically with new parameters.
CLEAR DATA COMMAND: Remove all previous run data from corresponding SWAT-MODFLOW folder.
OBS COMMAND: Import simulated data at observed location from MODFLOW package.
MON COMMAND: Convert daily data to monthly average as the observed values are monthly.
RESULTS COMMAND: 1) Refresh the result and represent GW level in each bore against the previous run which is similar to what PMWIN and GMS do; 2) calculate RMSE and R2 for each observation bore as well as overall and represented simulated result of GW level as "underestimated" (negative) and "overestimated" (positive) for each bore (red circles in Fig. 8) and whole aquifer as total; and 3) create plots from simulated and observed GW level automatically over the simulation period.

Figure 5. The GWCal interface.

SWAT-MODFLOW. Each time aquifer parameters are calibrated in this interface the corresponding files in SWAT-MODFLOW are replaced with the new input data. The main steps involved in the GWCal (summarized in Fig. 5) are as follows:

- The results of the coupled SWAT-MODFLOW model are considered as initial values in various zones of the aquifer (multiple polygons).
- The calibrated values from polygons in ArcGIS format were changed to raster.

- The raster files of hydraulic conductivity and specific yield were converted to ASCII file as MODFLOW input (UPW package) in SWAT-MODFLOW.
- Finally, SWAT-MODFLOW was run using the new calibrated values.

After calibration, the model output is represented in the GWhead in each observation well. The GWCal code is capable of presenting the results of model performances from SWAT-MODFLOW after each run. By comparing the model performance (RMSE) from the last run with the previous run the user would be able to track changes in the model performance result and improve the result. The SWAT result is then extracted and represented alongside the MODFLOW result to give the modeller a good understanding of whether the model performance is improving or not.

RESULT

The SWAT-MODFLOW results for runoff simulation at two points (Chenar Sokhteh (outlet 19) and Chenar Khoshk (outlet 16)) during simulation and validation period are presented in Fig. 6. Based on the results, the SWAT-MODFLOW model

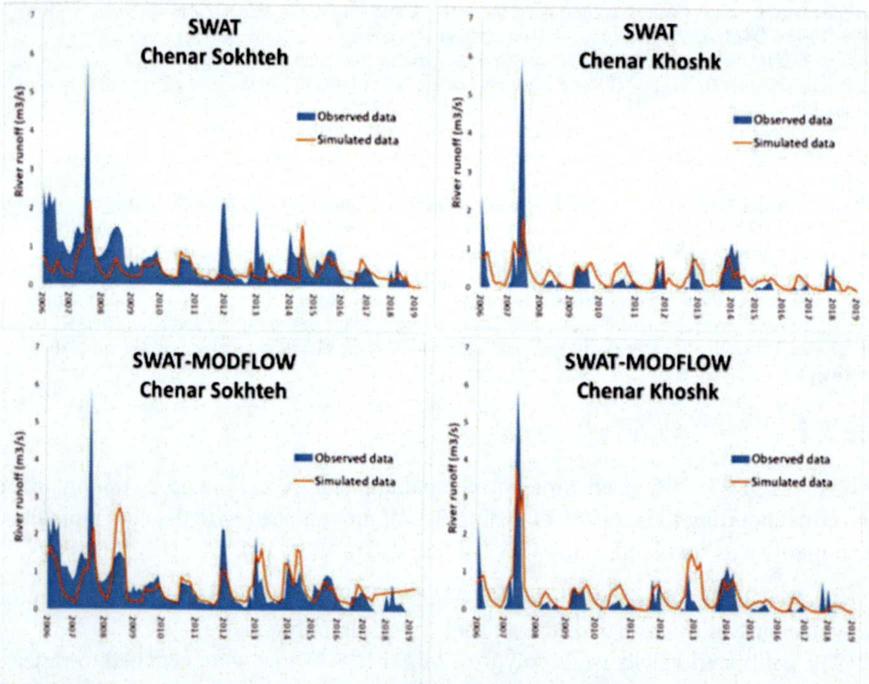

Figure 6. Observed and simulated river discharge using the SWAT-MODFLOW model during calibration and validation periods.

performed well in simulating runoff with NSE = 0.66 at outlet 16 and NSE = 0.68 at outlet based on the guidelines in Moriasi et al. (2007). The calibrated SWAT-MODFLOW values are given in Table 1. The results show parameters which are related to SW such as CN (curve number), LAT_TIME (LATeral flow travel time), ALPHA_B (base flow alpha factor), are close to their original reference values in SWAT, while parameters related to saturated zone and GW (such as GW_DELAY) are not close to their original values in SWAT. This highlights the impacts of

Table 1. SWAT parameters used for calibration process before coupling

SWAT Parameter	*Calibrated range for study area*	*Calibrated values (16)*	*Calibrated values (19)*	*Parameter Description*
ALPHA_BF. gw	0 - 1	0.2	0.12	Baseflow alpha factor (days)
PLAPS.sub	-100 - 100	12	30	Precipitation lapse rate (mm H_2O/ km)
TLAPS.sub	-10 - 10	3.5	1.5	Temperature lapse rate (°C/km)
CH_N(2).rte	-0.01 - 0.03	0.25	0.26	Manning's "n" value for the main channel (–)
HRU_SLP. hru	0 - 1	0.02	0.02	Average slope steepness (m/m)
LAT_TTIME. hru	0 - 100	85	92	Lateral flow travel time (days)
CH_K(2).rte	-0.01 - 30	185	240	Effective hydraulic conductivity in main channel alluvium (mm/hr)
CN(2).mgt	50 - 70	57.5	55.3	Initial Soil Conservation Service (SCS) runoff curve number for moisture condition II (–)
ESCO.hru	0 - 1	0.5	0.51	Soil evaporation compensation factor (–)
GWQMN.gw	0 - 500	4565	4855	Threshold depth of water in the shallow aquifer required for return flow to occur (mm)
GW_DELAY. gw	0 - 500	13	15	GW delay time (day)
GW_REVAP. gw	0.02 - 0.2	0.087	0.018	GW revap. coefficient[1] (–)
OV_N.hru	0.01 - 0.3	0.3	0.3	Manning´s roughness coefficient (n value) for overland flow (–)
REVAPMN. gw	0 - 500	185	205	Threshold depth of water in the shallow aquifer for revap. or percolation to the deep aquifer to occur (mm)
SOL_AWC. sol	0 - 1	0.6	0.9	Available water capacity of the soil layer (mm H_2O mm/ soil)
SOL_BD.sol	0.9 - 2.5	2.5	2	Moist bulk density (g/cm^3)
RCHRG_DP. gw	0 - 1	0.75	0.9	Deep aquifer percolation fraction (–)

MODFLOW on SWAT parameters and determines SWAT limitations in simulating saturated zone-related components in the water cycle.

Comparison between observed and simulated values in the selected observation wells (Fig. 7) shows that the model performance is good in the GW section. Moreover, the GW level calculated by SWAT-MODFLOW and observed GW level values in some piezometers are shown in Fig. 8. Simulated values in the aquifer are fitted to observation values where the observation point is located near to the river (Fig. 8).

The SWAT-MODFLOW performance in simulating aquifer level is assessed based on RMSE and MAE (Mean Absolute Error) values at observation wells (Fig. 9). The results show more than half of the observation wells including RMSE with less than 1 m. The water budget information was also used for model calibration. GW parameters in SWAT-MODFLOW were calibrated using GWCal. Based on the SWAT-MODFLOW water budget results (Fig. 10), total rainfall is 225.5 MCM. From total rainfall of 225.5 MCM/year, around 63% is evaporated. From the remaining water 24.7 MCM (11%) contributes to runoff while 58.3 MCM (26%) goes into GW system.

CONCLUSION

The SWAT-MODFLOW was used to simulate SW-GW interactions at Shiraz basin in Iran. To calibrate and validate SW and GW parameters a river runoff variable and a variable for GW level were used.

The results show that the calibration of SW parameters along with GW parameters in SWAT-MODFLOW significantly influenced the outputs of the model. The SWAT-MODFLOW calibrated values are considerably different to standalone SWAT in the SW parameters and standalone MODFLOW in the GW parameters. This difference is the result of the exchange happening between SW and GW. The results demonstrate that the GWCal tool, developed to enable the manual calibration of MODFLOW and SWAT, improves the efficiency and effectiveness of the calibration process which ultimately leads to improved performance of SWAT-MODFLOW. The results also demonstrate that properly calibrating both SW and GW components of the water cycle is critical. The key research findings are:

- An integrated approach that considers both SW and GW processes is required to realistically simulate SW-GW interactions.
- SWAT-MODFLOW produces more realistic results than standalone SWAT for SW-GW interactions and for simulating the effects of GW extraction on streamflow.
- SW parameters interact with GW parameters (and vice versa). Therefore, it is essential for realistic simulation of SW-GW processes to simultaneously calibrate SW parameters in SWAT based on insights obtained about GW from MODFLOW.

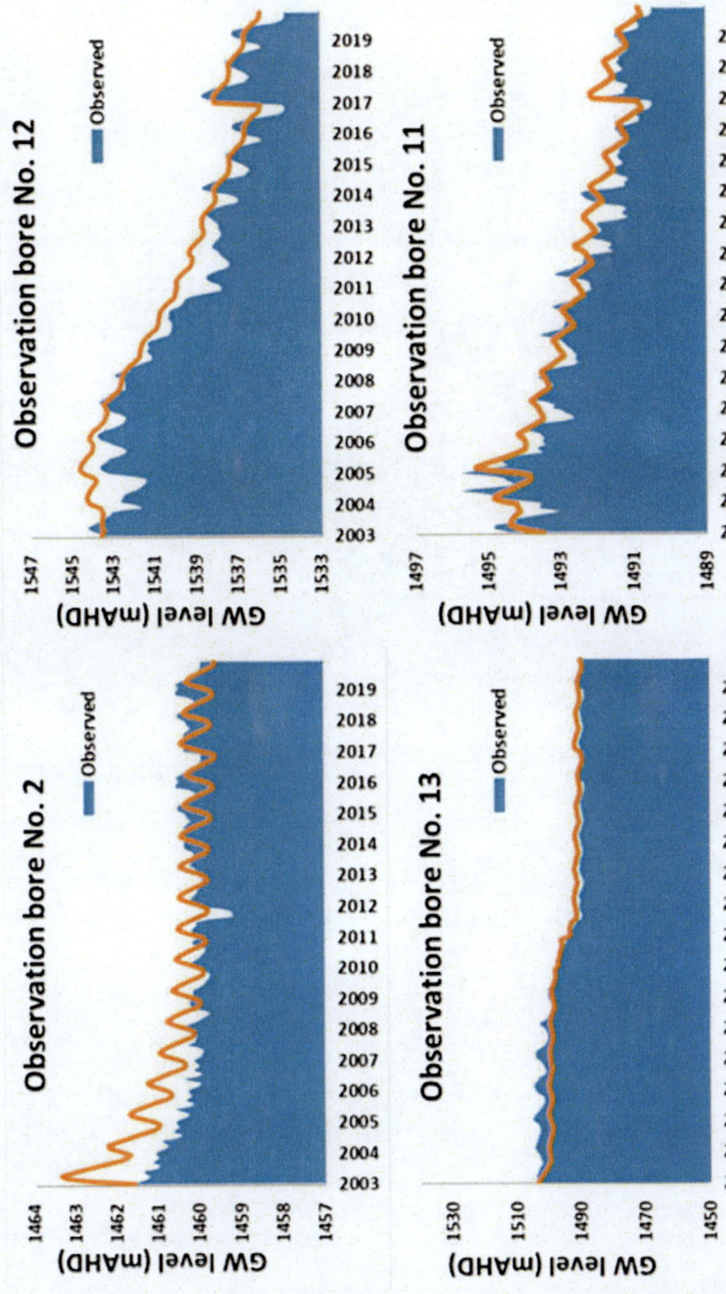

Figure 7. Observed and simulated GW level using SWAT-MODFLOW model in observation bores.

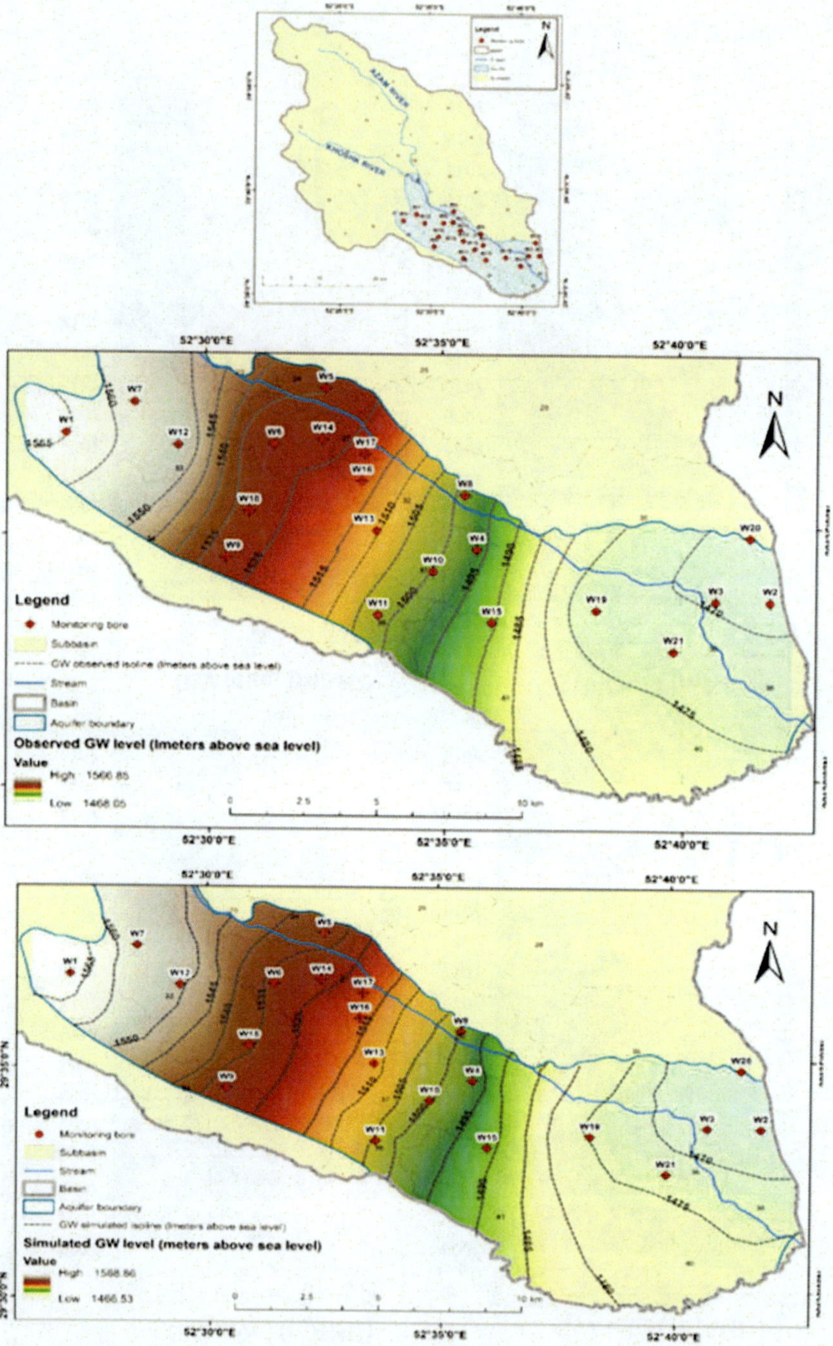

Figure 8. Simulated and observed GW level.

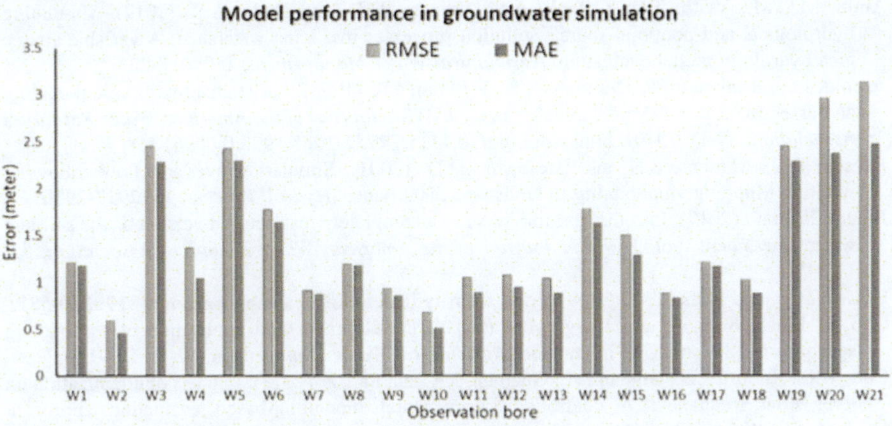

Figure 9. The RMSE and MAE values in SWAT-MODFLOW.

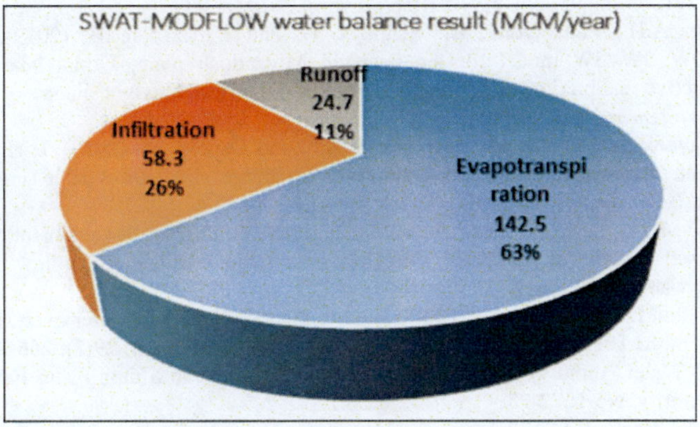

Figure 10. The SWAT-MODFLOW water balance.

The approach developed in this research, could be applied in other basins. Future work should be conducted to determine if automatic calibration of GW parameters can further accelerate and/or improve the calibration of SWAT-MODFLOW.

REFERENCES

Abbaspour, K. (2015). User Manual for SWAT-CUP: SWAT Calibration and Uncertainty Analysis Programs; Swiss Fed. Inst. of Aquatic Science and Technology: Dübendorf, Switzerland.

Aliyari, F., Bailey, R.T., Tasdighi, A., Dozier, A., Arabi, M. and Zeiler, K. (2019). Coupled SWAT-MODFLOW for large-scale mixed agro-urban river basins. *Environmental Modelling & Software*, 115: 200-210.

Amin, M.G.M., Veith, T.L., Collick, A.S., Karsten, H.D. and Buda, A.R. (2017). Simulating hydrological and nonpoint source pollution processes in a karst watershed: A variable source area hydrology model evaluation. *Agricultural Water Management*, 180: 212-223.

Arnold, J.G., Srinivasan, R., Muttiah, R.S., William, J.R. (1998). Large area hydrologic modeling and assessment part I: Model development. *JAWRA Journal of the American Water Resources Association*, 34(1): 73-89. https://doi.org/10.1111/j.1752-1688.1998.tb05961.x.

Beeson, P.C., Martens, S.N. and Breshears, D.D. (2001). Simulating overland flow following wildfire: Mapping vulnerability to landscape disturbance. *Hydrol. Process*. 15: 2917-2930.

Bieger, K. et al. (2017). Introduction to SWAT+, a completely restructured version of the soil and water assessment tool. JAWRA *Journal of the American Water Resources Association*, 53: 115-130.

Brown, M.E. and Funk, C.C. (2008). Food security under climate change. *Science* 319: 580-581.

Cho, J., Mostaghimi, S. and Kang, M.S. (2010). Development and application of a modeling approach for SW and GW interaction. *Agricultural Water Management*, 97(1): 123-130.

Chu, M.L., Knouft, J.H., Ghulam, A., Guzman, J.A. and Pan, Z. (2013). Impacts of urbanization on river flow frequency: A controlled experimental modelling-based evaluation approach. *J. Hydrol,* 495: 1-12.

Deb, P., Kiem, A.S. and Willgoose, G.R. (2019). A linked SW-GW modelling approach to more realistically simulate rainfall-runoff non-stationarity in semi-arid regions. Journal of Hydrology, 575: 273-291, doi:https://doi.org/10.1016/j.jhydrol.2019.05.039.

Dehghanipour, A.H., Zahabiyoun, B., Schoups, G. and Babazadeh, H. (2019). A WEAP-MODFLOW SW-GW model for the irrigated Miyandoab plain, Urmia lake basin, Iran: Multi-objective calibration and quantification of historical drought impacts. *Agricultural Water Management*, 223: 105704.

Eini, M.R., Javadi, S., Delavar, M., Monteiro, J.A.F. and Darand, M. (2019). High accuracy of precipitation reanalyses resulted in good river discharge simulations in a semi-arid basin. *Ecological Engineering*, 131: 107-119.

Eini, M.R., Javadi, S., Delavar, Gassman, M. and Jarihani, B. (2020). Development of alternative SWAT-based models for simulating water budget components and streamflow for a karstic-influenced watershed. *Catena,* 195: 104801

Esteller, M.V. and Diaz-Delgado, C. (2002). Environmental effects of aquifer overexploitation: A case study in the Highlands of Mexico. *Environmental Management*, 29(2): 266-278.

Fars Regional Water Authority (2015). Report of Water budget in Shiraz Plain, Fars Provincial Water Authority, No 1: 22-23.

Graaf, I.E., van Beek, R.L., Gleeson, T., Moosdorf, N., Schmitz, O., Sutanudjaja, E.H. and Bierkens, M.F. (2017). A global-scale two-layer transient GW model: Development and application to GW depletion. *Advances in Water Resources*, 102: 53-67.

Guzman, J.A., Moriasi, D.N., Gowda, P.H., Steiner, J.L., Starks, P.J., Arnold, J.G. and Srinivasan, R. (2015). A model integration framework for linking SWAT and MODFLOW. *Environmental Modelling & Software*, 73: 103-116.

Harbaugh, A.W. (2005). MODFLOW-2005, the US Geological Survey modular ground-water model: the ground-water flow process (pp. 6-A16). Reston, VA: US Department of the Interior, US Geological Survey.

Harbaugh, A.W., Banta, E.R., Hill, M.C. and McDonald, M.G. (2000). MODFLOW-2000, The U. S. Geological Survey Modular Ground-Water Model-User Guide to Modularization Concepts and the Ground-Water Flow Process. Open-file Report. U.S. Geological Survey, 92: 134.

Jafari, F., Javadi, S., Golmohammadi, G., Karimi, N. and Mohammadi, K. (2016a). Numerical simulation of GW flow and aquifer-system compaction using simulation and InSAR technique: Saveh basin, Iran. *Environmental Earth Sciences*, 75(9): 833.

Jafari, F., Javadi, S., Golmohammadi, G., Mohammadi, K., Khodadadi, A. and Mohammadzadeh, M. (2016b). GW risk mapping prediction using mathematical modeling and the Monte Carlo technique. *Environmental Earth Sciences*, 75(6): 491.

Javadi, S., Kavehkar, N., Mousavizadeh, M.N. and Mohammadi, K. (2011). Modification of DRASTIC model to map GW vulnerability to pollution using nitrate measurements in agricultural areas. *Journal of Agriculture Science and Technology*, 13(2): 239-249.

Kahinda, J.M., Taigbenu, A.E. and Boroto, R.J. (2010). Domestic rainwater harvesting as an adaptation measure to climate change in South Africa. *Physics and Chemistry of the Earth,* Parts A/B/C, 35(13-14): 742-751.

Lachaal, F., Mlayah, A., Bédir, M., Tarhouni, J. and Leduc, C. (2012). Implementation of a 3-D GW flow model in a semi-arid region using MODFLOW and GIS tools: The Zéramdine–Béni Hassen Miocene aquifer system (east-central Tunisia). *Computers & Geosciences*, 48: 187-198.

Martínez-Santos, P. and Andreu, J. (2010). Lumped and distributed approaches to model natural recharge in semiarid karst aquifers. *Journal of Hydrology*, 388: 389-398.

Molina-Navarro, E., Bailey, R.T., Andersen, H.E., Thodsen, H., Nielsen, A., Park, S., Jensen, J.S., Jensen, J.B. and Trolle, D. (2019). Comparison of abstraction scenarios simulated by SWAT and SWAT-MODFLOW. *Hydrological Sciences Journal*, 64(4): 434-454.

Moriasi, D.N., Arnold, J.G., Van Liew, M.W., Bingner, R.L., Harmel, R.D. and Veith, T.L. (2007). Model evaluation guidelines for systematic quantification of accuracy in watershed simulations. Trans. ASABE 2007, 50: 885-900.

Nash, J.E. and Sutcliffe, J.V. (1970). River flow forecasting through conceptual models part I — A discussion of principles. Journal of Hydrology, 10: 282-290.

Neitsch, S.L., Arnold, J.G., Kiniry, J.R. and Williams, J.R. (2011). Soil and Water Assessment Tool Theoretical Documentation Version 2009. Texas Water Resources Institute.

Pai, H. (2015). High resolution synoptic sampling and analysis for understanding GW-SW interactions in lowland rivers (Doctoral dissertation, UC Merced).

Pérez-Martín, M.A., Estrela, T., Andreu, J. and Ferrer, J., (2014). Modeling water resources and river-aquifer interaction in the Júcar River Basin, Spain. *Water Resources Management*, 28(12): 4337-4358.

Playan, E. and Mateos, L. (2006). Modernization and optimization of irrigation systems to increase water productivity. *Agr. Water Manage*. 80: 100-116.

Scanlon, B.R., Reedy, R.C., Stonestrom, D.A., Prudic, D.E., and Dennehy, K.F. (2005). Impact of land use and land cover change on GW recharge and quality in the southwestern US. *Glob. Change Biol.* 11: 1577-1593.

Sheikhipour, B., Javadi, S. and Banihabib, M.E. (2018). A hybrid multiple criteria decision-making model for the sustainable management of aquifers, *Environmental Earth Sciences*, 77 (19): 712.

Surinaidu, L., Rao, V.G., Rao, N.S. and Srinu, S. (2014). Hydrogeological and GW modeling studies to estimate the GW inflows into the coal Mines at different mine development stages using MODFLOW, Andhra Pradesh, India. *Water Resources and Industry*, 7: 49-65.

Tian, Y., Zheng, Y., Wu, B., Wu, X., Liu, J. and Zheng, C. (2015). Modeling SW-GW interaction in arid and semi-arid regions with intensive agriculture. *Environmental Modelling & Software*, 63: 170-184.

Wheeler, T. and von Braun, J. (2013). Climate change impacts on global food security. *Science,* 341: 508-513.

Winter, T.C., Harvey, J.W., Franke, O.L. and Alley, W.M. (1998). Ground water and SW – a single resource. US Geol Surv Circ 1139.

Chapter 8
Effect of Changing Climate on the Water Resources of Upper Jhelum Basin (UJB), India

Suhail A. Lone, Ghulam Jeelani, Akhtar Alam, M. Sultan Bhat, and Hakim Farooq

INTRODUCTION

Water resources originating from the alpine snow, glacier, recent precipitation and groundwater in the Kashmir Himalayas are important sources of fresh water to millions of people downstream. In the Upper Jhelum Basin (UJB) an increase in temperature has led to reduction in glacier cover and modification in the snowfall pattern (Jeelani et al., 2012; Lone et al., 2017) leading to the disturbance in timely availability of fresh water (Jeelani et al., 2012, 2013, 2017a). There is a lug in peak discharge from summer season to early spring season (Jeelani et al., 2017b, c), which can have strong bearing on the agriculture, horticulture and hydropower generation of the UJB, under prevailing climate change. It is, therefore, essential to comprehend the sources of water resources of UJB and to analyse and study the climate change impact on these resources. Climatic elements, such as precipitation, temperature, solar radiation, wind speed, humidity, etc. tend to have an effect on the direct input parameters for runoff, and alter it indirectly. For example, precipitation is received directly as a moisture input for runoff. Here, the form of input, e.g. rain or snow may generate different responses. Solar radiation, wind speed and temperature have an indirect relation with the runoff. Temperature increases the snow melting processes and thus would seem to enhance runoff; on the other hand, the evapotranspiration processes are also dependent upon the temperature of the region, which tends to retard the runoff volume. With respect to precipitation the intensity, duration and volume of a storm event affects the daily discharge and the peak flows (Subramanya, 2013). The stream flow in UJB show significant temporal and spatial variation exhibiting characteristic temporal/local variations, which may be dominantly due

S. A. Lone · G. Jeelani (✉)
Department of Earth Sciences, University of Kashmir, Srinagar, India

A. Alam · M. S. Bhat · H. Farooq
Department of Geography and Regional Development, University of Kashmir, Srinagar, India

A. Mukherjee (ed.), *Riverine Systems*, https://doi.org/10.1007/978-3-030-87067-6_8

to snow melt, glacier melt, rainfall and base flow sources. Stable water isotopes (^{18}O and ^{2}H) are ideal tracers for estimating the sources of stream flow and other water resources of a region due to their conservative characteristics that is inherent to the water droplets (Gat, 1996; Clark and Fritz, 1997; Jeelani et al., 2013; Lone et al., 2019). The stable water isotopic values ($\delta^{18}O$ and $\delta^{2}H$) of river and stream waters are often used to estimate the origin of water, through exploring the fraction among different reservoirs and evaporation processes (Dalai et al., 2002; Lambs et al., 2005; Bowen et al., 2011). In this study, we used natural tracers ($\delta^{18}O$ and $\delta^{2}H$) of source waters of UJB to estimate their contribution and to assess the impact of climate change on the water resources of the basin.

STUDY AREA

The transboundary UJB shares the boundary between India and Pakistan. In India, the upper Jhelum (Kashmir Valley) is NW-SE oriented alluvial basin bounded by higher and lesser Himalayas, respectively. It lies between the geographical coordinates of latitude 33°25′N and 34°32′N and longitude 74°0′E and 75°30′E (Fig. 1) covering an area of 12,757 km^2 with a mean altitude of 1850 m, amsl. The climate of UJB is strongly controlled by mid-latitude circulations (western disturbances

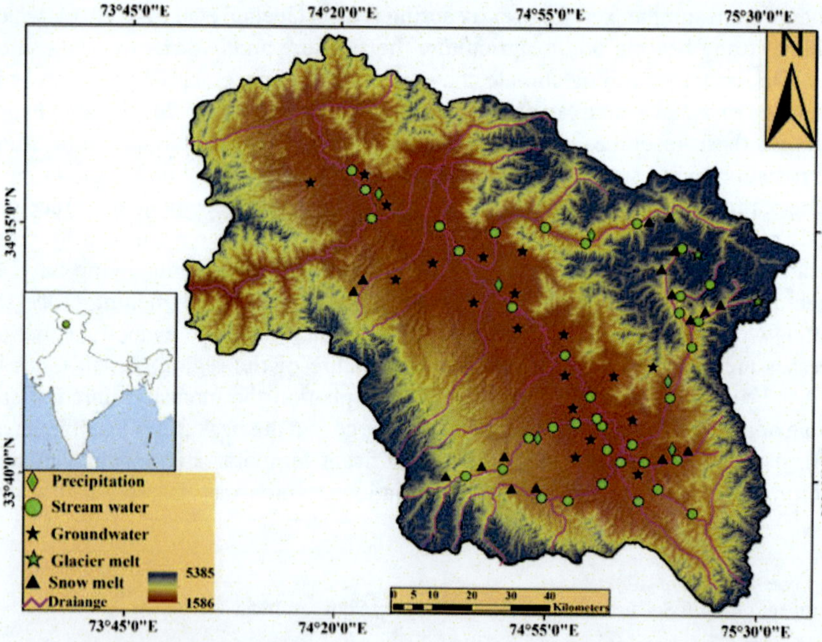

Figure 1. Upper Jhelum Basin (UJB) with sampling sites (base map was obtained from SRTM downloaded from https://earthexplorer.usgs.gov/).

(WDs)) and southwest monsoons (ISM, Lone et al., 2017; Jeelani et al., 2017a, 2018, 2020). The mid-latitude circulations (WDs) and southwest monsoons in UJB and other parts of western Himalayas are active from October to May and June to September respectively (Lone et al., 2016, 2017, 2019, 2020; Jeelani et al., 2017a). The UJB experiences a fairly cold and showery spring, extending from March through April and May. Unlike mainland India, Kashmir valley is a meager beneficiary of the southwest monsoon. In fact, summer monsoons hardly constitute one-sixth to one-fourth of the total precipitation received by Kashmir valley. This is for major part attributed to the obstructive nature of the Pir Panjal.

Autumn in the valley begins in September and is till November. Although autumn is almost altogether a dry season but rain or snow is not entirely unknown. Winter in UJB is rigorously cold with widespread snow and low temperatures and largely dominated by mid-latitude circulations. The onset of winter in UJB is by December and continues through till February. The UJB experiences moderate climate with a mean annual temperature and precipitation of 13.6°C and of 681 mm, respectively (1901-2015, Jeelani et al., 2017a, 2020). The mean minimum and maximum temperatures are observed in January and July while, mean minimum and maximum precipitation is received in March and November, respectively (Jeelani et al., 2020).

HYDROLOGICAL SETUP

The meltwater originating from winter accumulated snow and glacier cover constituting a significant proportion of discharge in rivers of UJB (Fig. 2) and downstream descending from high altitudinal terrains of Himalayas (Greater Himalayas and Pir Panjal). However, meltwater contributed by these winter accumulated snow and glaciers vary spatially from one watershed to the other depending upon a number of factors including extent of snow and glacier cover, topography, aspect etc. The precipitation received during winter and early spring seasons generates meltwater, which sustains the flow in Jhelum River throughout the year (Jeelani et al., 2017c). During late summer and autumn seasons, glacier melt and baseflow are the significant sources to regulate the flow in streams and river Jhelum (Jeelani et al., 2013, 2017c). Therefore, not only stream flow but also groundwater recharge is a mixture of snow, glacier melt and the summer precipitation, which are jointly affected by distinction in precipitation, glacier melt and winter accumulated snow cover (Jeelani et al., 2017c; Lone et al., 2017). Due to temperate climatic conditions, UJB receives significant amount of annual precipitation (avg: 681 mm year^{-1} (Jeelani et al., 2017a, 2020)). Rainwater is also an important source of river discharge in downstream areas of the basins at elevations less than 2000 m, amsl.

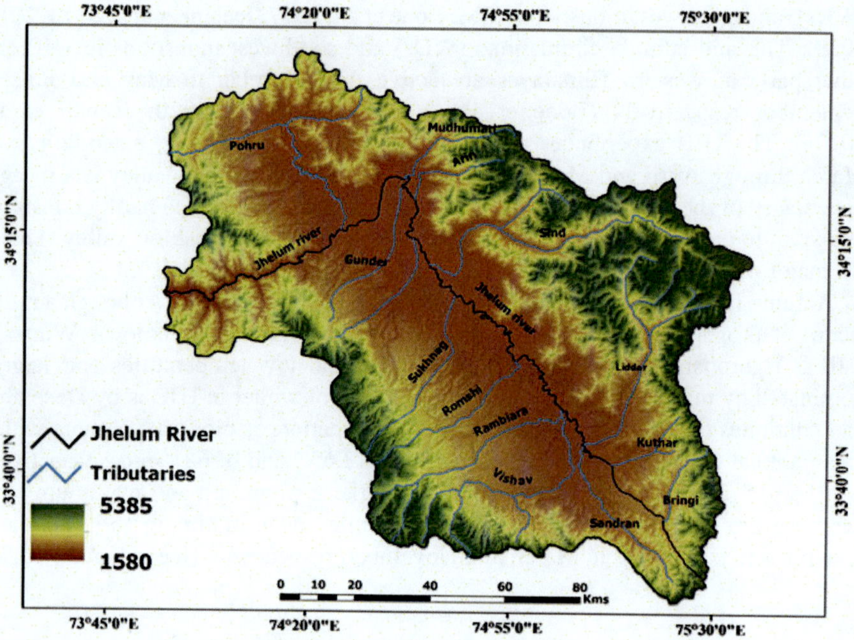

Figure 2. Drainage network of Upper Jhelum Basin. Black polyline shows Jhelum river and blue lines depict the tributaries.

METHODOLOGY ADOPTED

Water samples from Jhelum river ($n = 60$), snowmelt ($n = 36$) and glacier melt ($n = 23$) were collected for $\delta^{18}O$ and δ^2H analysis. The snowmelt (seasonally) was collected across valley in winter season and from snow packs in other seasons. Samples from glaciers were collected as solid ice at different altitudes and as melt at snout. Stream water samples were collected seasonally from river Jhelum. The samples were collected in 10 ml high-density polyethylene bottles and were properly labelled and coded at the spot to avoid contamination. The water samples were sent to Geosciences Division, Physical Research Laboratory (PRL), Ahmedabad, India, for stable water isotopic analysis. The isotope ratio mass spectrometer (IRMS, Delta V plus, Thermo Scientific) was used to measure the ratios of oxygen and hydrogen in continuous flow mode (Epstein and Mayeda, 1953). Results obtained were defined with respect to Vienna Standard Mean Ocean Water and reported in standard δ-notation. The precision of measurement is ± 0.1‰ and ± 1‰ for $\delta^{18}O$ and δ^2H, respectively.

END MEMBER-MIXING ANALYSIS (EMMA)

The seasonal deviation in proportion of different source waters to river Jhelum were estimated based on following assumptions:

(1) The contributing source waters (end-members) have constant stable water isotopic values; (2) the stable water isotopic values of end-members differ considerably. The seasonal contributions from rainfall, snowmelt and glacier melt to river Jhelum was evaluated using following equation (Singh et al., 2019):

$$Q_{st} = Q_p + Q_{gm} + Q_{sn} \tag{1}$$

where Q_p, Q_{gm} and Q_{sn} are the contribution from precipitation, glacier melt and snow melt to regional hydrology (Q_{st})

$$1 - f_p + f_{gm} + f_{sn} \tag{2}$$

$$\delta_{st} - \delta_p . f_p + \delta_{gm} . f_{gm} + \delta_{sn} . f_{sn} \tag{3}$$

δ_{st}, δ_p, δ_{gm}, δ_{sn} are $\delta_{18}O$ for the stream, glacier melt and snow melt, respectively.

$$\delta_{st} - \delta_p . f_p + \delta_{gm} . f_{gm} + \delta_{sn} . f_{sn} \tag{4}$$

d_{st}, d_p, d_{gm} and d_{sn} are *d*-excess tracer for the stream, glacier melt and snowmelt, respectively.

$$Q_p(\%) = \frac{(d_{st} - d_{sn})(\delta_{gm} - \delta_{sn}) - (d_{gm} - d_{sn})(\delta_{st} - \delta_{sn})}{(d_p - d_{sn})(\delta_{gm} - \delta_{sn}) - (d_{gm} - d_{sn})(\delta_p - \delta_{sn})} \times 100 \tag{5}$$

$$Q_{gm}(\%) = \frac{(d_{st} - d_{sn})}{(d_{gm} - d_{sn})} \times 100 - \frac{(d_p - d_{sn})}{(d_{gm} - d_{sn})} \times Q_p \tag{6}$$

$$Q_{sn}(\%) = 100 - Q_p - Q_{gm} \tag{7}$$

In this study, the Moderate Resolution Imaging Spectroradiometer (MODIS) Terra MOD10A2 snow product was used for the estimation of accumulated winter (November-February) snow cover from 2000 to 2018 to comprehend the temporal snow cover changes. Numerous spectral bands and superior spatial resolution of the MODIS data products permits a better ability to map snow and ice cover (Hall et al., 2001; Hall et al. 2016). The MOD10A2 data is 8-day composite snow cover observations with 500 m resolution (Pu et al., 2007). It contains maximum snow cover data in compressed Hierarchical Data Format-Earth Observing System (HDF-EOS), along with metadata sets. The snow cover information is extracted using an algorithm-Normalized Difference Snow Index (NDSI). The areas under snow always show the NDSI value > 0; however, not all surfaces with NDSI value of > 0 are snow cover (Riggs et al., 2015).

RESULTS AND DISCUSSIONS

Variability of Stable Water Isotopes ($\delta^{18}O$ and δ^2H) in Source Waters

The stable water isotopic values ($\delta^{18}O$ and δ^2H) in source waters showed a distinct spatial and temporal variation (Table 1). A substantial positive correlation exists between the mean monthly $\delta^{18}O$ (and δ^2H) of precipitation and the elevation of precipitation stations in UJB (Table 2). The estimated isotope altitudinal gradient in UJB is −0.15 to −0.73‰ and −1.2 to −2.56‰ per 100 m rise in elevation for $\delta^{18}O$ and δ^2H, respectively. The stable water isotopic values ($\delta^{18}O$ and δ^2H) of precipitation across UJB showed lighter/depleted values in winter and heavier/enriched values in summer (Fig. 3).

A significant positive correlation was observed between mean monthly temperature and the stable water isotopic values of precipitation. Stable water isotopic values become enriched from January to July and depletes from July to December associated with cycle of increase and decrease in temperature throughout the year. However, the abrupt decrease in stable water isotopic values in August indicates reversal of winds from western disturbances to Indian summer monsoons (Jeelani et al., 2017a).

Table 1. The isotopic values ($\delta^{18}O$ and δ-excess) of different end-members

	$\delta^{18}O$ (‰)			*δ-excess (‰)*		
End-Members	*Max*	*Min*	*Mean*	*Max*	*Min*	*Mean*
Precipitation	−2.9	−10.4	−7.5	20	1	14
Glacier	−9.3	−12.8	−10.9	20	14	17
Snow melt	−6.6	−8.9	−8.2	19	14	17
Groundwater	−7.6	−9.2	−8.5	26	14	19
Stream water	−7.4	−9.9	−8.7	23	17	19

Table 2. Correlation coefficient and significance level (*p*-value) of $\delta^{18}O$ and δ^2H of end-members with altitude

		Altitude (m, amsl)	
Components		*R^2*	*p-value*
Precipitation	$\delta^{18}O$ (‰)	0.88	>0.001
	δ^2H (‰)	0.84	=0.002
Glacier	$\delta^{18}O$ (‰)	0.81	=0.002
	δ^2H (‰)	0.75	=0.007
Snowmelt	$\delta^{18}O$ (‰)	0.79	>0.001
	δ^2H (‰)	0.71	<0.003

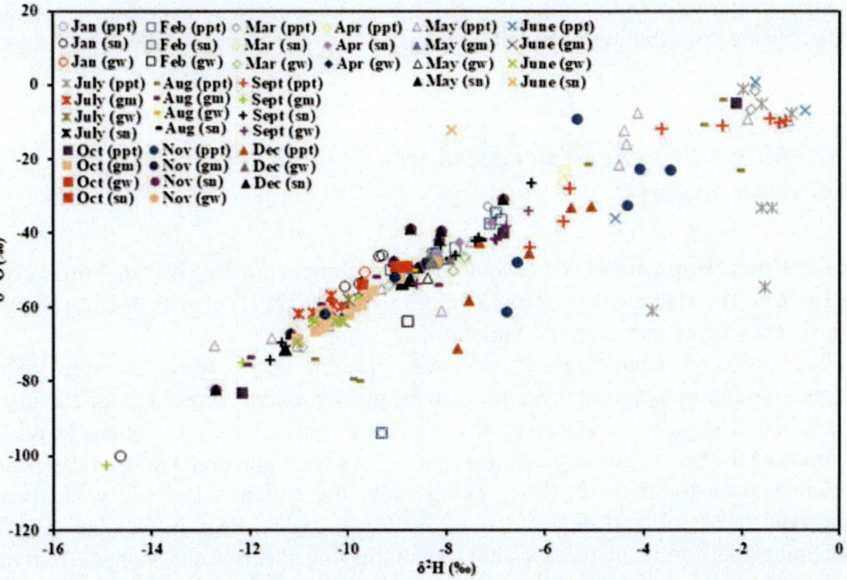

Figure 3. Stable water isotopic variation in precipitation, glacier melt, snowmelt and groundwater.

It was observed that glacier exhibit depleted isotopic values: $\delta^{18}O$ and/or $\delta^{2}H$ -12.8‰ and -74‰ at higher altitudes than lower altitudes, which is attributed to well-known altitude effect. A significant positive correlation observed between $\delta^{18}O$ (and $\delta^{2}H$) of ice and the elevation of sampling locations (Table 2). The stable water isotopic values of glacier melt showed a depleting trend temporally. The enriched isotopic values are attributed to the contribution of seasonal snow and rainfall (Fig. 3). Therefore, depleted isotopic values in autumn is real representation of glacier melt when snow is completely exhausted. Like precipitation and glacier melt, a significant positive correlation was observed between stable water isotopic values ($\delta^{18}O$ and $\delta^{2}H$) of snowmelt samples and elevation of sampling locations (Table 2). The calculated isotopic altitudinal gradient of snowmelt samples is -0.24‰ to -0.42‰ and -2.4‰ to -3.9‰ for $\delta^{18}O$ and $\delta^{2}H$. During the observational period, it was observed that the snowmelt showed depleting trend in stable water isotopic values. However, in August the enriched stable water values were observed suggesting contribution from rainfall, which enriched the isotopic values (Fig. 3).

The statistical summary of stable water isotopic composition of groundwater is given in Table 1. The stable water isotopic composition of groundwater does not indicate any relationship with altitude due to intermixing of waters from different sources. Like streams, groundwater was depleted in September (-10 ‰) and enriched in May (-6.8 ‰). The depleted stable water isotopic values observed is attributed to melting of accumulated winter snowfall. However, enriched stable water isotopic values observed in May reflects the signatures of recent rainfall.

Insignificant snow and/or glacial melting due to subzero temperatures in winter indicates the contribution of baseflow (Fig. 3).

Variability of Stable Water Isotopes (^{18}O and ^{2}H) in Stream Water

The statistical summary of stable water isotopic composition of river Jhelum is given in Table 1. The stable water isotopic values ($\delta^{18}O$ and $\delta^{2}H$) of river Jhelum showed significant spatial and temporal variation.

The different tributaries/streams have different stable water isotopic values depending upon physiography, snow cover, glacier extent, aspect etc. of the micro watershed. The isotopic values of Jhelum River is affected by the complex physiography of the basin. The depleted isotopic values were perceived in headwaters and enriched towards the basin floor. Temporally, the enriched isotopic values were perceived in March and depleted stable water isotopic values in October (Fig. 4) indicating the dominant role of snow in modifying the isotopic values of stream water. The isotopic composition of winter accumulated snowfall is reflected in stream flow in March ($\delta^{18}O$: −7.5‰ and $\delta^{2}H$: −53‰) through April ($\delta^{18}O$: −10‰ and $\delta^{2}H$: −59‰) up to October ($\delta^{18}O$: −9.8‰ and $\delta^{2}H$: −65‰). This is attributed to the increasing ambient temperature, which increases melting while as decrease in ambient temperature with increase in altitude, which hinder melting at higher altitudes. In the early melting season water with enriched (^{18}O and ^{2}H) stable water isotopic values are released from lower altitudes (>2200 m, amsl), which

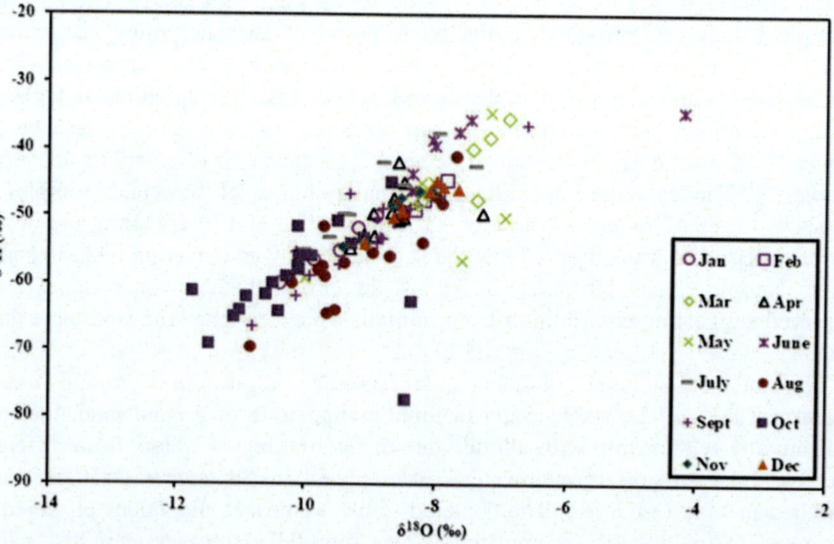

Figure 4. Temporal variation of $\delta^{18}O$ and $\delta^{2}H$ in stream water.

progressively melts snow pack with depleted isotopic values as the melting season advances. In autumn (September/October) most of the snow at lower altitudes is exhausted and snow pack at the higher altitudes, with stable water isotopic values similar to glaciers release melt water. However, in winter the negligible snow and glacier melting due to low temperature and less discharge represents the contribution of groundwater to stream flow.

Contribution of Rain, Snow and Glacier Melt to River Jhelum

The cyclic contribution from snowmelt, glacier melt and groundwater to river Jhelum were estimated using three-component mixing analysis. The results suggest that (Fig. 5) snowmelt (29 to 58%) is a dominant contributor to River Jhelum followed by rain (18 to 47%) baseflow (13 to 38%) and glacier (1 to 6%). The contribution of snowmelt to river Jhelum in spring is 29%, which tended to increase in summer (58%) and decreases in autumn season (38%). The glacier melt contribution in spring season to river Jhelum was >1% but it increases in summer (3%) and autumn seasons (6%), respectively. Contribution of baseflow to river Jhelum is 23% in spring, however, it decreases in summer (13%) but increases in autumn (38%). The rainfall contribution to river Jhelum was higher in spring (47%), however, it decreases in summer (26%) and autumn (18%), respectively.

The significant influence of snowmelt to stream flow in summer season (Fig. 5) dominantly controls the economic growth of the UJB, which largely depended upon the agriculture, horticulture and ecotourism activities (Jeelani et al., 2017d). Besides, large amount of water is required during summer for agricultural and horticultural

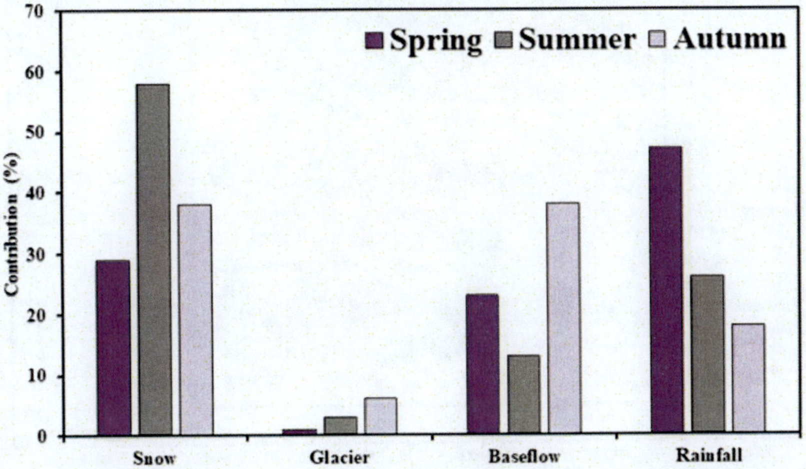

Figure 5. Seasonal contribution of snow, glacier, baseflow and rainfall to river Jhelum.

activities and the gushing waters in headwaters of river Jhelum are used for water sports and other commercial activities.

RELATIONSHIP BETWEEN DISCHARGE AND PRECIPITATION REGIMES

Among the climatic controls of river discharge, the precipitation is the most important variable factor having direct relationship with the discharge. Thus, investigating the inter-relationship between discharge and the stream flows can be very important in understanding the overall hydrology and aid in efficient water management.

The relationship was explored using Pearson correlation coefficient (r). Its value varies between 1 to +1 with negative values representing inverse relationship whereas positive values reflect direct relationship. The correlation coefficient was worked out between the precipitation and the stream flow volumes for different sub-watersheds at annual scales. Average precipitation for a particular sub-watershed was estimated using the Thiessen polygon method (Subramanya, 2013). Basic assumption is that a precipitation gauge station best represents the area, which is close to it. Thiessen polygons for the Jhelum basin were computed in the ArcGIS environment using meteorological data (Fig. 6). For a watershed encompassing n Thiessen polygons, the areal precipitation over the watershed is computed as:

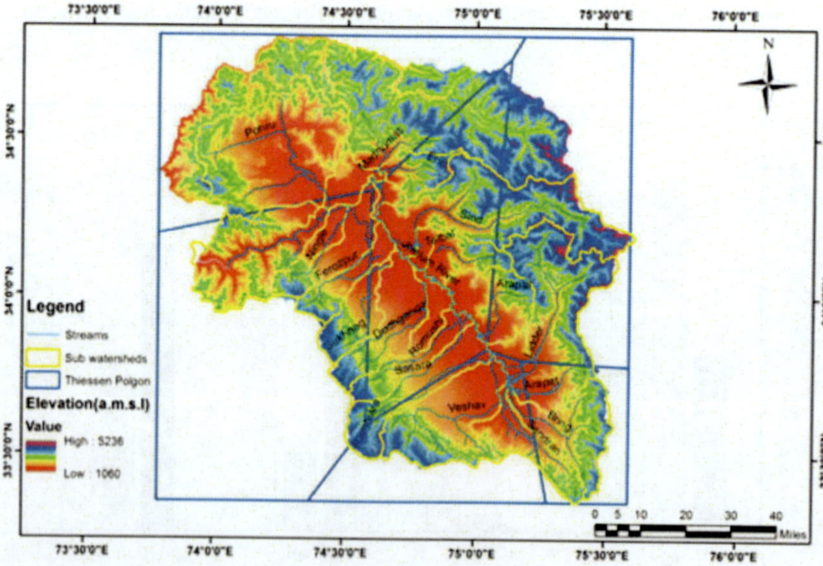

Figure 6. Map showing Thiessen polygons plot for Jhelum Basin.

$$P_{avg} = \sum_{i=1}^{M} P_i \frac{A_i}{A}$$

where P_i is the observed rainfall at the centroid of the *i*th polygon; A_i the area defined by the intersection of the Thiessen polygon and watershed and A the total area of watershed. The correlation analysis reveals positive and statistically significant relationship between the two variables (precipitation, discharge). Furthermore, the value of (r) varies considerably between the sub-watersheds.

Aripath shows a maximum correlation ($r = 0.73$) whereas Ferozpur shows lowest correlation ($r = 0.35$). Moreover, the analysis reveals that sub-watersheds forming the head waters of river Jhelum tend to show relatively higher correlation coefficients of the order 0.73 (Aripath), 0.67 (Bringi), 0.69 (Lidder), 0.69 (Vishow), 0.49 (Sandran) and 0.46 (Rambiara). Among the other sub-watersheds located in middle and lower reaches Sukhnag and Dakil-Nallah exhibits r-values of 0.71 and 0.62, respectively, while for other sub-watershed, values range between 0.35 and 0.55. On the regional scale there exists a noteworthy positive relationship between the stream flow and precipitation ($r = 0.55$). The trend analysis of annual precipitation data series indicates, an overall decline in the precipitation amount in the region from 1980 to 2016 (Fig. 6). Moreover, relationship between the stream flows and temperature variation was also assessed on the regional scale. There exists an inverse correlation between the stream flow and mean maximum temperature ($r = -0.52$). The temperature change largely determine the amount of evapotranspiration and tend to reduce the surface runoff. In addition, there exists a significant inverse relationship between the precipitation and temperature ($r = -0.65$) in the region implying that the region will experience precipitation deficit as a consequence of increasing temperature (Fig. 7). The long term temporal changes in the stream flow, precipitation and temperature were also explored by subjecting them to linear trend analysis (Fig. 8a). The linear trend shows the annual precipitation is declining with 5.81 mm year^{-1} in the basin. The temperature in the region is also showing increase in magnitude 0.035°C year^{-1} (T_{max}) and 0.019°C (T_{min}). The decrease in precipitation and the rising temperature has severely affected the stream flows, which can be observed from decreasing stream flow trends in the Jhelum basin from 1980 to 2016 (Fig. 8b). The annual average discharge is showing decrease of the order 82.3 cusecs year^{-1}. Thus, variation in climatic conditions are tightly coupled with stream flow and have been a primary factor in explaining the decreasing trends in stream flow.

LONG-TERM CHANGE IN SNOW COVER

The amount of snow received during the winter months in high altitude mountainous basin (Kashmir Himalayas) governs the availability of water for drinking and other recreational uses throughout the year (Jeelani et al., 2013). In this study, two contrasting patterns were observed in the snow cover over Kashmir basin from

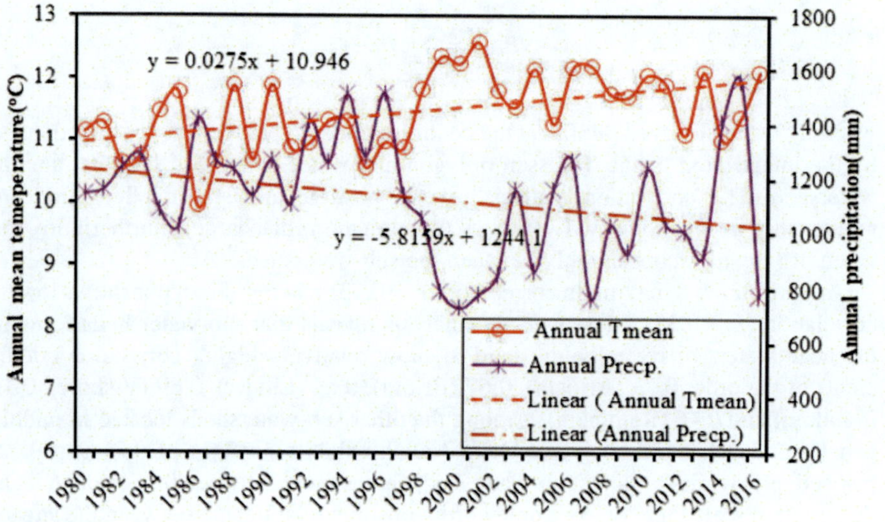

Figure 7. Relationship between regionally averaged-annual mean temperature and annual precipitation.

2000 to 2018 (Fig. 9). In general, snow cover reveals an increasing trend during November and December and a declining pattern in January and February (Fig. 10). The overall trend of snow cover mapping in UJB suggests snowfall decrease. Trends observed in the snow cover are in agreement with the declining summer flow patterns (gauge records) exhibited by the different tributaries of the Jhelum River suggesting the dominance of winter accumulated snow cover in maintaining the annual runoff.

Due to climate change effect the natural cycle of snowfall, accumulation and melting are adversely affected which in turn have led to increase in liquid precipitation and decrease in solid precipitation. This change trend had adversely affected the water resources of UJB.

INFLUENCE OF CHANGING CLIMATE ON THE WATER RESOURCES OF UJB

The impacts of climate change are extensive and cut across different natural and human systems. Water resources are one of the key sectors with a high degree of vulnerability from the climate change due to their direct exposure and dependence on the climatic variables. The IPCC AR4 places high confidence with the fact that observed changes in the fresh water resources are mainly due to variations in temperature and precipitation. These changes are more pronounced in the mountain ecosystems like Himalayas where the surface hydrology is driven by the seasonal snowmelt and glacial melt (Jeelani et al., 2012). The signals of climate change are

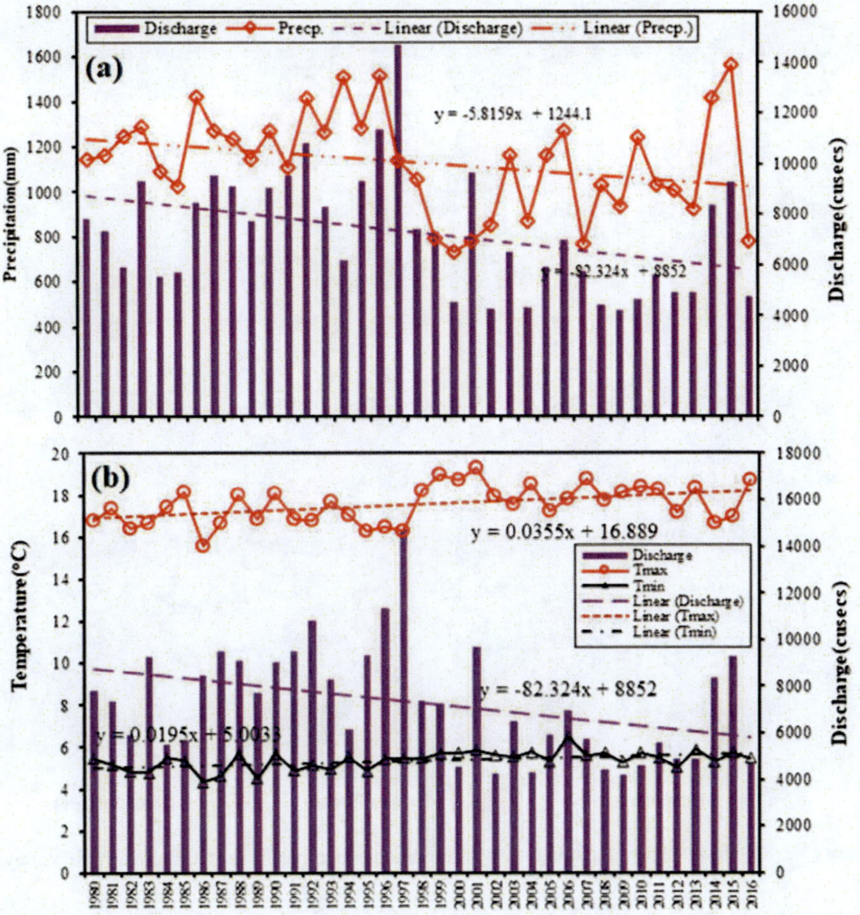

Figure 8. Annual variability of precipitation, temperature and discharge for entire Jhelum Basin.

well evident in the Kashmir Himalayas as has been analyzed through the recent trends in climate variables over the region. The region has witnessed a rise in mean maximum and minimum temperature associated with reduction in precipitation and hosts rich fresh water resources in term of snow and glaciers having tremendous regional significance. In wake of the ongoing climate change, glaciers in the area have observed a substantial recession and the snow cover area is showing a significant decreasing trend. Noting the observed and projected changes in the climatic variables, it is clear that water resource base of the region tends to remain sensitive to the vagaries of climate change. Other sectors, which are highly dependent on the water resources will also be affected. Agriculture, hydropower, tourism, which are the backbone of the economy not only in upper Jhelum basin but also downstream would be worst hit due to untimely availability of water, e.g., shifting of river flow peak towards early spring season, when its requirement is lesser in comparison to

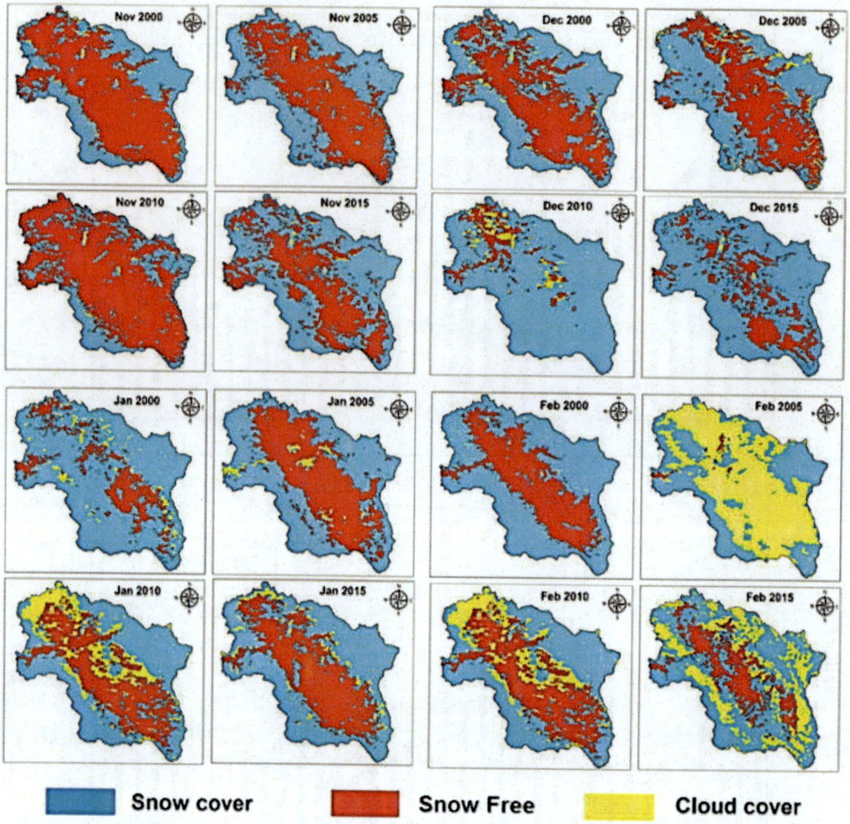

Figure 9. Satellite imagery depicting snow cover change in upper Jhelum basin (Kashmir Valley).

summer when its demand is highest. Crop shifting, which some people favour, is not a viable option, as the temperature shows large variability. Changes in cropping pattern from agriculture to horticulture is a newly adopted trend, is a clear indication that food security of the region is under pressure. Among other sectors, tourism also depends upon the health of water resources. As UJB has been a major and popular tourist destination, the change in climate would not only affect it severely but also would damage the fragile mountain ecosystem due to decreasing carrying capacity of the basin/catchments.

CONCLUSION

The isotopic values of source waters and stream water showed marked spatial and temporal variation influenced by altitude and meteorology of the basin. The precipitation, snowmelt, glacier melt and surface water exhibited depleted stable water

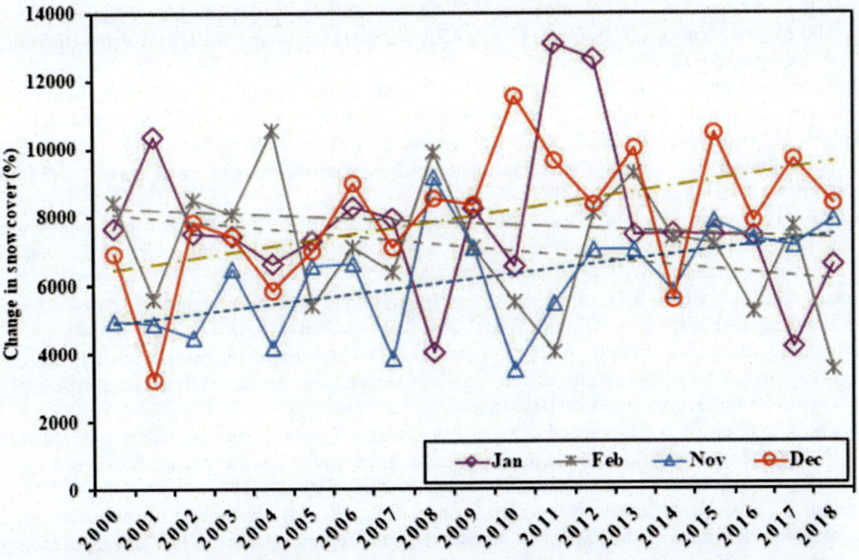

Figure 10. Plot showing the variation in snow cover in upper Jhelum basin (Kashmir Valley).

isotopic values at higher elevation and enriched at lesser elevation. The hydrograph separation mixing models suggests that the snow is the dominant contributor to river Jhelum (29 to 58%), followed by rainfall (18 to 47%), baseflow (13 to 38%). The annual contribution of glaciers to river discharge is very low (1 to 6%). The study suggests that the basin has observed a rise in mean maximum and minimum ambient temperatures with reduction in solid precipitation and also suggests that dissimilarities in amount, timing, precipitation type and differential melting behaviour of winter-accumulated snow might have severe consequences on river hydrology and consequently on the economy of the basin in headwaters and downstream.

References

Bowen, G.J., Kennedy, C.D., Liu, Z. and Stalker, J. (2011). Water balance model for mean annual hydrogen and oxygen isotope distributions in surface waters of the contiguous United States. *Journal of Geophysical Research: Biogeosciences*, 116: G04011.

Clark, I.D. and Fritz, P. (1997). Environmental Isotopes in Hydrogeology. Lewis Publishers, Boca Raton.

Dalai, T.K., Bhattacharya, S.K. and Krishnaswami, S. (2002). Stable isotopes in the source waters of the Yamuna and its tributaries: Seasonal and altitudinal variations and relation to major cations. *Hydrol. Proc.*, 16: 3345-3364.

Epstein, S. and Mayeda, T. (1953). Variation of $\delta^{18}O$ content in waters from natural sources. *Geochem Cosmochim Acta*, 4: 213-224.

Gat, J.R. (1996). Oxygen and hydrogen isotopes in the hydrologic cycle. *Annu. Rev. Earth Planet. Sci.*, 24: 225-262.

Hall, D.K., Riggs, G.A. and Salomonson, V.V. (2016). MODIS/Terra Snow Cover Daily L3 Global 500 m Grid, Version 6. Boulder, CO: NASA National Snow and Ice Data Center Distributed Active Archive Center.

Hall, D.E., Incropera, F.P. and Viskanta, R. (2001). Jet impingement boiling from a circular free-surface jet during quenching: Part 1—single-phase jet. *J. Heat Transfer*, 123(5): 901-910.

Jeelani, G., Feddema, J.J., Veen, C.J. and Stearns, L. (2012). Role of snow and glacier melt in controlling river hydrology in Liddar watershed (western Himalaya) under current and future climate. *Water Resour. Res.*, 48(12).

Jeelani, G., Kumar, U.S. and Kumar, B. (2013). Variation of $\delta^{18}O$ and δD in precipitation and stream waters across the Kashmir Himalaya (India) to distinguish and estimate the seasonal sources of stream flow. *J. Hydrol.*, 481: 157-165.

Jeelani, G., Deshpande, R.D., Shah, R.A. and Hassan, W. (2017a). Influence of southwest monsoons in Kashmir Valley, western Himalayas. *Isot. Environ. Health Stud.*, 53: 400-412.

Jeelani, G., Shah, R.A., Fryar, A.E., Deshpande, R.D., Mukherjee, A. and Jerome, P. (2017b). Hydrological processes in glacierized high altitude basins of western Himalaya. *Hydrogeol. J.* https://doi.org/10.1007/s10040-017-1666-1.

Jeelani, G., Shah, R.A., Deshpande, R.D., Fryar, A.E., Perrin, J. and Mukherjee, A. (2017c). Distinguishing and estimating recharge to karst springs in snow and glacier dominated mountainous basins of the western Himalaya India. *Hydrol.*, 550: 239-252.

Jeelani, G., Shah, R.A., Jacob, N. and Deshpande, R.D. (2017d). Estimation of snow and glacier melt contribution to Liddar stream in a mountainous catchment, western Himalaya: An isotopic approach. *J. Isot. Environ. Health Stud.*, 53(1): 18-35.

Jeelani, G., Deshpande, R.D., Galkowski, M. and Rozanski, K. (2018). Isotopic composition of daily precipitation along the southern foothills of the Himalayas: Impact of marine and continental sources of atmospheric moisture. *Atmos. Chem. Phys.*, 18 (12): 8789–8805.

Jeelani, G., Lone, S.A., Nisa, A.U., Mukherjee, A. and Deshpande, R.D. (2020). Sources and processes of groundwater arsenic mobilization in upper Jhelum basin, Western Himalayas. *Journal of Hydrology*, 591: 125292.

Lambs, L., Balakrishna, K., Brunet, F. and Probst, J.L. (2005). Oxygen and hydrogen isotopic composition of major Indian rivers: A first global assessment. *Hydrological Processes*, 19: 3345-3355.

Lone, S.A., Lone, A.A. and Jeelani, G. (2016). Characterization of groundwater potential of Sindh Watershed Western Himalayas. *Journal of Research & Development*, 16(2016): ISSN 0972-5407.

Lone, S.A., Jeelani, G. and Deshpande, R.D. (2017). Evaluating the sensitivity of glacier to climate by using stable water isotopes and remote sensing. *Environ. Earth Sci.*, 76: 598. https://doi.org/10.1007/s12665-017-6937-6.

Lone, S.A., Jeelani, G., Deshpande, R.D. and Mukherjee, A. (2019). Stable isotope ($\delta^{18}O$ and δD) dynamics of precipitation in a high altitude Himalayan cold desert and its surroundings in Indus river basin, Ladakh. *Atmospheric Research*, 221: 46-57.

Lone, S.A., Jeelani, G., Mukherjee, A. and Coomar, P. (2020). Geogenic groundwater arsenic in high altitude bedrock aquifers of upper Indus river basin (UIRB), Ladakh. *Applied Geochemistry*, 113: 104497.

Pu, Z., Xu, L. and Salomonson, V.V. (2007). MODIS/Terra observed seasonal variations of snow cover over the Tibetan Plateau. *Geophysical Research Letters*, 34(6).

Riggs, G.A., Hall, D.K. and Román, M.O. (2015). MODIS snow products collection 6 user guide. National Snow and Ice Data Center: Boulder, CO, USA, p. 66.

Singh, A.T., Rahaman, W., Sharma, P., Laluraj, C.M., Patel, L.K., Pratap, B., Gaddam, V.K. and Thamban, M. (2019). Moisture Sources for Precipitation and Hydrograph Components of the Sutri Dhaka Glacier Basin, Western Himalayas. *Water*, 11(11): 22-42.

Subramanya, K. (2013). Engineering Hydrology, 4th edn. Tata McGraw-Hill Education.

Chapter 9
Geogenic and Anthropogenic Impacts on the Water Quality of Cauvery River

R. RamyaPriya and L. Elango

INTRODUCTION

Human civilisation has evolved generally along the banks of rivers in most of the regions around the world. Over the years, the population living along the rives have grown rapidly and several such regions have been highly urbanised. Further, several industries requiring a large amount of water are also located along the rivers. These activities along the rivers have led to the deterioration of their water quality. The groundwater in the nearby regions has also been severely affected due to the poor quality of river water. The consumption of poor-quality water has affected more than one-third of the world's population. Water quality issues have been regarded as the major environmental issue globally and are caused due to variety of natural and human influences. Natural factors include geological, topographical, meteorological, hydrological and biological characteristics of the drainage basin (Bartram and Ballance, 1996). The effect of natural factors in influencing the water quality largely depends on the seasonal variation of runoff volumes, climatic conditions and water levels.

Though rivers are not much affected by geogenic sources, the groundwater is prone to severe contamination by geogenic sources. The human activities impact significant effects on water quality and these effects are much pronounced due to the changes in hydrological conditions of the rivers, such as the construction of dams, changes in riverine and wetlands ecosystem and flow diversion. The discharge of domestic, industrial, urban waste waters into the rivers along with the runoff from agricultural land into the drainage basin are important human influences that pollutes the river basin. In addition to this, climate change also contributes indirectly to water pollution in both rivers and groundwater such as sea-level rise in coastal environments inducing seawater intrusion in coastal aquifers and changes in flow patterns in

R. RamyaPriya · L. Elango (✉)
Department of Geology, Anna University, Chennai, India

A. Mukherjee (ed.), *Riverine Systems*, https://doi.org/10.1007/978-3-030-87067-6_9

rivers. To understand the nature of contamination of the freshwater system, it is essential to study the sources and types of pollutants and the factors controlling the contamination in rivers. Regional basin studies are thus necessary for a better elucidation and this chapter describes the factors, sources, causes of contamination in rivers and its surrounding environments. A study carried out on these aspects on Cauvery river flowing in southern peninsular India is also presented.

FACTORS CONTROLLING CONTAMINATION OF RIVER WATER

Contamination of the river water is controlled by several factors such as meteorological, geographical, geological, hydrological, landuse/land cover conditions of the basin etc. The major environmental factors are discussed below.

Meteorological and Geographical Factors

The geographical and meteorological factors of the basin area of rivers dominate the flow in the river which in turn influences the quality of water. In temperate zones, the rainfall received is usually stored in the form of ice or snow during winter and the water infiltrates into the subsurface through recharge which then flows into rivers during summer. Tropical rivers are defined by distinct annual flow cycles with contrasting dry and wet seasons. The tropical rivers are mostly dry during summer or dry season and receive abundant rainfall in the wet season (Bartram and Ballance, 1996). The seasonal effects due to changes in wet and dry cycle are more pronounced in tropical rivers. Time period of the dry season also affects the nature of tropical river. In such arid regions, evaporation also plays a major role in controlling the riverine chemistry. Hence, this indicates that the local conditions and position of rivers within the catchment determines the flow in the river, which is a representation of a mixture of hydrological regimes.

Hydrological Factors

To interpret the water quality of rivers, it is essential to understand the discharge regime of the river. Discharge of a river reflects the environment of its catchments. The physical, chemical and biological characteristics of rivers are chiefly affected by flow variation in rivers (Zeiringer et al., 2018). Changes in discharge also influences the habitat changes in the river, since thriving of aquatic species in rivers depends on the seasonal changes in river flows (Poff et al., 1997). The residence time of the water bodies is an important concept in water contaminant studies since it is related with the time taken for recovery of an aquatic body from contamination. Rivers have a short residence time, due to rapid flow which aids in faster recovery of the aquatic system. The unidirectional flow in the river with good lateral and vertical mixing attributes to the river's self-cleansing mechanism due to rapid dispersion and transport. However, deep lakes and aquifers have long residence times that results in the slow recovery of the system from contamination, which generally ranges from years to decades.

Lithological Factors

The influence of lithology in determining the quality of river water is related to the mineralogy of the rocks, weathering of minerals and flow of river present in the basin (Haidvogl, 2018). Soil and rock leaching are also responsible for the solubilization of some elements such as cations, aluminium and iron. The erosion process in the river is responsible for sediment transported to the sea. The process of erosion varies depending on various conditions such as amount and pattern of rainfall, changes in river flow, slope, lithology, vegetation, soil type and temperature. The rates of erosion are extremely variable, where highest erosion rates occur usually in mountain streams. This is due to the fact that human intervention in such pristine locations has resulted in immense damage to vegetation, thus speeding up the erosion process. The erosion also highly depends on the lithology of basin where the susceptibility of rocks to weathering is lower for hard rocks such as granite and higher for sedimentary rocks like limestone. In smaller watersheds, the local geological conditions highly influence the variations in concentration of ions in dissolved and particulate form.

Landuse/Land Cover

The spatial variation in water quality has been highly influenced by changes in landuse/land cover pattern where the urban and agricultural areas are hotspots of anthropogenic contamination. Although the landuse change was indirect, it is one of the most severe human impact on the rivers. The large-scale shifts of land use from forests to cultivable land or urbanised land have triggered rapid surface runoff resulting in erosion, reduced the evapotranspiration, and have also increased the flow in rivers (Haidvogl, 2018). Terrestrial vegetation has also contributed for pollutant load where the production of terrestrial plants and its decomposition in the soil increases the organic carbon and nitrogen in water. In addition, growth and decomposition of aquatic vegetation and planktons will also impact the pH, dissolved oxygen, nutrients and other chemicals in water that are sensitive to oxidation/reduction conditions.

The landuse changes occurring during the urbanisation, industrialisation and agricultural activities have changed the surface and subsurface characteristics of watersheds which thereby affect the quality and quantity of water. In general, areas subjected to high land use changes that are associated with human are often found to have high concentration of water pollutants, whereas the pristine areas such as forest have good water quality (Camara et al., 2019). However, understanding of such relationships on a catchment scale during various seasons is not yet carried out due to difficulties in monitoring over large areal extent (Rodrigues et al., 2018).

TEMPORAL VARIATION IN WATER QUALITY

The variability in flow cycles is a fundamental characteristic of rivers and is essential for proper functioning of their ecosystems (Poff et al., 1997). Flow variation of the river depends on time scales, from hours to days, seasons, years and for longer time scales. The water cycle and climatic conditions of each catchment acts as a boundary condition for the hydrological regimes. These in turn would define the distinct seasonal and daily flow patterns in rivers. Water quality differences in the rivers results due to internal or external processes in riverine system. The internal processes are cyclic and for a shorter duration, which occurs either daily or seasonal whereas the external processes, for example, contamination of the riverine system, depends on the flow regimes and longer water residence times.

The quality of water in a system differs spatially as well as temporally. The influence of time period in water quality is categorised into five types (Bartram and Ballance, 1996):

- Variations in water quality during minute to minute time period occurs due to fluctuations in pollutant inputs which would result in irregular water mixing. These type of water quality variations are much more pronounced in smaller water bodies.
- Diurnal variations in water quality results due to biological cycles and day/night cycles of the water system. These diurnal patterns also arise due to the cyclic nature of discharge of sewage and effluents from domestic and industrial sources.
- Irregular variations in quality of water arises from the discharge of agricultural runoff which consists of fertilisers, pesticides and herbicides and food processing wastes. These types of variations in water quality are evident for a few days to months.
- Seasonal biological and hydrological cycles changes depend on the flow regime of the river and also on wet and dry conditions of water body. This effect is much significant in riverine systems.
- Year-to-year trends in water quality occur due to the increased anthropogenic activities in the watershed. Similarly, the differences in water quality at different times in a year would be much pronounced for a stream than for a large river.

The sampling frequency is much essential to interpret the water quality of the water. Sampling in adherence to sampling protocols would minimize the error and would produce accurate results. Hence, the sampling frequency follows the order: rivers > streams > lakes.

SOURCES AND TYPES OF CONTAMINATION

Natural Source

Geogenic contamination is more pronounced in groundwater rather than river water due to the increased period of rock water interaction. The river water carries the eroded weathered particles, which in turn also increases the dissolved concentration of river water. The increase of dissolved constituents depends on the type of weathered rocks. In groundwater, increased salinity could be due to leaching of ions from the host rock. Groundwater contamination due to fluoride, arsenic and uranium is well-known example of geogenic contamination. Other than these, the presence of microorganisms also affects the river water quality to a greater extent. Though the presence of pathological components such as cyanobacteria and microcystins are common in rivers, increase of their counts degrades the quality of water causing algal blooms.

Anthropogenic Sources

Anthropogenic sources include agricultural activities, urbanisation, industrialisation and mining. Agricultural activities are the most common anthropogenic activity in the river basins especially for developing countries like India that is heavily dominant on agricultural practices for livelihood and, hence, the number of crop lands is high. Pesticides and insecticides are common pollutants originating from agricultural activities. The run-off water from such croplands carries a high amount of salt which also increases the dissolved ionic concentration in surface water. Disposal of untreated or partially treated sewage effluents along with dumping of solid wastes in the rivers is the most common after-effect of urbanisation. Untreated disposal of industrial effluents to the river would increase the dissolved load of river thereby deteriorating the river water quality. Mining activities also enhance the weathering of the rocks and the erosion of soil increases the dissolved concentration in rivers.

SOURCES OF ANTHROPOGENIC CONTAMINATION

The anthropogenic contamination is further divided as point source and non-point source which is classified based on the entry of contaminants into the river.

Point Source

Point source water pollution refers to the contaminants that enters the water bodies from a single distinguishable source. Sources could be specific locations of anthropogenic activities, which would result in discharge of sewage and industrial effluents, dumping of solid wastes, animal feedlots and quarries. The effect of point source pollution on the receiving water body mostly depends on the population of the area or size and nature of activity in the area, quantum of waste discharged and also the self-cleansing capacity of water body.

Non-point Source

Non-point source pollution is the contamination which does not have single discrete source as an origin. It is mostly a cumulative effect, which usually consists of smaller amounts of various contaminants that are gathered from a large area. The non-point source arises from diffuse sources such as run off carrying pesticides and fertilisers from agricultural fields which thereby contaminate the rivers and groundwater. Nutrient run off in stormwater in the form of "sheet flow" over an agricultural field or forest is the best example of non-point source. The sewage from latrines and septic tanks in rural, semi urban and urban areas are considered as diffuse sources due to monitoring difficulties.

CLIMATE CHANGE

Climate change impacts the hydrological nature of rivers, which limits the carrying capacity of chemical load into the oceans. The decreasing trends of rainfall have decreased runoff in the rivers, which subsides the dilution effect in rivers increasing the major ion concentrations in the rivers (Khan et al., 2017). Increasing temperature caused by climate change also enhances the rate of rock weathering which leads to release of large amount of major ion concentrations in river water (Miriyala et al., 2017). Also, rise in sea levels has impacted the groundwater near the coastal regions causing seawater intrusion in the aquifers.

TYPES OF POLLUTANTS

Chemical pollutants in water are grouped as inorganic and organic pollutants. These pollutants are further divided upon their concentration of pollutants. A proper assessment of pollution in natural water depends on the following conditions:

(a) type and origin of pollutants, (b) availability of analytical methods to identify the chemical concentrations, (c) understanding the processes that determines the transport and fate of chemicals, (d) transport and fate mathematical models which would predict the future contamination scenario and (e) quantification on the adverse effects of chemicals (Schwarzenbach et al., 2010).

Inorganic Pollutants

Inorganic pollutants are classified as macro and micropollutants. Macropollutants are in which the concentration is at a higher range of milligrams per litre whereas micropollutants are in micrograms per litre. The most common macropollutants in river are the major ions such as calcium, magnesium, sodium, potassium, bicarbonate, sulphate, chloride, nitrate and phosphate. Other than nitrate, the major ions could have been derived from weathering of rocks present in the basin as well as anthropogenic activities whereas nitrate originated mainly from anthropogenic activities. High nitrogen and phosphate nutrient loads would increase production of biomass causing algal blooms and depletion of oxygen. Increase in salinity of water could be due to agricultural runoff and leaching of road salts, domestic waste sludge etc.

Inorganic micropollutants such as heavy metal and metalloids e.g., Cr, Ni, Cu, Zn, Cd, Pb, Hg, U, Se, As) could have originated from the geogenic and anthropogenic sources such as release of untreated industrial effluents from dyeing and tanning industries. These could also have been originated from the solid hazardous wastes dumped in the riverbanks and from municipal wastewater. The main challenge of environmental risks regarded to these pollutants is that they depend on the redox conditions. Most of the elements do not undergo biodegradation and their fate and transport depends on chemical reactions such as oxidation, reduction, complexation, absorption, adsorption, precipitation and dissolution (Schwarzenbach et al., 2010).

Organic Pollutants

Organic pollutants are diverse in nature and are originated from various sources. The common organic pollutants are persistent organic pollutants originating from multiple sources like dumping of waste sites, spills, agriculture and urban wastewater. Some of the common examples of persistent organic pollutants are polychlorinated biphenyls, DDT, polyaromatic hydrocarbon, triazines. Pharmaceutical wastes, hormones such as dicyclophene, ethinylestradiol, personal hygiene products (PPCPs) like cosmetics, flame retardants, perfumes, waterproofing agents, plasticisers and endocrine-disrupting chemicals (EDCs) are other types of organic contaminants originating from urban wastewater and industrialised effluents (Schwarzenbach et al., 2010). The chemical structure, especially the reactivity of functional group or substituent regulates the transformation of the organic micropollutants (Schwarzenbach et al., 2010). Presence of such pollutants in the environment leads to adverse toxic effects such as mutagenicity, estrogenicity, and genotoxicity. Prevalence of antibiotic-resistant organisms in water is also one of the major risks to microbial ecosystems.

Microbial Pollutants

Bacteria are significant indicators of microbial pollution in river water since they actively respond to environmental changes. Salmonella species, *E. coli* and Vibrio species are important bacteria species found in water. Faecal coliforms and intestinal enterococci are excellent indicators of faecal pollution. These microbial pollutants are found in the aquatic systems due to the discharge of untreated/partially sewage originating from domestic households, dumping of solid wastes and biomedical and also run off from agricultural land. Microbial contamination of rivers by faecal pollutant poses serious risks to human and animal health if consumed as they carry numerous pathogens (Reischer et al., 2008). Knowledge about the hydrologic nature and the sources of pollution in the catchment is more essential to identify microbial pollution. Long-term monitoring of the sources of microbial pollution and the quality of water bodies is essential to identify the extent of microbial pollution (Reischer et al., 2008).

WATER QUALITY PARAMETERS

The complete assessment of the water quality involves the evaluation and assessment of chemical and physical characteristics of the water (Gorde and Jadhav, 2013).

- Chemical analyses of water involve the determination of major/minor ions, trace elements, nutrients, inorganic and organic pollutants, biochemical oxygen demand (BOD), chemical oxygen demand (COD)
- Physical measurements of water include measurement of temperature, pH, conductivity, turbidity, ORP, dissolved oxygen
- Microbiological analysis – estimation of microbial species

Physical Parameters

pH

The pH of water is one of the major limiting factors that governs solubility and bioavailability of the chemical constituents. pH is defined as the logarithmic scale of acidity of a solution or water. The pH scale normally ranges from 0 to 14. The liquids with pH value of 7 such as pure water said to be neutral and where the pH is below 7.0 are considered acidic. Liquids with pH greater than 7.0 are considered basic or alkaline.

Electrical conductivity

Electrical conductivity (EC) is a numerical expression on the ability of an aqueous solution to transmit electric current and is generally used to estimate the total amount of dissolved ions present in water. The dissolved ions in water are composed of inorganic elements and some organic matter and hence EC acts as a good indicator of water quality. The EC of water is influenced by factors such as total concentration of ions, and their valence states, mobility, and temperature of water.

Turbidity

Turbidity is a measure of the degree of water at which it loses its transparency. Turbidity of water is because of the presence of suspended particles in water. Higher the amount of total suspended solids in water, higher would be the turbidity of water. Turbidity is also used as an indicator for light penetration in aquatic systems.

Oxidation-reduction potential

Oxidation-reduction potential (ORP) measures the ability of a river to self-cleanse. In other words, it measures the oxidising nature of contaminants and biological matter. High ORP value indicates higher oxygen presence which would enable rapid oxidation process.

Temperature

Temperature of the water is also an important parameter in controlling the water chemistry. Higher temperature generally fastens the rate of chemical reactions in water. Increase of water temperatures rapidly dissolves more minerals from the host rocks leading to increase in electrical conductivity of water.

Dissolved oxygen

Dissolved oxygen (DO) represents the amount of oxygen dissolved in water. It is an important representation of the quality of water since high DO in water causes degradation of contaminants and facilitates better aquatic health of the riverine system. The dissolved oxygen levels vary at diurnal intervals and also during seasonal time periods. Water temperature and altitude are the major factors which controls dissolved oxygen concentration in water. Cold water holds more oxygen than warm and less oxygen is observed at higher altitudes.

Chemical Parameters

Major ions, minor ions and trace elements

The river water is known to have dissolved and particulate inorganic matter which is derived from precipitation, mineral weathering and groundwater. Anthropogenic activities also contribute some amount of the dissolved and particulate ions rather than natural sources. Some of the major ions are calcium, magnesium, sodium, potassium, chloride, sulphates, bicarbonate and carbonate and nutrients like nitrate and phosphate. Fluoride and arsenic are the most common minor ions, where their concentrations are high in groundwater of specific regions. Their concentrations are less in river in comparison with the groundwater due to lesser rock water interactions. Heavy metals such as Ni, Cu, Pb, U are among the common trace elements which occurs due to geogenic and anthropogenic activities.

Alkalinity

It is a measure of the buffering capacity of water which neutralises acid and bases, thus maintaining the pH of water. Higher alkalinity levels in surface water will buffer the acidity in surface water thus maintaining pH which would avoid the harmful effects to aquatic ecosystem. The common chemicals present in water that increases alkalinity are carbonates, bicarbonates, phosphates and hydroxides.

Nutrients

Nitrogen and phosphate are important nutrients present in fresh water, especially in rivers. Though nitrogen and phosphorous are present in rivers through natural sources such as decomposition of organic matter, rainfall and soil, the anthropogenic contribution is very high. Phosphorus and nitrogen are largely derived from runoff from agricultural land due to high usage of fertilizers, sewage, yard waste, etc. Both nitrogen and phosphorus are essential elements for both terrestrial and aquatic plant growth but higher concentrations of phosphorous causes algal blooms and higher concentration of nitrogen are converted to ammonia and has degradation effects on aquatic life.

Biochemical oxygen demand

Biochemical oxygen demand (BOD) is the quantification of amount of dissolved oxygen required by aerobic biological organisms to metabolize the organic material in water at a certain temperature over a specific time period. BOD is indicator of organic pollution in streams and is usually considered for 5 days. Greater the BOD values, the depletion of oxygen in river water is rapid.

Chemical oxygen demand

Chemical oxygen demand (COD) is defined as the amount of oxygen consumed for the decomposition of organic matter and also for the oxidation of inorganic chemicals such as ammonia and nitrite. The quantification of oxidizable pollutants in water is understood from COD.

METHODS TO IDENTIFY ANTHROPOGENIC AND GEOGENIC CONTAMINATION IN RIVERS

Microbial Identification

Bacteria such as total coliforms, faecal coliforms, *Escherichia coli* and intestinal enterococci found in excreta of humans and warm-blooded animals are considered as the indicators of faecal pollution. When partially treated sewage enters into the rivers, these bacteria survive even in those conditions preserving their pathogenicity. *E. coli* is one of the significant indicators of faecal contamination in water, especially in tropical and temperate regions and hence thorough assessment of the bacterial density of water is essential (Páll et al., 2013). These microbial pathogens are the most suitable indicators of anthropogenic contamination especially faecal contamination (Páll et al., 2013).

Isotopic Identification

Tracers have been used for various purposes in riverine environments, viz. (1) identification of contaminant origin (2) analysis of the fate and transport of contaminant and (3) evaluation of biogeochemical cycling of an element in the aquatic systems. Isotopes are the excellent tracers used in different aspects of environmental chemistry. Isotopes such as lead, copper, zinc, bromide, boron and several heavy metals act as excellent proxies for differentiating the anthropogenic and natural sources in the environment (Chetelat and Gaillardet, 2005; Chen, Gaillardet and Louvat, 2008;

Sivry et al., 2008). Isotopic fractionations occur during biological process, adsorption of metal, chemical diffusion, and column elution.

IDENTIFICATION USING MAJOR IONS

Mass balance approach can be used to differentiate anthropogenic sources and natural sources. Chloride is a conservative ion and could be used as tracer. The residual chloride can be calculated by the method proposed by Meybeck (Meybeck, 1983). Residual chloride is the remaining chloride obtained by subtracting the rainwater's chloride concentrations from chloride concentrations of the river. These could be used to estimate the atmospheric corrected concentrations of other major ions which would help in tracing the source of ions. Nitrate ions are also used for such anthropogenic identification, where a high amount of nitrate in aquatic system could be originated from agricultural water and urban waste water (Huang et al., 2018).

STATISTICAL ANALYSIS

Correlation analysis, analysis of variance, principal component analysis, regression analysis, cluster analysis, discriminant analysis, redundancy analysis are the commonly used statistical techniques for distinguishing the anthropogenic contamination from geogenic contamination (Wang et al., 2007; Ouyang et al., 2006). The clusters and different components by statistical methods are derived using chemical composition, which depicts differences in chemical indices differentiating anthropogenic and geogenic contamination.

Case study 1: Assessment of groundwater and river water quality along the Cauvery river

Description

The Cauvery river is one of the most important river of southern India, which originates in the Western Ghats and flows along the length of about 800 km. The river forms a delta before draining into the Bay of Bengal (Fig. 1). The western part of the river part mainly receives rainfall during south-west monsoon (July-September) and the eastern regions receive rainfall during north-east monsoon (October-December). The river flows entire course mostly during October to January and hence is non-perennial. Harangi, Hemavati, Shimsha, Akkravati, Kabini, Suvarnavathi are the tributaries which join Cauvery in upper region, while Sarabangaha, Bhavani, Noyyal and Amravathi join at the middle region, where the lower region has 7 principal distributaries (India WRIS, 2015).

Though the water distribution has been a problem for decades, water quality of the river is of greater concern due to pollution load carried by the river. Cauvery and

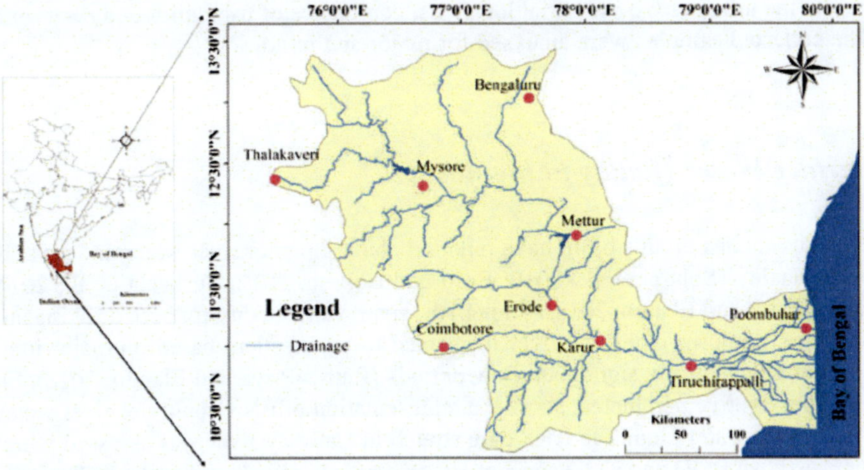

Figure 1. Cauvery river basin.

its tributaries, Noyyal and Sarabangaha have been regarded as one of the most pollutant rivers in India (CPCB, 2014). Several water quality assessment studies in surface water and groundwater in the basin (Susheela et al., 2014; Basu and Lokesh, 2012; Solaraj et al., 2010; Brindha and Kavitha, 2015; Suresh et al., 2010; Vetrimurugan et al., 2017a, but these are carried out only in certain smaller sections of the Cauvery river. Hence, it is essential to evaluate the water quality for the entire stretch of the river for a better understanding of the factors influencing the water quality and identification of locations of anthropogenic contamination, on a larger scale, but only few studies (Pattanaik et al., 2013; RamyaPriya and Elango, 2018) have carried out such exercise in the river.

Sampling Methodology

Sampling was carried out from 2013 to 2019, approximately once in four months. The exact period of sampling was decided based on the monsoonal conditions of the basin. As this basin receives rainfall during two different monsoons (south-west monsoon) and (north-east monsoon), each of about three months duration. It was decided to carry out sampling during these monsoons as well as during the non-monsoon period in a year. Hence, sampling was carried out three times every year considering the changes in the monsoon pattern. The water samples were collected from 28 locations along the river where the distance between each sampling location was fixed at an interval of 18 to 25 km. Care was taken to select the sampling locations such that it represents the natural hydrogeochemistry of the river. Sampling locations were determined based on the landuse of the region such that it covers the urban areas, industrial locations, confluence of tributaries and reservoirs. The collected samples were analysed for major and minor ions.

Surface Water Quality in the Basin

Previous studies on this basin have reported about the poor river water and groundwater quality (Solaraj et al., 2010; Basu and Lokesh, 2012; Susheela et al., 2014; RamyaPriya and Elango, 2018). Though the river water quality for both drinking and irrigation purposes along the river is deemed to be excellent based on major ionic concentrations during significant flow periods (RamyaPriya and Elango, 2018), the non-conservative parameters show the representation of poor quality of river water. The surface water along the KRS dam stretch in Cauvery river was assessed which indicated that BOD and COD concentrations are high, clearly indicating pollution in Cauvery river. High values of Cl is also observed around the KRS dam stretch along with high levels of total and faecal coliform indicating anthropogenic pollution. The pollution in river Cauvery at the upper stretches, especially at KR Nagara (location 23, and location 24, Fig. 2) was also escalating over the years due to indiscriminate human activities and agricultural uses along the river (Basu and Lokesh, 2012). The river water quality at the deltaic regions was also evaluated which indicates that during monsoonal periods where the phosphate concentration is greater. The high concentration of phosphate in water samples indicates that pollution have been

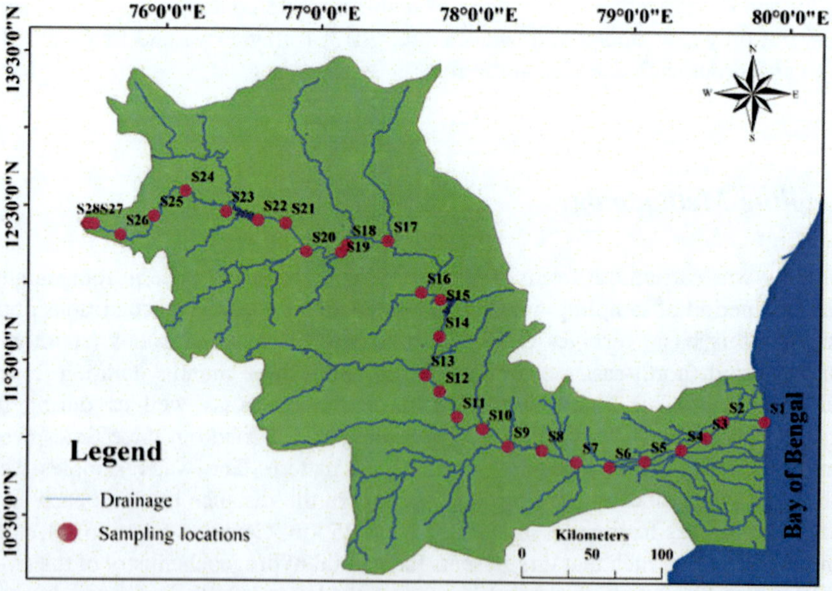

Figure 2. Surface water and groundwater sampling locations.

originated from runoff from agricultural fields, sewage and other non-point sources. Surface water in Cauvery river deltaic was also highly contaminated due to high TDS during non-monsoon seasons (Solaraj et al., 2010; Brindha and Kavitha, 2015).

Groundwater Quality in the Basin

Groundwater in locations near the urbanised, agricultural land and near the regions surrounding the principal tributaries Noyyal and Amaravathi have been severely contaminated due to the release of high sewage and effluent pollutant load to the river continuously (RamyaPriya and Elango, 2018). Due to low flow in rivers, the groundwater in deltaic part has also been severely affected. Along with high major ionic concentration, high levels of heavy metals like silver, lead, nickel, aluminium, boron, cadmium, copper, iron and manganese were also determined in the groundwater of delta regions of the basin. Poor-to-unsuitable water quality was found in most of the regions. The human exposure risk due to groundwater consumption was high due to presence of chemicals such as silver, lead, manganese, cadmium and lithium (Vetrimurugan et al. 2017a, b). The concurrence of all these elements in groundwater indicated that both geogenic processes and anthropogenic activities such as agricultural runoff, industrial effluent discharge and sea water intrusion have caused contamination.

Impact of river water recharge on groundwater quality

Since the quality of groundwater in nearby wells to the river was deteriorated (RamyaPriya and Elango, 2018), the study was carried out in the Cauvery river to identify the impact of recharge from river water to the adjacent groundwater quality. Major ion concentration in river water was found to vary spatially. The concentration of ions, in general, was found to increase along the flow direction of the river except in certain regions of anthropogenic activities. This increase in ions towards the flow direction could be due to the changes in precipitation chemistry, ocean spray effects along with weathering and anthropogenic disturbances (Fig. 3). The concentration of river water during full flow conditions of the river remains fairly constant and low due to dilution effect. Groundwater shows an irregular pattern, and the major ion concentration in groundwater were higher than surface water in all locations due to rock water interaction. But, temporally, similar variation in the ionic concentration is observed between the surface water and groundwater indicating the influence of river recharge on groundwater. For example, Fig. 4 explains the surface water-groundwater interactions at location 14 during various time periods. The recharge of groundwater from river water improves the groundwater storage and also the quality since the concentration of groundwater is low during months of significant river flow. This could be attributed due to the surface water-aquifer

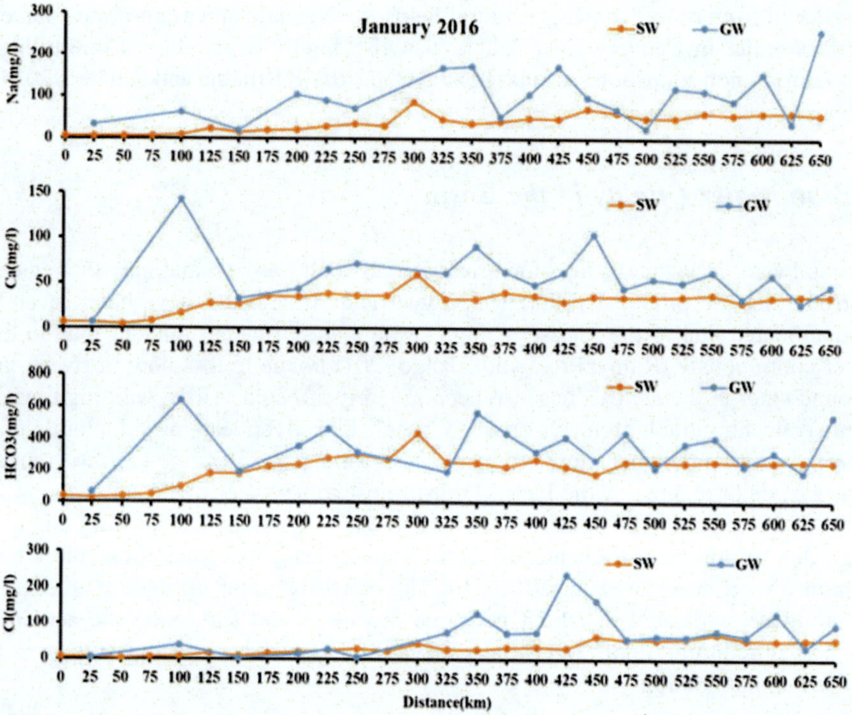

Figure 3. Spatial variation of major ions.

exchange which depends on topography, elevation, lithology and flow characteristics of the aquifer and river.

FACTORS INFLUENCING RIVER WATER QUALITY IN CAUVERY BASIN

Meteorological Conditions

The basin lies in the tropical zone and experiences the contrasting monsoonal changes. The western part of the basin receives rainfall during south-west monsoon and the eastern part of the basin receives rainfall during north-east monsoon and the middle region receives rainfall during both the seasons. The highest rainfall in the basin is usually received during the months of June-September due to prevalent south-west monsoon wind (India Wris, 2015). Hence, the river is dry during summer and flows completely during August-January. Though the quantum of rainfall during north-east monsoon is lesser than the south-west monsoon, it is important especially to the lower part of the basin since it is one of the major source of water for the

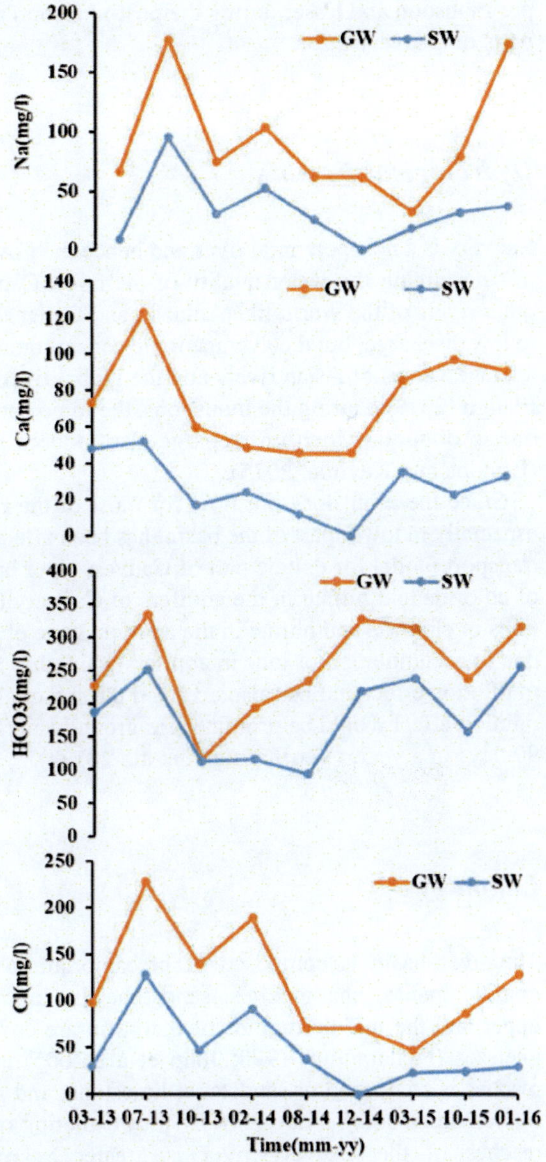

Figure 4. Temporal variation of major ions in surface water and groundwater.

agricultural activities in the deltaic region. Hence, the flow of Cauvery river is intermittent and thus the seasonal climatic wet and dry periods of the basin severely impacted the water quality of Cauvery river. The major ion concentrations of the river water and its adjacent groundwater are less during the river flow, whereas during the dry season their concentration is greater. (RamyaPriya and Elango, 2018). The conservative variables such as Na, K, Ca are found to be greater during pre-monsoon and lesser during the post-monsoon seasons in the river water of upper parts of Cauvery basin.

Hydrological Factors

Cauvery is a non-perennial river and hence the flow characteristics are major factors in determining the water quality of the river. Cauvery river, when compared with other rivers of the world like Amazon and Ganges, carries lesser dissolved load due to low discharge, but the average ionic concentrations of riverine are greater than the average values of Asian rivers and the global rivers as reported by RamyaPriya and Elango (2018). During the monsoons, the ionic concentrations in the river decreases due to dilution, which on recharge also reduces groundwater ionic concentration. (Brindha and Kavitha, 2015).

Since the river does not flow for most of the months, the groundwater quality, especially in lower part of the basin, has been affected to a greater extent. The solute transport model for deltaic part of Cauvery river basin indicated high concentration of chloride and nitrate in the aquifers of Cauvery deltaic region. Higher concentrations of chloride and nitrate in the aquifers were observed in the coastal area. This is due to accumulation of ions in aquifers which has happened because of the absence of dilution effect and has also declined the soil quality. The model predicted that the dilution effect would be prominent in groundwater only if the river flows for at least 90 days in a year. (Vetrimurugan et al., 2017b).

Lithology

The river basin is composed of precambrian rocks which are closepet granite, granitic gneiss, charnockites, sandstone and alluvium (Pichamuthu, 1978). The upper and the middle regions of the basin are covered by peninsular granites and gneisses (Pichamuthu, 1978; John et al., 2005) and the coastal regions at lower reaches is composed of sandstone, limestone, and shale (Sundaram and Rao, 1981; Subramanian and Selvan, 2001). The weathering of the gneissic rocks at the upper reaches of the Cauvery river is greater as reported by several researchers (Vaithiyanathan et al., 1988; Pattanaik et al., 2013). At higher temperatures, the rate of chemical weathering increases which increase the dissolved ion concentration in rivers. Silicate weathering is dominant in the basin and the riverine dissolved and suspended load reflects the dominance of weathering (Vaithiyanathan et al., 1992). The dissolved load carried by Cauvery river to sea was lower, but the dissolved load yield per area is higher than most of the rivers around the world indicating active chemical weathering in the basin (RamyaPriya and Elango, 2018).

LANDUSE/LAND COVER

The basin undergoes rapid urbanisation along with industrialisation and significant conversion of barren lands to agricultural land also occurs in the basin. Agriculture is widely practiced in this river basin. The deltaic regions are the most fertile regions of the basin and are extensively irrigated where major crops being rice and sugarcane. About 96 dams and 16 major and minor irrigation projects are present in the basin (India Wris, 2015). Forest lands also occupy significant portion of land use in the basin. About 36 districts are present in the basin consisting a population of about 3,18,89,280 (Census, 2011). Urbanisation has been rapid in this river basin, where many locations along this river over the past few decades has been transformed from rural areas to semi-urban and urban locations. The major urban cities along the river are Mysore, Erode, Karur, Tiruchirappalli and Kumbakonam. Industrial hubs were also present at many regions around the river, where several textiles, dyeing, cement, sugar, chemical and mineral processing industries are present. In addition to that, many small scale to medium scale industries are present in the basin. The changes in the land use have also made a greater impact on the surface water and groundwater quality of the river basin.

Studies on the river water quality in the river reported that the ionic concentration in river generally increases along the flow direction of the river. But few regions are exempted from such patterns where a high amount of ionic concentration was found in both river water and groundwater. The region of anomalies in water chemistry was found to be locations of industries, urban areas, agricultural lands and the confluence of tributaries. Mass balance studies using chlorine as the tracer were carried out using river water and rainfall chemistry to identify the regions of anthropogenic contamination. It is evident that the river and groundwater water quality in locations near to industries and confluence of the tributaries especially Noyyal and Amravathi and urban locations were severely affected by anthropogenic activities (RamyaPriya and Elango, 2018). It could also be deciphered from Fig. 5 that the locations 12, 6, 11, 9, 10 have been highly contaminated since these locations are urban (location 12 and 6), industrial areas (location 11) and confluence of tributaries (location 9 and 10).

CONCLUSION

Water quality studies for rivers are gaining momentum in recent years due to increasing concern for public and aquatic health. This article describes the sources and types of contamination in rivers, the factors which influence the contamination in rivers, identification of anthropogenic sources in riverine water chemistry, temporal variation of water quality with an emphasis on a major non-perennial tropical river of Southern India, Cauvery. River Cauvery flows across four states on southern India and has been extensively utilised for drinking and irrigation purposes. The

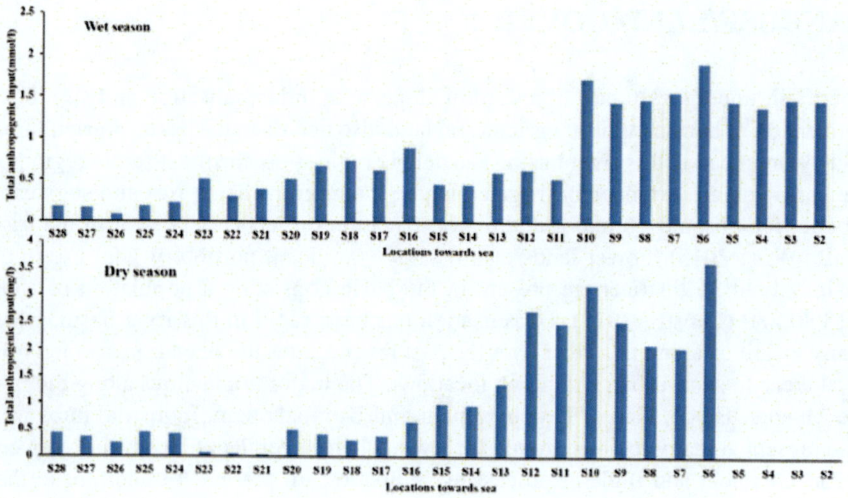

Figure 5. Spatial variation of concentration of ions from anthropogenic origin.

general scenario of the river water and groundwater quality in the basin is also described. Along with this, factors influencing the riverine water quality in the Cauvery basin is also discussed. Thus, this chapter emphasises that studies on assessment of water quality in such riverine environments are essential and continuous periodical long-term monitoring studies are needed to keep in check of the riverine quality.

References

Bartram, J. and Ballance, R. (Eds) (1996). Water Quality Monitoring: A Practical Guide to the Design and Implementation of Freshwater Quality Studies and Monitoring Programmes. CRC Press.

Basu, S. and Lokesh, K.S. (2012). Trend of temporal variation of Cauvery river water quality at KR Nagar in Karnataka. *International Journal of Engineering Science and Technology,* 4(8): 3693-3699.

Brindha, K. and Kavitha, R. (2015). Hydrochemical assessment of surface water and groundwater quality along Uyyakondan channel, south India. *Environmental Earth Sciences*, 73(9): 5383-5393.

Camara, M., Jamil, N.R. and Abdullah, A.F.B. (2019). Impact of land uses on water quality in Malaysia: a review. *Ecological Processes*, 8(1): 10.

Chen, J., Gaillardet, J. and Louvat, P. (2008). Zinc isotopes in the Seine River waters, France: a probe of anthropogenic contamination. *Environmental Science and Technology*, 42(17): 6494-6501.

Chetelat, B. and Gaillardet, J. (2005). Boron isotopes in the Seine River, France: a probe of anthropogenic contamination. *Environmental Science and Technology*, 39(8): 2486-2493.

CPCB (Central Pollution Control Board) (2014). Status of water quality in India—2007. https://data.gov.in/catalog/status-water-quality-india-2012.

Gorde, S.P. and Jadhav, M.V. (2013). Assessment of water quality parameters: a review. *Journal of Engineering Research and Applications*, 3(6): 2029-2035.

Haidvogl, G. (2018). Historic Milestones of Human River Uses and Ecological Impacts in River Hydrology, Flow Alteration, and Environmental Flow. *In:* Riverine Ecosystem Management, Springer.

Huang, H., Liu, M., Wang, J., He, J. and Chen, H. (2018). Sources Identification of Nitrogen Using Major Ions and Isotopic Tracers in Shenyang, China. *Geofluids*.

India Wris (2015). Cauvery. https://indiawris.gov.in/downloads/Cauvery%20Basin.pdf

John, M.M., Balakrishnan, S. and Bhadra, B.K. (2005). Contrasting metamorphism across Cauvery Shear Zone, South India. *Journal of Earth System Science,* 114(2): 1-16.

Khan, M.Y.A., Gani, K.M. and Chakrapani, G.J. (2017). Spatial and temporal variations of physicochemical and heavy metal pollution in Ramganga River—A tributary of River Ganges, India. *Environmental Earth Sciences,* 76(5): 231.

Meybeck, M. (1983). Atmospheric inputs and river transport of dissolved substances. *Dissolved loads of rivers and surface water quantity/quality relationships,* 141: 173-192.

Miriyala, P., Sukumaran, N.P., Nath, B.N., Ramamurty, P.B., Sijinkumar, A.V., Vijayagopal, B., Ramaswamy, V. and Sebastian, T. (2017). Increased chemical weathering during the deglacial to mid-Holocene summer monsoon intensification. *Scientific Reports*, 7: 44310.

Ouyang, Y., Nkedi-Kizza, P., Wu, Q.T., Shinde, D. and Huang, C.H. (2006). Assessment of seasonal variations in surface water quality. *Water Research*, 40(20): 3800-3810.

Páll, E., Niculae, M., Kiss, T., Şandru, C.D. and Spînu, M. (2013). Human impact on the microbiological water quality of the rivers. *Journal of Medical Microbiology*, 62: 1635.

Pattanaik, J.K., Balakrishnan, S., Bhutani, R. and Singh, P. (2013). Estimation of weathering rates and CO_2 drawdown based on solute load: Significance of granulites and gneisses dominated weathering in the Kaveri River basin, Southern India. *Geochimica et al Cosmochimica Acta*, 121: 611-636.

Pichamuthu, C.S. (1978). Archaean Geology Investigations in Southern India. *Geological Society of India*, 19(10): 431-439.

Poff, N.L., Allan, J.D., Bain, M.B., Karr, J.R., Prestegaard, K.L., Richter, B.D., Sparks, R.E. and Stromberg, J.C. (1997). The natural flow regime. *BioScience*, 47(11): 769-784.

RamyaPriya, R. and Elango, L. (2018). Evaluation of geogenic and anthropogenic impacts on spatio-temporal variation in quality of surface water and groundwater along Cauvery River, India. *Environmental Earth Sciences*, 77(1): 2.

Reischer, G.H., Haider, J.M., Sommer, R., Stadler, H., Keiblinger, K.M., Hornek, R., Zerobin, W., Mach, R.L. and Farnleitner, A.H. (2008). Quantitative microbial faecal source tracking with sampling guided by hydrological catchment dynamics. *Environmental Microbiology*, 10(10): 2598-2608.

Rodrigues, V., Estrany, J., Ranzini, M., de Cicco, V., Martín-Benito, J.M., Hedo, J. and Lucas-Borja, M.E. (2018). Effects of land use and seasonality on stream water quality in a small tropical catchment: the headwater of Córrego Água Limpa, São Paulo (Brazil). *Sci Total Environ*, 1553-1561.

Schwarzenbach, R.P., Egli, T., Hofstetter, T.B., Von Gunten, U. and Wehrli, B., 2010. Global water pollution and human health. *Annual Review of Environment and Resources*, 35: 109-136.

Solaraj, G., Dhanakumar, S., Murthy, K.R. and Mohanraj, R. (2010). Water quality in select regions of Cauvery Delta River basin, southern India, with emphasis on monsoonal variation. *Environmental Monitoring and Assessment*, 166(1-4): 435-444.

Subramanian, K.S. and Selvan, T.A. (2001). Geology of Tamil Nadu and Pondicherry, *Geological Society of India*, 1: 192.

Sundaram, R. and Rao, P.S. (1981). Lithostratigraphy of Cretaceous and Palaeocene rocks of Tiruchirapalli District, Tamil Nadu, South India. *GSI Rec*, 115(5): 9-23.

Suresh, M., Gurugnanam, B., Vasudevan, S., Dharanirajan, K. and Raj, N.J. (2010). Drinking and irrigational feasibility of groundwater, GIS spatial mapping in upper Thirumanimuthar sub-basin, Cauvery River, Tamil Nadu. *Journal of the Geological Society of India*, 75(3): 518-526.

Susheela, F.S., Srikantaswamy, F.S., Shiva Kumar, F.D., Gowda, F.A. and Jagadish, F.K. (2014). Study of Cauvery river water pollution and its impact on socio-economic status around KRS Dam, Karnataka, India. *Journal of Earth Sciences and Geotechnical Engineering*, 4(2): 91-109.

Sivry, Y., Riotte, J., Sonke, J.E., Audry, S., Schäfer, J., Viers, J., Blanc, G., Freydier, R. and Dupré, B. (2008). Zn isotopes as tracers of anthropogenic pollution from Zn-ore smelters The Riou Mort–Lot River system. *Chemical Geology*, 255(3-4): 295-304.

Vaithiyanathan, P., Ramanathan, AL. and Subramanian, V. (1988). Erosion, transport and deposition of sediments by the tropical rivers of India. *In: Sediment Budgets. IAHS Publication,* 174.

Vaithiyanathan, P., Ramanathan, AL. and Subramanian, V. (1992). Sediment transport in the Cauvery River basin: Sediment characteristics and controlling factors. *Journal of Hydrology,* 139(1-4): 197-210.

Vetrimurugan, E., Brindha, K., Elango, L. and Ndwandwe, O.M. (2017a). Human exposure risk to heavy metals through groundwater used for drinking in an intensively irrigated river delta. *Applied Water Science*, 7(6): 3267-3280.

Vetrimurugan, E., Senthilkumar, M. and Elango, L. (2017b). Solute transport modelling for assessing the duration of river flow to improve the groundwater quality in an intensively irrigated deltaic region. *International Journal of Environmental Science and Technology*, 14(5): 1055-1070.

Wang, X.L., Lu, Y.L., Han, J.Y., He, G.Z. and Wang, T.Y. (2007). Identification of anthropogenic influences on water quality of rivers in Taihu watershed. *Journal of Environmental Sciences*, 19(4): 475-481.

Zeiringer, B., Seliger, C., Greimel, F. and Schmutz, S. (2018). River Hydrology, Flow Alteration, and Environmental Flow. *In:* Riverine Ecosystem Management, Springer.

Chapter 10
Fluvio-dynamics and Groundwater System in the Narmada River Basin, India

Sudarsan Sahu

INTRODUCTION

Availability of water resources in any region depends on various factors like climate, precipitation, geologic and geomorphic setting. As such the surface and groundwater resource potentials of different river basins also differ. Climatic driven dynamics in the fluvial regimes in a river basin also affect the groundwater system, since both these spheres of water resources remain mutually dependent and do possess a give-and-take relationship. Variabilities in the discharge and sediment load along with changes in the river power modify the centres of erosion and deposition in the floodplain. Other than the autogenic controls of hydrologic origin, various tectonic regimes of allogenic origin also affect the river channel behaviour, its unit power and the morphometry of the river channel and floodplain as well by modifying the slope of land surface. The river channel dynamisms (shifting through migration and avulsion), the channel morphometry and the associated fluvial landforms (channels, point bars, braid bars, levees, floodplain etc.) carve out the geomorphology and the alluvail architecture in a river valley (Cant and Walker, 1978; Wells and Dorr, 1987; Bristow, 1996; Sinha et al., 2005). Spatial variation in lithofacies record in the sedimentary architecture bears clues to the statial and temporal variations in the fluvial processes over time (Bridge and Leeder, 1979; Blum and Tornquist, 2000; Bridge and Karssenberg, 2005). Thus, the availability of granular zones, their spatial extent and geometry in the sedimentary architecture which form aquifer systems of various potentials depends on the spatial variabilities in the fluvial dynamics in the river basin over time.

The fluvial dynamics during the processes of sedimentation and the resulted alluvial geomorphology give insights into the underlying stratigraphy and possible

S. Sahu (✉)
Central Ground Water Board, South Eastern Region, Bhujal Bhawan Khandagiri, Bhubaneswar, Odisha, India

A. Mukherjee (ed.), *Riverine Systems*, https://doi.org/10.1007/978-3-030-87067-6_10

heterogeneities in the sedimentary architecture (Allen, 1978; Straub et al., 2009). These can be interpreted for the aquifer bodies, their geometry and extent. The locations of palaeochannels, palaeochannel belts, palaeo-levees, palaeo-floodplains etc. depict the hydrogeological conditions in the alluvial plain. The palaeochannels width-to-depth ratio bears implications for the underlying aquifer body dimensions.

The Narmada River Basin (NRB) in the central parts of India is one of the water rich river basins in the country. However, very few works are available (Roy, 1971; GSI, 1995; Roy, 2009; Khan and Aziz, 2016) related to aquifers in the alluvial valley of Narmada River Basin. This article tries to gather the available information on the alluvial lithology and their spatial extent in the NRB in Madhya Pradesh (MP), where the major part of the basin is situated. Information related to the alluvial aquifer systems in the basin has been synthesized. In relation to the tectonic perturbations and the associated river channel dynamisms in the basin, the development of the alluvial architecture has been assessed. Available information on palaeochannels in the basin has been compiled and an attempt is made to analyze the groundwater potential of aquifer systems in the NRB.

THE NARMADA RIVER BASIN (NRB)

The Narmada River in central India (Fig. 1a) is the seventh largest river in the country preceded only by the Ganga, Brahmaputra, Godavari, Krishna, Indus and Mahanadi. The NRB that extends between 72°32′E and 81°45′E longitudes and 21°20′N and 23°45′N latitudes, drains an area of 98,796 km^2, out of which major portions of 86.18% and 11.6% fall in the states of Madhya Pradesh (MP) and Gujarat, respectively (Fig. 1b). The river originates at a height of 1056 m above sea level (asl) near Amarkantak in the Maikal mountain ranges of MP. Unlike the other peninsular rivers like the Godavari, Krishna, Cauvery and Mahanadi which flow into the Bay of Bengal, the Narmada is the largest west flowing river that falls into the Gulf of Cambay of Arabian Sea near Barouch of Gujarat. It traverses a distance of 1312 km through the states like MP (1077 km), Maharashtra (74 km) and Gujarat (161 km). Approximately 35% of the NRB is under forest cover, 60% under arable land and 5% are grassland, wasteland, etc (Gupta and Chakrapani, 2005). The approximate basin population is about 16 million; the urban share of the total population is only 20%, which indicates the sustenance of rural population in the basin (CPCB, 1994).

A humid tropical climate prevails in the basin, although extreme heat and cold are felt at certain places. Generally, the temperatures vary from 40°C to 42°C during summer (March-June) to the minimums of 8°C to 13°C during winter (October end to February end). The evaporation rates in the basin vary from 6 to 28 mm in summer (April to June) and 1 to 9 mm in winter (October to March). The average rainfall in the basin stands at 1180 mm, whereas the annual rainfall in the entire basin varies from 800 to 1600 mm (Bhandari et al., 2005). Southwest monsoon brings major portion (85-95%) of the annual rainfall in the basin. The cyclonic storms are often

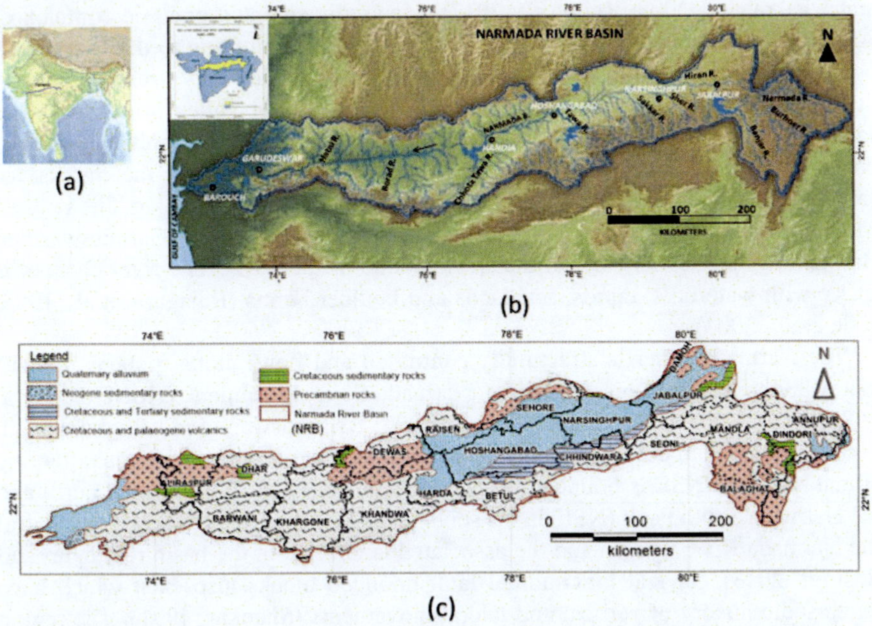

Figure 1. (a) Location of Narmada River in India. (b) Narmada River Basin (NRB) where the master drainage Narmada flows from east to west in an axial position and debouches into the Gulf of Cambay of Arabian sea near Barouch of Gujarat state. The map also depicts its major tributaries that join the river along its left and right banks. The inset shows a major part of the basin fall in the state of Madhya Pradesh. (c) The generalized lithology map of the NRB with the location of districts of Madhya Pradesh falling in the basin (redrawn from Rai et al., 2017).

associated with the southwest monsoon, which bring heavy rains and large floods in the river channels in the basin are a common phenomenon (Kale et al., 2003). During rains, the Narmada River becomes quite turbulent and is often in high spate, sometimes devastating, encroaching far inland on both the banks. The flood line goes up to ~25 m above the bed level of the river. The maximum flood discharge observed stands at around 53749 m^3/s. The average monsoon water discharge in the Narmada River varies from 48% to 78%, whereas in its tributaries, average water discharge in monsoon varies from 87% to 99% (Gupta and Chakrapani, 2005).

The Narmada River in the basin flows often close to the Vindhyans in the north. Most of the tributaries originate from the Satpura hills in south and join Narmada along its left bank (Fig. 1c). The major such tributaries are the Sher, Shakkar, Tawa, Ganjal and Chhoti Tawa (Fig. 1c). The only important tributary on the north (joining Narmada along the right bank) is the Hiran, which emanates from Vindhya hills. Most of these tributaries have short and precipitous course from the hills. The tributaries in south flow north-northwest, while those in the north flow south-southwest to meet the master drainage Narmada.

The Deccan basalts form the major lithology in the NRB (Fig. 1c), followed by sedimentary rocks of Vindhyan and Gondwana Super Groups (Roy, 1971; Gupta and Chakrapani, 2005; Rai et al., 2017). Metasediments and gneissic complex of Archean-Proterozoic age are the minor constituents contributing to the basin geology. Quaternary alluvium is restricted to the Narmada valley only, which have been deposited in narrow discontinuous patches along the river valley from the east of Jabalpur in MP to the west of Bharouch in Gujarat. The most significant alluvial coverage in the basin is observed in the eastern side falling in the districts of Jabalpur, Narsinghpur, Hoshangabad, Harda, Sehore and Raisen in the state of Madhya Pradesh. A second alluvium patch is observed between Garudeswar and Barouch in the Gujarat state in the western side of the NRB. The river channel is rocky with waterfalls, rapids, scablands and boulder berms (Rajaguru et al., 1995; Kale et al., 2003).

The Narmada River is structurally controlled and flows along grabens in a rift valley along the tectonically active Narmada-Sone Lineament (NSL) (Valdiya, 1984, 2010; GSI, 1995; Chamyal et al., 2002; Krishnaswamy and Raghunandan, 2005), which is the second most important tectonic feature of India after the Himalayas. The NSL is bounded by Narmada North Fault (NNF) in the north and the Narmada South Fault (NSF) in the south (Fig. 2). These two faults largely control the Narmada River system and the associated landforms in the basin (Kothyari and Rastogi, 2013). Several longitudinal fault bounded blocks also exist which have an episodic history of vertical and lateral movements (Shankar, 1991). The region has experienced two moderate earthquakes (in 1927 of magnitude 6.4 and during 1997 of magnitude 5.8) due to the movements along the NSL caused by the

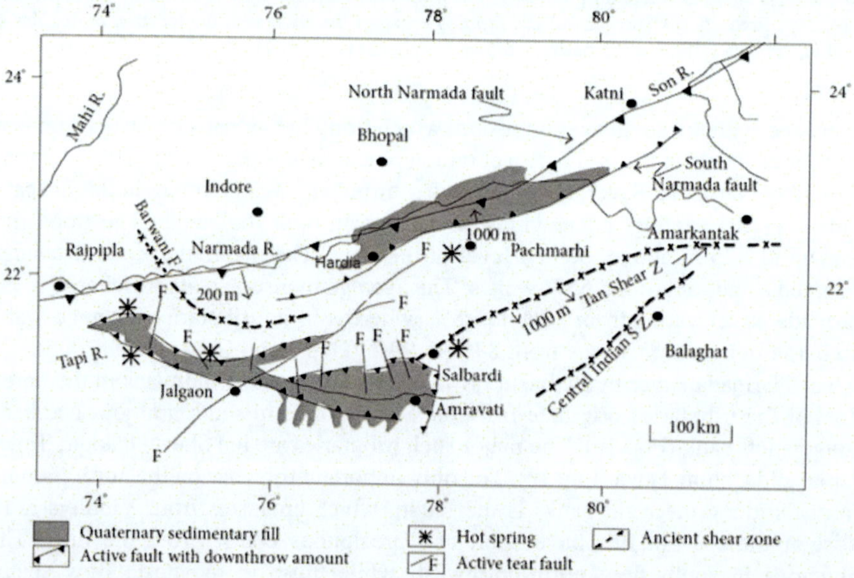

Figure 2. Geological map of NRB (taken from Kothyari and Rastogi, 2013) shows major lineaments in the area (GSI, 1995); active grabens, Narmada graben in the north, and Tapi graben in the south (Krishnaswamy and Raghunandan, 2005; Valdiya, 2010).

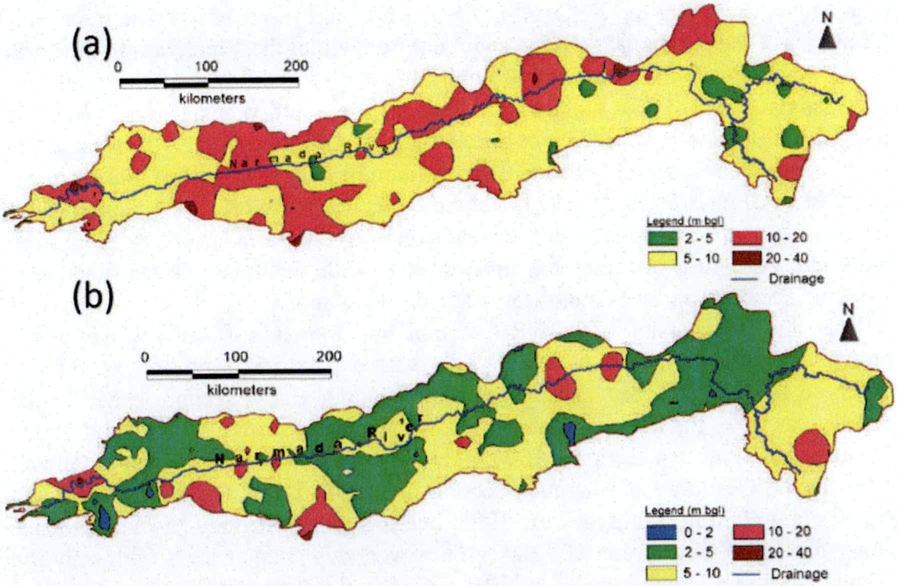

Figure 3. (a) Pre- (May 2016) and (b) Post-monsoon (Nov. 2016) depth to water level contours in the NRB (Govt. of India, 2019). The former map (a) shows deeper water levels aligned along the Narmada River, indicating the effluent character of the river (for details refer text).

prevailing N-S compression owing to the northward motion of the Indian plate (Rajendran and Rajendran, 1998).

Depths to water levels in the NRB during the pre-monsoon period (May) of 2016 varied largely in the range of 2-20 m below ground level (bgl) (Fig. 3a), where a major part of the basin remained in the range of 5-10 m bgl (Govt. of India, 2019). Whereas, during the post-monsoon period (Nov 2016), a significant part was covered by water levels in the range of 2-5 m bgl (Fig. 3b), indicating a good deal of recharge from the rainfall. A perusal of both the pre- and post-monsoon depth to water level maps (Fig. 3), indicates that the water levels along the axial river Narmada remained deeper (10-20 m bgl) during the former season, while during the post-monsoon period, significant patches along the river registered shallow water levels (2-5 m bgl). It suggests the effluent character of the Narmada River, where the river gets fed by groundwater during the lean periods. During the post-monsoon season, a good deal of bank storage of water takes place due to availability of granular zones in the river bank lithology.

NARMADA ALLUVIAL PLAIN

The Narmada Alluvial Plain (NAP) comprising pleistocene to recent sediment has been developed in a narrow valley in the NRB. Sedimentation has taken place in grabens forming troughs owing to down-faulting/dislocation and sinking of basement rocks along the NSL (GSI, 1995; Khan and Aziz, 2016). The transverse tributaries of the Narmada, deriving sediment from either the Vindhyans in the north or the Satpuras in the south, have contributed more in filling up the structural depressions (Roy, 1971). Gondwanas lying to the south of axial river Narmada might have yielded more sediment due to their greater susceptibility to weathering and erosion. The Quaternary deposits in the plain have been classified into Younger and Older Alluvium (Gupta, 1974; Rahate et al., 1985). While the Younger Alluvium occupies the floodplains and active channels of the Narmada and its tributaries, the Older Alluvium occupies the upland areas with floodplain facies of palaeodomain. The older terraces are observed in these uplands.

The alluvail deposits in the NRB, comprising mostly stiff, yellow, yellowish brown and grey clay with numerous intercalated bands of sand and gravel, are observed in two prominent patches: (1) the Jabalpur-Harda section in MP in the middle reaches of the river preserving the longest alluvial strip (~390 km) and (2) the Gurudeswar-Bharouch section (~100 km) in the lower reaches of the river in Gujarat state. In the Gurudeswar-Bharouch section, the basin-fill of Quaternary sediment reaches up to 800 m (Maurya et al., 1995) due to syn-sedimentary basin subsidence along the NSL (Chamyal et al., 2002). However, this chapter deals with only the Jabalpur-Harda alluvial corridor in NRB falling in the state of MP, which has been deposited in a peninsular upland basin.

The plain in MP covers an area of about 14500 km^2 between Jabalpur in the east and Handia in the west, a distance of about 325 km (GSI, 1995). It is almost flat with small undulations and isolated hillocks. Its general elevation of the plain varies between 200 and 350 m asl. In north of the plain, Vidhyan hills rise almost abruptly, rising to heights between 600 and 750 m asl, while the south of the valley is bounded by the escarpments of the Satpura Range, rising to heights of about 600 m asl. The alluvial plain is more or less flat with the average slope of the plain of around 60 cm per km towards west. The master drainage Narmada takes a course along the northern boundary of the plain close to the Vindhyans. It gives an asymmetric shape to the plain with slightly slanting towards the north. Though, the general width of the alluvial plain varies within 25-50 km only, the maximum width is observed in the Tonga (23°03′, 78°22′)-Dukrikhera section (22°30′, 78°22′). The plain attached to the right bank is only 5-20 km wide while that along the left bank is within 15-30 km wide. The plain is marked with good green vegetation and the fields are richly cultivated for paddy.

The Jabalpur-Harda alluvial plain in MP preserves the thickest Quaternary deposits in the peninsular India due to sedimentation in the troughs (Khan and Aziz, 2016). The plain in MP displays alluvial terraces which have been formed by valley deepening and incision in response to the allogenic forcing factors like climate and tectonics during late Quaternary (Khan, 2017). Due to river channel incision, the bank cliffs expose 15-50 m of the top alluvial layers all along the river in the valley. The channel incision has variously been attributed to basin uplift and climatic induced increased discharge in the river (Chamyal et al., 2002; Bhandari et al., 2005; Khan, 2017). The older and upper terraces exhibit erosional and non-depositional surfaces with rills and incised (20-25 m) gullies and ravines (Khan, 2017).

NEOTECTONICS AND DYNAMICS IN THE NARMADA RIVER SYSTEM

The fluvial regimes in the Narmada river have been affected significantly by both the climate and tectonics during the Quaternary period. It has been inferred from the alluvial sequences from lower Narmada valley that the river was a larger one earlier with higher discharge levels (than the present) and low sinuosity, and high lateral migration rates (Bhandari et al., 2005) during late Pleistocene. During this period, river was carrying large quantities of sand (Gupta et al., 1999) with period of large floods (Kale et al., 2003). The discharge regimes in the river were also stronger during the Holocene for stronger monsoon periods (Ely et al., 1996; Kale et al., 2003). Chamyal et al. (2002) from their study have also suggested that the active Narmada river appears to be a misfit river flowing within a larger valley bounded by incised cliffs. Even the largest flood discharge of the river is not able to fill the entire valley (Rajaguru et al., 1995). The larger size of the river with higher discharge and sediment budget is evident in the overbank sedimentary record from the existence of larger channel fills, horizontally stratified sands, massive sand sheets, crevasse splays and backswamp deposits (Bhandari et al., 2005). Isolated channel fills in the overbank sediment indicates the single channel character of the river.

Various geomorphic features in the Narmada plain such as the ravenous tracts with deep incised gullies, entranched meanders, alluvial cliffs along the banks of river and river terraces indicate neotectonic activities along the NSL. During Pleistocene also, the sedimentation took place in tectonically disruptive activities as evident from the dislocation and tilting in the sedimentary beds (GSI, 1995). Neotectonics in the NRB has induced dynamism in several rivers in the basin including the Narmada river. Narmada bears a persistent tendency to shift towards N-NNW, which suggests basin uplift in the south and subsidence in the north resulting in basin asymmetry (Kothyari and Rastogi, 2013; Khan and Aziz, 2016). Kothyari and Rastogi (2013) have interpreted uplift to the tune of a few hundred meters in the entire area south of NSF. The area south of NSF depicts minimum Quaternary deposits and uplift and erosion of basement rock. The terrain here is more rugged in comparison to the northern side. Thickness of the Quaternary deposits becomes more towards the graben area in the north. The vertical movements along the faults have given rise to strath terraces along the river in upper parts of the valley.

PALAEOCHANNELS IN NARMADA ALLUVIAL PLAINS

A palaeochannel is an old river course which no longer conduits water. It may happen owing either to dying of a river due to insufficient flow or change in the direction of flow of the entire/part of the river course. The river channel dynamisms

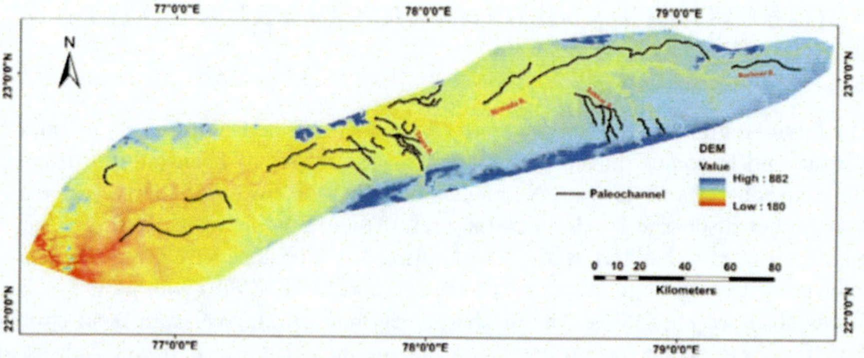

Figure 4. Upper Narmada river basin showing the palaeochannels as identified using the GIS techniques

such as meandering and migration, and avulsion are controlled by climate induced water/sediment budget and tectonics of intra/extra-basin origin. Alluvial plains of Narmada river in its upper reaches in MP indicate the presence of several palaeochannels of the stem channel Narmada as well as some of its tributaries (Fig. 4). The river channel of Narmada have migrated northward from south through meandering and cutoffs (Kothyari and Rastogi, 2013; Khan and Aziz, 2016). The imprints of such meander/chute cutoffs are visually traceable in the satellite maps on its left bank and in some cases on the right bank. The shifting of Narmada towards north has been more in the central parts around Tawa river in the districts of Narsinghpur and Hoshangabad, as evident from a wider alluvial stretch in this part. The tributaries from southern Satpura range sometime follow almost straight courses in the northward/northwestward direction to meet the Narmada. Important among them is the Tawa river that has remained quite dynamic in central parts of the NAP. The river is around 1.5-2.0 km wide and its palaeochannels are observed in a 10-12 km wide alluvial tract (Fig. 4). Some of the tributaries also show considerable meandering. Various chute and meander cut-offs are observed in the narrow valleys of the rivers such as Sakkar and Sher. Dimension of the palaeochannels varies a lot depending on the morphometry of the river. The length of palaeochannels varies within 5-30 km whereas the width is measured approximately between 400-1500 m.

ALLUVIAL HYDROGEOLOGY IN UPPER NARMADA RIVER BASIN IN MADHYA PRADESH

Earlier, Exploratory Tubewells Organisation (ETO), under its various groundwater exploratory programmes, had drilled 75 boreholes in the alluvial strecth between Jabalpur and Hoshangabad. In addition to it, under the Narmada valley groundwater project, the WTO (later known as Central Ground Water Board (CGWB)) drilled about 78 boreholes within the alluvial strecth. The lithological logs obtained from these boreholes alongside other geophysical sounding studies have established that the basement of the alluvium in the Narmada valley is quite uneven and marked within several highs and lows. Jain et al., on the basis of borehole data, have delineated the nature of the basement by preparing elevation map of the basement above the sea level (Fig. 5). It indicates the maximum numbers of lows are aligned along the axial zone where the thickness of alluvium has been noticed to be more, reaching to the maximum of 364 m in the southeast of Hoshangabad. The thickness of alluvium decreases both towards east and west.

The alluvial deposits cover significant parts of the districts Narsinghpur, Hoshangabad, followed by Harda, Raisen, Sehore and Jabalpur of MP. Several granular zones are encountered in the alluvium, comprising fine to medium to coarse grained sand, gravel and kankar separated by clay lenses. Kankar is present in abundance in the alluvium and pisolitic iron-ore granules are not infrequent.

In western parts of NAP, the thickness of alluvial fill goes up to ~100 m in depth. It increases towards the central parts around Hoshangabad and Narsinghpur, reaching up to ~250 m deep (Roy, 2009; CGWB 2013a, b, c, d; CGWB, 2015). This may be attributed to the convergence of fluvial activity of several palaeochannels including that of the Narmada, Tawa and others. The structural lows in this area might have induced such convergence. The thickness of the alluvial fill again decreases towards Jabalpur in the eastern parts of NAP. Maximum thickness of alluvium around 130 m has been explored in Jabalpur district (CGWB, 2013a). On the northern bank of the river in Jabalpur area, thickness of the alluvium ranges between 30 and 100 m. The cumulative thickness of the sand zones (fine to coarse and gravels) varies between minimum of 10 m to the maximum of ~50 m in the western parts. The frequency and thickness of sand zones within the framework of silt and clay increase towards the middle parts of the alluvial plain in the Hoshangabad and Narsinghpur districts. In Jabalpur district in east of NAP, the

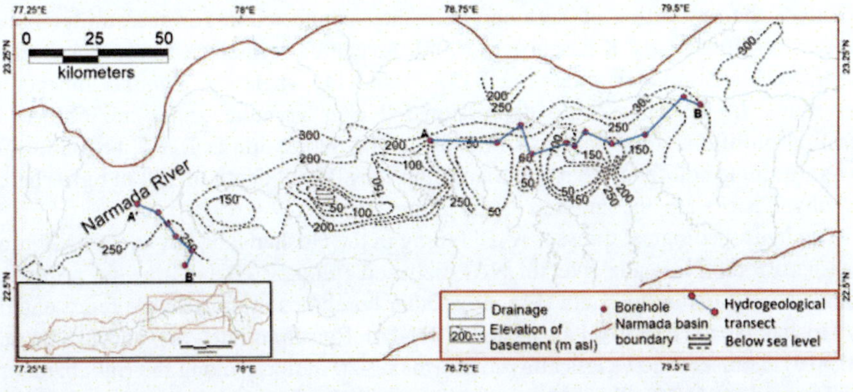

Figure 5. Elevation of the basement in parts of NRB (m asl) in its eastern parts around Jabalpur, Narsinghpur and Hoshangabad districts of MP (Source: GSI, 1995). Location of the map is given in the inset.

cumulative thickness of granular zones varies within ~10-60 m (CGWB, 2013e). The prolific alluvial aquifers are encountered in Hoshangabad, Narsinghpur and Jabalpur districts in the depth range of ~60-120 m bgl in the NAP. Average thickness of granular materials tapped by tube wells in the Narmada plains is around 30 m. The upper phreatic aquifer exists within the depth of ~20 m in general with sand thickness between 2 and 15 m.

Figure 6 (a) and (b) show two hydrogeological transects, redrawn from transects shown in different reports and publications (CGWB, 2015; Khan and Aziz, 2016). The transect A-B (Fig. 6a), in the NAP, falling in the Narsinghpur district of MP, runs from Jhiria (BH-1) in the west upto Bikrampur (BH-11) in the east along the left bank of Narmada river. The other boreholes are Navgaon (BH-2), Dungaria (BH-3):, Mothegaon (BH-4), Basedi (BH-5), Khairi (BH-6), Tindni (BH-7), Pansi (BH-8), Manegaon (BH-9), Noni (BH-10) and Bikrampur (BH-11). The Deccan traps, Gondwana sediments (limestones/sandstones/shales), Vindhyan shales and Precambrian granites/gneisses/schist kind of rocks form various basement formation over which the alluvium has been deposited. The transect depicts the phreatic aquifer within the depth of ~20 m below ground level (bgl), where kankars are also encountered with sand at various locations. The granular zones (comprising fine to coarse sand and gravels) are more in the west than in east. The deeper granular zones (beyond the depth of ~25-50 m below ground) form confined aquifer systems due to the presence of widely pervasive and thicker clay zones which act as confining layers. At Jhiria, beyond 20 m below ground, it seems, a single aquifer continues upto the alluvial depth of ~200 m below ground. In the western parts of transect A-B, the granular zones make ~40-60%, of the borehole depth in alluvium. Individual sand zones are found 5 m to ~50 m thick. Towards east, the granular zones diminish owing to appearance of several clay zones. It might be indicating the impact of tectonic perturbations was more towards the central parts of the alluvial corridor, rather than in the east and west. Depressions created owing to the down-throwing of crustal blocks might have attracted the fluvial activity towards it, thereby facilitating the deposition of sand bodies. Rates of river channel migrations and avulsions influence the proportion of sand and their inter-connectedness in the stratigraphic succession (Bridge and Karssenberg, 2005). As can be seen in the transect A-B, the basement highs are encountered at BH-3, 7 and 11, where the granular zones are minimum. In the intervening depressions, thicker granular zones are observed forming prolific aquifer systems in the area. The depression at BH-5, beyond ~40 m bgl, a single aquifer system continues upto the alluvial depth of 180 m bgl, where Vindhyan rocks are encountered as the basement rock.

The hydrogeological transect A′-B′ falling in the Hoshangabad district of MP run N-S across the Narmada river in NAP between Pilikarar (BH-12) in the north to Pathrota (BH-17) in the south (Fig. 6b). Other boreholes in transect are represented by Hoshangabad (BH-13), Pawarkheda (BH-14), Raisalpur (BH-15), and Sankheda (BH-16). Transect shows less alluvial fill (max. ~20 m thick) along the right bank of the Narmada (BH-12). Top phreatic aquifer is mostly clayey with minor sand lenses at a few locations (Khan and Aziz, 2016). At Hoshangabad (BH-13), the borehole (depth ~120 m) has not registered the basement. Sand thickness is appreciable at ~65

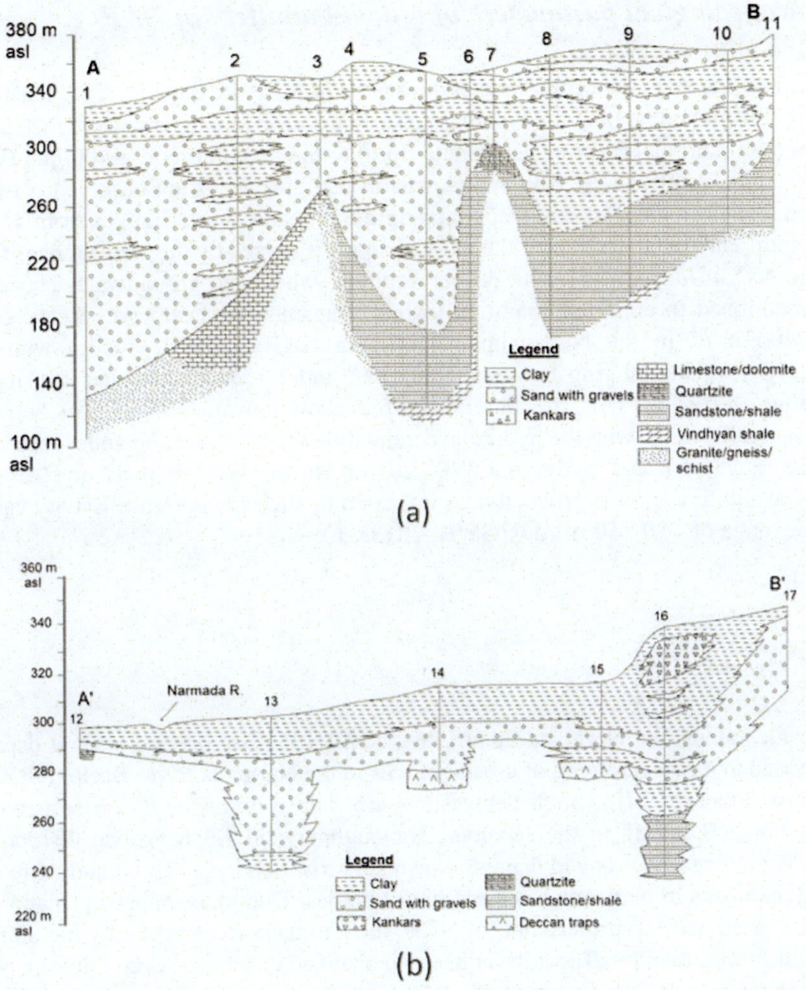

Figure 6. Hydrogeological transects in NRB: (a) A-B in Narsinghpur district and (b) A'-B' in Hoshangabad district depict the sub-surface lithology and the major aquifer systems in NRB (redrawn from CGWB, 2015; Khan and Aziz, 2016).

m. The deeper aquifer seems really extensive and continues towards south upto Pathrota (BH-17). Transect also shows various alluvial terraces in the valley along the left bank of the Narmada, as there has been sudden rise in surface elevations at BH-14 and BH-16 (Fig. 6b).

Hydrogeological parameters of alluvial aquifers in NAP

Aquifer systems in the central and eastern parts of NAP are quite prolific. The tube-wells constructed tapping the granular zones in the alluvium register yield that varies from minimum of 150 litres per minute (lpm) to the maximum of ~4400 lpm (Roy, 2009; CGWB, 2013c, d). The drawdowns in pumping wells are observed between 1 and 7 m. The yield of dug wells tapping the phreatic aquifer ranges from 450 to 750 lpm. The transmissivity of the alluvial aquifers tapped by dug wells ranges from 57 to 400 m^2/day (m^2/d). The deeper aquifers, where the groundwater occurs in semi-confined to confined condition, exhibit transmissivity values in the range of 23-2400 m^2/d in the Narsinghpur district in NAP (CGWB, 2013c) with the storativity values ranging between 2.01×10^{-6} and 1.15×10^{-3}. Further east in the Jabalpur district in NAP, the alluvial aquifers show transmissivity values between 300 and 3000 m^2/d, with the hydraulic conductivities of the granular zones estimated in the range of 6-204 m/day (CGWB, 2013d). In the western parts of NAP, the alluvial aquifers are less productive as indicated by their lower transmissivity values in the range of ~20-150 m^2/d (CGWB, 2013a, b).

CONCLUSION

Tectonics along the Narmada-Sone Lineament (NSL) has created several depressions due to down-throwing of crustal blocks in the Narmada River Basin (NRB) in Madhya Pradesh (MP). Such depressions are more prominent in the east-central parts of NRB falling in the Jabalpur, Narsinghpur and Hoshangabad districts of MP. These linearly arranged depressions in a narrow belt have significantly affected the dynamisms in Narmada river and its tributaries. Due to asymmetric location of the subsided belt in northern half of NRB, the Narmada river gradually has shifted towards north. Besides, Tawa River has also changed its course several times within an alluvial span of ~10-12 km width. Within the narrow valley of around 25-50 km width, the Quaternary alluvium has been deposited in an assymetric fashion. The plain attached to the left bank of Narmada is wider with width varying within 15-30 km. Alluvial deposits vary from a few tens of meters to the maximum of ~360 m in thickness. The thicker alluvial fills are observed in the Narsinghpur and Hoshangabad districts, which are attributed to the fluvial dynamisms within this stretch of alluvial plain. The granular zones comprising fine to coarse sand and gravels constitute ~40-60% of the alluvial thickness in parts of these districts. Towards Jabalpur district in eastern parts of the plain, granular zone diminish to ~30-40% of the alluvial column in boreholes.

The shallow phreatic aquifer lies within the depth of ~20 m below ground, where the granular zones of thickness between 5-15 m are found. Beyond the depth of ~25-50 m below ground, the multilayered aquifers behave as a single aquifer system in parts of Hoshangabad and Narsinghpur districts due to pinching out of intervening clays. Aquifer system exists in semi-confined to confined condition. The dug wells in phreatic zone yield in the range of 7.5 to 12 lps of groundwater, whereas the deeper aquifer system has the potential of yielding in the range of ~150-4400 lpm. The transmissivity values of the aquifer system widely vary in the range of ~20-3000 m^2/d. The confined nature of the aquifer system is indicated by the low storativity coefficient values falling in the range of 2.01×10^{-6} and 1.15×10^{-3}.

ACKNOWLEDGEMENT The author extends sincere thanks to Shri K. C. Naik, Chairman, Central Ground Water Board (CGWB), and Shri G. C. Pati, Member (CHQ), CGWB, for their encouragement to complete the study. Thanks are also extended to Dr. Utpal Gogoi, former Regional Director, CGWB, SER, Bhubaneswar for his constant advice and inspiration to complete the work. Views and opinions expressed in this chapter are of the author.

References

Allen, J.R.L. (1978). Studies in fluviatile sedimentation: An exploratory quantitative model for the architecture of avulsion-controlled alluvial studies. *Sed. Geol.*, 21: 129-147, https://doi.org/10.1016/0037-0738(78)90002-7.

Bhandari, S., Maurya, D.M. and Chamyal, L.S. (2005). Late Pleistocene alluvial plain sedimentation in Lower Narmada Valley, Western India: Palaeoenvironmental implications. *Journal of Asian Earth Sciences*, 24(2005): 433-444.

Blum, M.D. and Tornquist, T.E. (2000). Fluvial responses to climate and sea level change: A review and look forward. *Sedimentology*, 47: 2-48.

Bridge, J.S. and Karssenberg, D. (2005). Simulation of Flow and Sedimentary Processes, including Channel Bifurcation and Avulsion, on Alluvial Fans. 8th International Conference on Fluvial Sedimentology, Delft, The Netherlands, Abstracts, Delft, p. 70.

Bridge, J.S. and Leeder, M.R. (1979). A simulation model of alluvial stratigraphy. *Sedimentology*, 26: 617-644.

Bristow, C. (1996). Reconstructing fluvial channel morphology from sedimentary sequences. *In:* Carling, P.A., Dawson, M.R. (Eds.), Advances in Fluvial Dynamics and Stratigraphy, Wiley, New York, pp. 351-371.

Cant, D.J. and Walker, R.G. (1978). Fluvial processes and facies sequences in the sandy braided South Saskatchewan River, Canada. *Sedimentology*, 25: 625-648.

CGWB (2013a). District Ground Water Information Booklet. Raisen District, Madhya Pradesh. Central Ground Water Board, North Central Region, Bhopal. Ministry of Water Resources, Government of India. Cgwb.gov.in.

CGWB (2013b). District Ground Water Information Booklet. Sehore district, Madhya Pradesh. Central Ground Water Board, North Central Region, Bhopal. Ministry of Water Resources, Government of India. Cgwb.gov.in.

CGWB (2013c). District Ground Water Information Booklet. Narsinghpur district, Madhya Pradesh. Central Ground Water Board, North Central Region, Bhopal. Ministry of Water Resources, Government of India. Cgwb.gov.in.

CGWB (2013d). District Ground Water Information Booklet. Hoshangabad district, Madhya Pradesh. Central Ground Water Board, North Central Region, Bhopal. Ministry of Water Resources, Government of India. Cgwb.gov.in.

CGWB (2013e). District Ground Water Information Booklet. Harda district. Madhya Pradesh. Central Ground Water Board. Ministry of Water Resources. Government of India. www.cgwb.gov.in.

CGWB (2015). A report on deep groundwater exploration and hydrogeological studies (unpub.). Central Ground Water Board, North Central Region, Bhopal. Ministry of Water Resources, River Development and Ganga Rejuvenation, Government of India.

Chamyal, L.S., Maurya, D.M., Bhandari, S. and Rachna, R.R. (2002). Late quaternary geomorphic evolution of the lower Narmada valley, western India: Implications for neotectonic activity along the Narmada–Son Fault. *Geomorphology*, 46(3–4): 177-202.

CPCB (1994). Basin sub-basin inventory of water pollution. The Narmada Basin, Central Pollution Control Board, Delhi.

Ely, L.L., Enzel, Y., Baker, V.R., Kale, V.S. and Mishra, S. (1996). Changes in the magnitude and frequency of Holocene monsoon floods on the Narmada River, Central India. *Geol. Soc. Am. Bull.*, 108: 1134-1148.

Govt. of India (2019). Chapter 5: Groundwater level scenario in the country. *In:* National Compilation on Dynamic Ground Water Resources of India, 2017. Government of India Ministry of Jal Shakti Department of Water Resources, RD & GR Central Ground Water Board, Faridabad, pp. 40-45.

GSI (1995). Project Crumansonata: Geoscientific Studies of the Son-Narmada-Tapi Lineament Zone, Geological Survey of India, pp. 1-154.

Gupta, A., Kale, V.S. and Rajaguru, S.N. (1999). Varieties in Fluvial Form (eds. Miller A.J., Gupta, A.), John Wiley & Sons, New York, pp. 295-315.

Gupta, H. and Chakrapani, G.J. (2005). Temporal and spatial variations in water flow and sediment load in Narmada River Basin, India: Natural and man-made factors. *Environ. Geol.* 48: 579-589.

Gupta, S.K. (1974). Geomorphology and sedimentation in parts of Narmada basin, Sahore and Hoshangabad districts, M.P. (Unpubl.) Prog. Rep. of Geol. Surv. India.

Kale, V.S., Mishra, S. and Baker, V.R. (2003). Sedimentary records of palaeo floods in the bedrock gorges of the Tapi and Narmada rivers, central India. *Curr. Sci.*, 84(8): 1072-1079.

Khan, A.A. (2017). Geomorphology and neotectonics of quaternary deposits, Narmada valley, central India. *Int. J. Adv. Res.*, 5(4): 2230-2296.

Khan, A.A. and Aziz, M. (2016). Quaternary sedimentology tectonics and sedimentation Narmada rift valley, Central India. *Int. J. Adv. Res.*, 4(10): 1690-1719.

Kothyari, G.C. and Rastogi, B.K. (2013). Tectonic Control on Drainage Network Evolution in the Upper Narmada Valley: Implication to Neotectonics. *Geography Journal*, 2013: 1-9. Article ID 325808, https://doi.org/10.1155/2013/325808.

Krishnaswamy, V.S. and Raghunandan, K.R. (2005). The Satpura uplift and the palaeo climate of the Holocene and auxiliary evidence from the Valmiki Ramayana. *Journal of the Geological Society of India*, 66(2): 161-170.

Maurya, D.M., Chamyal, L.S. and Merh, S.S. (1995). Tectonic evolution of the Central Gujarat plain, Western India. *Curr. Sci.*, 69: 610-613.

Rahate, D.N., Khan, A.A. and Banerjee, S.N. (1985). Geomorphological and geological studies of Quaternary sediments in collaboration with project Crumansonata in parts of the Narmada basin, Sehore, Dewas and Hoshangabad districts (unpubl.). Geol. Surv. Ind. Progress Report.

Rai, P.K., Chaubey, P.K., Mohan, K. and Singh, P. (2017). Geoinformatics for assessing the inferences of quantitative drainage morphometry of the Narmada Basin in India. *Appl Geomat*, 9: 167-189. doi:https://doi.org/10.1007/s12518-017-0191-1.

Rajaguru, S.N., Gupta, A., Kale, V.S., Ganjoo, R.K., Ely, L.L., Enze, Y. and Baker, V.R. (1995). Channel form and processes of the flood dominated Narmada River, India. *Earth Surf. Process. Landforms*, 20: 407-421.

Rajendran, K. and Rajendran, C.P. (1998). Characteristics of the 1997 Jabalpur earthquake and their bearing on its mechanism. *Curr. Sci.* 74: 168-174.

Roy, A.K. (1971). Geology and groundwater resources of Narmada valley, Bull. Geol. Surv. Ind. No. 30.

Roy, A.K. (2009). Exploratory drilling for groundwater in the Narmada valley, Madhya Pradesh, India. *Hydrological Sciences Journal*. 2(4): 27-45, DOI: https://doi.org/10.1080/02626665709493086.

Shankar, R. (1991). Thermal and crustal structure of SONATA: A zone of mid continental rifting in Indian Shield. *J. Geol. Soc. India*, 37: 211-220.

Sinha, R., Tandon, S.K., Gibling, M.R., Bhattacharjee, P.S. and Dasgupta, A.S. (2005). Late Quaternary geology and alluvial stratigraphy of the Ganga basin. *Himal. Geol.*, 26: 223-240.

Straub, K.M., Paola, C., Mohrig, D., Wolinsky, M.A. and George, T. (2009). Compensational stacking of channelized sedimentary deposits. *J. Sed. Res.*, 79: 673-688, https://doi.org/10.2110/jsr.2009.070.

Valdiya, K.S. (1984). Aspects of Tectonics Focus on South-Central Asia. Tata McGraw-Hill, New Delhi. 310 pp.

Valdiya, K.S. (2010). The Making of Indian Geodynamic Evolution. Macmillan Publication House, New Delhi.

Wells, N.A. and Dorr, J.A. (1987). Shifting of the Kosi River, northern India. *Geology*, 15: 204-207.

Chapter 11
Groundwater Resources in Punjab and Bist-Doab Area: An Appraisal and Overview

Gopal Krishan, M. S. Rao, and Narayan C. Ghosh

INTRODUCTION

Punjab, a land of five rivers (in hindi Punch nada), namely, Beas, Sutlej, Ravi, Chenab and Jhelum are historically part of larger Punjab region comprising parts of Afghanistan and Pakistan on international front and states of Haryana and Himachal Pradesh (trifurcated into 3 states: punjabi speaking Punjab, and hindi speaking Haryana in 1966 and Himachal Pradesh in 1971), parts of Jammu & Kashmir and Delhi at national front.

Punjab is historically considered as the breadbasket and a role model for green revolution in India. However, it is presently under serious threat of groundwater level decline, salinity problem and also aquifer contamination.

To understand the issues, trends and challenges of groundwater resource, a comprehensive outlook on characteristic and property, which shape and are responsible for such abnormality to happen, is needed for deriving a technically sound sustainable solution. To appraise the issues and challenges, an overview covering hydrogeology, changing scenarios of groundwater resources, soils, trends in groundwater levels, quality of groundwater, etc. together with an indepth analysis of the Bist-Doab area, as a case study are described. The data and results compiled from different publications are duly acknowledged.

G. Krishan (✉) · M. S. Rao
National Institute of Hydrology, Roorkee, Roorkee, Uttarakhand, India

N. C. Ghosh
Bengal Institute of Technology, Kolkata, India

A. Mukherjee (ed.), *Riverine Systems*, https://doi.org/10.1007/978-3-030-87067-6_11

HYDROLOGICAL AND HYDROGEOLOGICAL FEATURES

The following sections describe the characteristic aspects, which have direct bearing on the occurrence and prospects of groundwater resource in an area.

Physiography

Punjab has five divisions covering 22 districts (Jalandhar 7, Patiala 5, Firozepur 4, and 3 district each in Faridkot and Ropar), 82 tehsils, 87 subtehsils and 147 blocks (Kaler, 2016) with Chandigarh as the state capital. Geographically, Punjab lies in 29.30° N to 32.32° N latitude and 73.55° E to 76.50° E longitude with a total area of 50,362 km^2 and shares international boundary with Pakistan on the west and national boundary with states of Jammu & Kashmir, Himachal Pradesh, Haryana and Rajasthan on the north, northeast and south, respectively.

Physiographically, Punjab is characterized in 8 sub-units, namely, Hills (Shiwalik), Tableland, Intermontane valleys, Piedomont Plains (Upper piedmont & Lower Piedmont), Sirowal Zone, Alluvial Plains, flood plains and Sand Dunes (CGWB, 2014). The Siwalik hills cover 900 km^2 in the northeastern part. These hills traverse NW–SE direction and are made up of predominating bands of clays alternated with sands of varying grade and form the boundary with the neighbouring Himachal Pradesh. Sediments of the eroded upper Siwalik have formed the flat tableland in the lower Shiwalik zone. This fertile zone supports the local agricultural activities (in Hopshiarpur district). Inter-montane Sutlej valley is 40 km long, 5 km wide and about 50 m thick filled sediment stretch between Nangal and Ropar towns along the Sutlej river.

The transitional area of width varying 6 to 10 km between the alluvial plains and Shiwalik hills is occupied by the Piedmont plains, locally known as *Kandi* belt. These Piedmont plains have distinct features and are divided into different zones on the basis of these features. Zone with extremely dissected large number of sub-parallel streams, loose to semi-consolidated very coarse materials; poor vegetation and rain fed condition prevailing is the *upper piedmont zone* while the zone having large fan shaped body; lesser coarser sediments; fine textured soils; parallel drainage pattern; flat to undulating slope; irrigated; higher tree population dense forests and orchards is the *lower piedmont zone*. Area towards more southwest of *Kandi* with gentle is the *Sirowal Zone*.

Alluvial plains formed by the alluvium transported by the rivers mark the main physiographic unit are: (i) Recent (or active): Flood Plains-composed of unconsolidated material deposited along the banks of the rivers, which are subjected to periodic or occasional flooding. (ii) Abandoned flood plains (or older flood plains): Fairly wide with flat surfaces; are less prone to flooding but are fertile; therefore, are under intensive cultivation. (iii) Upland areas.

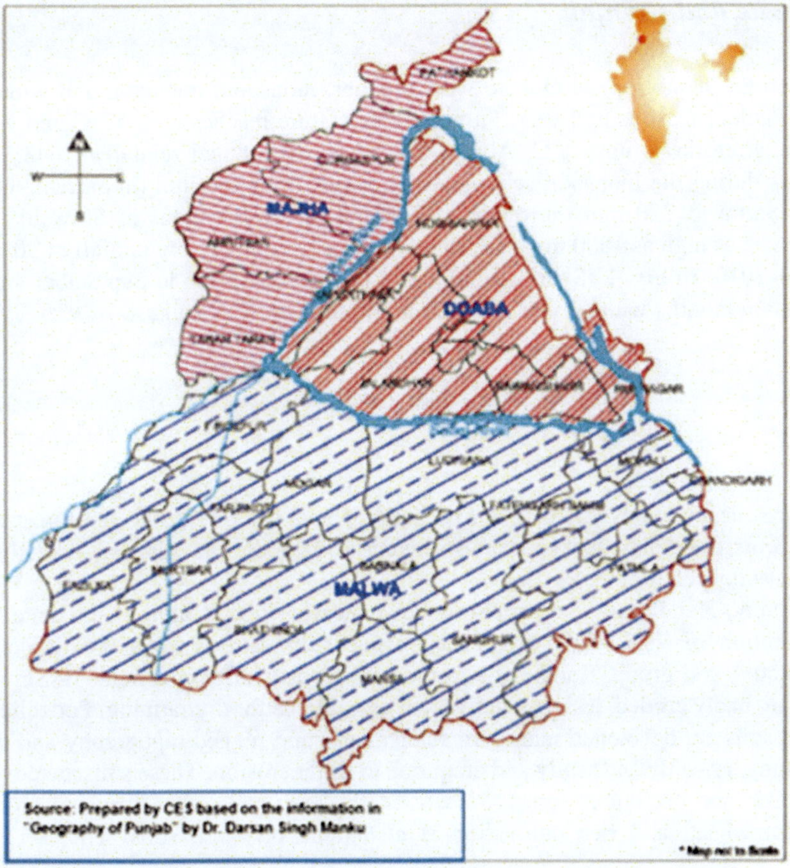

Figure 1. Physiography of Punjab showing three zone of Majha, Malwa and Doaba.

South-western part of the State comprising about 28% of the area, experiences semiarid type of climate and has isolated '*Sand Dunes*' of varying size and height.

Sutlej and Beas rivers divide the state into three zones, viz. Malwa, Doaba (land of two rivers *Do* means two, *Ab* means river) and Majha (Fig. 1).

Demography

As per Census (2011), Punjab had a population of 27.7 million that represented 2.29% of country's population with average population density of 550 persons/km^2, which is about 70% more than the national average of 327 persons/km^2. The trend of population growth showed a gradual fall in the last three decades from 20.26% (1981-1991) to 13.73% (2001-2011).

Climate and Rainfall

Punjab has a subtropical climate with summer, monsoon, transition and winter as four distinct seasons in a year. Summer temperature reaches upto 47°C and winter temperature drops upto 1°C. About 80% of normal annual rainfall of 648.8 mm occurs during the monsoon season, which varies from 260 mm (in extreme north-west parts) to 720 mm (northern region) and in few locations in Shiwaliks that increases as high as 1000 mm. The distribution of mean monthly rainfall of 30 years (1981-2010) in Fig. 2 (Sharma et al., 2017) shows that June to September are the monsoon months with July as the peak month followed by August.

Soils

Major soils are loamy (Ustochrepts of Ustic), Kandi soils, Podzolic and forest soils, sierozems, flood plain (bet soils or ustifluvent) soils, sodic and saline soils, sandy and desert soils (calciorthids).

Loamy, Kandi and sierozem soils cover nearly 70% of area of the state with distribution of 25%, 23% and 22%, respectively. Loamy soils are fertile and productive soil group. Kandi soils formed by deposits of river torrents of Shiwalik hills are badly eroded, less productive and are suitable for dry farming. Podzolic and forest soils are developed mainly in steep slopes and rugged topography and these are stony, gravelly and sandy and are prone to water erosion. These soils are covered by shrub and deciduous forest. *Sierozems,* the grey soils when irrigated produce highest wheat, and rice cultivation is also done on these soils. The Bet soils

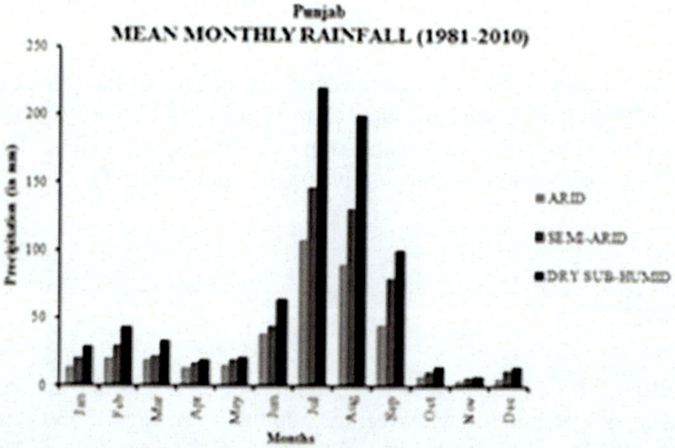

Figure 2. Distribution of mean monthly rainfall of Punjab (Source: Sharma et al. 2017 based on IMD, Pune and Directorate of Land Records, Jalandhar).

distributed on 7% area of the state are found on flood plain areas of rivers, streams or choes. Sodic and saline soils are developed mainly in waterlogging areas. Sandy and desert soils covering about 20% of the area occur mainly in south-western and south central Punjab and are favourable for cotton, citrus, oilseeds, wheat and fodder, cotton, moth, citrus, wheat, bajra and other kharif cultivation. The desert soils are prone to wind erosion especially in summer. Table 1 presents region and district-wise population, area, and soil types of the Punjab state.

Rivers of Punjab

Three rivers namely, Sutlej, Beas and Ravi, which are tributaries of Indus system, form the main drainage system of Punjab (Fig. 3). Two major rivers, Sutlej and Beas, traverse in the state while Ravi and Ghaggar—an intermittent and endorheic river—touch the northern and southern borders of the state, respectively. At Harike head-work, the two rivers Beas and Sutlej join and continue thereafter, as Sutlej. Besides these major rivers, the state is also drained by several small rivers like White Bein, Black Bein, Ghaggar river and seasonal revaluates. The area in Punjab in between the Beas and Sutlej is called Doaba.

Agriculture

Punjab is an agrarian state with 70% population engaged in agricultural activities. The state has intensified the agriculture activities using fertilizers, pesticides, machinery, irrigation intensification, etc. with the inception of green revolution in 1960's. As a result, the country was relieved from importing food grains and the state's economy increased rapidly. By the year 2000, area under cultivation reached to 83% of the geographical area (50.362 lakh hectares), cropping intensity increased to 189%, and 98% of cultivable area came under assured irrigation, use of fertilizer reached to 235 kg/ha (which is 1.84 times the national average of 128 kg/ha) and farming became mechanised. According to National Sample Survey Report (NSS Rep no 45, 54th round) of 1992, Punjab had 1024 tractors and 584 power tillers per 100 km^2. This was nearly 10 times more than the national average of 109 tractors and 41 power tillers per 100 km^2.

Water and Irrigation

Irrigation water requirement is supported by canal and ground water. Surface water and groundwater (GW) development in the state started with the onset of green revolution. By 1986-87, the state had 218 minor irrigation schemes operational per

Table 1. Region and district-wise population, area and soil cover

Region	*District*	*Population (Census, 2011)*	*Area*		*Soil cover*
			Sq. km.	*% of state*	
Doaba	Hoshiarpur	15,86,625	3365	6.68	Kandi soils, Sierozems, pod-zolic and forest soil
	Jalandhar	21,93,590	2632	5.23	Loamy, sierozems, Sodic and saline soils
	Kapurthala	8,15,168	1632	3.24	Loamy, Sierozems, Sodic and saline soils
	Shaheed Bhagat Singh Nagar (NawanShahr)	6,12,310	1267	2.52	Loamy, Kandi soils, podzolic and forest soil
Majha	Amritsar	24,90,656	2647	5.26	Sierozems, Sodic and saline soils
	Gurdaspur	16,21,725	2635	5.23	Kandi soils, podzolic& forest soil and Sodic and saline soils
	Pathankot	6,76,598	929	1.84	Kandi and loamy soils
	Tarn Taran	11,19,627	2449	4.86	Sierozems
Malwa	Ajitgarh (Mohali)	9,94,628	1093	2.17	Loam with sandy patches
	Barnala	5,95,527	1410	2.80	Loam and Sandy loam
	Bathinda	13,88,525	3385	6.72	Loamy, Sandy and desert soils
	Faridkot	6,17,508	1469	2.92	Sandy, sierozems, Sodic and saline soils
	Fatehgarh Sahib	6,00,163	1180	2.34	Sierozems
	Fazilka	10,63,737	3113	6.18	Sodic and saline soils
	Firozpur	9,65,337	2190	4.35	Sodic and saline soils and sandy and desert soils
	Ludhiana	34,98,739	3767	7.48	Sierozems, Loamy, Sandy and desert soils
	Mansa	7,69,751	2171	4.31	Loamy, desert soils, Sodic and saline soils
	Moga	9,95,746	2216	4.40	Loamy
	Patiala	18,95,686	3218	6.39	Sierozems, loamy and sandy soils
	Rup Nagar (Ropar)	6,84,627	1369	2.72	Kandi soils, podzolic and forest soil
	Sangrur	16,55,169	3610	7.17	Sierozems, laomy, sandy, desert soils and Sodic and saline soils
	Muktsar	9,01,896	2615	5.19	Loamy, Sandy, desert soils and Sodic and saline soils
		2,77,43,338	50,362	100	

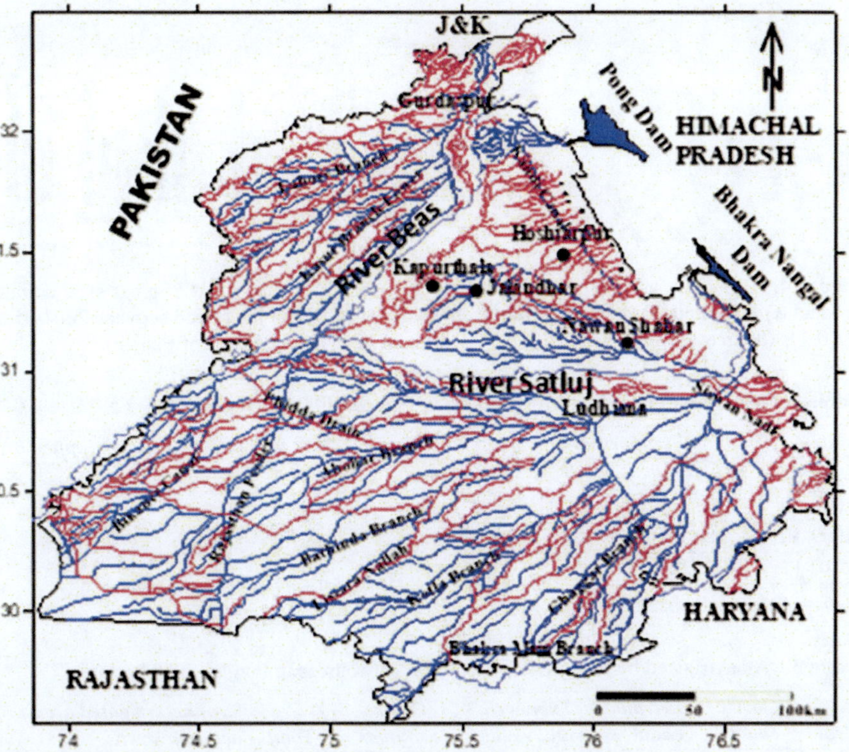

Figure 3. Drainage systems of rivers in Punjab.

100 km^2 area against the National average of 75. In 1993-94, the number of GW withdrawal structures (electric and diesel pumps) in the states rose to 18.8 pumps per 100 km^2 against the national average of 1.4 pumps (electric + diesel) per 1000 km^2 for the same period (MI Census 1993-94). In 2009-2010, 13.15 lakh tube wells were operational in the state. The surface water is distributed through network of 10 main canals and distributaries over a length of 14,500 km. However, post 90's, although the economy continued to increase but the share of the agriculture in the total GDP started decreasing with the progressive increase in the share from the secondary and tertiary sectors contribution in the total GDP (Fig. 4).

The state has intensive agricultural inputs (Table 2) showing 97-98% irrigated area of the total cropped area (in hectare) during 2007-2011. Area irrigated (in hectare) by using water from different sources is given in Table 3. Irrigation by the use of GW is nearly 2.6 times more than the other surface water sources including canal water.

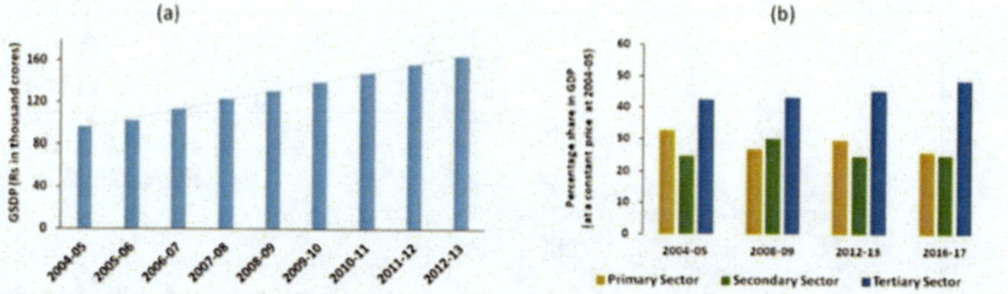

Figure 4. Economic growth of Punjab during 2004-11: (a) annual trend of gross state domestic product (GSDP), and (b) share of sectors in the total state GDP for the period between 2004-05 and 2010-11. (Source: Annual Plan 2011-12, Department of Planning, Govt. of Punjab).

Table 2. Status of cropped and irrigated area in Punjab ('000 ha.) during 2007-11

Year	*Total cropped area*	*Irrigated area*	*% of irrigated area to cropped area*
2007-08	7870	7689.3	97.7
2008-09	7912	7723.6	97.6
2009-10	7876	7714.2	97.9
2010-11	7882	7723.8	98.0

Source: Statistical Abstract, Punjab

Table 3. Area irrigated by using water from different sources in Punjab ('000 hectare)

Year	*Govt. canal*	*Private canal*	*Tube-well and well*	*Other sources*	*Total*	*% of net irrigated area to net area sown*
2007-08	1142	-	2922	4	4068	97.2
2008-09	1110	3	2950	1	4064	97.4
2009-10	1111	3	2955	2	4071	97.9
2010-11	1113	3	2954	-	4070	97.9

Source: Statistical Abstract, Punjab

Groundwater Aquifer

The seismic surveys by ONGC have indicated that the basement rocks dip gently towards the Himalayan foothills and correspondingly the sediment thickness increases towards the foothill. The aquifer basin is deep and wide in the north-western segment and gets narrower to the southwest, and the basement topography rises gradually in that direction. A high basement occurs in the subsurface corresponding to the present water divide between the rivers in the Punjab and Yamuna of the Ganga system. The crest of the buried ridge is ~ 450 m below ground level (bgl). Sediment thickness is about 450 m at Dasuya, Hoshiarpur which decreases to 154 m in south-western parts of Punjab. The alluvial deposits formed

Table 4. Aquifer extent, geometry and yield of different districts

District	*Total area (km^2)*	*Aquifer thickness*	*Average yield (m^3/hr)*
Faridkot, Moga, Bathinda, Mansa, southern parts of Sangrur and Ferozepur district	12,000	Upto 50 m with local discontinuity	50
South of Fazilka to north of Moga, north eastern parts of the district of Gurdaspur, Hoshiarpur, Nawanshahar, Patiala, Ropar and Anandpur Sahib valley	7,000	Upto 300 m	50-150
Amritsar, Kapurthala, Fatehgarh Sahib, Ludhiana, Patiala districts and parts of Sangrur, Gurdaspur, Ferozepur, Patiala, Nawan- shahar, Jalandhar and Ropar district	29,000	Upto 450 m	150
Hilly terrain in parts of Ropar, Gurdaspur, Hoshiarpur and Nawanshahar district	2,000	Semi-consolidated formations below	< 50
Plateau area (beet area) in Garhshankar block of Hoshiarpur district	750	100-370	100-200
In parts of Gurdaspur district and Anandpur Sahib block		40-150	Artesian flowing aquifers 1-70

Table 5. Block-wise groundwater development as assessed by the GWD of Punjab (total blocks = 138)

Category (GW draft in % of annual net recharge)	*Year: 2000* *No. of block (% of area)*	*Year: 2005* *No. of block (% of area)*	*Year: 2010* *No. of block (% of area)*
Over-exploited (Dark) (> 100%)	73 (52.90)	103 (75.18)	110 (79.71)
Critical (85-100 %)	11 (7.97)	5 (3.65)	3 (2.17)
Semi critical (65-85 %)	16 (11.59)	4 (2.92)	2 (1.45)
Safe (<65%)	38 (27.54)	25 (18.25)	23 (16.67)

Source: Jain, A.K., Department of Soil & Water Engineering, PAU, Ludhiana

by the mighty rivers of Punjab have developed extensive thick multiple freshwater aquifers throughout the state. More than 5 thick freshwater aquifers are encountered in Majha region at a depth within 500 m bgl. In the south-western part, freshwater aquifers are underlain by brackish/saline water. Table 4 presents district- and area-wise aquifer thickness and average yield of the respective aquifer. Major parts of the state have aquifer yield that ranges between 50 and 200 m^3/hr except parts of Gurdaspur district and Anandpur sahib block.

Groundwater Department (GWD) of the state together with the Central Groundwater Board (CGWB) assesses the status of GW exploitation every five years. The focus of such assessment is to examine the level of exploitation and categorising under different status with respect to the annual recharge. Table 5 presents the GW

Table 6. Groundwater Potential of Punjab (2006)

Annual replenishable groundwater resource (in BCM)	18.22
Provision for domestic, industrial and other uses (in BCM)	1.822
Available groundwater resources for irrigation (in BCM)	16.398
Utilizable groundwater resources for irrigation (in BCM)	14.755
Gross draft (in BCM)	20.307
Net draft (in BCM)	14.215
Balance groundwater resource for future use (in BCM)	2.179
Level of groundwater development (%)	86.71

Source: DWSS, Punjab, 2006*

exploitation status of block assessed in different years. Out of 138 assessment blocks, the number of over-exploited (annual GW extraction > annual recharge) blocks has continuously increased during 2000 to 2010, while the critical (annual GW extraction is between 85 and 100% of the annual recharge), semi-critical (annual GW extraction is between 65 and 85% of the annual recharge), and safe aquifer (extraction < 65%) have reduced during these period. The GW potential in Punjab (Table 6) shows that the average stage of GW development is ~86.7% against the annual replenishable GW resources of 1.821577 mham, i.e., 18.21577 BCM (billion cubic meter). Table 6 also gives distribution of GW for different sectoral usages.

Increased GW over-exploitation has been observed in the north and central part of Sutlej basin while water logging and increased salinity reported in the south-western part and in lower Sutlej basin, problems of high levels of fluoride, selenium and arsenic in the GW have also been reported. Over exploitation of GW, implying a negative balance of GW storages, withdrawal being more than the GW recharges has been experienced in most parts of the basin (NAPCC, 2011).

GROUNDWATER RESOURCES

An estimate of GW potential by NAPCC in 2011 (Table 7) shows that the net annual GW resource in the state was 21.44 BCM, of which the net annual draft was 31.16 BCM reflecting an annual deficit of 9.72 BCM. The GW assessment conducted by CGWB in 2004 showed that out of 138 blocks in the state, 103 were "over-exploited". The stage of GW development was estimated to 145%. This eventually put the state into an "over-exploited" category. Table 7 also gives status of groundwater in past years. The average water table depth was 7.32 m in 1998 and that had depleted to 12.79 m bgl in 2012, indicating an annual fall of 41.6 cm/yr. In central Punjab, the decline was between 0.11 m/year and 1.34 m/year (NAPCC, 2011). The north-eastern part of the state showed GW depth between 5 m and 10 m, in the north-central it was between 10 m and 15 m, and in areas around major cities, namely, Jalandhar, Ludhiana, Amritsar, Patiala, Fatehgarh Sahib, Nawashahar and Sangrur,

Table 7. Status of groundwater in Punjab in 2011 (Source: NAPCC, 2011)

Annual groundwater availability	21.44 BCM
Groundwater extraction	31.16 BCM
Average level of groundwater development	145%
Level of groundwater extraction in districts of Fatehgarh Sahib, Amritsar, Jalandhar, Kapurthala, Ludhiana, Mansa, Moga, Nawanshahar, Patiala and Sangrur .	144% to 254%
Over-exploited groundwater blocks	103 out of 137
Critical blocks	5
Over-exploited blocks in 1984	64
Over-exploited blocks in 2006-07	103
Decline in groundwater level reported by CGWB	4.5 to 13.5 m
Area identified for groundwater recharge estimated by CGWB	16450 km^2
Waterlogged area	200,000 ha
Salinity affected area	1,000,000 ha

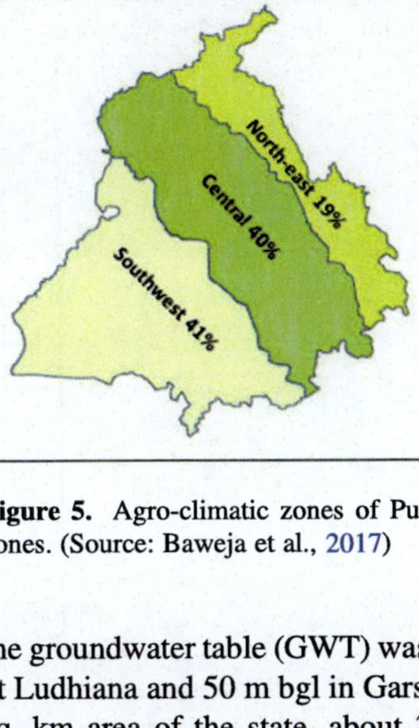

Proportion of area and the water table depth in different agro-climatic zones of Punjab

Year	Northeast zone			Central zone			Southwest zone		
	<3	3-10	>10	<3	3-10	>10	<3	3-10	>10
1998	0	81	19	0	65	35	12	84	04
2004	0	70	30	0	18	82	03	65	32
2010	0	61	39	0	12	88	12	48	40
2011	0	50	50	0	13	87	19	41	40
2012	0	58	42	0	21	79	08	44	48

Figure 5. Agro-climatic zones of Punjab and % of state geographical area falling under these zones. (Source: Baweja et al., 2017)

the groundwater table (GWT) was more than 20 m bgl. Depth to GW was at 33 m bgl at Ludhiana and 50 m bgl in Garshankar block of Hoshiarpur district. Out of 50,362 sq. km area of the state, about 39,000 sq. km (78%) showed a decline in GWT. However, the decline in GWT was not spatiotemporally uniform in the whole state. Figure 5 presents depiction of area under different zones and Table 8 describes ranges of GWTs in different years in those zones. Figure 6 portrays the yearly trend (decline) of GWTs (bgl) of two hydrograph stations; one in Gujarwala of Ludhiana district and the other in Nakodar block of Jalandhar district from 1973 to 2005. A continuous decline with no sign of GWT reversion was common in more than 75% area of the state. The fluctuation of GW levels (RL of ground surface—depth to GWT beneath the respective ground surface) of two hydrograph stations, namely,

Table 8. Trend of groundwater level in districts of Punjab (Source: NAPCC, 2011)

Zones	*District*	*Trend (decline) (in m/yr)*
Zone I: North-eastern area	Hoshiarpur	0.68-0.07
	Nawashahar	-
	Ropar	-
	Fatehgarh Sahib	0.025-0.58
	Patiala	0.5
Zone II: North-central area	Jalandhar	-
	Ludhiana	0.11-1.34
	Kapurthala	0.20-1.00
	Moga	0.20-1.00
	Mansa	-
	Sangrur	0.65
Zone III: South-western area	Faridkot	-
	Ferozepur	-
	Muktsar	-
	Bathinda	-

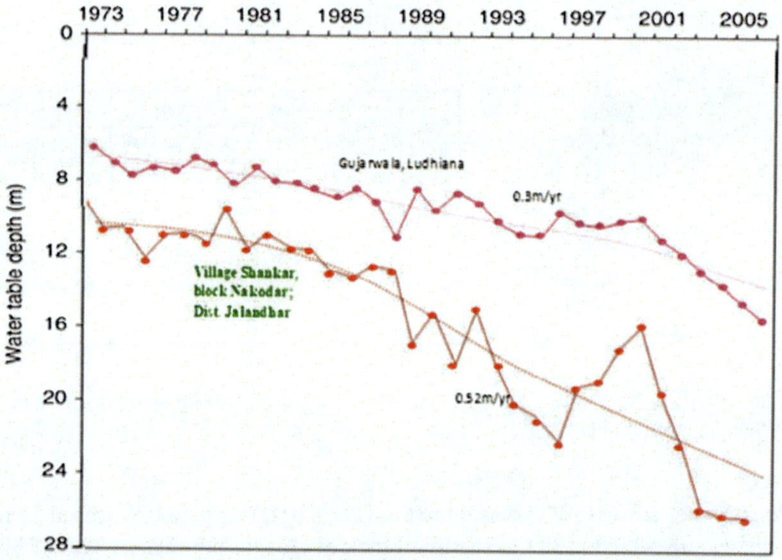

Figure 6. Depth to water level in Punjab during 1973-2005.

Bajwara and Noormahal (Fig. 7) for the period 1998-2011 in the month of July showed a continuous decline in GWL @ 0.64 m/year for the Bajwara and 0.79 m/year for Noormahal site. Some improvement in GWL in few locations like Noormahal (Fig. 7), was reported. However, the overall GWL followed a declining trend. This sign of improvement in local scale was either due to augmentation of groundwater by artificial recharge or due to reduced groundwater extraction or both.

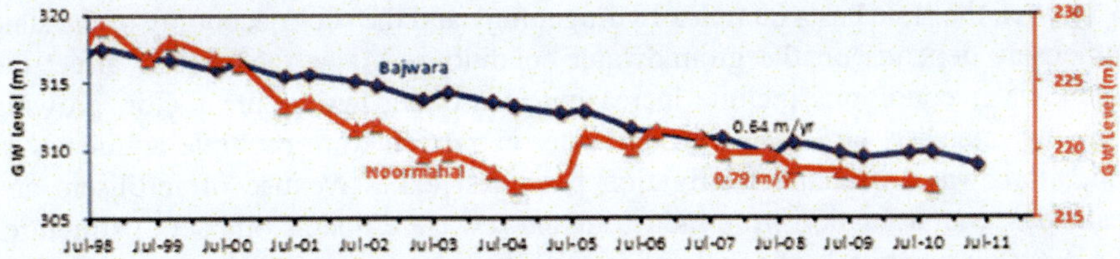

Figure 7. Groundwater fluctuation at Bajwara and Noormahal, Bist Doab during 1998-2010.

Table 9. Fluoride and nitrate contamination in GW of Punjab (Source: NAPCC 2011)

District	*Fluoride concentration (ppm)*	*Nitrate concentration (ppm)*
Bhatinda	-	61-401
Faridkot	-	60-287
Firozpur	1.63-3.4	69-241
Fatehgarh Sahib	1.54	-
Ludhiana		70-206
Mansa	1.58-8.3	70-206
Moga	1.96-5.36	-
Muktsar	-	83-940
Patiala	2.05-2.80	47-52
Sangrur	1.71-11.30	107-1180

Water Quality

About 75-80% of water for domestic requirement is fulfilled by groundwater (GW). Groundwater quality (GWQ) analysed by the state departments under various projects have shown that GWQ is good in northern and central zones, and is mostly saline in southern and south-western parts of the state. Isolated saline patches are also present in Mansa Moga, Sangrur and SAS Nagar districts. In general, there is an increasing trend of salinity in the direction from north-east to south-west (Chopra and Krishan, 2014; Krishan et al., 2013, 2014, 2017). CGWB (2013) reported (Table 9) presence of high levels of fluoride, iron, TDS (Krishan et al., 2015), chloride and nitrate in the unconfined aquifers in some blocks of Punjab. Out of 22 districts in Punjab, high level of fluoride content in GW were reported from 7 districts, viz. Firozepur, Fatehgarh Sahib, Mansa, Moga, Muktsar, Patiala and Sangrur. The maximum amount of fluoride (> 11 mg/l) was reported from the Sangrur district. The GW nitrate contamination was mostly reported from Bathinda, Faridkot, Firozpur, Ludhiana, Mansa, Muktsar, Patiala and Sangrur district. High contamination of nitrate and fluoride was mainly reported from Malwa region. In addition to fluoride and nitrate contamination, occurrence of heavy metals namely, chromium, uranium, cadmium, etc. was also reported from Malwa region. Recent studies had reported presence of heavy metal contaminants in the deeper aquifers. It is speculated that it might be due to induced flow carrying contaminants from shallow to deeper aquifer.

In brief, the state has a complex hydrogeology and the socio-economy of the state immensely depends on the groundwater conditions. Major factors that affect the groundwater conditions include increasing area under agriculture sector, growing cropping intensity, large scale groundwater extraction from multiple aquifers system, more water use under irrigation practices, excessive use of fertilisers and pesticides, contamination from industrial and sewage effluents, impact of structures like reservoirs, network of canal and its distributaries, etc. on GW.

The last one decade witnessed a new range of techniques (high resolution remote sensing and GIS analysis, isotopic techniques, automated field based systems etc.) in addition to the conventional technique for assessment of anthropogenic influence on groundwater system especially in the Bist-Doab region of Punjab. The following section provides an illustration of outcome acquired from these new techniques.

BIST-DOAB REGION: A CASE STUDY

The Bist-Doab region is between rivers Beas and Sutlej. This region is known as one of the most fertile area in the world in terms of agricultural productivity, and was the centre of the green revolution in India. Even today, it is recognized as one of the largest per capita producers of wheat in the world. A geomorphologic map of Bist-Doab region is shown in Fig. 8.

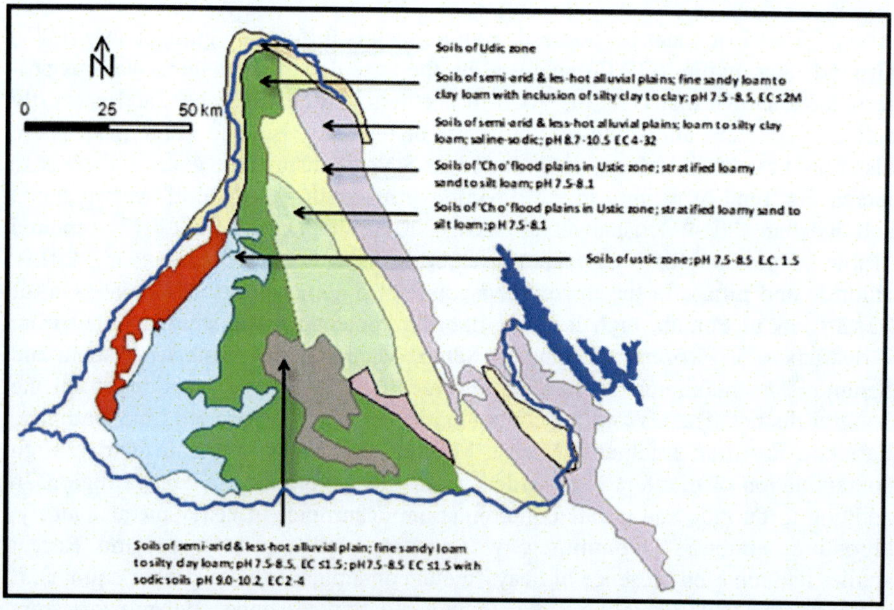

Figure 8. Geomorphological map of Bist-Doab tract, Punjab

Table 10. Aquifer groups in Bist-Doab region

Zones	*Texture*	*Thickness*
Aquifer Group I	Coarse sand beds separated locally by thin clay beds.	72-94 m
Clay Layer	A regionally extended confining clay layer that separates aquifer I from aquifer II.	It is 16 (at Kapurthala) to 32 m (at Hoshiarpur) thick.
Aquifer Group II	Alternating sequences of thin layers of sand, clay and gravel beds and, occasionally with kankar. The clay beds are thin and pinch out.	80-105 m thickness (81 m at Hoshiarpur district 85 m at Kapurthala district 87 m at Jalandhar district 105 m at SBS Nagar district.)
	Aquifer Group II is followed by a regionally confining layer upto the depth of 250 m bgl.	

The region is bounded by Shivalik hills in the north-east, the river Beas in the north-east and west, and by river Sutlej in the south (Fig. 8). Doab has the highest road density than all other areas in Punjab. The area is drained by perennial rivers, Sutlej and Beas and their tributaries. The two rivers join at village Harike and thereafter, continue as Sutlej river. The Bist-Doab region comprises districts Hoshiarpur, Kapurthala, Jalandhar and SBS Nagar of Punjab. Hydrogeologically, the region is characterised by two aquifer groups I and II extended upto a depth of 250 m bgl (Table 10). The two aquifer groups I and II are separated by regionally extended thick clay layer. Each aquifer group is composed of thick sand-gravel beds separated by small, thin clay beds that pinch out locally (Table 10). A fence diagram prepared by CGWB for the Bist-Doab area showing possible connectivity of the geological strata is shown in Fig. 9. This fence diagram will help understand the extent of different aquifers and their geometry and also recognises that Bist-Doab region has muti-aquifer systems with intermittent clay layers. No pumping test data were available for determining the aquifer parameters exclusively for these groups. However, on the basis of pumping tests data of wells tapping multiple aquifer groups the storage coefficient S values were estimated that ranges from 2.5×10^{-3} to 7.1×10^{-3} with an average of 3.85×10^{-3}.

Land-Use Changes (LUCs)

High resolution satellite data of 1975-2015 used for investigating the impact of social development on land-use changes (Table 11). showed that built-up area grew from 1.36% (1975) to 16.32% (2015), mainly due to conversion of 10.8% agricultural land into built-up land. The river area has increased from 2.33% (in 1975) to 2.52 % (in 2015) due to inclusion of additional 0.33% barren area and 0.23% agricultural area into the flood zone. The agricultural area increased from 58.87% (1975) to 64.73% (2015). This increase resulted from the change of 3% of shrub and 1.5% of sandy area into agricultural land. Reservoir area which was 0% in 1975

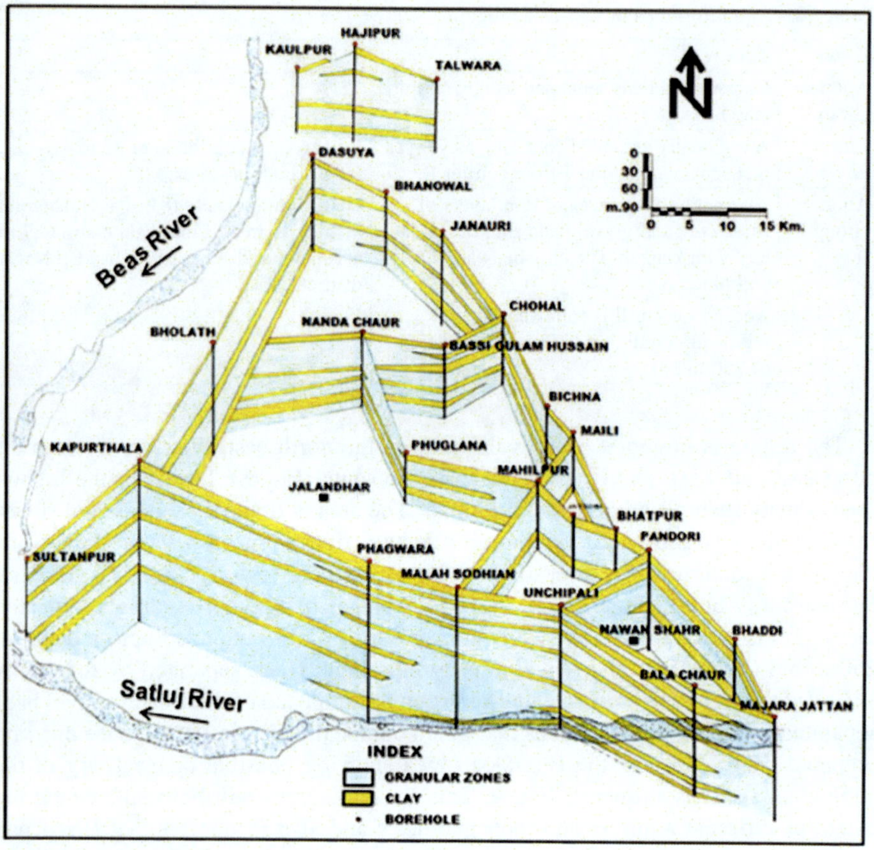

Figure 9. Bist-Doab fence diagram

increased to 0.22% because of 8 minor dams (Chohal, Dhamsal, Dholbaha, Janauri, Mailli, Patiari, Saleran and Thana), which came up during this period.

Groundwater Fluctuation

State and Central groundwater boards' historical data on groundwater levels involved measurement from bore-wells using rope and weight method on quarterly intervals in a year. Such data cannot be used to investigate diurnal changes, impact of high groundwater withdrawal on neighbouring pumps, impact of flood wave in the region, etc. In recent years, these manual methods of data monitoring are getting replaced by automatic water level recorders which can record water level by piezometers at desired intervals and for several months without being physically

Table 11. Percent Land-use Land cover change (source Kaundal et al., 2018)

1975	*2015*									
	River	*Reservoir/ Pond*	*Dense scrub*	*Open Scrub*	*Barren area*	*Built-up area*	*Agricultural Area*	*Current fallow*	*Sandy area*	*Total*
River	1.61	0.01	0.33	0.05	0.04	0.01	0.09	0.15	0.04	2.33
Reservoir/ Pond	0	0	0	0	0	0	0	0	0	0
Dense scrub	0.05	0.16	5.27	1.40	0.16	0.32	0.15	0.14	0.05	7.70
Open Scrub	0.01	0.01	0.21	0.35	0.08	0.33	3.40	0.45	0.06	4.90
Barren area	0.33	0.01	0.30	1.37	0.09	0.46	0.17	0.20	0.17	3.10
Built-up area	0.00	0.00	0.00	0.01	0.00	1.33	0.01	0.01	0.00	1.36
Agricultural Area	0.23	0.00	0.16	0.29	0.44	10.81	45.58	0.27	1.09	58.87
Current fallow	0.06	0.00	0.06	0.47	0.05	0.19	0.10	0.11	0.55	1.59
Sandy area	0.23	0.03	0.01	0.66	0.12	2.87	1.52	0.24	0.76	20.15
Total	2.52	0.22	6.34	4.60	0.98	16.32	64.73	1.57	2.72	100

Figure 10. Piezometer (Pz) development in Bist-Doab: (A) Location points of Pz, (B) lowering of Pz pipes, (C) cleaning of Pz, (D) lowering data logger, (E) Measurement of recorded water level after 3 months and (F) Pz strata chart (of Tanda).

attended (Krishan and Rao, 2017). The data from these systems can also be transmitted using telemetry system at the data recording station.

In order to monitor the deep groundwater response to withdrawal and recharge continuously; under Purpose Driven Study in Hydrology Project Phase II, six deep piezometers (~150 m depth) were constructed (Fig. 10) and change in water level at those sites were recorded at hourly intervals. Figure 11 shows the daily drawdown and subsequent recoup as diurnal fall and rise in water levels. The water levels fall generally during day and recover by wee hours. In case of seasonal variations, water levels fall rapidly in pre-monsoon season and recharge through monsoon mostly during post-monsoon season when extraction is also less. Seasonal difference in water levels is shown in Fig. 11 (a) to (c).

Groundwater level fluctuation depending on the extraction and recovery rates in deep piezometers varies from 3 to 8.6 m and major 2-3 withdrawal events take place. Groundwater level started rising after 2 months of rainfall and continued for 3 more months (Table 12). This is also the time taken for transmitting hydraulic pressure from surface to aquifer.

Relationship between rainfall and groundwater depletion is not prominent in the Bist-Doab region as 42% is recharge from rainfall and 58% comes from other

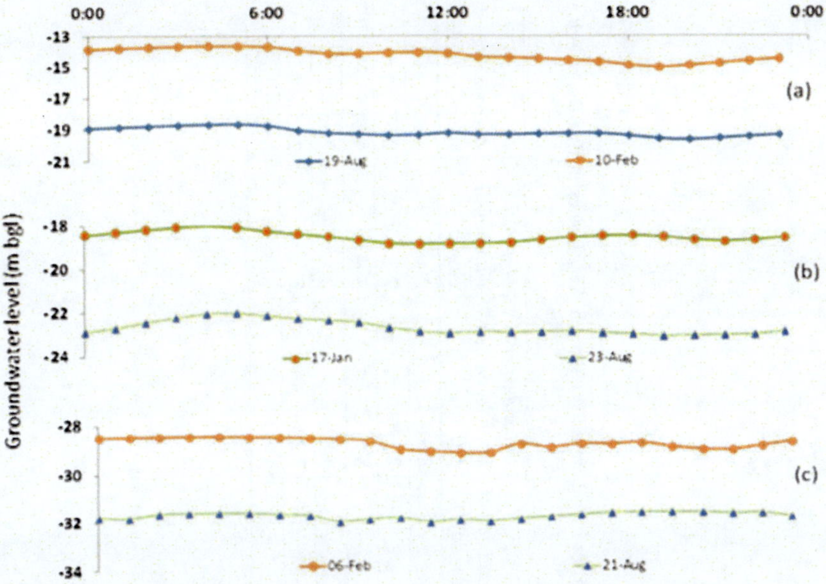

Figure 11. Trends of diurnal and seasonal groundwater levels in Bist Doab at sites: (a) Bhogpur, (b) Kapurthala and (c) Nakodar.

sources like leakage from reservoirs, tanks, ponds, water conservation structure, seepage from canal, irrigation return flow, etc. (Govt of Punjab, 2013).

Satellite based (GRACE data) GWL analysis

Groundwater depletion has been estimated by Rodell (2009) using satellite based GRACE data available from 2003. The GRACE derived results is also found to be well correlated with field based data (Fig. 12). The GRACE data based analyses showed the depletion in groundwater rate in northwest India (covering states of Punjab, Haryana, Delhi and Rajasthan) is about 20.4 Gt/yr averaged over the period 2003-2012 and 29.4 Gt/yr for 2003-2007 (Rodell, et al., 2009). Groundwater issues of Indo-gangetic basins were reported by researchers in detail (Bonsor et al., 2017; MacDonald et al., 2015, 2016)

ISOTOPE ANALYSIS

Isotopes have been used widely as tracers in hydrogeology for recharge estimation and source identification (Guay et al., 2006, Blasch and Bryson, 2007) and can also be used for studies of age depth profiles (Lapworth et al., 2014, 2015, 2016).

Table 12. Seasonal variation in groundwater level in response to monsoon and withdrawal for irrigation at the observed AWLR sites in Bist-Doab region during 2011–12. (Source: Rao, et al., 2017)

Site name	*Start of water level rise after monsoon onset (days)*	*Date of maximum water level in Feb.*	*Maximum water level and end of monsoon time difference (days)*	*Groundwater level fluctuation (m)*	*Major withdrawal events*	*Average rate of recovery (m/hr)*
Bhogpur	60	5th	142	8.01	3	4.16
Tanda	62	20th	162	2.74	2	0.57
Saroya	57	3rd	145	8.6	3	9.84
Kapurthala	59	3rd	141	4.89	2	2.35
SultanpurLodhi	57	20th	150	3.29	2	2.35
Nakodar	48	1st	143	3.9	1	0.85

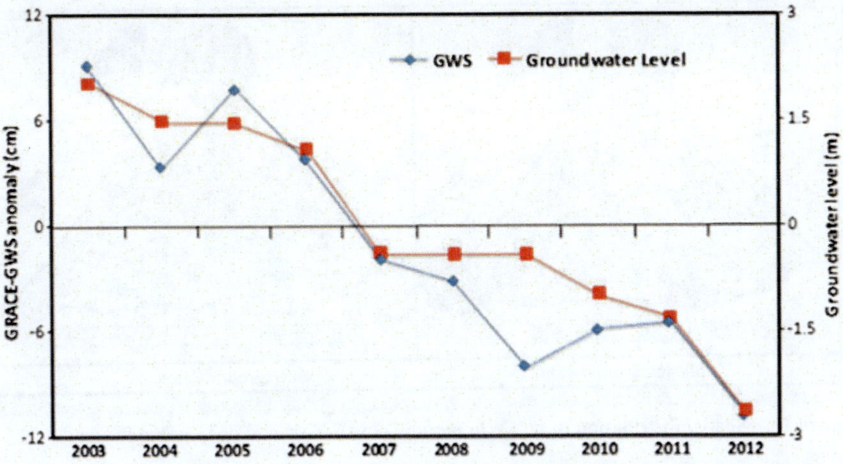

Figure 12. Comparison of anomalies in the GRACE-derived groundwater storage (GWS) and in situ groundwater level averaged from about 250 observation wells in Punjab based on data of 2003–2012.

Table 13. Equations of the characteristic isotopic lines and data range of source water

Water source	Characteristic isotopic equation	Observed data range	
		(dD, d^{18}O)min	*(d^{18}, dD)max*
Rainfall	$\delta D_R = 8.16\delta^{18}O+7.35$	(−12.9, −97.0)	(−0.02, +5.1)
Shallow groundwater	$\delta D_S = 7.26\delta^{18}O+1.06$	(−9.56, −73.07)	(−4.87, −32.06)
Deep groundwater	$\delta D_D = 6.92\delta^{18}O-0.84$	(−9.56, −73.1)	(−5.75, −36.0)
River Beas	$\delta D_{RB} = 6.31\ \delta^{18}O-0.81$	(−9.97, −65.27)	(−4.61, −34.07)
River Sutlej	$\delta D_{Rs} = 5.4\ \delta^{18}O-14.51$	(−11.9, −80.73)	(−8.1, −54.5)

The study conducted in the Bist-Doab region (Rao et al., 2017) demonstrated stable isotopic composition of deep and shallow groundwater, river water and rainwater and also dated the samples using environmental tritium analysis. Table 13 highlights the equation of the characteristic isotopic lines. Figures 13 and 14, respectively, present spatial distribution map of $\delta^{18}O$ and environmental tritium of groundwater. Figure 15 presents spatial distribution map of $MgHCO_3$, $CaHCO_3$ and $NaHCO_3$ in groundwater of Bist-Doab area.

The interpretation and description of isotopic composition are summarised as follows:

(i) Shallow groundwater showed $\delta^{18}O$ values between −6.8‰ and −10.5‰, whereas in deep aquifer values ranged between −6.8‰ and −8.4‰, that indicated recharge of shallow groundwater from that water source that contained depleted water (−8.4‰ to −10.5‰), such as the water that originates from canal and rivers (Sutlej, Beas etc.), and this type of water is less observable in the deeper aquifer.

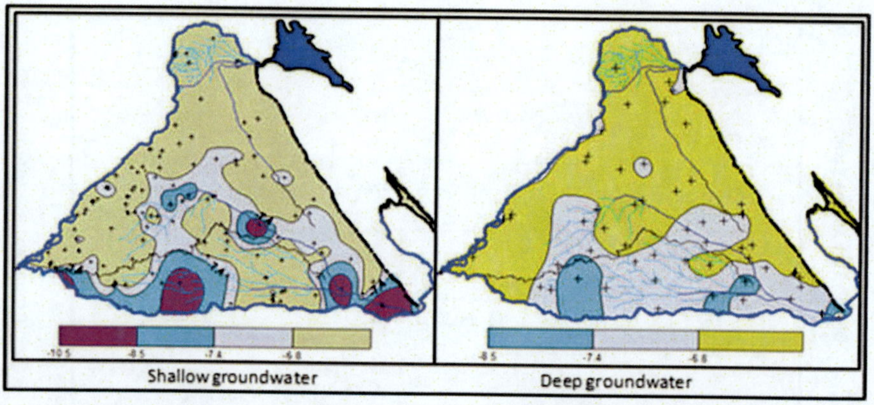

Figure 13. Spatial distribution of d[18]O (in per mill) in shallow and deep groundwater.

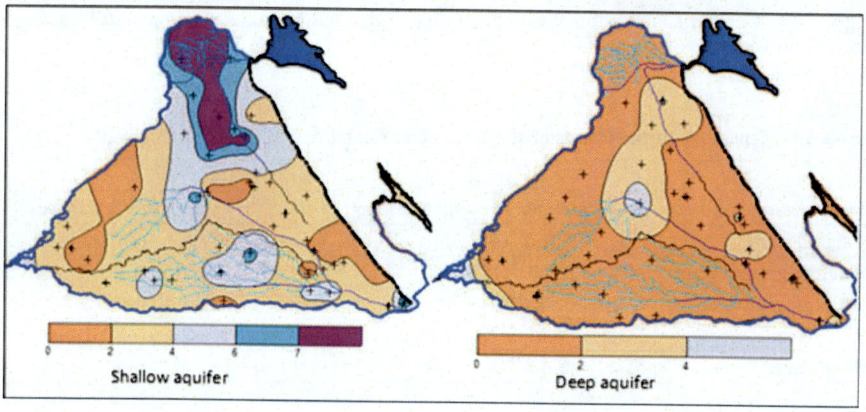

Figure 14. Spatial distribution of environmental tritium in shallow and deep groundwater.

(ii) Depleted groundwater ($\delta^{18}O < -6.8‰$) indicating recharge through canal to the shallow aquifer is very localised and this water is not moving down horizontal beyond a kilometre and therefore, not entering into the deeper aquifer.

(iii) Oxygen isotopic composition of groundwater (in shallow and deep aquifer) matches with the rainfall isotope value indicating rainfall is the main source of groundwater recharge, and recharge from other sources such as canals (originating from Sutlej and Beas rivers), river Beas and river Sutlej, is very small.

(iv) The isotopic mass balance has indicated that the shallow groundwater of the region is recharged through rain by 91%, through Sutlej canal by 8.6% and through the river Beas canal by 0.36% whereas, deep aquifer is recharged from canals' seepage by 5.1% (4.08% through Sutlej canal and 0.6% through Beas canal) and the remaining 94.9% is through rain.

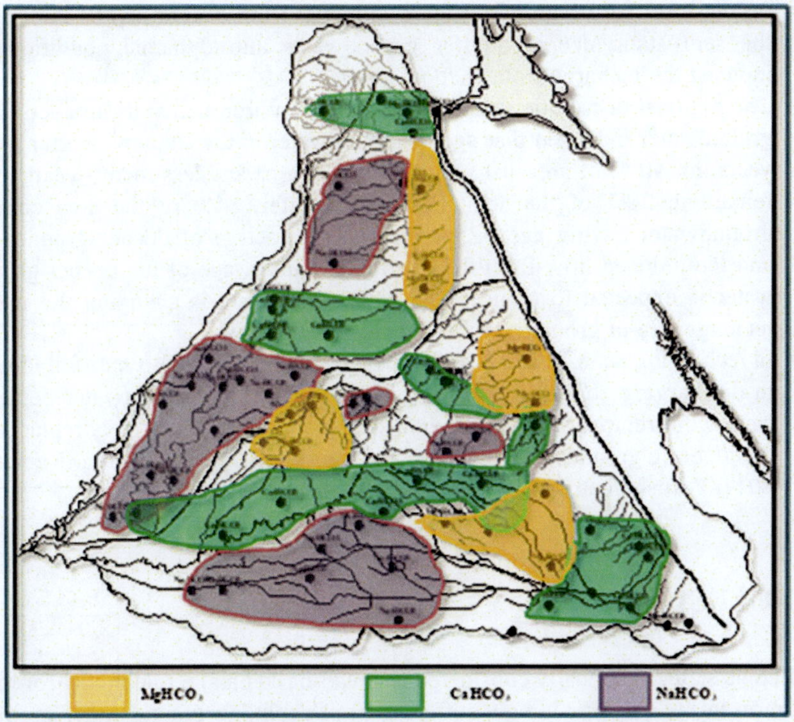

Figure 15. Spatial distribution of $MgHCO_3$, $CaHCO_3$ and $NaHCO_3$in groundwater.

(v) Groundwater is recharged mainly from rainfall in the foot-hills and higher altitude regions of Shiwalik hills. The recharged groundwater moves down to the plains and in the process the shallow aquifers and deeper aquifers are recharged by leakage. After a small amount of intermittent recharge from Sutlej canal, the composite groundwater is finally discharged into the rivers Sutlej and Beas as base-flow.

(vi) The seasonal variation in isotopic composition of shallow aquifer indicated that during rabi season (November-March), groundwater levels are deep and the canal discharge is high. Thus, the seepage from canal contributes the local shallow aquifer. During the monsoon period, canal discharge is low and land is filled-up with rain. During this season, recharge to groundwater occurring from rainfall makes the isotopic composition of groundwater similar to that of the rainfall.

(vii) Intense groundwater withdrawal over the past five decades has resulted into deepening of groundwater levels by several tens of meters. Therefore, the induced canal water isotopic signature is penetrating progressively to greater depths in the aquifer. With the progress of groundwater withdrawal from deeper depths and its replenishment through fresh (young age) water (including canal irrigated water), the average age of groundwater in the region is decreasing, that is, groundwater withdrawal from deeper depths is bringing the semi-static deeper aquifer groundwater into dynamic condition and induced the recharge from shallow aquifer.

(viii) The regional distribution and assessment of environmental tritium activity in groundwater indicated that age of groundwater in the shallow aquifer is ~16 years in ~40 % of area, 10 years in 20% of area and less than 6 years in the remaining 20% of the area. For deep aquifer, 84% of the area contains groundwater having age >25 years. The extraction of deeper groundwater and the induced flow from the overlying aquifer, age of the deeper groundwater is expected to reduce. The over-exploitation is changing the natural recharge rate of groundwater and its residential time.

(ix) In recharging zone, groundwater of $MgHCO_3$ and $CaHCO_3$ types found while in discharging zone it is $NaHCO_3$ type. Progress of groundwater depletion causes the removal of old mineralised groundwater which gets replaced by fresh/young groundwater of $MgHCO_3$ and $CaHCO_3$ composition (Rao et al., 2017; Purushothaman et al., 2013).

CONCLUSIONS

The Punjab state, historically considered as bread-basket and a role model for green revolution in the country, is now suffering from declining ground water problem, groundwater salinity and aquifer contamination. The Punjab basin is rich in water resources due to rich aquifers (thickness about 4.5 km in the northeast to about 400 m in the southern parts) and the five major rivers. Since the time of green revolution, the state has put all its efforts in maximising the food grain productivity. Towards this, vast areas of the state has intensified the rice and wheat cropping system, started using farming machines, chemical fertilisers, pesticides, herbicides, intensified irrigation practices through constructing huge number of groundwater extraction units, canal networking, reservoir constructions, promoting farming by providing cheap credit, free electricity, conversion of barren lands into agricultural, etc. As a result, irrigated area and the agricultural productivity doubled since 1965. However, indiscriminate extraction of groundwater for cultivating water intensive crops even in areas where groundwater recharge (e.g. Doab region) is low, led to falling of groundwater at a rate exceeding 1 m/yr. In areas where canal network is dense (like southwest Punjab), the canal irrigation is getting overused. Less extraction of groundwater, excessive use of surface water and poor groundwater transmission characteristic led to waterlogging and salinity problem in these areas (like south and southwest Punjab). Use of chemical fertilisers, pesticides, burning of crop residues for quick clearance of farms, uncontrolled disposal of urban waste on open surface, into open water bodies and rivers, poorly managed landfills and open dumps started contaminating soil, decreased soil fertility and polluted groundwater. Monitoring and compilation of water level and water quality over the past few decades has shown concentrations of cadmium, lead, cyanide and chromium, exceeding the permissible limits in the shallow aquifers of ground water in south west Punjab. Number of dark blocks started increasing in the Bist-Doab and adjoining region (e.g. groundwater at Ludhiana is 33.5 m bgl), the blocks with waterlogging, salinity and water contamination zones started increasing in the south and southwest parts of Punjab (e.g. groundwater at Muktasar is 1.12 m). In south-western zone flooding was obvious due to flood irrigation practice and presence of sandy-clay and sandy-silt type of soils which are not allowing water to penetrate into deep under the earth surface. Technological advancements indirectly promoted the consequences by replacing ordinary pumps with high power centrifugal pumps, high speed of well construction, developing hybrid and salt tolerant crops.

In the water declining zone, the only comfortable areas, where the water level did not lower much, happened to be in the vicinity of rivers or canals. However, in the waterlogging zone water table are rising due to the negligible draft of groundwater. Presently, total area under water logging conditions is 200,000 hectare on one hand and total area under salinity is 1,000,000 hectare on the other.

In recent days, new range of technologies have come up and these provide immense information on the changing groundwater resources and the factors influencing it. Some such technique includes remote sensing and GIS, satellite based system (e.g. GRACE) to estimate change in groundwater reserve at regional scale, isotope techniques, noble gas analysis, geophysical imaging and airborne survey systems, automated groundwater level (with other parameters like temperature, conductivity etc.) recorders with real time data acquisition system (RTDAS) etc. These systems find vide scope in groundwater monitoring and management. Remote sensing images in the Bist-Doab clearly indicated human influence on water and environment through change in landuse and land cover. GRACE data has reconfirmed the drying groundwater resource in Punjab during 2003-12. Isotopic data has provided vital information on groundwater dynamics, induced flow of groundwater from shallow to deeper aquifers and surface water groundwater interaction zones in the Bist-Doab region. Automated water level recorders installed at 6 sites in Bist-Doab clearly indicated groundwater level fluctuation from hourly scale to seasonal scale occurring from combined effect of monsoon recharge, irrigation withdrawal and daily water withdrawals. Thus, these techniques can be effectively utilised in understanding the hydrological processes from micro to regional scale and management methods for augmentation of groundwater reserve.

To protect and conserve groundwater resources, the state government has also initiated several programmes in recent years. To prevent waterlogging and groundwater salinity issues the state has developed programmes on canal linings, improving the surface drainage system, construction of tertiary canals and water courses. To protect the falling groundwater resource, the state has signed MoU with Israeli institutions to prepare master plan for projections of the water resources, alternate water supply schemes and economic analysis of water use. The government has also taken up comprehensive action plan to check pollution levels in the Beas and Sutlej rivers as well as to restore the water quality of state rivers. *To control the soil and water contamination issues the state has implemented* State Solid Waste Management Policy *from 2018.* The state has also notified action plan for reuse of treated waste water for various urban activities (Notification No.5/29/2017-1lg4/2717) and already implemented Punjab Preservation of Subsoil Water Act, 2009, that encourages paddy planting in consonance with onset of monsoon to avoid extraction of ground water if planted earlier.

The state has installed automated water level recorders in the state installed piezometers for continuous monitoring of water level in shallow and deep aquifers. Under the ongoing National Hydrology Project, the state is going to install 5 automated weather stations with real time data acquisition system. Punjab has five major perennial rivers of which Sutlej and Beas passes from area adjoining to the severely depleted groundwater zone. The state needs to develop *interlinking of these rivers and canal system in the depleted groundwater zone to reduce pressure on the groundwater resource.*

References

Blasch, K.W. and Bryson, J.R. (2007). Distinguishing sources of groundwater recharge by using δH and δ18O. *Groundwater*, 45: 294-308.

Baweja, A., Aggarwal, R. and Brar, M. (2017). Groundwater depletion in Punjab, India. Encyclopedia of Soil Science. Third Edition: Three Volume. https://doi.org/10.1081/E-ESS3-120052901.

Bonsor, H.C., MacDonald, A.M., Ahmed, K.M., Burgess, W.G., Basharat, M., Calow, R.C., Dixit, A., Foster, S.S.D., Gopal, K., Lapworth, D., Lark, R.M., Moench, M., Mukherjee, A., Rao, M. S., Shamsudduha, M., Smith, L., Taylor, R., Tucker, J., van Steenbergen, F. and Yadav, S.K. 2017. Hydrogeological typologies of the Indo-Gangetic basin alluvial aquifer, South Asia. *Hydrogeology Journal*, 25(5): 1377-1406. DOI: https://doi.org/10.1007/s10040-017-1550-z

CGWB (2004). Impact Assessment of Artificial Recharge to Ground Water Utilizing Runoff Generated in Patiala Nadi, Block Patiala, and Punjab. New Delhi: Central Ground Water Board.

CGWB (2013). Master plan for artificial recharge to groundwater in India. Central Groundwater Board. Ministry of Water Resources, Government of India.

CGWB (2014). Dynamic groundwater resources of India (As on 31st March 2011), Central Groundwater Board. Ministry of Water Resource, River Development & Ganga Rejuvenation, Government of India.

Chopra, R.P.S. and Krishan, G. (2014). Analysis of Aquifer Characteristics and Groundwater Quality in Southwest Punjab, India. *Journal of Earth Science and Engineering*, 4: 597-604.

Guay, B.C., Eastoe, R. and Long, A. (2006). Sources of surface and ground water adjoining the Lower Colorado River inferred by δ18O, δD and 3H. *Hydrogeol. J.*, 14: 146-158.

Kaler, Sunita (2016). Administrative structure of Punjab. *Advances in Social Research*, 2: 59-62.

Kaundal, A., Brar, K.K. and Kahlon, S. (2018). Land Use and Land Cover Change Detection through Remote Sensing and GIS: Bist Doab, Punjab (India). *Asian Resonance*, 7(4): 67-78.

Krishan, G. and Rao, M.S. (2017). Variability in groundwater level. *Geography and You*, 101: 23-28.

Krishan, G., Rao, M.S., Kumar, C.P., Kumar, S., Loyal, R.S., Gill, G.S. and Semwal, P. (2017). Assessment of Salinity and Fluoride in Groundwater of Semi-Arid Region of Punjab, India. *Curr. World Environ.*, 12(1): 34-41.

Krishan Gopal, Singh, R.P., Rao, M.S., Gupta, Sushil and Tiwari, P.K. (2015). Fluoride, Iron and Nitrate affected areas of Punjab. Suresh Gyan Vihar, *International Journal of Water and Research*, 1(1): http://www.gyanvihar.org/researchjournals/c3w_vol1_2.pdf

Krishan, Gopal, Rao, M.S., Kumar, C.P., Garg, Pankaj and Semwal, Prabhat (2014). Assessment of salinity and groundwater quality with special emphasis to fluoride in a semi-arid region of India. *Journal of Earth Science and Climate Change*, 5(6): 149. https://doi.org/10.4172/2157-7617.S1.016

Krishan, Gopal, Rao, M.S., Kumar, C.P., Semwal, Prabhat, G.S. (2013). Identifying Salinization Using Isotopes and Ionchemistry in Semi-Arid Region of Punjab, India. *Journal of Geology and Geosciences*, 2: 129, https://doi.org/10.4172/jgg.1000129

Lapworth, D.J., Krishan, G., Rao, M.S. and MacDonald, A. (2014). Intensive groundwater exploitation in the Punjab – An evaluation of resource and quality trends. Technical Report. NERC Open Research Archive, BGS-UK. Project report Groundwater Science programme, BGS, UK OR-14-068, 34 p.

Lapworth, D.J., MacDonald, A.M., Krishan, G., Rao, M.S., Gooddy, D.C. and Darling, W.G. (2015). Groundwater recharge and age-depth profiles of intensively exploited groundwater resources in northwest India. *Geophys. Res. Lett.*, 42, doi: https://doi.org/10.1002/2015GL065798.

Lapworth, Dan, Macdonald, A.M., Taylor, R.G., Burgess, Willy, Kazi Ahmed, Krishan, Gopal, Mukherjee, Abhijit and Zahid, Anwar. (2016). Hydrogeological responses to intense groundwater pumping in the Indo-Gangetic Basin – Evidence from environmental tracers. IAH International Congress "Groundwater and Society" Montpellier, France, 25-29 September 2016, 175 p.

MacDonald, A.M., Bonsor, H.C., Taylor, R., Shamsudduha, M., Burgess, W.G., Ahmed, K.M., Mukherjee, A., Zahid, A., Lapworth, D., Gopal, K., Rao, M.S., Moench, M., Bricker, S.H., Yadav, S.K., Satyal, Y., Smith, L., Dixit, A., Bell, R., van Steenburgen, F., Basharat, M., Gohar, M.S., Tucker, J. and Maurice, L. (2015). Groundwater resources in the Indo-Gangetic basin - Resilience to climate change and abstraction. *British Geological Survey Open Report*, OR/15/047, 63 p.

MacDonald, Alan, Bonsor, Helen, Ahmed, Kazi, Burgess, William, Basharat, Muhammad, Calow, Roger, Dixit, Ajaya, Foster, Stephen, Krishan, Gopal, Lapworth, Daniel, Lark, Murray, Moench, Marcus, Mukherjee, Abhijit, Rao, M.S., Shamsudduha, Mohammad, Smith, Linda, Taylor, Richard, Tucker, Josephine, Steenbergen, Frank van and Yadav, Shobha (2016). Groundwater depletion and quality in the Indo-Gangetic Basin mapped from in situ observations. *Nature Geosciences*, 9: 762-766.

NAPCC (2011). Final Report Appendix 2: Lower Sutlej Sub Basin. New Delhi: National Action Plan on Climate Change, Government of India.

Purushothaman, P., Rao, M.S., Rawat, Y.S., Kumar, C.P., Krishan, Gopal and Parveen, T. (2013). Evaluation of hydrogeochemistry and water quality in Bist-Doab region, Punjab, India. *Environmental Earth Sciences*, 72: 693-706.

Rao, M.S., Krishan, Gopal, Kumar, C.P., Purushotmanan, P. and Kumar, Sudhir (2017). Observing changes in groundwater resource using hydro-chemical and isotopic parameters: A case study from Bist Doab, Punjab. *Environmental Earth Science,* 76: 175. DOI: https://doi.org/10.1007/s12665-017-6492-1

Rodell, M., Velicogna, I. and Famiglietti, J.S. (2009). Satellite-based estimates of groundwater depletion in India. *Nature*, 460: 999-1002.

Sharma et al. (2017). A study on farm women awareness for climate variability and its effect on water resources in Punjab. *British Journal of Applied Science & Technology*, 21(5): 1-9.

Chapter 12
Environmental Flow Impacts on Water Quality of Peninsular River System: Tunga-Bhadra River, India

M. Rajesh, S. Rehana, and C. T. Dhanya

INTRODUCTION

Environmental flows play a major role in terms of quantity and quality for sustainable riverine ecosystems (Brisbane Declaration, 2007). Environmental Flows (EFs) have a variety of impacts in different regions of the world, including fisheries and other aquatic life, assimilative capacity, drinking water security, agriculture, transportation, navigation, industry, flood protection, recreation and tourism, and other cultural aspects (Iyer, 2005). The EFs are a measure of the amount and quality of water flowing in a freshwater river or stream over time. Estimation of EFs should be able to consider hydrologic, hydraulic, habitat and biodiversity, water quality, socioeconomic and cultural aspects with consideration of water regulation policies (Tennant, 1976). Estimation of EFs is generally practiced with consideration of one or multiple factors. Among these, the hydrological aspect with consideration of historical natural flow data is the most common and practiced by several river water management stakeholders (Zeiringer et al., 2018). Environmental flows based on hydrological criteria were conventionally estimated using flow indices based on a selected threshold level, which may be different for various river systems (Sharma and Dutta, 2020).

There have been many developments in calculating EFs based on hydraulics, hydrology, habitat based on the data availability (Chen and Wu, 2019; Jain and Kumar, 2014; Kumar and Kv, 2019; Zhao et al., 2020). Most of the studies considered low flow indices based on specified thresholds of discharge to identify the low flow regimes in the river systems. A low flow can be a measure of the flow in

M. Rajesh · S. Rehana (✉)
Lab for Spatial Informatics, International Institute of Information Technology, Hyderabad, India
e-mail: rehana.s@iiit.ac.in

C. T. Dhanya
Department of Civil Engineering, Indian Institute of Technology Delhi, Hauz Khas, Delhi, India

A. Mukherjee (ed.), *Riverine Systems*, https://doi.org/10.1007/978-3-030-87067-6_12

a stream during dry weather conditions. Low flow can be considered as a seasonal phenomenon of a river system Smakhtin (2001). One of the most widespread informative tools in the characterization of riverine systems low flows based on the discharge is the flow duration curve (FDC) which is very popular in illustrating the low flow response for any river system. The most conservative threshold, which is considered in the low-flow hydrology is Q50. This is also considered as one of the criteria to define streamflow drought analysis to represent the extreme low flow values (Fleig et al., 2006). The stream flows within the range of 70-99% probability of exceedance are mostly used as design low flows (Smakhtin, 2001). In this context, several studies made efforts for the assessment of low flow events based on the data availability and level of accuracy (Jha et al., 2008; Tharme, 2003; Zhou et al., 2017).

Another phenomenon which defines the deficit of water in rivers is hydrological drought which is observed to commence after a meteorological drought (Geng and Shen, 1992). A low-flow period is the annual cycle of streamflow, which can occur once or twice a year depending on the climatic conditions, unlike the hydrological droughts that are generally associated with low flow concepts, where a single hydrological drought can have multiple low flow events (Zelenhasić and Salvai, 1987). The most conventional way to represent hydrological drought based on the indices to capture the occurrence of water availability below average (Van Lanen et al., 2004). Given that, hydrological drought indices also try to capture the water availability anomalies that can be a better measure to identify the extreme low flow events. Low flows indices and hydrological drought indices are always closely related. For example, lowest annual flow of various durations such as 1, 7, 15 and 30 days is one of the ways to interpret the hydrological drought. Low flows can result due to hydrological drought phenomenon which is conventionally related with meteorological drought which originates based on the climatological anomalies (Monish and Rehana, 2019). The hydrological droughts and low flow indices can be valuable to study the occurrence of low flow events and to define low flow durations which are critical for river water quality as most of the indicators gets affected due to reduced water volumes and consequent impact on the riverine aquatic system.

Increasing frequency of low flow events and hydrological droughts in the river water systems necessitates understanding the impacts on river water quality variables. Extreme low flow conditions were always a major concern towards changes in river water quality. For example, longer water residence time periods under low flows can result in higher river water temperatures affecting the self-purification capacity of the river systems. Higher river water temperature can reduce the dissolved oxygen (DO) concentrations and potential for algal blooms (Caruso, 2002). Extreme low flow events and corresponding deterioration of river water quality can have adverse impact on socio-economic conditions and riverine aquatic biodiversity system. The pollutants from various sources such as domestic, industrial, agricultural developments and human activities can introduce significant nutrients into the river systems and intensifies the eutrophication process affecting aquatic life. The organic degradable material from municipal and industrial effluents into the river systems may decrease in DO concentrations. This is evident during low-flow periods where pollutant transport and dilution capacity gets affected severely with a consequent decrease in the assimilative capacity of the river system.

Given the significance of understanding the river water quality under low flows, the present study emphasized on identification of low flow events and durations based on low flow indices based on threshold basis and hydrological anomaly basis. Overall, the study has considered the concept of FDCs and hydrological drought indices to characterize the low flow events and durations. The study has demonstrated to relate the river water quality impacts during the low flow event commencement. The main emphasis is to understand how different river water quality parameters such as temperature, DO, BOD, pH, nitrates, etc. can affect during a low flow event with respect to a wet reference period. Considering a peninsular river system, the present study tested hypothesis that (1) river water quality variables have significant impact under low flow conditions; and (2) the changes in river water quality concentrations were mainly driven due to reduced water volumes. The present study made efforts to asses the EFs with a focus on hydrological low flows with use of FDCs and hydrological drought index in the assessment of low flow events and impacts on river water quality.

DATA AND METHODOLOGY: CASE STUDY

Tunga-Bhadra is a perennial river and one of the major tributaries of Krishna river which is the fifth largest river system of India. Tunga-Bhadra river flows through Karnataka and Andhra Pradesh with an area of 71,417 sq km (Jain et al., 2007). Tunga, Bhadra, Kumudvati, Varada and Hagari (Vedavthy) are the contributing tributaries to the Tunga-Bhadra river. The major tributaries of Tunga and Bhadra originate at Gangamoola hills, western Ghats of Kuduremukh in Chikkamagaluru district, Karnataka, India. Tunga and Bhadra rivers are about 147 km and 178 km, respectively, and join at Koodli to form Tunga-Bhadra river and joins Krishna river at Sangamaleshwaram after flowing for about 382 km through Karnataka and Andhra Pradesh. The mean annual rainfall of Tunga-Bhadra river basin is about 884 mm (Jain et al., 2007). The Bhadra river stretch is highly polluted by major industries such as paper, pulp, rayon, steel and municipal effluents (CPCB, 2020). Major industries along the Tunga-Bhadra river stretch are Mysore Paper Mill, Visveshvarya Industrial Steel Limited and Harihar poly fibre.

The river location considered for the calculation of hydrological based environmental flows is Hosaritti station along the Tunga-Bhadra river (Fig. 1). The hydrological and water quality data was obtained from Advanced Centre for Integrated Water Resources Management (ACIWRM), Bengaluru, Karnataka, India. Daily discharge data at Hosaritti station for a period of 2005 to 2017 was considered in the analysis. Water quality data were obtained for one station, namely, downstream of Haralahalli bridge which is near Hosaritti discharge station for a period of 2005-2017 from ACIWRM, India (Table 1).

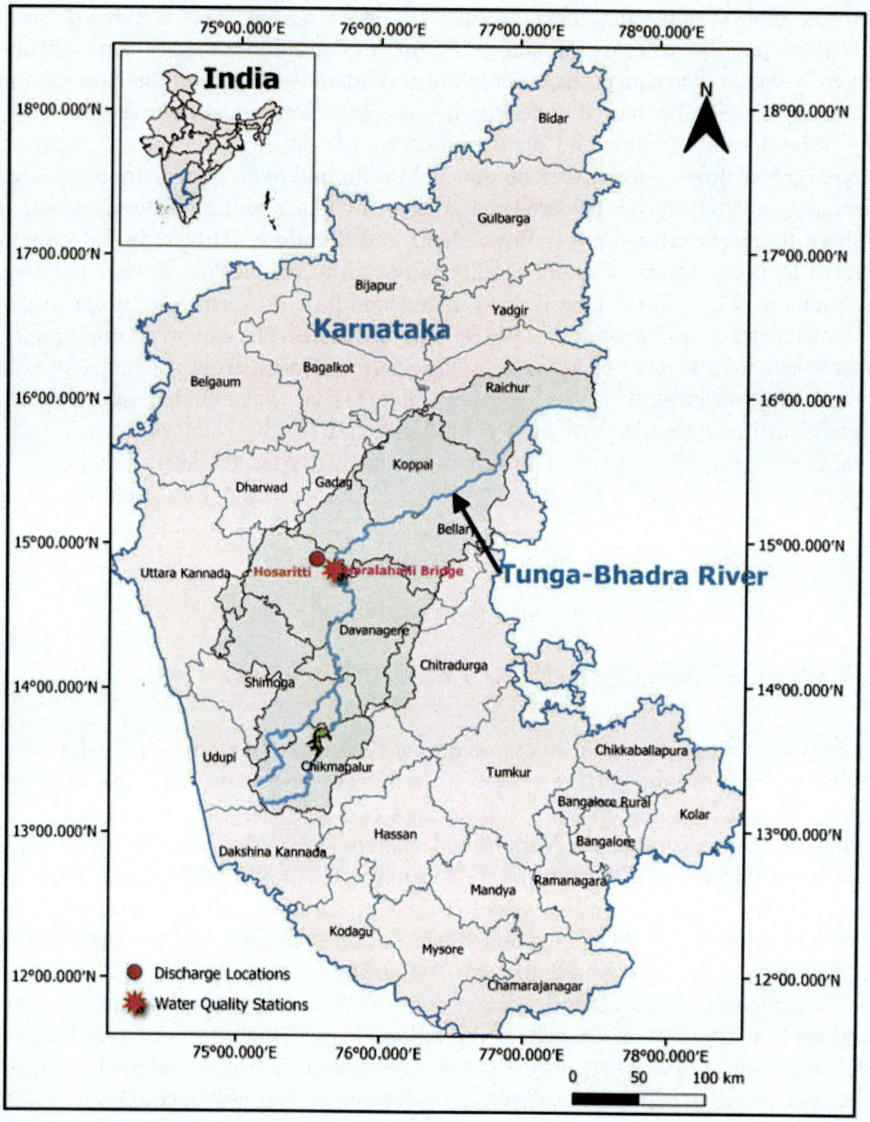

Figure 1. Tunga-Bhadra river basin along with the discharge location and water quality monitoring station, Karnataka, India.

Table 1. Details of selected monitoring stations, Tunga-Bhadra river, Karnataka, India.

Station Type	*Station Name*	*River*	*Time Period*
Discharge Station	Hosaritti	Tunga-Bhadra	2005 -2017
Water Quality Station	Haralahalli Bridge	Tunga-Bhadra	2005 -2017

The present study demonstrates how low flow events can affect the river water quality parameters and used conventional method to define the low flow events based on FDC threshold (Q75) and hydrological anomaly with respect to hydrological drought index, Standardized Runoff Index (SRI). The study identified one common low flow event based on FDC and SRI indices and river water quality was assessed under low flow conditions and compared various hydro-climatology, low flows and river water quality parameters with one low flow duration and reference to wet duration.

Threshold indices for low flows

Environmental flows were estimated using flow indices and flow duration curve (FDC) analysis, and it is a plot between the flow values and the percentage of time that flow is likely to be equal or exceeded.

The step-wise procedure to estimate the low flows using flow duration curves is as follows:

- The monthly stream flows are sorted and ranked from the largest values to the smallest values.
- Each flow is assigned a rank M starting with 1 for the largest value.
- The exceedance probability (*P*), where *P*(*Q* £ *qi*) is then calculated as:

$$P = \frac{M}{n+1} \tag{1}$$

where *P* is the probability that a given flow will be equalled to or exceeded, as a percentage of time and *n* the number of flow events.

The probability of exceeding a given flow can be estimated as:

$$P_i = P(Q > q_i) = 1 - P(Q \leq q_i) \tag{2}$$

- The flow duration curve is now plotted using flow on a logarithmic scale as the ordinate and the associated exceedance probability *P* as the abscissa.

Standardized Runoff Index (SRI)

The Standardized Runoff Index (SRI) is a widely used index to assess the hydrological drought events due to its simplicity and data availability, which identifies the hydrological anomalies over a river system (Bayissa et al., 2018; Shukla and Wood, 2008). The SRI is calculated by considering monthly streamflow value $Q_{i,j}$, where i is the year and j is months within the year. Monthly streamflow values and a

Table 2. Characterization of dry and wet conditions based on the Standardized Runoff Index (SRI) and with the cumulative probabilities (Tao et al., 2014; Wang et al., 2015)

Description	*Criterion*	*Cumulative Probability (%)*
Extremely wet	SRI $\geq$ 2.0	2.28
Moderately wet	1.5 $\leq$ SRI < 2.0	6.68
Slightly wet	1.0 $\leq$ SRI < 1.5	15.87
Near normal	$-1.0 \leq$ SRI < 1.0	50.00
Mild drought	$-1.5 \leq$ SRI <-1.0	84.13
Moderate drought	$-2.0 \leq$ SRI <-1.5	93.32
Extreme drought	SRI < -2.0	97.72

historical time series for the streamflow gauge were considered as input variable. More precise results can be obtained with a longer streamflow data available. Shukla and Wood (2008) mentioned that SRI incorporates hydrologic processes that determine seasonal lags in the influence of climate on streamflow.

- For estimating SRI index, the cumulative flow values were estimated individually for each month, then the SRI values for various time scales. The cumulative streamflow $V_{i,k}$, was calculated based on the equation

$$V_{i,k} = \sum\nolimits_{j=1}^{3k} Q_{i,j} \quad k = 1, 2, 3, 4 \tag{3}$$

Equation (3) gives $V_{i,k}$ values for 3-, 6- and 12-month periods, respectively. The SRI is described with cumulative streamflow volumes $V_{i,k}$ for each reference period k of the ith hydrological year as

$$SDI_{i,k} = \frac{V_{i,k} - \overline{V}_k}{S_k} \quad k = 1, 2, 3, 4 \tag{4}$$

where and S_k are the mean and the standard deviation of cumulative streamflow for a period k, respectively. The values of SRI greater than zero can be considered as wet conditions and the SRI values greater than zero are classified as dry conditions. The characterization of wet and dry conditions based on SRI values are given in Table 2.

Environmental Flows and Water Quality

Five river water quality parameters such as water temperature, pH, DO, BOD and nitrates based on the data availability were considered to study the effect of low flow events on the river water quality. Monthly values of water quality parameters were taken to study the significance between them during low flow and reference wet periods. Mean, minimum and maximum values of water quality indicators were computed, and the p-values obtained from the two-sample t-test was used to study the statistical difference in water quality indicators between a dry and wet period. The two-sample *t*-test is a statistical test used in the assessment of the importance of significant differences between groups, which may be related in certain features (Levy, 1967). A paired two-sample *t*-test to presume whether the difference between the sample means is statistically distinct from a hypothesized difference (Cressie and Whitford, 1986).

RESULTS AND DISCUSSION

Low Flow Analysis

The hydrology of river was studied by plotting the mean monthly precipitation and discharges for the Hosaritti station from 1st June 2005 to 31st May 2017. The annual average rainfall for the study location was about 1168 mm and streamflow about 440 m^3/s estimated for the period 2005 to 2017. Figure 2, shows that precipitation and discharge values for Hosaritti station were decreased from 2010 onwards. The Tunga-Bhadra river has experienced a decline in precipitation in recent years along with high fluctuations in the rain patterns based on Karnataka State Natural Disaster Monitoring Centre (KSNDMC). Karnataka state receives an annual rainfall of 1,135 mm in which the south-west monsoon accounts for about 73%. But, since 2011, the state has received an average rainfall of 1,033 mm, which is 10% less than the normal rainfall (The Times of India, 2019). The variations in the rainfall patterns have affected the streamflows severely. The annual streamflow at Hosaritti station has shown a decreasing trend of about 26.1 m^3/s per decade from 2005 to 2017 (Fig. 3). Such decrease of annual streamflows has been found for other river locations along the Tunga-Bhadra river such as Shimoga with about 3.1% and Honnalli with 12.26% based on the study of (Rehana and Mujumdar, 2011.)

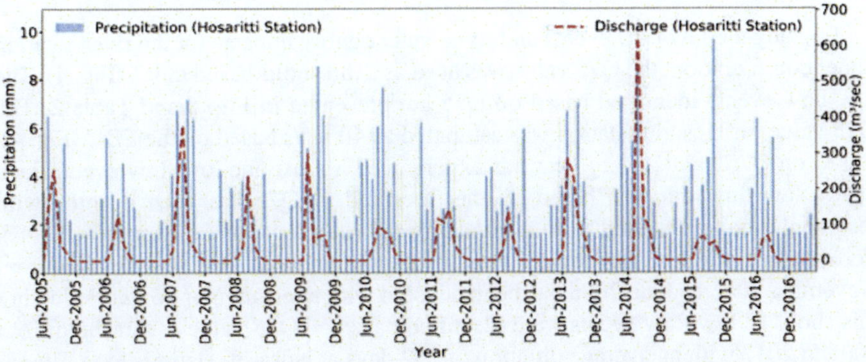

Figure 2. Monthly average precipitation and discharge of Tunga-Bhadra river at Hosaritti station, India (2005–2017).

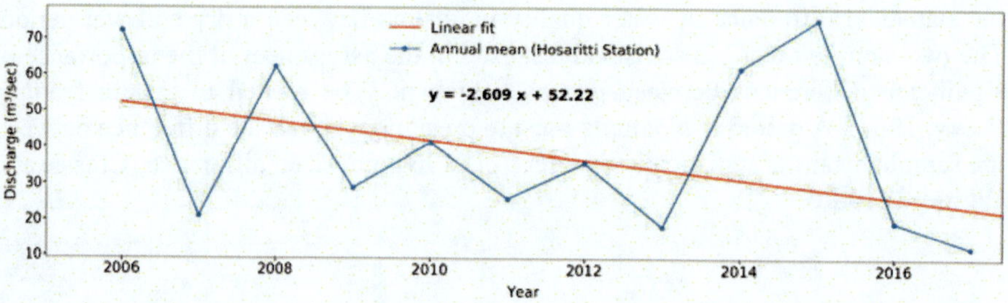

Figure 3. Annual discharge at Hosaritti station along with linear trend line from 2005 to 2017

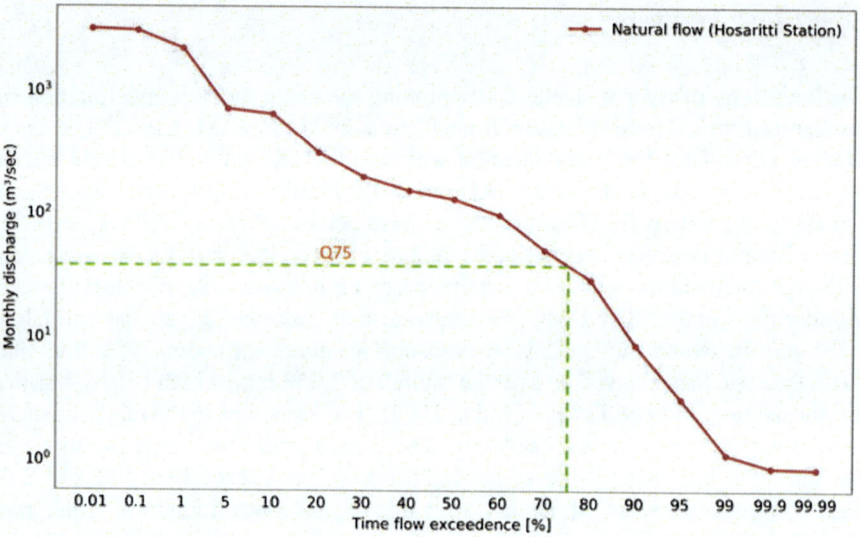

Figure 4. Mean monthly flow duration curve (flow on a logarithmic scale) of Tunga-Bhadra river at Hosaritti station, India (2005–2017).

To study the low flow events and river water quality impacts for the river location under consideration, the Q75 value was used as a threshold-based index (Fig. 4). The low flow events identified based on Q75 are presented in Fig. 5 and Table 2. The Q75 value for Hosaritti station was estimated as 40 m^3/s based on the FDC (Fig. 4). Based on Q75 value, every year has experienced almost one low flow event. Each year's low flow duration based on the threshold of Q75 has been denoted with rectangular boxes labelled from *a* to *l* along the river location of Hosaritti (Fig. 5). For example, the low flow event with a duration of about 240 days starting from November 2005 to June 2006 has been denoted with a rectangular box *a*. Following this about 12 low flow events were identified based on Q75 value as threshold from 2005 to 2017 with an average duration of 252 days at Hosaritti station along Tunga-Bhadra river.

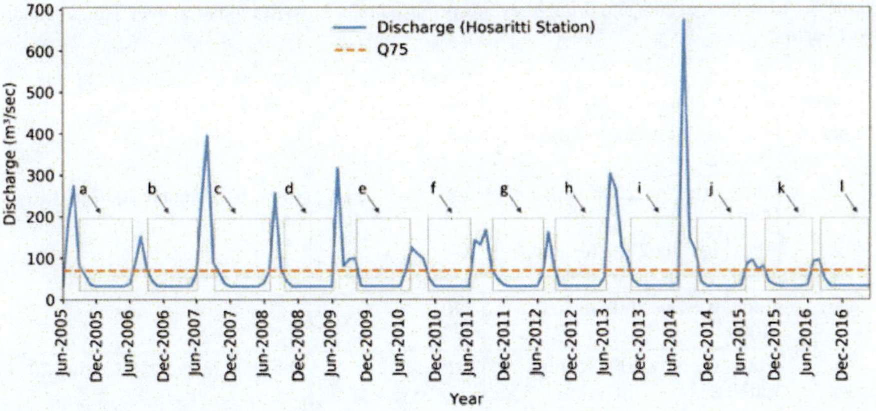

Figure 5. Low flow events and durations along Tunga-Bhadra river at Hosaritti station, India (June 2005-May 2017).

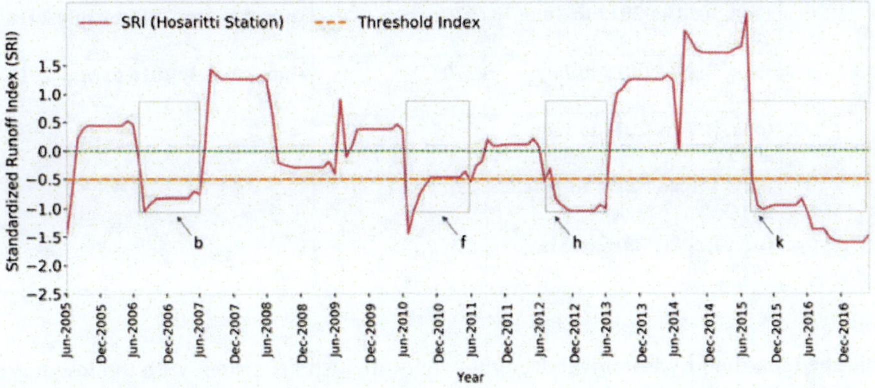

Figure 6. The SRI index on a 12-month scale at Hosaritti station, Tunga-Bhadra river, India, estimated (2005-2017).

The monthly values of SRI were calculated for 12-month time scale for a period ranging from 1st June 2005 to 31st May 2017, and the SRI time series was plotted for Hosaritti station as (Fig. 6). The intensity of drought was explained with reference to the low flow events estimated based on Q75 values (Fig. 5; Table 2). The study considered the extreme low flow events based on both threshold-based criteria of Q75 and hydrological anomalies-based criterion drought index of SRI. For this, the study considered –0.5 as SRI threshold to pick the hydrological low flow events which are also consistent with Q75 criterion. By considering the SRI threshold as –0.5 Hosaritti station had experienced drought between August 2006 to June 2007 (Fig. 6: box (b)), June 2010 to May 2011 (Fig. 6: box (f)), June 2012 to June 2013 (Fig. 6: box (h)) and August 2015 to May 2017 (Fig. 6: box (k)). It can be noted

Table 3. Low flow events and drought events occurred over Hosaritti location over Tunga-Bhadra river basin

Location	*Low flow events (Fig. 5)*	*Occurrence (days)*	*Hydrological Drought events (Fig. 6)*
Hosaritti	(box, a): Nov 2005 to June 2006	240	
	(box, b): Nov 2006 to June 2007	240	**(box, b): August 2006 to June 2007**
	(box, c): Oct 2007 to June 2008	270	
	(box, d): Oct 2008 to June 2009	270	
	(box, e): Nov 2009 to June 2010	240	**(box, f): June 2010 to May 2011**
	(box, f): Nov 2010 to June 2011	240	**(box, h): June 2012 to June 2013**
	(box, g): Nov 2011 to June 2012	240	
	(box, h): Oct 2012 to June 2013	270	**(box, k): Aug 2015 to July 2016**
	(box, i): Nov 2013 to July 2014	270	**(box, k): July 2016 to May 2017**
	(box, j): Nov 2014 to July 2015	270	
	(box, k): Nov 2015 to July 2016	270	
	(box, l): Nov 2016 to May 2017	210	

that the identified hydrological drought events are also consistent with the low flow events and durations identified based on Q75 values are shown in Fig. 5 and Table 2.

The study focused on one of such common low flow periods which was identified based on Q75 and SRI to analyze the river water quality impacts under low flows. For example, from Figs. 5 and 6, and Table 3, a recent and yet common low flow period from December 2015 to May 2016 (Fig. 5: box, (k)) was considered to study the river water quality impacts at Hosaritti station along Tunga-Bhadra river. A comparison conducted for various water quality parameters such as air and river water temperatures, pH, DO, BOD, and nitrates. The river water quality parameters during one of the extreme low flow period was investigated and compared with the reference period which is the preceding period (June 2013 to November 2013). The study compared low flow period with the wet reference period before the occurrence of a low flow period. The river water quality indicators under low flow period of December 2015 to May 2016 was compared with a wet reference period of June 2013 to November 2013.

Before understanding river water quality impacts under low flows, the basic climatology of the study area was compared for low flow period and referenced

Table 4. Annual total precipitation and maximum air temperature at discharge stations for low flow period, reference period, and averaged over the entire period 2005 - 2017

Discharge location	*Parameter*	*Low flows* *Dec 2015-May 2016*	*Reference* *June 2013-Nov 2013*	*Average* *2005-2017*
Hosaritti	Annual total precipitation (mm)	354	805	1168
	Air temperature (°C)	35.75	28.44	31.64

wet period. The climatology of a low flow period and reference period were compared by comparing the precipitation and temperatures for Hosaritti station. Annual total precipitation and maximum air temperatures were compared for low flow period with reference period (Table 4). Comparison of annual total precipitation, mean and maximum air temperatures for low flow period and for reference period demonstrates drier and warmer conditions during the low flow period. Annual total precipitation was substantially lower (354 mm) during the low flow period, while the maximum air temperature was incredibly higher (35.75 °C) when compared to the reference period (Table 4). The long-term average annual rainfall was noted as 1168 mm over the basin, whereas, average annual rainfall during low flow and reference periods were noted as 354 mm and 805 mm, respectively. Table 4, shows that the annual precipitation was low during low flow event duration, i.e., December 2015 to May 2016 compared to the preference period (June 2013 to November 2013) for Hosaritti station along Tunga-Bhadra river. While the annual rainfall is decreased, the maximum air temperatures were observed to be increased during low flow period for Hosaritti station. The low flows in current study are observed to be a combined effect of both lower precipitation and higher air temperatures.

The study compared the hydrological aspects for a low flow period with the reference period by comparing the discharge values. For this, the study compared the low flow hydro-climatology of Hosaritti station for low flow period and reference period. The lower precipitation values, higher temperatures and therefore consequent decrease in stream flows were observed under low flows compared to hydrological period over Tunga-Bhadra river basin.

River Water Quality Analysis

Impact of low flows on river water quality was studied by observing variation in the water quality parameters under low flow and wet reference periods. The variations within each water quality parameter were studied using the statistical test, such as a two-sample *t*-test (Table 5). The *t*-test was conducted on each water quality parameter for the low flow period and reference period (Table 5).

Table 5. Mean and significant values of surface water quality variables at Haralahalli bridge in low flow period (Dec 2015 to May 2016) and reference period (June 2013 to Nov 2013). Values indicate significant relationships with a low flow period ($P < 0.05$).

Water quality station	*Water quality variable*		*Low flow period*	*Reference period*	*P-Value*
Tungabhadra at Haralahalli bridge	Water Temperature	Mean	28.33	25.33	**0.02**
		Minimum	26	22	
		Maximum	31	27	
	pH	Mean	7.80	8.23	
		Minimum	7.6	7.7	**0.03**
		Maximum	8.1	8.7	
	DO	Mean	7.73	7.25	
		Minimum	7.2	6.6	**0.10**
		Maximum	8.4	7.8	
	BOD	Mean	2.97	2.33	
		Minimum	2.3	2.0	0.20
		Maximum	5.0	3.0	
	Nitrate	Mean	0.49	0.14	
		Minimum	0.27	0.10	**0.00**
		Maximum	0.59	0.20	

The river water temperatures were observed to differ significantly at the Haralahalli bridge station ($P < 0.05$) during low flow period. For example, average air (water) temperature for low flow period was noted as 35.75 (28.33) and 28.44 (25.33) °C for the low flow and hydrological periods, respectively. The increased water temperatures under low flows can be correlated with the increase in air temperature (Fig. 6) (Rehana, 2019). An increase in river water temperature can result in decrease in DO levels (Fig. 7), which may lead to anaerobic conditions affecting the aquatic life and self-purification capacity of the river system (Rehana et al., 2019). River water temperature is an important quality parameter affecting the physical, chemical and biological characteristic of a river system and aquatic life of riverine environment (Webb et al., 2003). Linear regression models relating to air and water temperatures become popular in the prediction of river water temperature (Neumann David W. et al., 2003; Rehana and Mujumdar, 2011). Air and water temperatures were observed (Fig. 8a) to follow a linear trend at Hosaritti station. River water temperatures were observed to increase with the increase in maximum air temperatures during the study period (Fig. 7). Research findings of the current study are in agreement with the results of Rehana and Mujumdar (2011) for the same case study at stations Shimoga (Tunga river) and Honnalli (Tunga-Bhadra river) in terms of linear dependency of air and water temperatures (Rehana and Dhanya, 2018).

But unlike direct relation between air and water temperatures, the discharge is observed to follow a quadratic relationship with the river water temperature (Fig. 8 (b)). The river water temperature decrease to a point and then increase at Hosaritti station. Such an inverse relationship between river water temperature and discharge

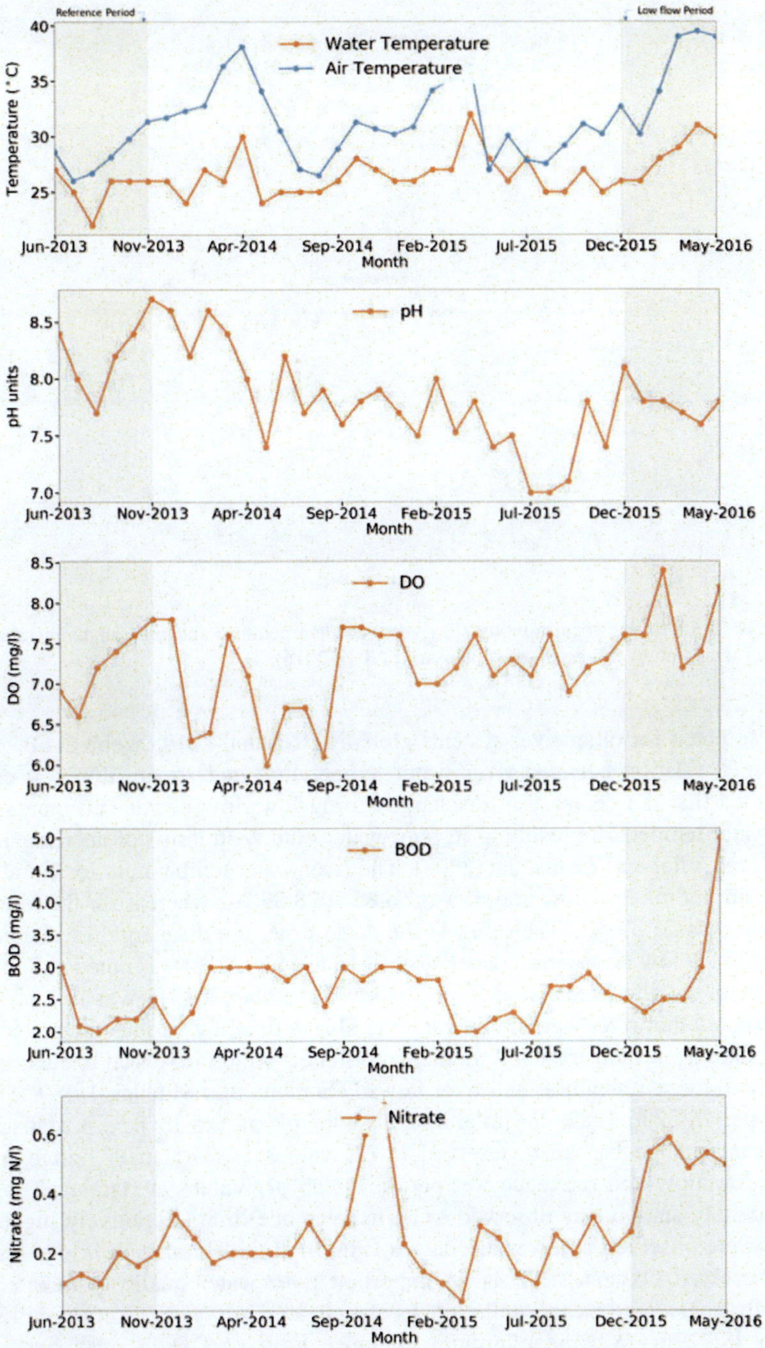

Figure 7. Water temperature, max air temperature, pH, DO, BOD and Nitrate (2013-2016)

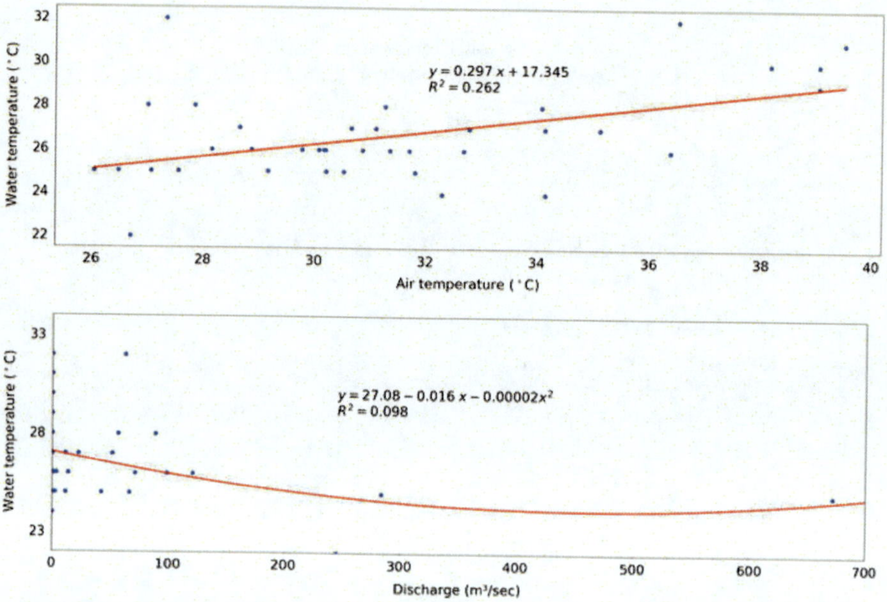

Figure 8. Relation between monthly (a) river water temperature and max air temperature and (b) water temperature and discharge at Hosaritti (2013-2016).

was also noted for other river systems globally (Rehana, 2019; Webb et al., 2003). With such relation between river water temperature and stream flows, it can be concluded that the decrease in discharge during low flow events can increase the river water temperature resulting in poor water quality in terms of decrease in DO levels (van Vliet and Zwolsman, 2008). The river water temperature for Haralahalli station during the low flow period was noted at 28.33 °C, whereas for the reference period it was 25.33 °C (Table 5). At the same time, the discharge for Haralahalli station for the low flow period (December 2015 to May 2016) was noted as 7 m^3/sec, whereas for the reference period (June 2013 to November 2013) it was 122 m^3/sec. It can be noted that river water temperature is also defined by various factors such as excess heat from industries and municipal effluents which may lead to such inconsistency in the temperatures and river flows (Thomann and Mueller, 1987).

As per IS:2296-1982, the tolerance limit of pH is 6.5 to 8.5. A statistically significant change has been observed in pH values at Haralahalli bridge station during low flow and reference wet period. Mean pH values at Haralahalli bridge water quality station was observed to be in tolerance limits. Relatively higher pH values were observed at lower discharges (Fig. 9) at the Haralahalli bridge station. The Dissolved Oxygen (DO) is an important river water quality parameter and generally considered as the pollution level indicator in water quality monitoring. As per IS:2296-1982, the minimum tolerance limit of DO in river systems is 4 mg/l. The DO values at Haralahalli Bridge station was observed to be above the tolerance limit (Fig. 7). For example, the DO level for Haralahalli bridge station in

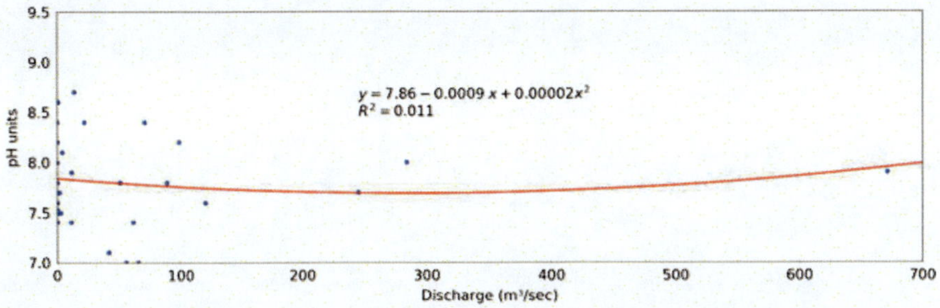

Figure 9. The relation between monthly discharge and pH at Haralahalli bridge (2013-2016).

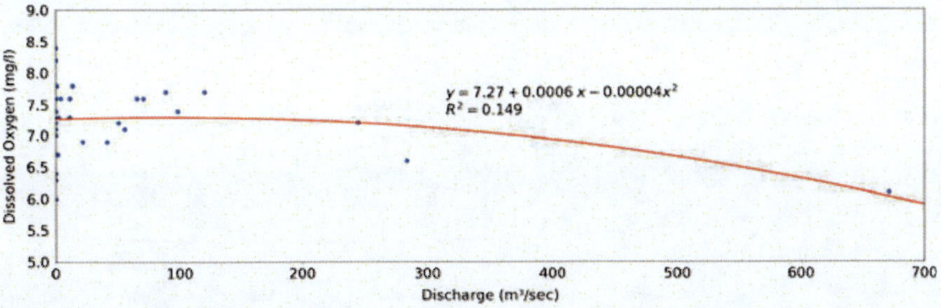

Figure 10. Relation between monthly discharge and dissolved oxygen at Haralahalli bridge (2013-2016).

the low flow period was noted as 7.73 mg/l, whereas, the DO levels for the reference period was noted as 7.25 mg/l. The higher DO values at Haralahalli bridge station during the lower discharges (Fig. 10) can be because of an increase in primary production in the river water (Zwolsman and van Bokhoven, 2007).

The next important river water quality indicator analyzed for the low flow period is BOD. As per IS:2296-1982, the maximum tolerance limit of BOD is 2 mg/l for tolerance limits for inland surface waters, class - A, with the use as a drinking water source without conventional treatment but after disinfection. The BOD values at Haralahalli station was observed to be more than the tolerance limit (Fig. 7). High BOD values indicate that the river is heavily polluted. The higher BOD value of about 2.97 mg/l was noted under low flow period, while for wet period the BOD value was noted as 2.33 mg/l (Table 5). Note that the high BOD values at the lower discharge values at the Haralahalli bridge station (Fig. 11). Higher BOD values during low flows may have severe impact on the riverine aquatic system (Sharma and Dutta, 2020).

As per IS:2296-1982, the maximum tolerance limit of nitrate is 20 mg/l for tolerance limits for inland surface waters, class - A, with the use as drinking water source without conventional treatment but after disinfection. Though the nitrate differs significantly ($P < 0.05$) at Haralahalli bridge, the values were observed to be below the desirable limit (Fig. 7; Table 5). For example, the average nitrate

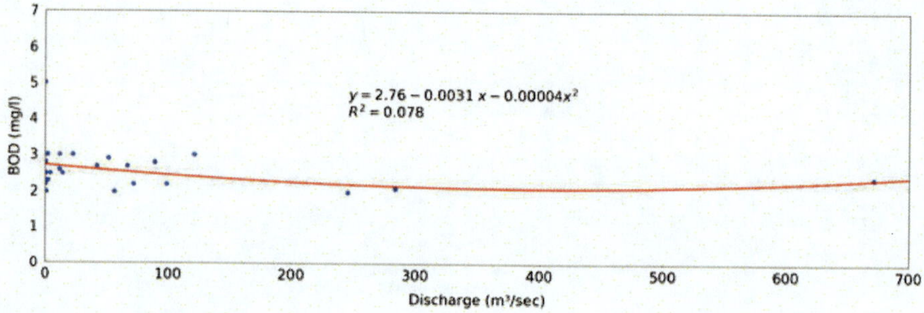

Figure 11. Relation between monthly discharge and BOD at Haralahalli bridge (2013-2016).

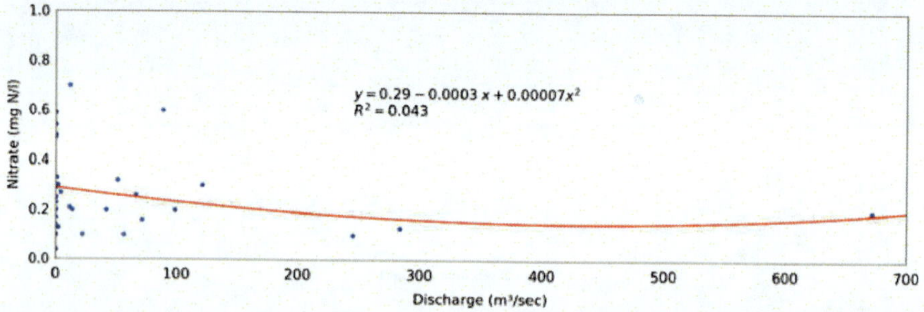

Figure 12. Relation between monthly discharge and nitrate at Haralahalli bridge (2013-2016).

concentration during the low flow and wet referenced durations was 0.49 mg/l and 0.14 mg/l, respectively, with higher maximum values during low flow periods. Although the nitrates concentrations are within the desirable limits, higher concentrations were expected during low flow periods compared to the reference periods (Fig. 12). Increase in the nitrate concentrations under low flow conditions can be due to the reducing dilution capacity of the water body (Prathumratana et al., 2008; van Vliet and Zwolsman, 2008; Zwolsman and van Bokhoven, 2007). Furthermore, the decrease in nitrate concentration can be expected due to the reduced agricultural runoff and drainage (Muchmore and Dziegielewski, 1983).

CONCLUSIONS AND FUTURE DIRECTIONS

Environmental flows are important in the context of river water quality management. Sufficient hydrological based environmental flows should be maintained along the river stretches to maintain the river water quality parameters within desirable limits. Flows are very much essential for maintaining the river regime in terms of its self-purification capacity, sustainable aquatic life and vegetation, well-being of livelihoods etc. River water systems have to be protected in terms of both quantity and quality. In India, the low flow management is challenging due to several aspects including water regulation policies and other socio-economic and cultural factors. In India, rivers have great religious significance during holy dips along the banks of rivers, necessitates maintaining minimum quantity and quality of flow (Jain and Kumar, 2014). Given the emerging importance of flows in the riverine systems, it is essential to develop an appropriate environmental flow estimation strategy for Indian context given the seasonal hydroclimatological variability. Further, a detailed understanding about the river water quality impacts under low flow events in the present study is of very much relevance for the water resources management under low flow events.

The present study made efforts to understand the river water quality responses under low flow event compared to referenced wet periods along a peninsular river system, Tunga-Bhadra river, India. The emphasis of the study was to assess the impacts of low flow events on the river water quality parameters such as water temperature, BOD, DO, pH, nitrates etc. of Tunga-Bhadra river, India. The research findings of the present study demonstrate that low flows significantly impacted river water quality of Tunga-Bhadra river system. The lower precipitation values, higher temperatures and therefore consequent decrease in stream flows were observed during a low flow compared to hydrologically wet period over the Tunga-Bhadra river. Results showed significant effect of low flow durations with respect to the river water temperature when compared to the reference wet period. The pH is observed to be nominal, while DO and BOD are offsetting desirable limit at Haralahalli bridge station, indicating deterioration of water quality due to industrial effluents. The nitrate concentration is observed to be in desirable limits. As per standards given by the CPCB (CPCB, 2020) for the designated best use of water, the river water quality at Haralahalli station is not meeting any of the criteria due to high amounts of BOD present. Higher concentrations of BOD and nitrates were observed during low flow events compared to reference hydrological period. Whereas, lower concentrations of pH and DO were noted under low flow conditions compared to reference wet period. The most affected river water quality variable during low flow events is river water temperature along Tunga-Bhadra river. Higher river water temperatures were noted for low flow duration period compared to reference wet period, which can directly impact DO levels. The river water quality concentration responses of major water quality parameters are mainly determined by its behaviour with the fluctuations in the discharge.

This study demonstrates how threshold based low flow index of Q75 using FDC and hydrological drought index of SRI could be scaled-up and used for assessing and identification of low flow events. The study considers extreme low flow events based on both threshold-based criteria of Q75 and hydrological anomalies-based criterion drought index of SRI.

One of the major limitations is the non-availability of discharge data for the possible implementation and understanding of the river low flow analysis and its response towards river water quality. Specifically, estimating low flows for ungauged catchments is a challenging factor for water resources management and policy making. Due to limitation of the data availability, the present study demonstrates the environmental flow impacts on river water quality at low flow events and duration with one river location, which can be extended to other stations based on data availability. Since, environmental flow is a multifaceted phenomenon, therefore, coupling of other flow indices may generate enhanced results for analysts and policy makers. To conclude, this study gives an overview of hydrological low flow events and its potential impacts on water quality parameters.

ACKNOWLEDGEMENTS The authors sincerely thank the Advanced Centre for Integrated Water Resources Management (ACIWRM), Bengaluru, Karnataka, India, for providing Tunga-Bhadra river data. The research work presented in this study is funded by Ministry of Science & Technology, Department of Science & Technology, TMD (Energy, Water & Others), Water Technology Initiative (WTI), Project no. DST/TMD-EWO/WTI/2K19/EWFH/2019/306.

References

Bayissa, Y., Maskey, S., Tadesse, T., Van Andel, S.J., Moges, S., Van Griensven, A. and Solomatine, D. (2018). Comparison of the Performance of Six Drought Indices in Characterizing Historical Drought for the Upper Blue Nile Basin, *Ethiopia. Geosciences,* 8: 81. https://doi.org/10.3390/geosciences8030081

Brisbane Declaration (2007). The Brisbane Declaration: Environmental flows are essential for freshwater ecosystem health and human wellbeing. *In:* 10th International River Symposium and International Environmental Flows Conference, Http://Www.Eflownet.Org/Download_documents/Brisbanedeclaration-English.Pdf.

Caruso, B.S. (2002). Temporal and spatial patterns of extreme low flows and effects on stream ecosystems in Otago, New Zealand. *Journal of Hydrology,* 257: 115-133. https://doi.org/10.1016/S0022-1694(01)00546-7

Chen, A. and Wu, M. (2019). Managing for Sustainability: The Development of Environmental Flows Implementation in China. *Water,* 11: 433. https://doi.org/10.3390/w11030433

CPCB (2020). Central Pollution Cotrol Board. Wikipedia, The Free Encyclopedia.

Cressie, N.A.C. and Whitford, H.J. (1986). How to Use the Two Sample t-Test. *Biometrical Journal,* 28: 131-148. https://doi.org/10.1002/bimj.4710280202

Fleig, A.K., Tallaksen, L.M., Hisdal, H. and Demuth, S. (2006). A global evaluation of streamflow drought characteristics. *Hydrology and Earth System Sciences,* 10: 535-552. https://doi.org/10.5194/hess-10-535-2006

Geng, H. and Shen, B. (1992). Definition and significance of hydrological droughts. *Agricultural Research in the Arid Areas,* 4: 91-94.

Iyer, R.R. (2005). The notion of environmental flows: A caution. *In:* NIE/IWMI Workshop on Environmental Flows. New Delhi.

Jain, S.K., Agarwal, P.K. and Singh, V.P. (2007). Krishna and Godavari Basins, *In:* Jain, S.K., Agarwal, P.K., Singh, V.P. (Eds.), Hydrology and Water Resources of India, Water Science and Technology Library. Springer Netherlands, Dordrecht, pp. 641-699. https://doi.org/10.1007/1-4020-5180-8_14

Jain, S.K. and Kumar, P. (2014). Environmental flows in India: towards sustainable water management. *Hydrological Sciences Journal,* 59: 751-769. https://doi.org/10.1080/02626667.2014.896996

Jha, R., Sharma, K.D. and Singh, V.P. (2008). Critical appraisal of methods for the assessment of environmental flows and their application in two river systems of India. *KSCE J Civ Eng,* 12: 213-219. https://doi.org/10.1007/s12205-008-0213-y

Kumar, U. and Kv, J. (2019). Assessment of environmental flows using hydrological methods for Krishna River, India. *Advances in Environmental Research,* 7: 161-175. https://doi.org/10.12989/aer.2018.7.3.161

Levy, P. (1967). Substantive Significance of Significant Differences Between Two Groups. *Psychological Bulletin,* 67(1): 37-40. https://doi.org/10.1037/h0020415

Muchmore, C.B. and Dziegielewski, B. (1983). Impact of Drought on Quality of Potential Water Supply Sources in the Sangamon River Basin. *JAWRA Journal of the American Water Resources Association,* 19: 37-46. https://doi.org/10.1111/j.1752-1688.1983.tb04554.x

Monish, N.T. and Rehana, S. (2019). Suitability of distributions for standard precipitation and evapotranspiration index over meteorologically homogeneous zones of India. *J Earth Syst Sci,* 129: 25. https://doi.org/10.1007/s12040-019-1271-x

Neumann, David W., Rajagopalan, Balaji and Zagona, Edith A. (2003). Regression Model for Daily Maximum Stream Temperature. *Journal of Environmental Engineering,* 129: 667-674. https://doi.org/10.1061/(ASCE)0733-9372(2003)129:7(667)

Prathumratana, L., Sthiannopkao, S. and Kim, K.W. (2008). The relationship of climatic and hydrological parameters to surface water quality in the lower Mekong River. *Environment International,* 34: 860-866. https://doi.org/10.1016/j.envint.2007.10.011

Rehana, S. (2019). River Water Temperature Modelling Under Climate Change Using Support Vector Regression, *In:* Singh, S.K., Dhanya, C.T. (Eds.), Hydrology in a Changing World: Challenges in Modeling, Springer Water. Springer International Publishing, Cham, pp. 171-183. https://doi.org/10.1007/978-3-030-02197-9_8

Rehana, S. and Dhanya, C.T. (2018). Modeling of extreme risk in river water quality under climate change. *Journal of Water and Climate Change,* 9: 512-524. https://doi.org/10.2166/wcc.2018.024

Rehana, S. and Mujumdar, P.P. (2011). River water quality response under hypothetical climate change scenarios in Tunga-Bhadra river, India. *Hydrological Processes* 25: 3373-3386. https://doi.org/10.1002/hyp.8057

Rehana, S., Munoz-Arriola, F., Rico, D.A. and Bartelt-Hunt, S.L. (2019). Modelling Water Temperature's Sensitivity to Atmospheric Warming and River Flow, *In:* Sobti, R.C., Arora, N.K., Kothari, R. (Eds.), Environmental Biotechnology: For Sustainable Future. Springer, Singapore, pp. 309-319. https://doi.org/10.1007/978-981-10-7284-0_12

Sharma, U. and Dutta, V. (2020). Establishing environmental flows for intermittent tropical rivers: Why hydrological methods are not adequate? *Int. J. Environ. Sci. Technol.* 17: 2949-2966. https://doi.org/10.1007/s13762-020-02680-6

Shukla, S. and Wood, A.W. (2008). Use of a standardized runoff index for characterizing hydrologic drought. Geophysical Research Letters, 35. https://doi.org/10.1029/2007GL032487

Smakhtin, V.U. (2001). Low flow hydrology: A review. *Journal of Hydrology,* 240: 147-186. https://doi.org/10.1016/S0022-1694(00)00340-1

Tao, H., Borth, H., Fraedrich, K., Su, B. and Zhu, X. (2014). Drought and wetness variability in the Tarim River Basin and connection to large-scale atmospheric circulation. *International Journal of Climatology,* 34: 2678-2684. https://doi.org/10.1002/joc.3867

Tennant, D.L. (1976). Instream Flow Regimens for Fish, Wildlife, Recreation and Related Environmental Resources. *Fisheries,* 1: 6-10. https://doi.org/10.1577/1548-8446(1976)001<0006:IFRFFW>2.0.CO;2.

Tharme, R.E. (2003). A global perspective on environmental flow assessment: emerging trends in the development and application of environmental flow methodologies for rivers. *River Research and Applications,* 19: 397-441. https://doi.org/10.1002/rra.736

The Times of India (2019). Drought, floods new normal as Karnataka rain pattern changes | Bengaluru News - Times of India [WWW Document]. The Times of India. URL https://timesofindia.indiatimes.com/city/bengaluru/drought-floods-are-new-normal-in-karnataka/articleshow/69191638.cms

Thomann, R.V. and Mueller, J.A. (1987). Principles of surface water quality modeling and control. Harper & Row, New York.

van Vliet, M.T.H. and Zwolsman, J.J.G. (2008). Impact of summer droughts on the water quality of the Meuse river. *Journal of Hydrology,* 353: 1-17. https://doi.org/10.1016/j.jhydrol.2008.01.001

Van Lanen, H.A.J., Fendekova, M., Kupczyk, E., Kasprzyk, A. and Pokojski, W. (2004). Flow generating processes. *In:* L.M. Tallaksen and H.A.J. van Lanen (Eds.), Hydrological Drought. Processes and estimation methods for streamflow and groundwater. *Developments in Water Science,* 48: 53-96.

Wang, Y., Li, J., Feng, P. and Chen, F. (2015). Effects of large-scale climate patterns and human activities on hydrological drought: a case study in the Luanhe River basin, China. *Nat Hazards,* 76: 1687-1710. https://doi.org/10.1007/s11069-014-1564-y

Webb, B.W., Clack, P.D. and Walling, D.E. (2003). Water–air temperature relationships in a Devon river system and the role of flow. *Hydrological Processes,* 17: 3069-3084. https://doi.org/10.1002/hyp.1280

Zeiringer, B., Seliger, C., Greimel, F. and Schmutz, S. (2018). River Hydrology, Flow Alteration, and Environmental Flow, in: Schmutz, S., Sendzimir, J. (Eds.), Riverine Ecosystem Management: Science for Governing Towards a Sustainable Future, Aquatic Ecology Series. Springer International Publishing, Cham, pp. 67-89. https://doi.org/10.1007/978-3-319-73250-3_4

Zelenhasić, E. and Salvai, A. (1987). A method of streamflow drought analysis. *Water Resources Research,* 23: 156-168. https://doi.org/10.1029/WR023i001p00156

Zhao, C.S., Yang, Y., Yang, S.T., Xiang, H., Ge, Y.R., Zhang, Z.S., Zhao, Y. and Yu, Q. (2020). Effects of spatial variation in water quality and hydrological factors on environmental flows. *Science of The Total Environment,* 728: 138695. https://doi.org/10.1016/j.scitotenv.2020.138695

Zhou, Z., Ouyang, Y., Qiu, Z., Zhou, G., Lin, M. and Li, Y. (2017). Evidence of climate change impact on stream low flow from the tropical mountain rainforest watershed in Hainan Island, China. *Journal of Water and Climate Change,* 8: 293-302. https://doi.org/10.2166/wcc.2016.149

Zwolsman, J.J.G. and van Bokhoven, A.J. (2007). Impact of summer droughts on water quality of the Rhine River – a preview of climate change? *Water Sci Technol,* 56: 45-55. https://doi.org/10.2166/wst.2007.535

Part II
Hydrosocial

Chapter 13
Relationships First: Introducing Phenomenological Research Methodology in Hydrosocial Scholarship

Kalpita Bhar Paul

INTRODUCTION

Increasing water-related disasters, scarcity in accessing water for everyday needs across social strata, prevailing water-intensive patterns of development, and rising water pollution are some of the issues that brought water at the center of attention in this century. Although water is a natural resource, these water-related issues arise at the intersection of socio-political and cultural dimensions. By acknowledging this fact, the current research on water is evolving as researchers are adopting interdisciplinary or transdisciplinary frameworks. Hydrosocial framework is an example of that shift, similar to socio-hydrology. Wesselink et al. (2017) thoroughly tease out in what manner these two frameworks differ in their epistemology, ontology, and axiology. The hydrosocial framework emphasizes on reorienting our approach of studying water as an object, to understanding water as being integrally embedded in its social context that gives its meaning (Budds et al., 2014). Capturing this place-specific meaning of water that transcends the scientific definition is one of its major objectives. According to Wesselink et al. (2017), the second objective is to understand how this meaning leads to establishing different relationships with water and the subsequent choice of water management system (Wesselink et al., 2017). The choice of water management system becomes the focal point as in the hydrosocial framework, researchers are not only keen on implementing a scientific intervention for better management, but also embrace a pro-poor stand and consider all different facets of a solution comprehensively (*ibid*).

Mostly, hydrosocial scholarship explores human agency, power asymmetries, and the need for fostering a reflective attitude to offer socially-just solutions. To attain this, narratives act as the departure point in hydrosocial scholarship.

K. B. Paul (✉)
Krea University, Sri City, Chittor, India
e-mail: Kalpita.paul@krea.edu.in

A. Mukherjee (ed.), *Riverine Systems*, https://doi.org/10.1007/978-3-030-87067-6_13

Researchers acknowledge that narrative is one of the effective ways through which developing a rich understanding of complexity is possible. This scholarship also adopts Actor-Network Theory (ANT) to capture the internal relationship that water and social power possesses. Introduction to hydrosocial cycle, as Linton and Budds highlight, is a step toward

> [t]ranscend (ing) the dualistic categories of 'water' and 'society', and employ a relational-dialectical approach to demonstrate how instances of water become produced and how produced water reconfigures social relations. (Linton and Budds, 2013, p. 2)

Scholars employ ANT to tease out all the transient networks, and also to unveil the privileged worldview that shapes the knowledge about water and guides water management. Both ANT and narrative approach work with one basic hypothesis, that is, different systems interact with each other, and the complexity that arises from these interactions can be captured through these.

The transition in water research from hydrology to hydrosocial is a shift from the understanding of one system (bio-physical system) to understanding how the interaction of different systems gives rise to the meaning of water. It acknowledges that meaning of water is place and community-specific, and subsequently argues that these meanings aid in forming the relationship with water. To understand this place-specific and community-specific meaning at first, hydrosocial emphasizes how various structural factors interact with each other and what is the dominating worldview that plays a crucial role in producing knowledge of water. Once the power dynamics between structural forces come to the forefront, and the meaning of water gets revealed, hydrosocial scholarship claims that these meanings guide individuals' relationship with water. Narrative as a tool helps in capturing these varied relationships with water. In that same vein, the hydrosocial framework presented by Linton and Budds comprises three components, viz. H_2O or biophysical and chemical property of water, technology, and social power or structure. By focusing on the interactions among these structural components, this scholarship attempts to unveil the holistic underneath picture (Boelens, 2013; Falkenmark, 1997). This approach of starting from the parts or components and then considering the relationships/interactions to be able to gain an in-depth understanding, I term as a disengaged mode of inquiry. I will clarify the reason behind terming it as a disengaged approach as I contrast it against an engaged mode of inquiry.

Hydrosocial employs narratives as a tool to gain an in-depth understanding of water. While analyzing narratives, the structural forces and interactions remain a focal point driven primarily by political-ecological understanding and ANT. Thereby, I argue though the narrative approach is vested with significant importance, we fail to explore the full potential of this approach in the hydrosocial framework. Hydrosocial adopts a problem-oriented approach and use of narrative is merely a tool to find out an appropriate solution. In another article, I present a detailed philosophical critique of this problem-oriented approach. As hydrosocial follow this problem-oriented approach, similar to the other existing paradigms, it also gets caught into the calculative ways of thinking, in which humans as subjects think about the object water from a managerial perspective. However, one of the main aims of hydrosocial is to transcend this subject/object binary. Precisely for this, hydrosocial gives primacy to other structural factors, but limited use of narratives makes it a rather shallow disengaged mode of inquiry. On the contrary, I propose a methodology that vests primacy to the relationship between an individual and water by focusing on phenomenological narratives.

In everyday affairs, individuals live amidst the relationships that water is part of. Thereby, the relationships and interactions, I argue, first play out in our everydayness or we dwell completely being soaked into that milieu and often remain oblivious of the presence of these relationships. We are always-already dwelling in an engaged reality which is simultaneously shaped by as well as shaping these relationships and interactions. Therefore, relationships must rightfully be vested with higher ontological primacy than the components. I argue against the claim by Wesselink et al. (2017) that the meaning of water decides individual's relationship with water; rather, I see it is other way round. From individuals' everyday relationships with water, the meaning of water emerges. Hence, if we can get to these relationships from everyday experiences, then only we will arrive at a more engaged or situated understanding of the hydrosocial nexus. To capture these relationships, we need to focus on individuals who are being embedded in the hydrosocial nexus, live and witness the playing out of these relationships in their everydayness. In this context, this chapter introduces phenomenological research methodology that can capture experiential accounts of individuals and can also serve as a theoretical bedrock to understanding hydrosocial changes in an engaged mode.

Phenomenological Research Methodology is a powerful methodology that accepts narratives as its departure point. Through phenomenological narratives, we can attain the necessary experiential account for appropriately capturing the meaningful encountering of water. This methodology could illuminate the enmeshed relationship within which an individual belongs and in turn, could shed light on various components of hydrosocial nexus. Focusing on the relationship is the core for both phenomenology and hydrosocial—while phenomenology emphasizes individuals' experiences of these relationships, hydrosocial accentuates the need for understanding the relationship among various structures (social, political, and economic). My attempt here is to demonstrate how individuals' experiences provide an avenue to understand the relationship among various structures as well as makes us aware of those relationships through which the meaningful encountering of water becomes at all possible in our everydayness.

PHENOMENOLOGICAL RESEARCH METHODOLOGY

The dualistic nature of our thinking that divides subject and object to study the latter was problematized with the development of the stream of phenomenology which focuses on ordinary experiences to know about any phenomenon. To transcend the subject-object duality, the departure point of phenomenology is the first person's account of any conscious experience. Conscious experiences are those which not only one observes or engages with, rather which one experiences and lives through or performs. By taking into account these conscious experiences, phenomenology studies the intentionality of the experience which analyses the way experience is directed toward revealing the phenomenon. The basic epistemological shift that phenomenology brings with itself is that it does not agree with the fact that reality exists out there. Indeed, it focuses on individuals' experiences and more importantly, their intentionality, and through that, it attempts to decipher the meaning of things that emerge in our experiences of a phenomenon. Manen (1997) points out that upshot of phenomenology is the moments of seeing meaning or "in-seeing" which is only plausible through thoughtful relation to our involvement with things of our world in everydayness. This mindfulness of our everyday involvement with our world provides an inherent understanding of a phenomenon by illuminating our relation to it.

The Phenomenological Research Methodology (PRM) is widely accepted in the discipline of psychology, education, nursing, and consumer research. In environmental studies and humanities literature, employing PRM is quite limited[1]. There are examples that have employed phenomenology as a methodology to capture embodied experiences. Here we need to remember Manen's (1997) suggestion that the role of phenomenologists is to unveil the region from where meaning arises and which in-turn leaves an impression on us, instead of capturing the mere experiences. In the following, I illustrate the methodology in detail with all its nitty-gritty.

The first step of PRM is to identify the specific phenomenon that would be the concerned matter of inquiry. Phenomenon is a mere experience/event/happening or occurrence. There should not be any problematization or value judgment involved in deciding the phenomenon. The only thing one needs to keep in mind is that there should be a possibility of obtaining direct human experiences of the phenomenon. To elaborate, in the context of hydrosocial, an example of a phenomenon could be one of the lived-experiences like changes in water bodies, water availability, the experience of water quality, landscape change, or access to water. Here, we should not consider any particular problem/concern like water crisis, water pollution, submergence of land, etc. as a phenomenon to begin with.

The next important step is to choose the co-researchers or narrators, who have prolonged and in-depth experiences of the phenomenon. As co-researchers' reflective descriptions of their experiences about the phenomenon are the basis for further exploration, the success of such study is entirely contingent upon the participants (Creswell, 2007). Hence the choice of participants depends on three basic criteria, viz. prolonged experience of the phenomenon, the willingness to spare adequate time to share experiences, and possessing the necessary articulation ability. An essential characteristic of a phenomenological narrative is that it might not be a definitive account of the phenomenon itself; rather, it is a description of the awareness of the condition in which the phenomenon can manifest itself (Wrathall, 2006)

[1] See my other attempts to integrate Phenomenological Research Methodology into Environmental Humanities: [Kalpita Bhar Paul and Meera Baindur, 2016], and [Kalpita Bhar Paul, 2017].

who also points out that "the end goal of description is to guide the reader to the practical orientation for the world in which the phenomenon can show itself" (Wrathall, 2006).

This methodology does not follow the active questioning technique; rather a researcher's role is quite limited here as being just a facilitator. As phenomenological narrative emerges out of self-reflection, the interview process is entirely unstructured. A researcher's sole responsibility is to create the space or moment that leads a co-researcher to elaborate on the experiences of the phenomenon or the occurrence of the phenomenon in his/her everydayness in a mode of 'self-talk'. The best way to create such space is to conduct the research at the site of the phenomenon (Moustakas, 1994). It enables the researcher to engage with the world of experience, which eventually ensures that he/she is completely immersed in the context. Particularly, in the context of hydrosocial research, I feel this is quite important as the visualization of a phenomenon offers a sense of belongingness with the place.

The discussion with co-researchers revolves around three major themes: "What does an individual experience in terms of the phenomenon?", "What contexts or situations have typically influenced or affected one's experience of the phenomenon?", "How does it affect the narrator?" (Englander, 2012; Groenewald, 2004; Kornhaber, 2009; Bhattacherjee, 2012). As this methodology is quite fluid, a researcher should maintain the necessary space for himself/herself to contemplate and reflect throughout the entire process so that the researcher can refrain himself/herself from asking any causal explanatory questions and in-turn, could guide a co-researcher to reach to the mode of 'self-talk'. If a researcher is able to reach this stage of 'self-talk' then, it has been observed that saturation can be achieved with 5-25 narratives from a homogeneous group. Saturation indicates that a new narrative is not providing any further insights into the concerned phenomenon and becomes redundant[2].

Interpretive Approach in PRM

As per literature, two of the broad objectives of hydrosocial paradigm are that of exploring how the meaning of water emerges from our relationship, and how water gets meaningfully appropriated to a community. PRM guides us to reveal these two questions in depth. The interpretive approach of narrative analysis over the descriptive[3], would be more appropriate in this context. Kafle (2011) highlights the normative difference between these two approaches is that researchers who follow the

[2]For further details about the methodology see Meera Baindur and Kalpita Bhar Paul, 2015.

[3]According to DPRA, a researcher should bracket her own belief regarding the pre-existing conceptual framework about the concerned phenomenon before beginning the research work. In this process, a researcher's aim should be to look at the phenomenon from the descriptions given by the individuals who have directly experienced it. In other words, it suggests that one must bracket out her natural attitude, which is thought to be contaminated by the prevailing scientific paradigm.

former believe that philosophy should not be carried out from a detached, objective, and disengaged standpoint, precisely what the latter proposes. Indeed, in the interpretive stream, researchers attempt to interpret what it means to experience the same that is described by the narrators. This method tries to capture the underlying mindset of narrators to get to their experiences. More importantly with the help of that, it attempts to reach the underlying condition through which the phenomenon appears. From this point of view, Interpretive Phenomenological Research Approach (IPRA) is a dynamic process that demands a persistent effort from a researcher's end to attain the state of experience in an 'as if' mode.

IPRA is known as a double hermeneutic process (Pietkiewicz and Smith, 2014) which points toward a dual interpretation. The first layer of interpretation comes from the narrator, and to understand that a researcher tries to extract the essence of narratives from the narrator's point of view. This process is called an emic approach (Pietkiewicz and Smith, 2014). Next layer of interpretation is of researcher's understanding of the narrative through notes, explanatory comments, and multiple readings of a narrative. It is called an etic approach. Thorough readings help a researcher to break narratives down for grouping them under various themes. The explanatory comments noted down during readings as well as during fieldwork, guide a researcher to develop new themes. By clustering such themes, major themes emerge. Once these themes are in place, the task of the researcher is to proceed with the analysis of narratives based on the major themes. Following this structure, final step is to elaborate on each theme by drawing examples from narratives. This double hermeneutic process, on the one hand, demonstrates the way interpretation happens in an interviewee's mind; on the other, it also provides a scope to capture a researcher's interpretation of the same.

While analyzing a narrative concerning water availability, for example, a researcher should at first try to discuss those situations in which the question of availability has emerged, and when the narrator felt the lack of availability or excess of it and most importantly how the narrator is defining the changes in water availability. Answers will come from the first level of hermeneutic, i.e., emic approach. In the next level, the multiple readings of the narratives will help researcher to develop various themes, which may not always directly offer an understanding of water availability, it will rather show how this is intrinsically tied with other socio-political, ecological, and economic issues. Themes like industrial pollution, groundwater depletion, water resource management, impact on biodiversity, and women's drudgery etc. will inherently provide us the essence of the phenomenon of water availability. It will reveal how the availability of water is not only an issue related to water; it is indeed the tip of the iceberg.

In the following, as an illustration of how PRM can be employed to study water-individual society relationships, I am going to capture the nexus between water and changes in the environment in the Sundarbans within the framework of hydrosocial.

Furthermore, descriptive phenomenology also borrows from Husserl's eidetic analysis, which attempts to reach the universal from the particular, to establish the truth.

The phenomenon under consideration here is how people of this area perceive changes in the environment. We all are aware that in the Sundarbans rivers, creeks, and land create a cobweb-like structure which essentially forms the Sundarbans that is neither a waterscape nor a landscape region; indeed, it stands at the threshold of being both of these, concurrently. Hence, the environmental changes in the Sundarbans cannot entirely be captured phenomenologically, barring the discussion on water. Water acts as an active force in this region not only in determining the landscape but also in shaping the lives and livelihoods of this area.

PHENOMENOLOGICAL NARRATIVES ON CHANGE IN ENVIRONMENT AND LANDSCAPE OF SUNDARBANS

Following the double hermeneutic process, at first, it is important to grasp how the biophysical changes in the landscape and the environment come into the narrators' purview or how narrators are defining these changes. Reading the narratives in an 'as if' mode or from the narrators' standpoint will enable us to understand when the changes are being felt by the narrators and which changes have the most bearings on them. Here, the primary focus is on to capture those moments that induce a narrator to experience changes in landscape and waterscape.

Emic Approach: In Sundarbans, the islanders' lives and livelihoods revolve around various environmental phenomena like timing of tides, seasonal changes, lunar cycle, etc. For this dependency, there is an integral connection between the narratives of environmental change and their livelihoods or everyday endeavours. As they spoke about their livelihoods and household chores, changes in the landscape and waterscape surfaced. However, this contextualized reference of change is crucial as it determines whether or not any particular change in this 'ever-changing' world gets acknowledged as a 'change' in the narrator's everydayness. For example[4], a veteran boatman narrates, how before the introduction of GPS technology, he used to notice the submergence of the existing landmasses on the sea or rivers and the creation of new islands while taking his boat deep into Bay of Bengal. Being a boatman, the person acknowledges the submergence and the creation of new islands as these provide him with necessary landmarks for navigating at sea. Hence, we can see these subtle landscape changes appear to boatmen not merely as changes in the landscape; indeed, these possess higher significance to them as landmarks. And their work-world enables them to notice these changes.

However, young boatmen are hardly aware of these changes probably because the new age technologies like GPS, wireless phone, have made their journey so safe that they do not feel the need to remember any external landmarks for navigation.

[4] This field work was part of my PhD dissertation research. Portions of these narratives have been used in my different published articles. However, the interpretation and most importantly the contextualization are entirely distinct in each of these works.

Technological devices entirely guide them at sea. It renders that these landmarks are not necessary for them, and thus, mostly they remain ignorant about changes in those. It is an example of how structural modifications like the introduction of new-age technologies make people indifferent to changes in their surroundings. Nevertheless, these young boatmen have agreed on the fact that they cannot rely on technology entirely as there is always a scope that in front of nature's fury these technologies will fail, and in these moments their intuition can be their only savior. As one of them says:

> No one can teach how to ride a boat in the sea. At most one can tell us the know-how of driving a boat. However, true learning happens through our years-long experiences as an assistant of an experienced boatman. We gather how to make the right decision, how to decipher natural clues, etc. Sitting beside him only we truly learn how to make our journey safe at sea. It is much easier to learn the technical know-how, but that would not be enough for life [FLDN][5].

It is a perfect example of how changes in livelihood needs (i.e. context) subsequently transformed an individual's observation ability, and as a result, changes in the environment which earlier was significant becomes insignificant. Technological interventions, on the one hand, create a safe environment for dwellers to cope with natural elements. On the other hand, due to these interventions inhabitants become reluctant to observe the environmental cues that provide them an early sign of extreme natural events, as well as changes that are perpetually taking place and shaping their surroundings. No doubt, this lack of awareness makes them oblivious of their capacity of gathering intuitive knowledge of this land-waterscape.

Although narrators speak about changes in the coastline vividly, in their narration, they hardly mentioned about landscape changes. They were even quite reluctant to acknowledge it as a change. They mentioned modifications of the landscape as if it is the very nature of the place where they live. According to them, one of the primary characteristics of Sundarbans' rivers is that they shift their courses over time, and that, in turn, induces changes in the landscape. However, islanders do not acknowledge these landscape modifications as 'environmental changes'; they consider these as an integral characteristic of the place they inhabit. The attitude of accepting landscape changes as the feature of these islands, as we can understand, makes islanders skeptic to any policies and interventions that, in principle, attempt to safeguard land from water. The prevailing land-water binary is missing in their phenomenological experiences. Their lives on island lead them to acknowledge that living in the Sundarbans means living in a land that is liminal. Here, on the one hand, land emerges from water and on the other, "land remains until water takes it over" [FLDN].

With the help of these narratives, one can understand that there is a significant difference between outsiders' and insiders' acknowledgement of changes in land and waterscape. While these seem to be threatening to outsiders, the dwellers' lives flow in harmony with these changes. The first layer of analysis of phenomenological

[5] FLDN stands for Field Narrative. Henceforth, FLDN will indicate extract from narratives.

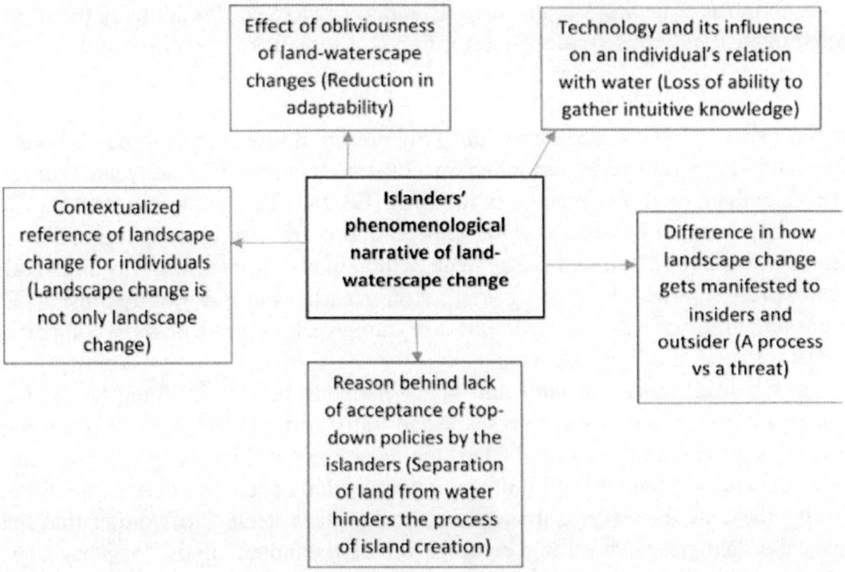

Figure 1. Various dimensions of hydro-social framework emerged out of islanders' phenomenological narratives.

narratives helps us to capture inhabitants' perception of changes in waterscape. It enables us to acknowledge how depending on the waterscape the landscape is being formed and consequently, how society thinks about this kind of liminal landscape. It also illuminates how a community, who lives in close proximity to such a liminal landscape, remains at ease with such integral land-water mingling. Sundarbans islanders for their living depend on water not only for their livelihoods or for their everyday needs but also for the continued existence of the very land itself, and eventually, they find themselves in an intrinsic relationship with water. The following diagram (Fig. 1) conclusively portrays how the first level of phenomenological narrative analysis unfolds various aspects of the hydrosocial nexus.

Etic Approach: The next step of narrative analysis is etic approach. From my reading of narratives, two themes emerge, viz. the first is obviously the landscape change and the second is change in biota. Following illustrates how these two themes help in understanding the hydrosocial nexus in this region. There are some sub-themes which also emerge within these two broad themes, for example, property loss, changes in livelihoods fall under the first theme, and risk perception, effects of institutionalized/top-down interventions come under the second theme. Following will provide the account of their acknowledgement of change; in addition, I will highlight the context of that acknowledgement.

Landscape change: Although the narrators have not explicitly mentioned or have not acknowledged several alterations in the landscape as change, phenomenological descriptions of the place implicitly contains a significant account of changes in land and water-scape. These phenomenological accounts offer a sense that land and water create an intricate network in this area. Islanders have experienced many floods with varied intensities. Those experiences of floods, I see, make them realize the essence of the place which is, the land here is always susceptible to the invasion of water. Islanders here live within a space constituted simultaneously with anxiety and ease. An old fisher in his narration mentions "my present house is the thirteenth house of mine, and I have moved as and when the embankment shifted. The recent one is the seventh embankment I witnessed in my life" [FLDN]. To capture constant transformation between landscape and waterscape narrators frequently use phrases like: *age chilo, ekhon nei/age chilo na ekhon ache*, which means, something was there earlier but not present now/or something was not there earlier but is currently present. This expression highlights the presence-absence framework to prove how the Sundarbans always remains in a state of flux.

The transition between land and water happens here quite fluently, land can change into water-body or a waterscape can transform into the land, as if the entire area is in a process of 'becoming'.[6] Due to this process of 'becoming', changes in the landscape are encountered so often and in almost daily activities that the inhabitants identify these as the very characteristics of the place itself. No wonder that these landscape changes have strong bearings on the islanders' lives. Property loss is common upshot of such change. As mentioned above, many islanders have shared their experiences of shifting their home from one place to another. And also this water invasion on land hampers farming to a large extent, e.g. islanders explain, after the cyclone Aila, in 2009 to cope with the saline water invasion, many farmlands are being converted to salt-water ponds for fishing a few particular species of fish. Farmers either have sold or leased the land to fishers or without having any alternatives they have moved from their primary occupation. Hence, fact is that though the islanders are habituated to landscape changes, their lives and livelihoods do get hampered by such change to a large extent. For this reason, in their phenomenological narratives landscape changes intrinsically appear as significant.

Change in biota: The connection between hydrosocial nexus and the biota of the Sundarbans region is intricate. All the narrators explicitly recognized changes in biota in some form or the other as livelihoods of the islanders directly depend on the biota. The most common observation in this regard came from fishers as they discussed the fish stock depletion and reduction in the varieties of fish. Local fishers undoubtedly accept that the introduction of trawler fishing in this region has aggravated fish stock depletion. Opening up Sundarbans waterscape for industrial trawlers not only hampers traditional practices of sustainable fishing but also harms the biophysical condition of the sea. One of the fisherman shares:

> The new trawlers come from outside, have a different mechanism to catch fish. They place their net so deep into the sea that it even scratches grass and fish seeds, everything from the seabed. [FLDN]

[6]For more details about phenomenological understanding of landscape change see [Kalpita Bhar Paul and Meera Baindur, 2016].

This deterioration in the seabed as per the fisher community of the islands puts a massive toll on fish reproduction and that in turn severely affects their livelihood.

Many narrators have alluded to the decline in the forest cover; however, in this regard, there was hardly any mention of extinction of any plant species in their account. On a few occasions, to explain the nature of the forest, some of the narrators described how, in spite of Sundarbans getting its name from the *Sundari* (*Heritiera fomes*) tree, at present the tree can hardly be found in the forest, rather the forest is mostly covered by Goran (*Ceriops decandra*) tree. Except this, narrators have mostly described the loss of forest cover along the river. Changes in the course of rivers and increasing demand for settlement area, together led to the depletion of the mangrove that in turn make this region significantly vulnerable to the brute force of storms and cyclones. Apart from this, it further impacts the availability of firewood. Mostly women's accounts capture how at present gathering sufficient fuelwood is a struggle, and their drudgery has increased significantly. In earlier days firewood availability was never a concern for them, but at present, on one hand their access to forest is highly restricted, on the other, trees alongside riverbanks are declining; therefore to collect firewood they are now dependent on water. They need to go to the river at dawn to collect logs which come floating downstream.

Small scale fishers, who regularly visit narrow creeks deep inside the forest, as well as honey gatherers[7] mentioned that the number of tigers have significantly increased in the last 10-15 years. The threat of tiger attack is a deterrent to their forest-based livelihoods. Their accounts sketch a vivid picture of how institutionalized conservation and preservation programmes have severely restricted their access to the forest. In the same vein, tiger protection and conservation programmes undoubtedly increase threat to their life inside the forest or even making their stay in the proximity to forest more unsafe. Forest conservation and preservation are considered as the backbone of ensuring the Sundarbans' sustainability and disaster risk reduction; however, the same pose threats to inhabitants whose lives and livelihoods are integrally dependent on forest and creeks.

Farmers shared how the introduction of monoculture practices led to the abolishment of different native paddy varieties. They also explain how land-shaping and concretization of embankments increase the risk of flood. As a result, their choices of paddy have changed.

> Earlier we used to cultivate native varieties paddy, which yielded less but those paddies could withstand saline water, massive storm, and even flooding. Now we focus more on yield and could achieve that only through hybrid varieties of paddy. But it has increased the risk by many folds. Hybrid varieties cannot stand any slightest of natural calamity. That is why during Aila [cyclone that hit this region in 2009] all our crops failed miserably [FLDN].

It shows how development in terms of separating land from water have induced some significant alterations in farming practices. These seem effective at the outset

[7] Mostly people from indigenous communities of this area go deep inside the protected forest, generally, in a group for collecting honey from beehives.

but practically makes farmers more vulnerable to the onset of any natural calamities like floods, cyclones.

The second level of narrative analysis with the help of theme-based analysis showcases the interconnection among various processes and events. It shows how structural changes induce alterations in traditional practices and how those, in turn, influence the biophysical environment of the Sundarbans, impacting the inhabitants' lives and livelihoods. The following diagram comprehensively illustrates the same.

The two-step analysis offers clarity on how to capture nuanced hydrosocial changes through detailed analysis of phenomenological narratives. I conclude, PRM is quite appropriate not only for deciphering the effect of hydrosocial changes in individuals' lives and livelihoods but also for teasing out existing interactions among structural forces and how that in-turn, creates this hydrosocial nexus. It also indicates how far institutional strategies are effective in managing the hydrosocial nexus in this region, and how it shapes a community's future. This methodology helps us to realize the bigger picture by illuminating the relationship between water and each individual and also water and the society/community at large. In this methodology, narratives are not only employed to come up with an appropriate solution of the problem at hand, instead it helps in teasing out the situation in-depth and all the interconnected issues that emerge out of this situation (Fig. 2). Especially, in area like the Sundarbans, where water plays a pivotal role in shaping every

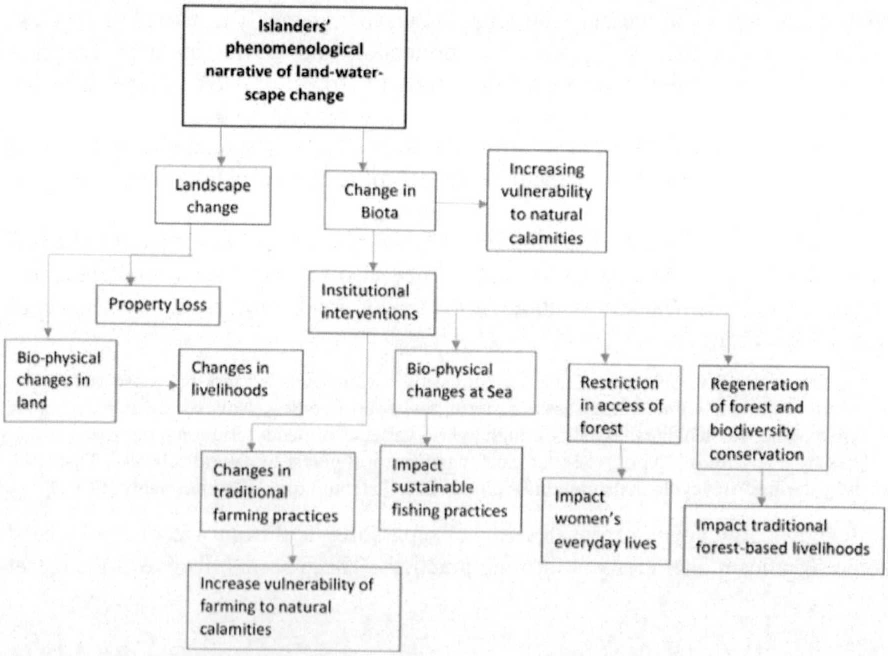

Figure 2. Illustration of various aspects of hydro-social framework emerged out of theme-based analysis of phenomenological narratives.

factor—tarting from landscape to livelihoods, to eventually the flow of life—it is imperative that PRM is employed to tease out various facets of the hydrosocial framework.

CONCLUSION

This is an attempt to demonstrate how the current hydrosocial scholarship in-spite of being able to induce a shift from hydrology to hydrosocial, gets limited by following a top-down approach or a disengaged mode of inquiry. This scholarship highlights the presence of different components of hydrosocial nexus as its point of departure and then attempts to capture the interactions among these components to be able to shed light on the relationship between water and society holistically. I argue the approach of employing an interdisciplinary perspective to grasp a comprehensive picture falls short of bestowing enough emphasis on the relationship as such. Here, keeping in mind that the primary objective of hydrosocial nexus is to unfold how water and society relate to each other and, in-turn, shape themselves, I introduce an alternative methodology named Phenomenological Research Methodology. It begins with everyday experience of individuals and through that intends to arrive at a more nuanced understanding of the complexity of water-society relationship. I vouch that this is an apposite one to capture the hydrosocial nexus as it adequately accentuates the need for an engaged mode of inquiry and opens up different components of the hydrosocial nexus as these appear in an individual's field of experience. The relationship between water and society is intrinsically present in each of our lives and could be possible to grasp by paying attention to our every-dayness. Therefore, I propose instead of various structural forces and components, individuals' everyday experiences should be the departure point for hydrosocial to reach a holistic understanding that it envisages. Through the case study of the Sundarbans, I demonstrate, how the islanders' phenomenological narratives provide a detailed understanding of various structural components of hydrosocial as well as illustrate the context in which it plays out in individual's situatedness.

This is a preliminary attempt that contributes meaningfully to the hydrosocial scholarship by not only providing a case study of the Sundarbans to illustrate the playing out of hydrosocial framework there, but also it is to bring an epistemological shift in the water-society relationship. By introducing the Phenomenological Research Methodology, I accentuate the role of individuals in this entire hydrosocial nexus. In other words, it highlights why individuals matter in understanding the water-society relationship. As we have seen individual stands in the milieu of social processes and also encounters bio-physical changes in water or water bodies at first. Only in everyday dwelling of human beings, changes in water and waterbodies first get manifested. Therefore, narrative of an individual's life experiences does not remain limited to being an affect-based narrative; indeed, it manages to highlight how various so-called changes get appropriated in his/her place-specific situatedness.

This attempt to introduce a new methodology can significantly help the hydrosocial scholarship to move away from the prevailing disengaged approach. Even though I have restricted myself in delineating phenomenology as a methodology, it also has the potential to philosophically enrich this scholarship. If phenomenology is adopted as a philosophical bedrock then it can strengthen the hydrosocial framework as far as the exploration of the relationship remains its central theme. I foresee more philosophically enriched version of Phenomenological Research Methodology not only will illuminate the unexplored dimensions of water-society relationship but also it will emphasize individuals' relation to social processes and natural systems which eternally remain in constant flux. Even I see phenomenology could radically change the conceptualization of this framework by rightfully highlighting the role of individuals' experiences. For establishing the same, nevertheless, we need more exploration and further deliberation on this line.

References

Bhattacherjee, A. (2012). *Social Science Research: Principles, Methods and Practices*. USF Open Access Textbooks Collection. Book 3. http://scholarcommons.usf.edu/oa_textbooks/3

Boelens, Rutgerd (2013). Cultural politics and the hydrosocial cycle: Water, power and identity in the Andean highlands. Geoforum, https://doi.org/10.1016/j.geoforum.2013.02.008

Budds, Jessica, Jamie Linton and Rachael McDonnell (2014). The hydrosocial cycle. *Geoforum,* 57: 167-169.

Englander, M. (2012). The Interview: Data Collection in Descriptive Phenomenological Human Scientific Research. *Journal of Phenomenology and Psychology*, 43: 13-35.

Falkenmark, M. (1997). Society's interaction with the water cycle: a conceptual framework for a more holistic Approach. *Hydrological Sciences-Joumal-des Sciences Hydrologiques*, 42(4): 451-466.

Groenewald, T. (2004). A Phenomenological Research Design Illustrated. *International Journal of Qualitative Methods*, 3: 42-55.

Kafle, N.P. (2011). Hermeneutic Phenomenological Research Method Simplified. *Bodhi: An Interdisciplinary Journal*, 5: 181-200.

Kalpita Bhar Paul and Meera Baindur (2016). Leopold's Land Ethic in the Sundarbans: A Phenomenological Approach. *Environmental Ethics,* 38(3): 307-325.

Kalpita Bhar Paul (2017). Introducing Interpretive Approach of Phenomenological Research Methodology in Environmental Philosophy: A Mode of Engaged Philosophy in the Anthropocene. *International Journal of Qualitative Methods,* 16: 1-10.

Kornhaber, R.A. (2009). A Lived Experience of Nursing Severe Burns Injury Patients: Phenomenological Enquiry (PhD diss.) The University of Adelaide, Australia.

Linton, Jamie and Jessica Budds (2013). The hydrosocial cycle: Defining and mobilizing a relational-dialectical approach to water. *Geoforum.* https://doi.org/10.1016/j.geoforum.2013.10.008.

Manen, Max Van (1997). Phenomenology of Practice. Phenomenology & Practice, 1, 11–30.

Creswell, J.W. (2007). Qualitative Inquiry & Research Design: Choosing among Five Approaches.

Meera Baindur and Kalpita Bhar Paul (2015). Mapping the observer in the observation in Anthropocene: A Methodological Exploration. *Humanities Circle,* 3(2): 61-81.

Moustakas, C. (1994). *Phenomenological Research Method*. Thousand Oak, CA: Sage Publication Ltd.

Pietkiewicz, I. and Smith, J.A. (2014). A Practical Guide to Using Interpretative Phenomenological Analysis in Qualitative Research Psychology. *Czasopismo Psychologiczne – Psychological Journal*, 20: 7-14.

Wesselink, Anna., Michelle Kooy and Jeroen Warner (2017). Socio-hydrology and hydrosocial analysis: toward dialogues across disciplines. *WIREs Water* 4.

Wrathall, M.A. (2006). Existential Phenomenology. *In:* H. Dreyfus and M. Wrathall (Eds.), *A Companion to Phenomenology and Existentialist* (pp. 31-47). USA: Blackwell Publishing.

Chapter 14
Socio-hydrology: A Holistic Approach to Water-Human Nexus in Large Riverine Islands of India, Bangladesh and Vietnam

Pankaj Kumar, Rajarshi Dasgupta, and Ram Avtar

INTRODUCTION

Considering the finite volume of freshwater resources, its sustainable management is a global challenge. Due to various factors ranging from natural disturbance in hydrological regime to poor governance, about 33% of global population currently live in water stress conditions and this number will go upto 67% by 2050 if no adaptation and mitigation measures are timely considered (Vörösmarty et al., 2010). On the other hand, rapid population increase, economic growth, and urbanization are putting further stress on fresh water availability (Kumar et al., 2018). Variability of climatic conditions in the tropical Indian and Pacific Oceans are also dominated by the Indian Ocean Dipole (IOD) and El Nino Southern Oscillations (ENSO). They both have huge impact on Asian region through inducing hydro-meteorological hazards and its induced health impacts (from water borne diseases like cholera). Since last three decades, approximately 3.2 million people killed in countries like India, Bangladesh and Vietnam only because of ENSO and IOD induced hazards (Masahiro et al., 2010; Tamaddun et al., 2019). With increasing nature and anthropogenic interferences and management efforts, hydrology of different landscapes around the world has significantly changed, and which has significant impact on both water resources and socio-cultural attributes deeply associated with water (Blair and Buytaert, 2016).

Despite world's largest river systems being located in Asia such as Ganges, Brahmaputra, Mekong etc., water availability per capita in Asia is behind rest of

P. Kumar (✉) · R. Dasgupta
Natural Resources and Ecosystem Services, Institute for Global Environmental Strategies, Hayama, Kanagawa, Japan
e-mail: kumar@iges.or.jp

R. Avtar
Faculty of Environmental Earth Science, Hokkaido University, Sapporo, Japan

A. Mukherjee (ed.), *Riverine Systems*, https://doi.org/10.1007/978-3-030-87067-6_14

the world (WWAP, 2015). Impacts of rapid global change include extreme weather condition, hydro-meteorological hazards, interruption in food production and water supply, human morbidity and mortality, disrupt the ways that people live and interact with their environment (IPCC, 2014, Mishra et al., 2017). Additionally, the total cost due to hydrological disasters in 2016 was estimated as 59 billion USD, which represents 74% of the annual average of all financial loss due to all hazards for the period (2006-2015) (Mukate et al., 2017). Among these damages caused by hydrological disasters, flood was estimated about 98.8% in 2016 (ADB, 2013). The most vulnerable group due to water hazards are communities from Riverine Island because of limited resources/infrastructure as well as institutional setup. Conventional models may help identify desirable future status of water environment, but they cannot guarantee their attainment because of adaptive responses by humans and management decisions which might have unintended consequences (Sivapalan et al., 2014; Palmer and Smith, 2014; Jamero et al., 2017). Therefore, an integrated perspective in analyzing water related risk through socio-hydrological pathways deemed essential for better understanding the action research and policy implication for sustainable water management (Baldassarre et al., 2015; Wesselink et al., 2017).

Baldassarre et al. (2019) has shown global hot spots of water crises based on the survey among water experts carried out by the International Association of Hydrological Sciences (IAHS) (Fig. 1). Finding of this survey emphasized that water crises at most of the places are due to poor governance and lack of understanding, or

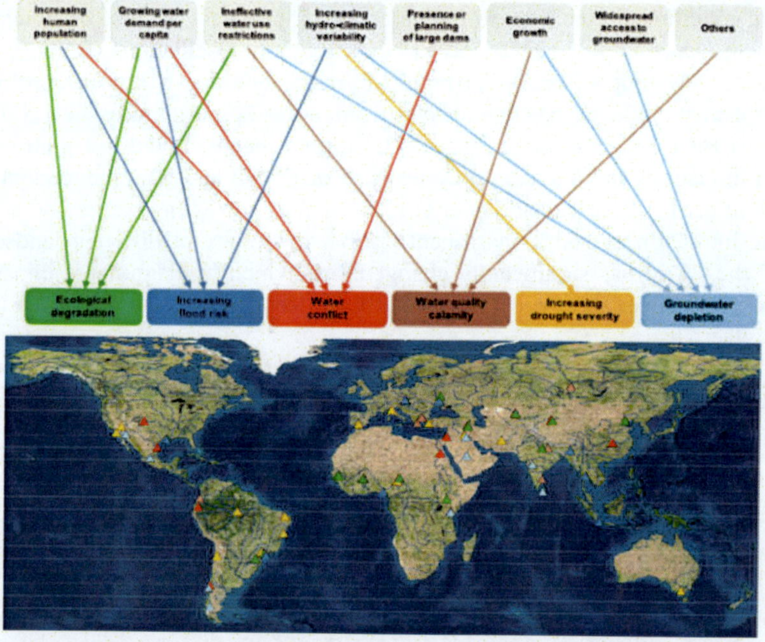

Figure 1. Global hot spots of water crises (Source: Baldassarre et al., 2019).

neglect by different stakeholders and decision makers about economic and socio-cultural perspectives of water resources and their interwoven nexus.

So far, most of the water management decision making are dealt with physical science behind hydrological processes, numerical simulations, scenario based approaches, which are good for providing short-term solutions but cannot lead to long-term sustainable solutions due to very limited consideration of dynamic feedbacks and nexus between the natural, technical, and social dimensions of human-water systems. That is how the role of socio-hydrology comes into existence. It has the potential to minimize the gaps identified before to understand the nexus between water and human society. In order to be proven as robust approach, socio-hydrology should address hydrology with sociology by engaging social scientists along with natural scientists for considering social heterogeneity, power relations, trust, cultural beliefs and cognitive biases, which strongly influence the way in which people alter, and adapt to, changing hydrological regimes. As this is a new venture in the scientific field, it needs to develop new methods to formulate and validate different hypothesis, which can explain the close interaction between water and human society (Baldassarre et al., 2019).

Globally, rivers are exhibiting significant changes in annual runoff as well as quality deterioration along with the global warming, decrease in snowfall, increase in glacier melt, sea level rise, groundwater salinization, rapid urbanization and socio-cultural practices ultimately causing water shortage/scarcity (Arnell and Gosling, 2013). Particularly, communities in the isolated riverine islands are worse affected due to their poor adaptive capacities, which is well documented in the contemporary literature (Srinivasan et al., 2017). Despite having plenty of water, sharp reduction in the usability of the available water became a serious problem since the past decades. In many such cases, the focal point for the vulnerability of these people lies in the shrinking water resources (i.e. drinking water availability, agricultural water quality, salt-water intrusion, flooding etc.) and the future interaction between human and water systems is a critical concern for sustainability.

While much progress has been made in making quantitative assessments of different hydro-meteorological risk, there remains a lack of fundamental understanding of the interplay between physical and social processes. As a result, the current analytical frameworks somehow failed to capture (or explain) the dynamics emerging from this interplay (Baldassarre et al., 2015). To understand this interwoven issue of water and societal interaction, there is a swift change in the direction of transdisciplinary research approach, which study water resources from humanities, social science and natural science perspectives that seek to address the hydrological and social challenges related to the complex human-water interactions (Wesselink et al., 2017). Although, many conceptual framework have been developed, very few of these research works have tried to implement in the field.

With abovementioned gaps, under Asia Pacific Network (APN) funding, we have initiated a project on study water resource management from the lens of socio-hydrological approach. This study will emphasize on different scale of human interaction with the water cycle, along with the coupling effect present between social and hydrological systems. In addition, as an adjoined system, it will explore decisions that affect water as well as its users, i.e. local communities. Socio-hydrological models will be used to quantify the feedbacks between water resources and society at multiple scales with aim to expedite stakeholder participation for its sustainable management.

More precisely, this work will try to achieve following objectives: (i) identification of the key parameters/indicators for quantification of losses and damages (in terms of income, life, health) due to hydro-meteorological disasters and extreme weather conditions in last three decades in each target sites, (ii) stocktaking of societal measures taken for mitigation and adaptation during the past disaster events for water resource management. Also, it will identify the gaps in facilitating better adaption or mitigation strategies for water resource management, (iii) future hydrological risk prediction incorporating different key drivers and pressures. Here, future projection (year 2050) for water quality and quantity will be made using key factors, namely, population change, lowering of water table, groundwater extraction rate, salt water intrusion, climate change and land use land cover change etc., (iv) pathway analysis for co-designing different adaptive countermeasures based on socio-hydrological simulation. Here, model will consider the biophysical sciences and in particular societal responses to and influence on extreme hydrological events. For bio-physical components, mainly structural measures like Wastewater Treatment Plants, retention ponds etc., will be considered. For societal components, different variables like socio-political attributes, indigenous knowledge, culture, crop types, water consumption patterns, practices and monitoring efforts to be made to reach the regional/ national targets in line with the sustainable development goals (SDGs) (i.e. goal 6) will be considered, finally, (v) best practices for co-designing the effective management/adaptation strategies for water resource as well as harmonizing it with human well-being will be show-cased to all the people/ organization concern for water resources management.

This model will consider the biophysical sciences and in particular societal responses to and influence on extreme hydrological events. For societal component, different variables like socio-political attributes, indigenous knowledge, culture, agricultural practices, water consumption pattern and monitoring efforts to be made to reach the targets, will be considered. The project will focus on three different large riverine islands in Asia, namely, Sagar (Ganges River, India), Dakshin Bedkashi (Padma River, Bangladesh) and Con Dao Island (Mekong River, Vietnam).

The expected result will be helpful to sketch projection of alternatives that explicitly account for plausible and co-evolving trajectories of socio-hydrological system, which will yield both insights into cause–effect relationships and help stakeholders to identify safe functioning space. It will try to evaluate key components like the societal measures taken as mitigation means for sustainable water resource management against the rapid global changes, and the societal characterization and monitoring efforts made for choosing management targets and checking the effects of measures taken to reach the targets.

METHODOLOGY

The proposed research aims to build on resilience to hydrological hazards and enhancing climate change adaptation while optimizing water resource management with the help of socio-hydrology as shown in Fig. 2 which presents interlinkage between physical science and social dynamics, while detail methodology is divided into following three steps (shown in Fig. 3):

(1) Participatory mapping for getting community perception about different recent past hydro-meteorological hazards, loss and damage and their action on adaptation and mitigation based on the existing socio-economic resources in three different riverine islands, namely, Sagar (Ganges River, India), Dakshin Bedkashi (Padma River, Bangladesh) and Con Dao Island (Mekong River, Vietnam) as shown in Fig. 4. This will be accomplished by focus group discussion (FGDs) (with policy makers, regulating bodies), key informants interviews (KIIs) and households questionnaire survey of different communities living in the study area to get a detail insight on local/indigenous adaptive capacity. For each study site, the collected survey database will outline the induced changes in diverse socio-political attributes including changes in demography, agricultural practices, water consumption pattern for personal use and irrigation, institutional framework etc.; that govern and regulate water management policies and practices over the last couple of decades. This will be accomplished through statistical analysis (principal component analysis, cluster analysis and factor analysis). Findings from abovementioned analysis will be validated through historical satellite images and documented information like various government reports. Finally, relation between water scarcity and its relation with human well-being over last two decades and its future trend can be estimated.

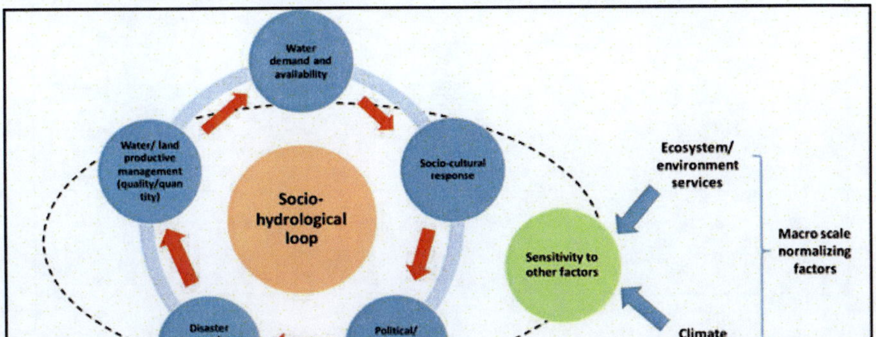

Figure 2. Concept of research framework.

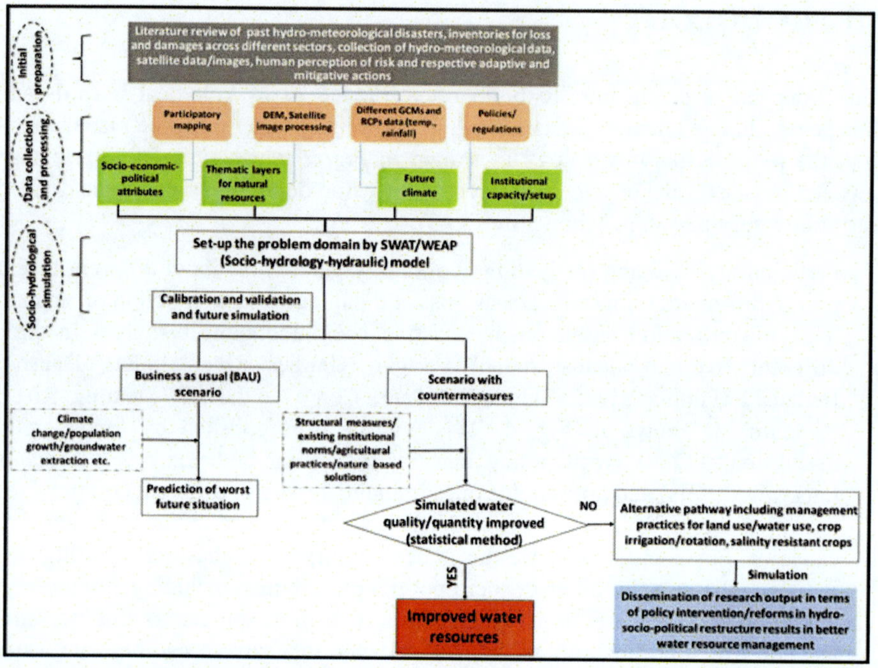

Figure 3. Research methodology.

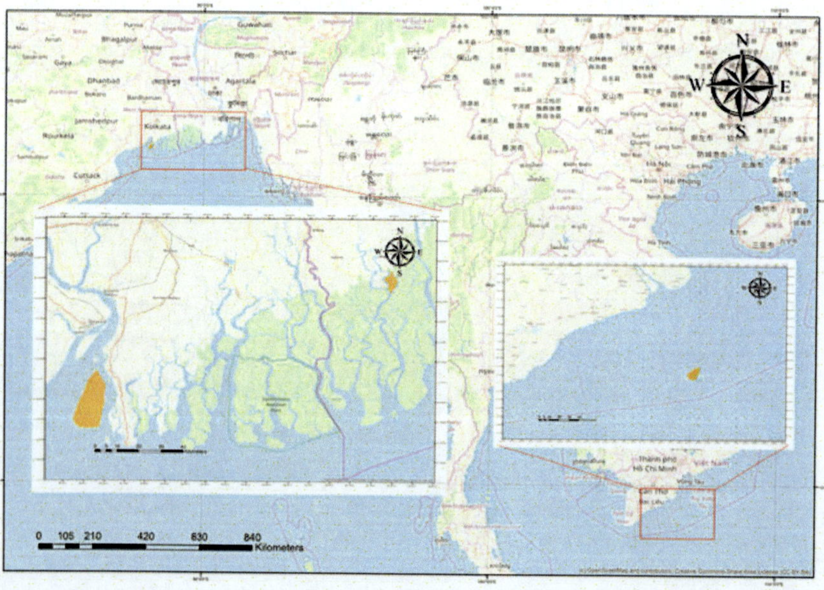

Figure 4. Study area map.

(2) Using socio-hydrological numerical tools (WEAP/SWAT), two different scenarios will be developed, i.e. business as usual (BAU) and scenario with countermeasures. For business as usual scenario, hydrological modelling will be carried out to estimate future risk and challenges on water security (both flooding and water quality deterioration) considering key drivers, namely, climate change, population growth, land use/land cover change, groundwater level depletion, groundwater extraction rate. It will also identify the gaps in terms of both soft and hard mitigation/adaptation measures (e.g. retention ponds, recharging zone, artificial wetlands, waste water treatment capacity, nature-based solutions etc.), which can be considered to make the hydro-sociological system more resilient to any possible future external/internal disturbances. Different set of input data will be used for different hydrological models to predict future water environment, e.g., different downscaled Global Climate Models (GCMs) output will be used to simulate climate prediction through different climate variables such as temperature, precipitation etc. Future population projection will be generated through global reports like UNDESA (United Nations Department of Economic and Social Affairs). Land use/land cover changes will be simulated by Land Change Modeler, an Arc-GIS component.

(3) Above activity will be followed by developing scenarios with countermeasures. Here development of different possible retrofitting adaptive mechanism will be proposed with regular feedback mechanism from socio-political components. Following possible options of adaptation/mitigation measures will be considered:

(i) Domestic wastewater management (like increase in domestic sewerage collection and treatment (Waste Water Treatment Plant capacity and sewerage network)).

(ii) Groundwater salinity management (such as change in groundwater extraction rate/cropping pattern/saline resistant crops/crop rotation/seed sowing time/homestead gardening/crop and food insurance/rainwater harvesting/recharging zone etc.).

(iii) Flood inundation management (e.g. retention ponds, artificial wetlands, nature-based solutions etc.). Since this is a retrofitting model with inputs from different stakeholder is required, it can provide different possible adaptation and mitigation measures of water resource management in synergy with human well-being.

(4) In case some of the socio-economic components cannot be incorporated in above models like cultural practices, GDPs etc. then we will try to add them through loop model.

(5) Finally, best possible adaptation/mitigation strategies will be suggested, which can help optimizing water resources (both quality and quantity). It will help to outline what is the best possible way to use indigenous knowledge/traditional agricultural practices along with existing scientific information to minimize water quality deterioration as well as to revive the economy. Since this is a

co-designing process, most of the stakeholders will be able to know "what if" situation and they can alter the design.

EXPECTED DELIVERABLES/OUTPUTS

(a) After assessing the information on loss and damage during past hydro-meteorological hazards, a regional database on gaps and opportunities for future research for all three areas will be created.
(b) With different predictive hydrological simulation tools and using the "what if" scenarios (business as usual and scenario with mitigation measures), the study will pin-point the extent and severity of water quality deterioration (primarily considering salt-water intrusion, and flood inundation etc.). Vulnerability maps for water scarce and hazard prone areas will be created. For business as usual scenario, different key factors, namely, population growth, climate change, groundwater level depletion, groundwater extraction rate etc. will be taken into account. For scenario with counter-measures, different adaptation/mitigation measures for sewerage management (such as increase in domestic sewerage collection and treatment (Waste Water Treatment Plant capacity and sewerage network)) and groundwater salinity management (like change in groundwater extraction rate/rainwater harvesting/cropping pattern/saline resistant crops/crop rotation/seed sowing time/homestead gardening/crop and food insurance etc.) will be considered. Therefore, the study can quantify the implications of possible adaptation and mitigation measures of water resource management and their relationship with human well-being.
(c) With different possible socio-hydrological adaptation/mitigation integration pathways, this study will facilitate to live with change and uncertainty in best possible way through nurturing various types of hydrological, social and political diversity. This will widen the range of knowledge for learning and problem solving skills as well as creating opportunities for self-organization. Furthermore, it would help to understand the macroscale effects of microscale drivers of human behaviour, culture, economic status. Finally, it will depict the role of co-evolutionary processes of socio-hydrological systems for managing complex issues of water security in vulnerable riverine islands.
(d) From the research perspective, we will publish comprehensive research report; peer reviewed articles, policy-brief, policy relevant documents and conference proceedings which will outline the shared learning from the case studies will be shared to policy makers in the target countries. In addition, networking with government experts in the target countries will be expected for future research collaborations.

RELEVANCE

APN's Strategic Plan

From the inception, this proposed research aims to contribute to all the goals outlined in the APN's Strategic Plan. This research is built on the gaps identified by recently published global climate reports (special IPCC report on global warming of 1.5°C and 6th APAN Forum), where more emphasis on adaptive measures/communication against climate change was reiterated. Since most of the environmental modelling focusses on scenario based snapshot of the future world but policy based intervention hardly get implemented because of the understanding gap between scientific community and decision makers. Therefore, proposed work intends to work on socio-hydrological interactions, which involves public investments with long-ranging impacts on hydrological cycle management. This will use attributes of existing water environment and social dynamics (cultural, political, ethical etc.) for improving adaptation communication in sustainable water management (water quality, flooding etc.). Here, for the modelling purpose, biophysical parameters will be given importance, however, greater emphasis will be placed on the role of politics and culture in shaping them. Since this work will be based on the concept of "with the people rather than for the people", the co-designed knowledge base and insights gained in this research will be shared with the relevant institutions, stakeholders and society in terms of reports, guidelines and policy briefs and also through capacity building activities.

Policy Processes and Sustainability

In implemented holistically, socio-hydrology can potentially contribute towards meeting Sustainable Development Goals as shown in Fig. 5 (Source: Baldassarre et al., 2019). Socio-hydrology is conceptually different than the previous widely adopted concepts of water resource management across the globe like (a) Water Resource System (WRS) (where goal of analysis is to combine hydrology and economics to design and operate optimal infrastructure projects); (b) Integrated Water Resource Management (IWRM) (which prescribes how to manage water resources in specific contexts); and (c) social-ecological systems which deals with interaction between human perturbances and its relation with ecosystem and ecosystem services. In contrast, the focus of socio-hydrology is to understand why certain water management outcomes arise rather than proposing actual management solutions. It also tries to analyze actual water management processes and outcomes to develop generalizable understanding. It has a more explicit focus on water, and on the specifics of the hydrologic cycle in space and time, including the role of water infrastructure.

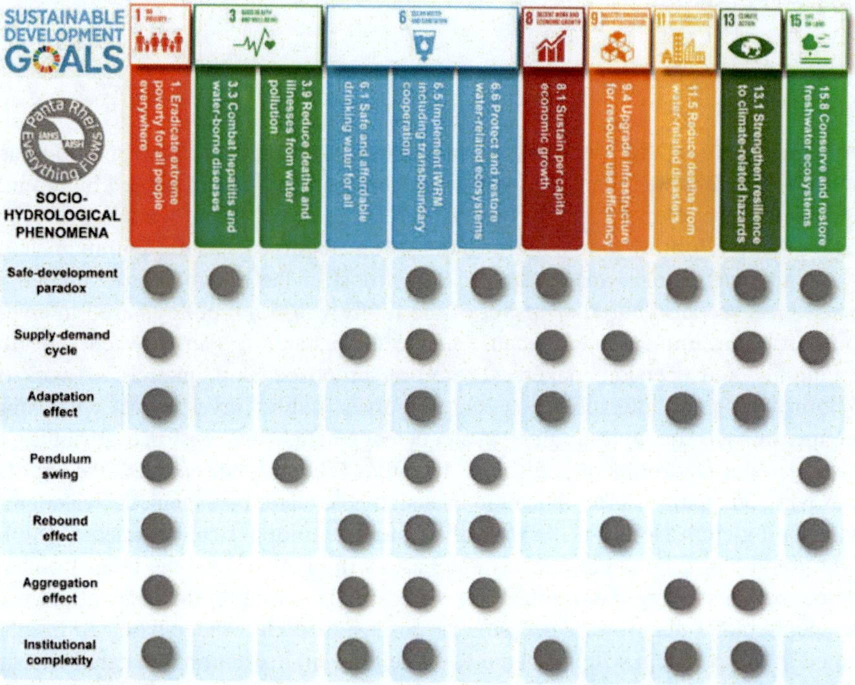

Figure 5. Socio-hydrology and its role in achieving Sustainable Development Goals (SDGs) (Source: Baldassarre et al., 2019).

While the project targets bridging these knowledge gaps effectively by doing pathway analysis for co-designing different adaptive countermeasures based on retrofitting socio-hydrological simulation. Here, model will consider the biophysical sciences and in particular societal responses to and influence on extreme hydrological events. For societal component different variables like institutional set-up, indigenous knowledge, culture, practices and monitoring efforts to be made to reach the regional/national targets in addition to the sustainable development goals (SDGs). Within the regional pretext, this work will target particularly for the goal 13 (climate action), 6 (clean water and sanitation), 3 (good health and well being), and 15 (life on land). The research will contribute towards better understanding of hydro-meteorological disaster risks and their best possible mitigation measures based on local socio-political context in the largest riverine island environment in three Asian countries. Thus, it will also contribute towards proactive, evidence-led policy planning under the priority 1 and priority 3 of Sendai Framework for Disaster Risk Reduction (SFDRR); which foresees better picture of the disaster risks and promotion of non-structural/structural measures for risk reduction. Finally, it will promote adaptation communication in best possible way to advocate local people to use their local resources for managing water resources for both short and long time scale.

Scientific Capacity Development for Global Change Research

Developing pathway analysis for co-designing different adaptive countermeasures for hydro-meteorological disaster management as well as managing local water needs based on socio-hydrological simulation is a very new domain of applied hydrological research. It is getting popularity after the concept of transdisciplinary research and is widely accepted as an only viable solution to solve complex issue of water scarcity. Despite a handful, scientific work is carried out in Europe and USA; there is a significant lack of such research work from the Asia-Pacific. In recent global climate reports (special IPCC report on 1.5°C, 6th APAN Forum summary output and Japanese new Adaptation Law), need for adaptation communication and action for mitigation along with adaptation was reiterated for better understanding of hydro-meteorological hazards and sustainable water resource management. This will provide a strong science-policy interface against the regional backdrop to solve the issues as thirst area of water scarcity. The study is, therefore, a modest attempt to fill this existing gap and to enrich the scientific understanding for pathway analysis for co-designing different adaptive countermeasures based on socio-hydrological simulation in Asia-Pacific region. Apart from its strong policy relevance in solving the interwoven issues of water scarcity, it will also help knowledge-transfer within and outside the project collaborators, including governmental sector and research institutions. This also promote "3P concept", i.e. for the people, by the people and of the people; where different section of the society will be able to co-design the mitigation/adaptation measures for sustainable water resource management.

CONCLUSION AND WAY FORWARD

Recently, scientific communities around the world are shifting their focus from multidisciplinary to transdisciplinary research with integration of physical science and socio-economics through scenario analysis to predict most likely risk for different natural resources (Pahl-Wostl, 2009). Scenario exercises have only recently become an important decision-making tool in the field of natural resources, soon after the IPCC's Special Report on Emission Scenarios (SRES) (Nakicenovic et al., 2000). However, in an increasingly globalized world like Anthropocene, these models must account for broader economic, social and cultural influences on the system of interest. In other words, these models must/should simulate outcomes society actually cares about, so they can facilitate stakeholder participation and steer societies onto better trajectories (Ferguson et al., 2018). Considering the significant gap, different scientific research works try to conceptualize idea of socio-hydrological simulation with the intention of bridging the gap of science-policy interface (Sivapalan et al., 2014; Palmer and Smith, 2014; Jamero et al., 2017). One of the studies tried to implement socio-hydrological approach for incorporating gender into biophysical models and implications for water resources research, where they particularly focused on how different gender people interact with water resources and landscapes in different ways (Baker et al., 2015). However, the number of socio-hydrological research in Asian region is almost none.

Henceforth, recently different adaptation related global platforms and events univocally called for more inclusive research work, which endorse the idea of scenario exercises with inputs from different social attributes at different stage to design action based mitigation measures for water resources management. Thus, the study provide a unique opportunity to fill this important research gap and will provide alternative, plausible and co-evolving trajectories through use of socio-hydrological models, rather than generate scenarios that present a snapshot of the world at some future date. Here, through socio-hydrologic modeling approach, this research will model the co-evolutionary dynamics of coupled human, water and ecological system.

ACKNOWLEDGEMENT This work is supported by Asia Pacific Network (APN) Collaborative Regional Research Programme (CRRP) project with project reference number CRRP2019-01MY-Kumar.

References

ADB (Asian Development Bank). The Rise of Natural Disasters in Asia and the Pacific: Learning from ADB's Experience; ADB: Mandaluyong, Philippines, 2013.

Arnell, N.W. and Gosling, S.N. (2013). The impacts of climate change on river flow regimes at the global scale. *Journal of Hydrology*, 486: 351-364, doi: https://doi.org/10.1016/j.jhydrol.2013.02.010.

Baker, T.J., Cullen, B., Debevec, L. and Abebe, Y. (2015). A socio-hydrological approach for incorporating gender into biophysical models and implications for water resources research. *Applied Geography*, 62: 325-338.

Baldassarre, D., Viglione, G.A., Carr, G., Kuil, L., Yan, K., Brandimarte, L. and Bloschl, G. (2015). Debates—Perspectives on socio-hydrology: Capturing feedbacks between physical and social processes, *Water Resour. Res.*, 51: 4770-4781, doi:https://doi.org/10.1002/2014WR016416.

Baldassarre, D., Sivapalan, M., Rusca, M., Cudennec, C., Garcia, M., Kreibich, H. et al. (2019). Socio-hydrology: Scientific challenges in addressing the sustainable development goals. *Water Resources Research*, 55: 6327-6355. https://doi.org/10.1029/2018WR023901.

Blair, P. and Buytaert, W. (2016). Socio-hydrological modelling: a review asking "why, what and how?" *Hydrol. Earth Syst. Sci.*, 20: 443-478.

Ferguson, L., Chan, S., Santelmann, M.V. and Tilt, B. (2018). Transdisciplinary research in water sustainability: What's in it for an engaged researcher-stakeholder community? *Water Alternatives*, 11(1): 1-18.

Intergovernmental Panel on Climate Change (IPCC) (2014). Climate Change 2014: Synthesis Report. Contribution of Working Groups I, II and III to the Fifth Assessment Report of the Intergovernmental Panel on Climate Change [Core Writing Team, R.K. Pachauri and L.A. Meyer (eds.)]. IPCC, Geneva, Switzerland.

Jamero, M.L., Onuki, M., Esteban, M., Billones-Sensano, XK., Tan, N., Nellas, A., Takagi, H., Thao, N.D. and Valenzuela, V.P. (2017). Small-island communities in the Philippines prefer local measures to relocation in response to sea-level rise *Nature Climate Change*, 7: 581-586.

Kumar, P., Masago, Y., Mishra, B.K. and Fukushi, K. (2018). Evaluating future stress due to combined effect of climate change and rapid urbanization for Pasig-Marikina River, Manila. *Groundwater for Sustainable Development*. 6: 227-234.

Masahiro, H., Faruque, A. S. G., Terao, T., Yunus, M., Streatfield, P.K., Yamamoto, T. and Moji, K. (2010). The Indian ocean dipole and cholera incidence in Bangladesh: A time series analysis, *Environmental Health Perspectives*, 119(2): 239-244.

Mishra, B.K., Rafiei Emam, A., Masago, Y., Kumar, P., Regmi, R.K. and Fukushi, K. (2017). Assessment of future flood inundations under climate and land use change scenario in Ciliwung river basin, Jakarta. *Journal of Flood Risk Management*. Taylor and Francis Publication. Doi: https://doi.org/10.1111/jfr3.12311

Mukate, S., Panaskar, D., Wagh, V., Muley, A., Jangam, C. and Pawar, R. (2017). Impact of anthropogenic inputs on water quality in Chincholi Industrial area of Solapur, Maharashtra, India. Groundw. Sustain. Dev. https://doi.org/10.1016/j.gsd.2017.11.001.

Nakicenovic, N., Alcamo, J., Grubler, A., Riahi, K., Roehrl, R.A., Rogner, H.H. and Victor, N. (2000). Special report on emissions scenarios (SRES), a special report of Working Group III of the intergovernmental panel on climate change. Cambridge University Press.

Pahl-Wostl, C. (2009). A conceptual framework for analyzing adaptive capacity and multi-level learning processes in resource governance regimes, 19: 354-365.

Palmer, P.I. and Smith, M.J. (2014). Earth systems: Model human adaptation to climate change. Nature, 7515: 365-366.

Sivapalan, M. et al. (2014). Socio-hydrology: use-inspired water sustainability science for the Anthropocene. *Earth's Future*, 2 (4), 225–230. doi:https://doi.org/10.1002/2013EF000164

Srinivasan, V., Sanderson, M., Garcia, M., Konar, M., Blöschl, G. and Sivapalan, M. (2017). Prediction in a socio-hydrological world, *Hydrological Sciences Journal*, 62(3): 338-345, DOI: https://doi.org/10.1080/02626667.2016.1253844

Tamaddun, K. A., Kalra, A., Bernardez, M. and Ahmad, S. (2019). Effect of ENSO on temperature, precipitation and potential evapotranspiration of North India's Monsoon: an analysis of trend and entropy, *Water*, 11: 189, doi:https://doi.org/10.3390/w11020189.

Vörösmarty, C.J., McIntyre, P.B., Gessner, M.O., Dudgeon, D., Prusevich, A., Green, P., Glidden, S., Bunn, S.E., Sullivan, C.A. and Liermann, C.R. (2010). Global threats to human water security and river biodiversity. *Nature*, 467: 555 doi:https://doi.org/10.1038/nature09440.

Wesselink, A., Kooy, M. and Warner, J. (2017). Socio-hydrology and hydrosocial analysis: Toward dialogues across disciplines. WIREs, *Water*, 4: e1196

WWAP (United Nations World Water Assessment Programme) (2015). The United Nations World Water Development Report 2015: Water for a Sustainable World. Paris, UNESCO.

Chapter 15
Interlinking of Indian Rivers: Floated Myths, Flouted Realities

Nilanjan Ghosh and Sayanangshu Modak

INTERLINKING RIVERS: A CONCERN

The gravity of concern about interlinking rivers (ILR) in India is huge and quite evidently has been associated with acrimonious debates. While various dispensations in the new millennium have generally been in favour of this mega-scale infrastructure project, no other infrastructure development project has perhaps received so much flak from the scientific community and the NGOs than the ILR project. Opposing discourses started flowing in ever since Suresh Prabhu took up the chairmanship of the task force of ILR in 2002—a position which he relinquished in 2006. The idea of interlinking is old enough to have invited the debate even in the last century. However, neither adequate scientific knowledge, nor appropriate technological know-how existed during that time to really take the debate to the public forum. The idea of linking rivers of India has its roots in the thoughts of Visveswarya, the stalwart engineer of yore. The idea was further extended by K.L. Rao, the legendary irrigation minister of India, and Captain Dastur, a pilot. Both Rao and Dastur envisaged the Ganga-Cauvery Link Canal and the Garland canals, respectively (Fig. 1). Rao felt that the solution to India's water problems lie in transferring water from "water surplus" to "water deficit" basins. Even today, many believe that "robbing Peter to pay Paul" (RPBP) is a great idea to resolve the water problems.

Shah (2010) attributes the origin of this massive engineering marvel to the historical development of South Asian irrigation especially to the era of *constructive imperialism* (1800 to 1970) that witnessed a quantum increase in irrigated agricultural acreage by manipulating large untapped rivers and reconfiguring basin hydrology and morphometry. This was chiefly pioneered by Sir Arthur Cotton in South

N. Ghosh (✉) · S. Modak
Observer Research Foundation Kolkata, Kolkata, India
e-mail: nilanjanghosh@orfonline.org

A. Mukherjee (ed.), *Riverine Systems*, https://doi.org/10.1007/978-3-030-87067-6_15

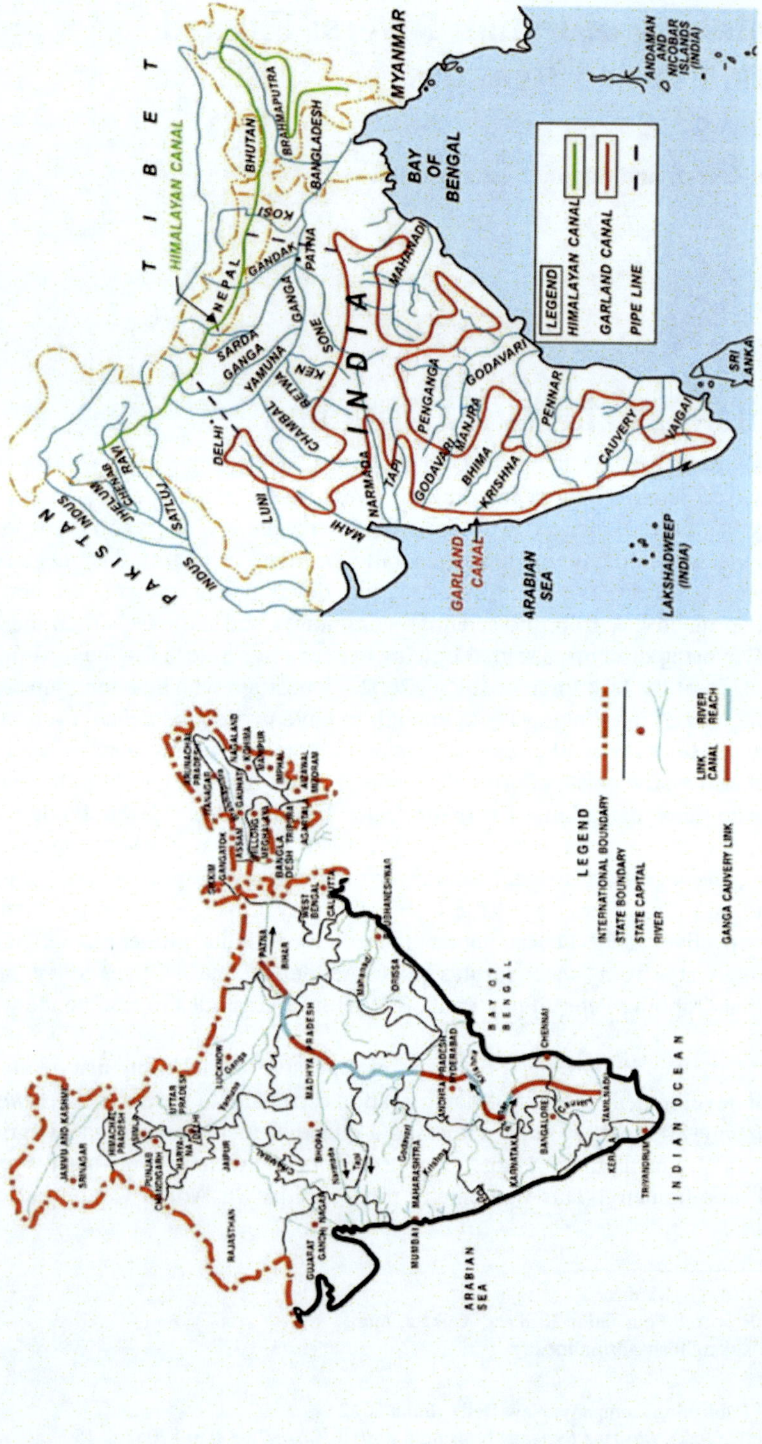

Figure 1. Map showing K.L. Rao and Captain Dastur's proposed links, respectively.

India and Major Proby Cautley in North India. However, despite the clear emphasis on irrigation, it is rather surprising to note that linking the rivers of the subcontinent was envisaged to create a single navigable water route across the length and breadth of the subcontinent. D'Souza (2008) pointed that for a while, this plan was even in the reckoning as a more viable alternative to the railways which Cotton had dismissed as an 'inferior mode of conveyance' in his attempt to syphon capital investments coming from London and into the colonies seeking high profits and quick returns. In his '*Report on the Mahanuddy*', which sought to apply the same treatment on the Orissa delta as required for every other delta in the subcontinent, Cotton had proposed a mix of irrigation and navigation canals along with a series of embankments to regulate the river and harness its flow (Cotton, 1858). This 'Orissa scheme' was a mere segment in the grander scheme of larger grid that would connect Karachi in northwest with Madras in the south. However, much to Cotton's dismay, not only did the railways continue to gain favour over his bizarre alternative, but he would also have to see the Orissa scheme (D'Souza, 2003) and several other irrigation ventures turning into financial disasters.

Subsequently, ILR remained a forgotten idea for more than a century—almost waiting for the right moment to wake up from its deep slumber. Meanwhile, the country had gained independence and was well on its course towards self-sufficiency through the Soviet-styled Five-Year Plans with a major emphasis on large infrastructure projects—particularly dams and canals. Much like his counterparts in Asia and Africa that had just emerged from the grips of colonialism, India's first Prime Minister Pandit Jawaharlal Nehru had made his opinion clear by remarking that dams were the new temples of modern India in 1954 and revering with optimism its potential role in the national development processes (Biswas and Tortajada, 2001). He would soon subsequently admit that *'we are suffering from what we may call a disease of gigantism'* owing to the suffering of displacement and the frustration of rehabilitation by those who had to cede their land for such large projects (Nehru, 1988).

However, the cart of gigantism had already gained significant momentum and it was not long before K.L. Rao, the then Union Minister of State for Power and Irrigation, came up with the idea of linking the Ganga in the north with the Kaveri in the south through a 2,640 km long canal in the 1960s. By the 1970's the idea had taken a concrete shape and Dr Rao proposed it as the National Water Grid in 1972, which would draw 60,000 cusecs of flood flows of the Ganga near Patna for about 150 days in a year and transfer it to the Cauvery (Radhakrishna, 2003). The project would have further required 5 to 7 million kW of power to pump the water over a head of 1800 ft. with no flood control benefits whatsoever (NWDA, 2020b).

Another equally fanciful idea was floated by Captain Dastur in 1977, suggesting two canals: (i) The Himalayan Canal at a level of 1100 to 1500 feet and (ii) the Garland Canal at 800 to 1000 feet. These canals were further stated to have lakes one mile wide and 100 feet deep along their course and the total project would have incurred a cost of 120 lakh crores (Ministry of Irrigation, 1980). Both K.L. Rao's plan and Captain Dastur's proposal were subsequently assessed by the National Commission for Integrated Water Resources Development Plan (NCIWRDP) which further remarked that the former was 'very costly and lower cost alternatives were available' while simply denouncing the latter as 'prima facie impractical' (SANDRP, 2003).

Yet, the pursuit for ILR to connect the 'water surplus' region with 'water deficit' did not stop. The ghost of IBWT had been truly unshackled and it would continue to occupy the mind-space of policymakers in the years to come. The National Water Development Agency (NWDA) was set up in 1982 with the mandate of undertaking detailed studies, surveys and investigations for preparing the feasibility report of the links that had been conceptualised in a previous National Perspective Plan (NPP) formulated by the Ministry of Water Resources in 1980 (Singh, 2008). For the next 20 years, the NWDA worked assiduously by involving all the states concerned to develop a mutually acceptable scheme and prepare pre-feasibility reports for the same. By 2002, the National Water Policy of the country had been revised and it carried an explicit mention of ILR through the insertion of subsection 3.5 within section 3 (Water Resources Planning) which read, "*Water should be made available to water short areas by transfer from other areas including transfers from one river basin to another, based on a national perspective, after taking into account the requirements of the areas/basins.*" It further sought for the creation of necessary guidelines for facilitating future agreements amongst states (subsection 21.1) and the promotion of frontier research and development for utilising water through such non-conventional methods (subsection 3.2) (MOWR, 2002).

In the same year, on the country's Independence Day eve, the then President of India, Dr. A.P.J. Abdul Kalam remarked that interlinking of rivers was 'inescapable' in solving India's flood and drought problems (Vombatkere, 2016). This prompted an advocate to file a Public Interest Litigation (PIL) for the cleaning of the Yamuna and regarding the dispute over Cauvery between Karnataka and Tamil Nadu by attaching Kalam's Independence Day speech (Singh and Singh, 2004). This act brought the issue out from the closets of public offices and put it out for deliberation right at the Apex Court of the nation. The then Chief Justice of India, Justice B.N. Kirpal, was heading the bench and he responded with unprecedented enthusiasm to convert the PIL into an independent writ petition, issuing notices to the Centre and the States for their opinion on the issue. Despite the absence of a response from all other states except Tamil Nadu and the Union Government, both of whom endorsed the idea, the bench assumed that the absence of affidavits from States meant that they did not oppose the plan and that there was a consensus amongst them for ILR (Singh and Singh, 2004). This paved the way for the issuing an all-important order by the Supreme Court in support of ILR despite significant apprehensions raised by observers regarding the 'defensibility' of this instance of judicial activism (Iyer, 2002). However, for all that matters, this set-off an immediate chain reaction that led the Union Government to appoint a Task Force under Suresh Prabhu to draft an action plan for implementing the project (Singh and Singh, 2004).

Therefore, it is worthwhile to ponder what makes the idea such a resilient force, emerging time and again as if a certain panacea for all the country's water woes and promising to restore both the quantitative imbalance as well as a serious fall in quality. To decipher this, there is a need to look at what is being sold as the proposed benefits of the project and contextualise it according to the challenges of water governance which have emerged.

ILR AS A REMEDY FOR INDIA'S WATER WOES

ILR in its present form is much more elaborate and well-laid out than its two previous predecessors. It draws its strength from the NPP of the Ministry of Irrigation (1980) and has two main components—Himalayan Rivers Development and the Peninsular Rivers Development (Fig. 2). The Himalayan component entails transferring the surplus water of the eastern tributaries of Ganga to the West and linking the main Brahmaputra with the Ganga. The peninsular component is divided into four parts. Part-I involves interlinking of Mahanadi-Godavari with Krishna-Pennar-Cauvery that would allow for the conveyance of surplus water from the former to the latter basins. Part-II seeks to interlink short length and surplus West Flowing Rivers located North of Mumbai, and South of Tapi like Ulhas, Vaitarni, Damanganga etc. to reduce the water demand from the Tapi in the north and also channelise supplies to the Mumbai Metropolitan area. Part-III requires interlinking of Ken with Chambal and increasing the irrigation coverage in the Malwa Plateau, Bundelkhand and Baghelkhand regions—a major drought-prone area in the region. Part-IV seeks to construct a canal along the west coast confirming to the contour of 500 feet above mean sea level to channelise water to Kerala and another tunnel to divert water over the Kallada river ridge to the east (Ministry of Irrigation, 1980).

Confirming to the now globally abandoned 'supply-side hydrology' (D'Souza, 2008), ILR is touted to be an answer to the deepening water crisis in the country that has been a direct consequence of mismanagement and resultant over-exploitation of the resource in the first place. By and large, the mammoth initiative tries to artificially reduce the regional imbalance of precipitation endowments by transferring water from a 'surplus' basin to a 'deficit' one (Fig. 2). Suresh Prabhu was the chairman of the Task Force that was appointed by the NDA government to examine the ILR after the Supreme Court directive of 2001. In a book chapter titled "The Vital Links", Prabhu has shared accounts of the work of the Task Force and his rich experiences of professional engagement within it. The chapter presents an important and, in many ways, a unique perspective into how this massive engineering feat is being proposed as an inevitable path for a water-secure nation in the 21st century. For our analysis, we use this text as our primary document while also considering all other aspects that emphasise the desirability of ILR.

Co-managing 'Menace' of Floods and 'Curse' of Drought

Unlike the precipitation regime of the temperate regions, India receives the bulk of its precipitation, more than 75 to 80%, through the South Asian Summer Monsoon

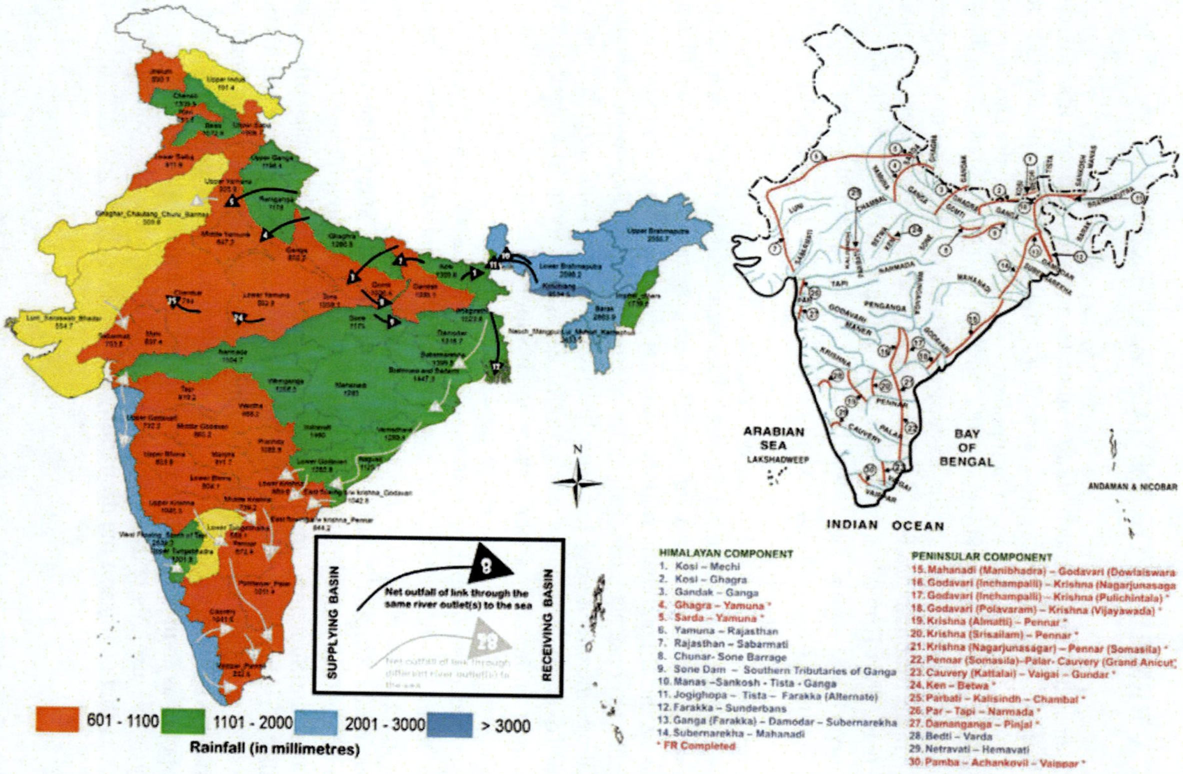

Figure 2. (left) River Basin Rainfall Normals based on data between the periods 1951 and 2000 (*Source:* IMD – modified by author), (right) Proposed Inter Basin Water Transfer (IBWT) (*Source:* NWDA).

(Nandargi et al., 2010). Apart from being highly concentrated, the precipitation is also immensely variable over different temporal scales—inter-seasonal, seasonal and yearly scale. Similarly, the spatial distribution of precipitation is also highly variable, and this can be largely attributed to the influence exerted by the topography of the land. The eastern, north-eastern and the coastal parts, particularly the western coast, experiences very heavy rainfall while other parts, particularly the Deccan hinterland and the western parts of India, experience acute water shortage which aggravates during the summer (Fig. 2).

Linking it to water availability for human use, a rather crude way to look at this disparity in water resource endowment is by looking at the Potential Utilisable Water Resource (PUWR) available on a per capita basis. PUWR per capita for Himalayan river basins (like Ganga[1] 1,044 m^3, Brahmaputra 1,448 m^3) and some easterly flowing (like Mahanadi 2,341 m^3, Godavari 1,431 m^3) and westerly flowing rivers (Narmada 2,448 m^3, Tapi 1,183 m^3) of Central India are significantly high. Compared to them, the other rivers basins (Sabarmati 797 m^3, Subarnarekha 568 m^3, Pennar 762 m^3 and Cauvery 852 m^3) (Amarasinghe et al., 2004).

This is stated to create a certain dichotomy in the country on an annual basis where parts of the country are regularly plagued by floods that severely damage infrastructure and impact livelihoods, while other parts of the country have to routinely grapple with drought and low availability of water that makes raising even one crop an impossible task. ILR has been projected to resolve this crisis of extremes once and for all! Citing the mounting costs of flood damages, Rs 645 million in the decade between 1953 and 1963 and rising to Rs 28,810 million between 1990 and 2000, Prabhu has opined that only the 'surplus flood water' may be transferred to deficit areas after all the in-basin requirements have been met.

Interlinking for Food- and Water-secure Future

The management of extremes has been ably matched by the argument that the links would allow for an increase in the irrigated area in the country, the total quantum of which is set to increase by 35 million hectares and additional water supply to domestic and industrial sectors if all the links were to reach fruition. Broadly, three specific concerns, or on the obverse—expectations, that buttress the argument for 'self-sufficiency' of food grains (Amarasinghe, 2012).

First, India has a large population that is set to grow further in the coming decades. Irrigation has remained the most important factor in boosting food grains production in the country from a low of 89.36 million tonnes in 1964 to 1965 to a

[1] The sub-basins comprising left-bank of the Ganga (bulk of which will be linked) like the Ghaghara, Gandak and Kosi with the sub-basins of the right bank tributaries and distributaries like the Yamuna, Son and Bhagirathi/Hooghli, have a much higher PUWC per capita than the average for the entire Ganga Basin and also much higher average annual volume of discharge than their right bank counterparts.

high of approximately 211.32 million tonnes in 2001 to 2002, making the country self-sufficient in food grain production. Prabhu has noted that India needs to expand its irrigation potential to 160 Mha for all crops by 2050 and that the maximum which can be attained if only reliant on conventional sources is 160 Mha. Out-of-the-box thinking is crucial and that is precisely where ILR is said to fit in quite well with its promise of an expansion of irrigated area by 35 Mha. Apart from that, the second expectation is that a boost in agriculture through an expansion of irrigation services will improve the rural economy and strengthen the rural livelihood, both of which are inextricably linked with agriculture. And finally, it has been argued that low foreign exchange reserves do not allow for the import of food grains (Amarasinghe, 2012).

Likewise, the sectoral demand for this precious resource is only going to increase (Fig. 3) which will lead to stiff competition over sectoral resource allocation. In terms of sheer volume, the total requirement for all other uses will remain a small percentage of the total demand since irrigation water requirement will continue to hold a larger share of the pie. However, in terms of sheer access, drinking water needs and industrial water requirements will remain as important as irrigation water needs since the proposed links are supposed to provide water to the parched landscapes of the country. This is linked further with the larger socio-economic development of large parts of the country beset with the problems of extremes—flood and drought. Prabhu has firmly put his weight behind addressing the burning problem of water for both India's economic and human development. Quite

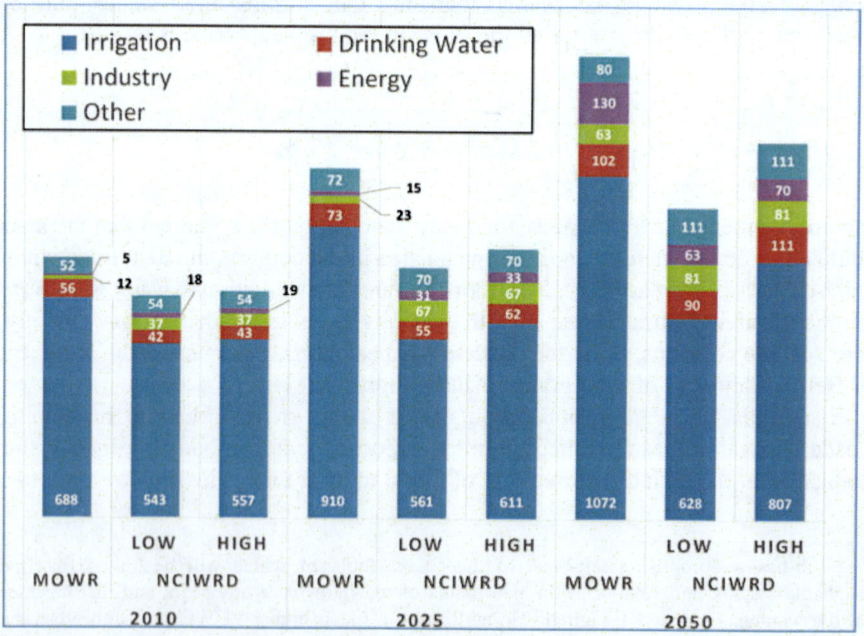

Figure 3. Projected Water Demand in India (Values in Billion Cubic Meters). *Source*: CSO (2018)

unassumingly, Krishna Nand Yadav (2007) remarks that ILR will "connect the different culture and parts of the country in a strong chain", thereby hinting at the belief held by many of the potentials of this mega project in nation-building.

Hydropower and Other Incidental Benefits

ILR also promises to quench India's thirst for energy. Despite standing at only 30% of the world's average per capita energy consumption at 0.44 tonnes of oil equivalent (toe) per capita, the energy demand has steadily increased over the past decades and will continue to increase even further as the country embarks on economic growth (IEA, 2020). The energy system is largely dependent on the use of coal for power generation. In 2017, the country's total primary energy supply (TPES) was 882 Million toe with coal contributing 44% of the total TPES and the black gold has accounted for almost half of the total growth in energy supply between 2007 and 2017 (IEA, 2020).

ILR promises to be an alternative to its dependence on non-renewables. Hydropower has been packaged as a 'clean' and 'green' alternative to the crisis of an economy dependent on hydrocarbons and the ensuing rise in carbon emissions contributing to global warming. If completed, ILR is expected to provide the nation with 34,000 MW of energy, which is almost 40% of the total potential of hydropower generation in the country (Misra et al., 2007). The bulk of it will be generated within the Himalayan component, leveraging the hydraulic head that exists in the mountains and utilising the year-round availability of flow owing to glacier-melt and snowmelt (see Annexure for a detailed account). Hydropower continues to remain the most important renewable source of energy for the country, accounting for 9% of the total energy mix in 2017 and being the second-largest domestic source of power (IEA, 2020).

ILR is also stated to have some incidental benefits such as navigation facilities for trade and commerce, fisheries, salinity and pollution control, recreation facilities and employment generation. These benefits would require very little additional investments, if any, and contribute to the development of the regions through which these links would pass. These have not been quantified and, therefore, not accounted for in the Benefit-Cost Ratio and Financial Returns from each of these links. For instance, the Godavari (Inchampalli)– Krishna (Nagarjunsagar) link lists several indirect benefits due to assured irrigation enroute stretch of the link. It specifies that agricultural labourers and other groups engaged in the allied activities would get employed for longer durations. Moreover, agro-based industries, dairy and poultry farms and other industries that would cater to agricultural inputs such as pesticides and fertilisers, would also find a base in the region after an extension of irrigation facilities. Other benefits listed are better connectivity and infrastructure development, plantation of trees along canal banks, groundwater recharge and, finally, a better living condition for those who would be rehabilitated. The Benefit-Cost Ratio for this particular link stands at 3.41 (NWDA, 2020a).

Highlighting the entire range of benefits of ILR, Prabhu issued a clarion call for "the vital links" calling them an urgent need for the development of the agriculture sector. Further, bulwarking the plan from the "big vs small" debate, Suresh Prabhu had remarked that the approach should not be a dogmatic, but a need-based one. He clarified that both are important, but a mega-project remains desirable because of the plethora of benefits that it provides. He further impressed upon the readers that considering the huge and ever-increasing population of the country, a trade-off has to be between 'playing with nature' and 'harnessing the water' for all the benefit that it would provide. However, a staunch rationalist that he is, Prabhu has maintained that any decision on ILR must be based on scientific reasoning and the execution must be transparent. In the subsequent section, we use the same lens of scientific reasoning to reassess the links considering emerging realities, particularly the ecological concerns associated with such large-scale water transfers.

UNDERSTANDING THE CRITICISM FOR ILR

A steady repository of literature has emerged ever since the Union Government started pursuing the link in all earnest at the beginning of the last decade. Broadly, the arguments against ILR can be grouped into two categories, viz. those who question the underlying assumptions of undertaking the ILR as well as the stated benefits, while others believe that various cost-effective alternatives exist to the behemoth of ILR. When seen in combination, the case against undertaking such a large-scale IBWT certainly gets strengthened.

Flawed Assumptions and Miscalculated Benefits

To begin with, questions have been raised against the entire scheme of identifying certain basins as 'surplus' and some as 'deficit'. Bandyopadhyay and Perveen (2008) have pointed that the estimated water requirements for classifying river basins does not consider the diverse ecosystem services provided by river flows. They identify the conceptual framework of reductionist engineering to be behind this major flaw that only recognises the irrigation, domestic and industrial needs without any mention for the needs of the ecosystems. From a holistic perspective, there is no 'surplus' since each drop of water performs an ecological service and is not really a water going to waste. Another assumption has been that the excess water that cause floods in the Himalayan river basins can be diverted to other basins and, therefore, the imminent loss from floods in those basins can be averted. However, this will require massive diversion of flows and the creation of storage structures to store the additional water; the excess water will not be required even in the so-called deficit basin during the monsoons! Therefore, flood moderation seems to be quite misplaced and unfounded in real terms. Iyer (2012) has observed that the flow in the Ganga during monsoons can exceed 2,000,000 cusecs while the link canals that have been envisaged will only be able to divert 1500 cusecs. This is hardly any flood moderation and if storage structures are to be constructed for storing and diverting all the excess water during the monsoon, then the feasibility reports need to be reworked.

As far as hydropower generation is concerned, the plan is riddled with questionable assumptions. For one, the assumption that hydropower is 'clean and green' is extremely problematic. Ahlers et al. (2015) in their assessment of drivers, risks, and tensions around hydropower development in the Eastern Himalayas have noted that while hydropower may be renewable, the process of generating energy may leave the waterscape irreversibly damaged. Another assumption has been that it would reduce the dependence on coal (Platt et al., 2008). Considering the trend in the decade between 2007 and 2017 (IEA, 2020), it may well be the case that use of coal may continue growing with the hydropower not being able to mitigate whatsoever. Lastly, despite the claims that the links will largely make use of gravity and no lift exceeding 120 m will be required, Iyer (2012) raises doubt at a prima facie basis for all the 30 projects that have been planned. He further questions if the entire project would be a net generator of large quantities of energy as publicised.

Questions have also been raised about the overall sustainability of hydropower in the Himalayas considering the uncertainties that will be expressed due to climate variability. Future projections of power generation and/or flood control (since many a times such projects are sold as multi-purpose projects) may grossly fail to meet their targets (Ahlers et al., 2015). Similarly, the plan for IBWT has been prepared based on the assumption that the flow in the rivers will confirm with the past trends—not to forget that the initial planning and pre-feasibility reports considered data from the 80s or even earlier! Lal (2008) noted that the current policies affecting water use, management and development in India do not take into consideration how climate change will impact the Indian summer monsoon and the broader hydrological system.

Overlooking Alternatives: A Case for Local Solutions

A major criticism has been the entire focus of the project as supply-side solution to the challenge of water availability and the issues of water quality. D'Souza (2008) has noted that supply-side hydrology is a product of a specific historical and political movement. He further mentions that its main proponents such as engineers, private construction companies and bureaucracies like Central Water Commission fundamentally ignore the inherent complexities in water governance from the ecosystemic approach. This led to a search for a quick fix of supply augmentation instead of taking the more arduous route of demand management. As soon as water crisis is looked at as a problem of ineffective demand management, the horizon widens, and a lot of other possibilities emerge.

For instance, Amarasinghe (2012) showed that water productivity improvements could reduce the water demand and subsequently the surface and groundwater overdraft. He pointed out that the total Consumptive Water Use (CWU) from irrigation for cultivating food grains in the year 2000 was only about 155 km^3 while the withdrawals for food grains were 430 km^3. Therefore, better water management alone can take care of the need to increase the production of food grains with any need for additional irrigation. This can also help to bridge the gap between actual and maximum attainable yield. The spatial variation in water productivity is also quite skewed (Fig. 4). Even a quick glance would reveal that large clusters of districts in the central and western parts of the country, home to the bulk of the drought-affected area in the country, has poor water productivity regarding food grains. Another way to improve water use and also simultaneously the nutrient intake is by cultivating alternate cereals like maize, finger millet, pearl millet etc. than rice and wheat. Davis et al. (2018) have shown that if rice as a staple crop could be replaced by the other cereals mentioned before, then the irrigation water demand can be replaced by 33%

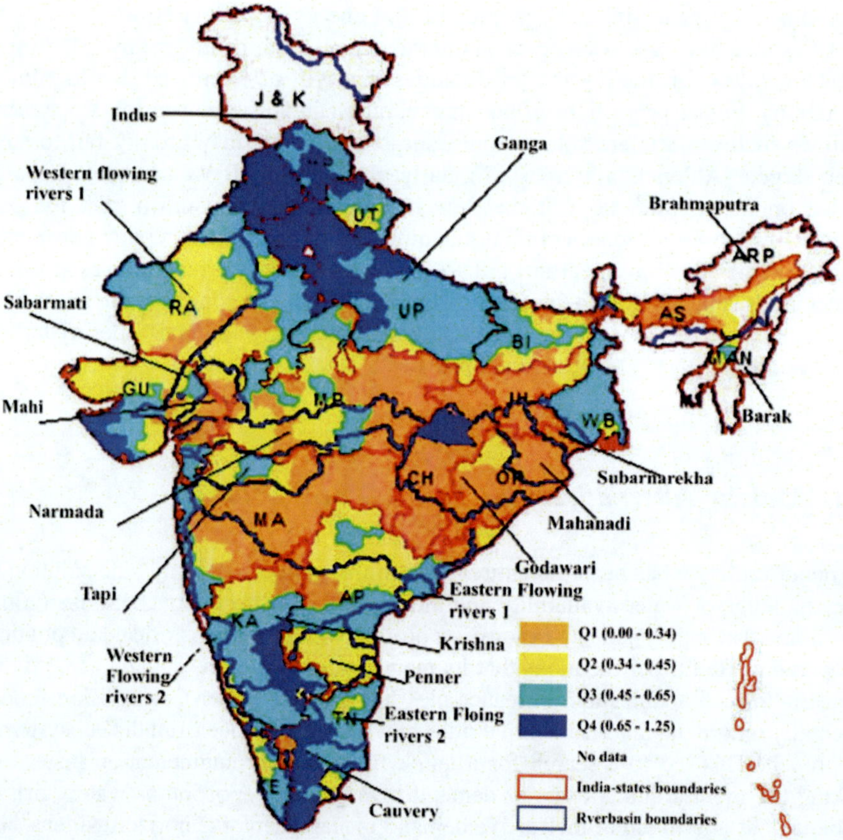

Figure 4. Variation of water productivity across districts in India. *Source*: Amarasinghe (2012)

while increasing the intake of protein, iron and zinc by 1%, 27% and 13%, respectively. Moreover, what is even more encouraging is that local knowledge to produce and consume such cereals already exists.

The need for ILR has also been questioned based on cost-effective alternatives that are tried and tested and better suited to meet the local requirements. Gopal and Marothia (2016) have positioned the traditional tanks, village ponds, and farm ponds as quick, cost-effective, and environmentally benign solutions to tackle the needs for water in the drought-prone regions such as Bundelkhand. There is also a very strong historical basis for constructing such structures as evidences suggest that several thousand large and small tanks had been constructed by the Chandela and Bundela rulers. However, due to neglect and encroachment, the time-tested solution for drought has been lost to the pages of history. Rajendra Singh (2008) (also known as 'waterman of India'), had also made a strong pitch for Indigenous Knowledge Systems as an effective way for society to live in a sustainable manner with nature and adapt to any situation of hydrological extremes.

Emerging Ecological Concerns

Since later half of the last century, a steady body of literature has emerged globally that shows the presence of flows in the rivers as critical for the functioning of riverine ecosystems—even high flows that may appear to be damaging and disruptive. Without the flow in the rivers, the rivers would not just starve but, in the long term, people and livelihoods dependent on the rivers in the downstream would be adversely and inequitably impacted. Calling the current proposal for addressing the twin challenges of floods and water scarcity by interlinking rivers as outdated and dangerous, Bandyopadhyay (2012) asserted that hydrological obscurantism has to make way for modern holistic water science. Bandyopadhyay further stated that the paradox is not 'surplus' and 'deficit' but rather it is the coexistence of scarcity and inefficient use of water. Therefore, hydro-luxury must not be extended to some while robbing others of their resource bases and livelihoods.

In a later piece, Bandyopadhyay (2019) has strengthened the foundations of the interdisciplinary Integrated Water Systems Governance (IWSG) by proposing an initial framework for a synergy-based perspective to the governance of flows in river basins. This framework is a diversion from the current reductionist perspective of a volumetric understanding of water in the river channel without considering the bio-geophysical constituents of flow or its linkages such as that with the subsurface water. It postulates an interdisciplinary perspective that considers the important constituents of flows—abbreviated as WEBS for Water, Energy, Biodiversity and Sediments. The interrelationship between the constituents of flow and the flow regime can be best understood through the concept of connectivity of rivers. Due to the heavy influence of South Asian Summer Monsoons in generating the flow in the rivers, the geomorphic functions in the channels such as erosion, transportation and deposition of sediments are most active during this period (Fig. 5). Variability in

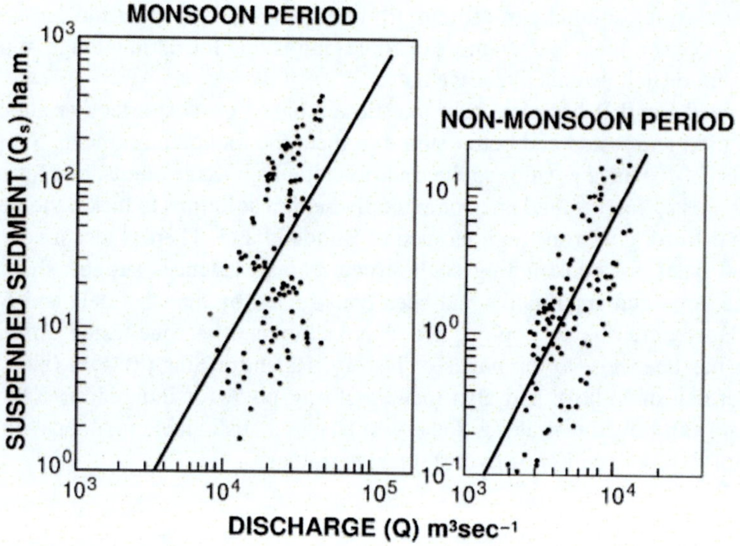

Figure 5. Power function relationship between flow (Q) and sediment transport (Q_s) for the monsoon (log Q_s = 1.79*log Q – 6296) and the non-monsoon (log Q_s = 2.018*log Q –7307). *Source*: Kale (1998)

inter-annual discharge and intra-monsoonal high flow magnitude is an integral part of the fluvial regime of Indian rivers (Kale and Gupta, 2001). The high flow in the rivers allow for the connectivity to be established wherein matter (water, solutes, sediment, organic matter) and organisms can move among patches in a landscape or ecosystem (Wohl, 2017). This movement is essentially tri-directional—lateral, longitudinal, and vertical.

Water is a conduit for the movement of matter and organisms. With a diversion of flow, as envisaged through ILR, the riverine ecosystem will be severely impacted and irreversible changes can occur. For example, there is a high degree of correlation between flow and sediment transport in the river (Fig. 5). Structural interventions on the rivers have already impeded to supply of sediments to regions in the downstream. Any diversion of flow would further aggravate the problem. This will be particularly disastrous for the Himalayan rivers. For example, according to Rahman et al. (2018), it has been estimated that the sediment load in the combined flow of the Ganga and Brahmaputra is decreasing at a rate of 4-10 MT/year. From a previous estimation of mean annual sediment load of 1.0 to 2.4 billion tonnes per year, the sediment load in the GBM system had declined to 500 million tonnes per year in 2015. However, as of now, this is still sufficient to offset the combined effects of land subsidence and sea-level rise but the unsustainable practices in the upstream and middle reaches could soon push the balance beyond a certain threshold. The diversion of flow would qualify as one such unsustainable practice.

ILR IN LIGHT OF EMERGING NEW PARADIGM OF INTEGRATED WATER GOVERNANCE

Water governance is going through a paradigm change across the world. The new emerging paradigm of governance recognises that the traditional reductionist engineering thinking of supply development through structural interventions over flow regimes is unsustainable. The traditional paradigm of "arithmetic hydrology" views water in terms of a numerical measure of availability without any concern of the ecosystems, and its linkages with livelihoods. This vision is dominated by the thinking of "water for food security" associated with some use of water in industry, hydropower, and urban sectors, for which supply augmentation plans through engineering interventions were thought of as offering the solutions. The new emerging paradigm of Integrated Water Resources Management (IWRM) or Integrated Water Governance (IWG) adds a new dimension to the reductionist thought process, by proclaiming the notion of "water for ecosystems" (Falkenmark and Rockström, 2004). In the process this new paradigm recognises the human society as a subsystem in the biosphere in which water is a key element (Ghosh, 2018a). The question is 'does river interlinking comply with the broad contours of integrated water governance'. This has largely been stated in the previous sections. Here we articulate it with some of the tenets of the new emerging paradigm. Following Ghosh (2018b) these tenets may be summarised as follows:

(a) *Water is viewed as an integral part of the global hydrological cycle, and not as a stock of material resource to be used for the satisfaction of human requirements.* In an integrated governance approach, water is viewed in the context of the totality of the global hydrological cycle. Therefore, the lack of acknowledgement of ecological cost due to interventions over streamflows elsewhere is an inbuilt subsidy to use water for economic purposes at will (Flessa, 2004). Unfortunately, ILR is way away from this thinking. Rather, with the thinking that water is a stock of resource, the idea of transfer of water from "surplus" to "deficit" basins as identified in an unpublished document by Mohile (1998) is a clear subscription to the arithmetic hydrology paradigm.

(b) *Supply of ever increasing volumes of water is not a pre-requisite for continued economic development. Hence, solutions to the problems of water resource development need not be searched in supply side management alone.* ILR is exactly the opposite. It is a mega-supply-augmentation project based on massive engineering interventions with no consideration of the negative externalities.

(c) *Clear and strict prioritization of various types of needs and demands for water is needed, including those of the ecosystems.* The new and interdisciplinary paradigm assigns clear priorities to the various competing requirements of water. The competing needs primarily involve two levels, viz., between the needs of ecosystems and the needs of human societies. This is completely missing in ILR. Rather, while the paradigm of integrated water governance talks of competing needs of economy and ecosystem (or the short and long-runs), ILR stands out as a myopic human intervention to address short-term economic needs at the cost of the long-run sustainability concerns.

(d) *There is a need for comprehensive assessment of the water development projects keeping the integrity of the full hydrological cycle.* This will necessitate the creation of an interdisciplinary knowledge base for objective and comprehensive assessments of the pros and cons of water infrastructure projects. In that sense, concern of the economy, society and ecosystem needs to be integrated. The integrated development paradigm has shifted to reconciling between the irreconcilable trinity of equity, efficiency, and sustainability. Unfortunately, assessments of ILR have so far been piecemeal and fragmented. There is no evidence of any comprehensive assessment even at individual basin link levels. The concern of social conflicts as also possibilities of incurring social costs related to transboundary water conflicts have been totally ignored.
(e) *A transparent and interdisciplinary knowledge base for the understanding of the social, ecological and economic roles played by water resources is required.* At the very outset, ILR is an opaque project. Independent scientific assessments on the projects are conspicuous by their absences. At least two cherished institutions justified the feasibility of this project based on such large-scale economic benefits. One such is the National Council for Applied Economic Research (NCAER) which in a 2008 study of the economic impacts of the interlinking project, justified the same on grounds of agricultural benefits. The other is International Water Management Institute (IWMI), Colombo. While the NCAER study is bereft of anything beyond economic benefits (without any consideration of the larger costs), thereby making it narrow and uninformed, IWMI has based their contentions on some flawed presumptions of river basin governance that have been challenged worldwide. IWMI assessments even talk of maintaining "environmental flows" (eflows) giving the impression as if such "minimum flow requirements" can conserve the river ecosystem. This very delineation of "maintaining eflows" reducing it to a specific number is also flawed from the perspective of global paradigm changes. None of these are adequately informed from a trans-disciplinary perspective.
(f) *Droughts and floods are to be visualized in the wider context of the ecological processes associated with them.* There is no doubt that ILR proposes exactly the opposite with its intentions of tampering with the ecological processes of droughts and floods.

Given the above arguments, it is clear that river interlinking in no way is in conformity with the contours of the new emerging discourse of Integrated Water Governance.

CONCLUDING REMARKS

The idea of interlinking and the objective with which this project has been conceived is indeed noble— promotion of "water equity" and "water security." For somebody, it might have been a dream project, but in reality, it is a myth. Given the constraints presented, it does not seem that a project of this mammoth proportion will really be sustainable. Indian water policymakers are still bogged in the traditional engineering framework, without having any consideration for social and environmental problems that has to enter into everyday strategic and economic decision-making. They are clearly oblivious of the fact that a lot of water has flown down the drain, ever since the U.S. embarked on the "dam building decades." The U.S. has witnessed the economic benefits, but has also counted on the ecological costs. Ever since the realization of the loss in ecosystem services, and how ecosystem services affect economic well-being, they have incorporated ecosystem concerns into economic decision-making. As a result, the U.S. and Europe have readily realized that often large water infrastructure projects need to be decommissioned for the greater common good. The fact remains that even if India intends to materialize a project as big as the ILR, it has to place adequate data and information in the public domain for independent scientific analysis and informed debate. Quite unfortunately, very little information on ILR has been placed in the public domain, and this has actually led to further enragement and agitation by civil society. In any case, Indian policymakers need to understand that they have to think of demand management of water resources, rather than relying on purely supply side interventions that are costly from economic, social, and ecosystem perspectives.

Merely confining water governance to traditional engineering paradigms is an outmoded policy. Water resource management has steadily emerged as a transdisciplinary paradigm. While transdisciplinary interactions, as argued by Falkenmark et al. are extremely important to evolve with better practices on water management, one may safely conclude that ILR is thoroughly devoid of such an interaction. Rather, it falls in the realm of the old paradigm, where disciplines were not intersecting with each other in a way that truly helped understand the potential contributions of other areas of competence, not even from closely neighbouring disciplines. However, under the changed scenarios of demand, the new paradigm requires a real understanding of the nature of water resources, their complex links and interrelations with other systems, and how societies manage them. Such complex interactions can no longer remain the domain of compartmentalized sector and single-disciplinary approaches. That is why an independent scientific assessment of ILR from a transdisciplinary perspective becomes important. Ecological economics, social sciences, livelihood problems, developmental studies, and ecosystem sciences have to be integrated with traditional civil engineering to create a framework of holistic engineering. Such a holistic ecosystem engineering framework should be embedded in the evaluation of all large water infrastructure projects, including that of ILR. Before such a holistic evaluation takes place, it would really be unwise to link the "dream with the myth"!

ANNEXURE

Peninsular Component

Sl. No.	Name	States concerned	States benefited	Annual irrigation (Lakh ha)	Domestic and industrial supply (MCM)	Hydro-power (MW)	Status
1	Mahanadi (Manibhadra)- Godavari (Dowlaiswaram) link	Orissa, Maharashtra, Andhra Pradesh, Karnataka, & Chattisgarh,	Andhra Pradesh & Orissa	0.91+3.52 =4.43	802	445	Feasibility report completed
2	Godavari (Inchampalli)- Krishna (Pulichintala) link	Orissa, Maharashtra, Madhya Pradesh, Andhra Pradesh, Telangana, Karnataka &Chattisgarh,	Telangana & Andhra Pradesh	1.09+5.04 =6.13	413	--	Feasibility report completed
3	Godavari (Inchampalli)- Krishna (Nagarjunasagar) link	-do-	Telangana	2.87	237	975	Feasibility report completed
4	Godavari (Polavaram)- Krishna (Vijayawada) link	Orissa, Maharashtra, Andhra Pradesh, Karnataka, & Chattisgarh	Andhra Pradesh	5.82	162	--	Feasibility report completed
5	Krishna (Almatti)- Pennar link	-do-	Andhra Pradesh & Karnataka	1.90+0.68 =2.58	56	--	Feasibility report completed
6	Krishna (Srisailam)- Pennar link	Maharashtra, Andhra Pradesh, Karnataka & Telangana	--	--	--	17	Feasibility report completed

7	Krishna (Nagarjunasagar)-Pennar (Somasila) link	Maharashtra, Andhra Pradesh & Karnataka,	-do-	5.81	124	90	Feasibility report completed
8	Pennar (Somasila)-Cauvery (Grand Anicut) link	Andhra Pradesh, Karnataka, Tamil Nadu, Kerala & Puducherry	Andhra Pradesh, Tamil Nadu & Puducherry	0.49+4.36 +0.06 =4.91	1105	--	Feasibility report completed
9	Cauvery (Kattalai)-Vaigai - Gundar link	Karnataka, Tamil Nadu, Kerala & Puducherry	Tamil Nadu	3.38	185	--	Feasibility report completed
10	Ken-Betwa link a) Ken-Betwa Link Phase-I b) Ken-Betwa link Phase-II	Uttar Pradesh & Madhya Pradesh - do-	Uttar Pradesh & Madhya Pradesh Madhya Pradesh	2.66 + 3.69 =6.35 0.99	496	78	DPR Phase-I completed in April 2010 & DPR Phase-II completed in January 2014.
11	Parbati-Kalisindh-Chambal link	Madhya Pradesh, Rajasthan & Uttar Pradesh (UP requested to be consulted during consensus building)	Madhya Pradesh & Rajasthan	*Alt.I: 2.05+0.25 =2.30 (Alt.II : 1.77+0.43 =2.20)	13.2	--	Feasibility report completed
12	Par-Tapi-Narmada link	Maharashtra & Gujarat	Gujarat	2.32	76	22	DPR completed in August, 2015
13	Damanganga- Pinjal link (As per DPR)	Maharashtra & Gujarat	Maharashtra (only water supply to Mumbai)	--	895	--	DPR completed in March 2014
14	Bedti-Varda link	Maharashtra, Andhra Pradesh & Karnataka	Karnataka	0.60	--	4	Pre-feasibility report completed
15	Netravati- Hemavati link	Karnataka, Tamil Nadu & Kerala	Karnataka	0.34	--	--	Pre-feasibility report completed
16	Pamba- Achankovil-Vaippar link	Kerala & Tamil Nadu,	Tamil Nadu	0.91	--	508	Feasibility report completed

(continued)

Himalayan Component							
Sl. No.	Name of the Link	States/Country concerned	States benefited	Annual irrigation (Lakh ha)	Domestic and industrial supply (MCM)	Hydro-power (MW)	Status
1.	Manas-Sankosh-Tista-Ganga (M-S-T-G) link	Assam, West Bengal, Bihar & Bhutan	Assam, West Bengal & Bihar	2.08 + 1.82 + 2.64 = 6.54	--	5287	FR in progress
2.	Kosi-Ghaghra link	Bihar, Uttar Pradesh & Nepal	Bihar & Uttar Pradesh	8.17+ 0.67 + 1.74 (Nepal) = 10.58	48	--	FR in Indian portion in progress
3.	Gandak-Ganga link	-do-	Uttar Pradesh	37.99+2.41 (Nepal) = 40.40	700	--	Draft FR completed (for Indian portion)
4.	Ghaghra-Yamuna link	-do-	Uttar Pradesh	25.30 + 1.35 (Nepal) =26.65	1391	10884	FR completed (for Indian portion)
5.	Sarda-Yamuna link	Bihar, Uttar Pradesh, Haryana, Rajasthan, Uttarakhand & Nepal	Uttar Pradesh & Uttarakhand	3.45 + 0.30 = 3.75	6250	3600	FR completed (for Indian portion)
6.	Yamuna-Rajasthan link	Uttar Pradesh, Gujarat, Haryana & Rajasthan	Haryana & Rajasthan	0.435 + 2.442 = 2.877	57	--	Draft FR completed
7.	Rajasthan-Sabarmati link	-do-	Rajasthan & Gujarat	5.35 + 2.04 = 7.39	282	--	Draft FR completed
8.	Chunar-Sone Barrage link	Bihar & Uttar Pradesh	Bihar & Uttar Pradesh	0.30 + 0.37 = 0.67	--	--	Draft FR completed
9.	Sone Dam-Southern Tributaries of Ganga link	Bihar & Jharkhand	Bihar & Jharkhand	2.99 + 0.08 = 3.07	360	95	FR in progress

10.	Ganga(Farakka)-Damodar- Subernarekha link	West Bengal, Orissa & Jharkhand	West Bengal, Orissa & Jharkhand	7.63 + 0.30 + 0.55 = 8.47	484	-	Draft FR completed
11.	Subernarekha-Mahanadi link	West Bengal & Orissa	West Bengal & Orissa	0.18 + 0.365 = 0.545	--	9	Draft FR completed
12.	Kosi-Mechi link	Bihar, West Bengal & Nepal	Bihar	2.99 + 1.75 (Nepal) = 4.74	24	3180	PFR completed. FR to be taken up Entirely lies in Nepal
13.	Farakka-Sunderbans link	West Bengal	West Bengal	1.50	184	--	Draft FR completed
14.	Jogighopa-Tista-Farakka link (Alternative to M-S-T-G)	-do-	Assam, West Bengal & Bihar	----	216	1115	Alternative to M-S-T-G Link. Not to be taken up.

References

Ahlers, R.J. Budds, D. Joshi, V. Merme and M. Zwarteveen (2015). "Framing Hydropower as Green Energy: Assessing Drivers, Risks and Tensions in the Eastern Himalayas." *Earth Syst. Dynam.* 6 (1): 195–204. https://doi.org/10.5194/esd-6-195-2015.

Amarasinghe, U., B.R. Sharma, N. Aloysius, C. Scott, V. Smakhtin, C. de Fraiture, A.K. Sinha and A.K. Shukla (2004). Spatial Variation in Water Supply and Demand across River Basins of India. IWMI Research Report 83. *International Water Management Institute*. Colombo.

Amarasinghe, Upali (2012). The National River Linking Project of India: Some Contentious Issues. *IWMI-TATA Water Policy Research Highlight*. Vol. 16. https://hdl.handle.net/10568/34700.

Bandyopadhyay, Jayanta (2012). Water Science in India: Hydrological Obscurantism. *Economic and Political Weekly* 47 (16): 45–47. http://www.jstor.org/stable/23214595.

Bandyopadhyay, Jayanta (2019). Towards a Synergy-Based Approach to River Basin Governance | ORF. Observer Research Foundation. August 2019. https://www.orfonline.org/research/towards-a-synergy-based-approach-to-river-basin-governance-54786/.

Bandyopadhyay, Jayanta and Shama Perveen (2008). The Interlinking of Indian Rivers: Questions on the Scientific, Economic and Environmental Dimensions of the Proposal. In *Interlinking of Rivers in India: Issues and Concerns*, edited by M. Monirul Qader Mirza, Ahsan Uddin Ahmed, and Q.K. Ahmad, 53. London: CRC Press. https://doi.org/10.1201/9780203894576.

Biswas, Asit K. and Cecilia Tortajada (2001). Development and Large Dams: A Global Perspective. *International Journal of Water Resources Development* 17 (1): 9-21. https://doi.org/10.1080/07900620120025024.

Cotton, Arthur (1858). Report on the Mahanuddy River. *Calcutta, May*. Calcutta.

CSO (2018). EnviStats-India 2018: Supplement on Environmental Accounts, Central Statistics Office, Ministry of Statistics & Programme Implementation, Government of India, New Delhi. http://mospi.nic.in/sites/default/files/reports_and_publication/statistical_publication/EnviStats/EnviStats_India_27sep18.pdf

Davis, Kyle Frankel, Davide Danilo Chiarelli, Maria Cristina Rulli, Ashwini Chhatre, Brian Richter, Deepti Singh and Ruth DeFries (2018). Alternative Cereals Can Improve Water Use and Nutrient Supply in India. *Science Advances* 4(7): eaao1108. https://doi.org/10.1126/sciadv.aao1108.

D'Souza, Rohan (2003). Canal Irrigation and the Conundrum of Flood Protection: The Failure of the Orissa Scheme of 1863 in Eastern India. *Studies in History* 19 (1): 41-68. https://doi.org/10.1177/025764300301900103.

D'Souza, Rohan (2008). River-Linking and Its Discontents: The Final Plunge for Supply-Side Hydrology in India. *In: Water First: Issues and Challenges for Nations and Communities in South Asia,* edited by Kuntala Lahiri-Dutt and Robert J. Wasson, 99-121. SAGE Publications India Pvt Ltd., New Delhi.

Falkenmark, Malin and Johan Rockström (2004). Balancing water for humans and nature: The new approach in ecohydrology. Earthscan.

Flessa, K.W. (2004). Ecosystem services and the value of water in the Colorado River delta and Estuary, USA and Mexico: Guidelines for mitigation and restoration. *International Seminar on Restoration of Damaged Lagoon Environments*, 79-86. Matsue, Japan.

Ghosh, Nilanjan (2018a). From Reductionist to Holistic Paradigm: Combining Ecology, Economics, Engineering, and Social Sciences in a Transdisciplinary Framework for Water Governance, *Ecology, Economy and Society—the INSEE Journal*; 1(2): 69-72.

Ghosh, Nilanjan (2018b). Water, Ecosystem Services, and Food Security: Avoiding the costs of ignoring the Linkage, in Kathuria, R., S. Ray and K. Bandyopadhyay (Eds.) *Low Carbon Pathways for Growth in India,* 161-176 (New Delhi: Springer).

Gopal, Brij and Dinesh K. Marothia (2016). Seeking Viable Solutions to Water Security in Bundelkhand. *Economic and Political Weekly*. https://www.epw.in/journal/2016/44-45/commentary/seeking-viable-solutions-water-security-bundelkhand.html.

IEA (2020). India 2020 Energy Policy Review. https://niti.gov.in/sites/default/files/2020-01/IEA-India 2020-In-depth-EnergyPolicy.pdf.

Iyer, Ramaswamy R. (2002). Linking of Rivers: Judicial Activism or Error? *Economic and Political Weekly* 37(46): 4595–96. http://www.jstor.org/stable/4412834.

Iyer, Ramaswamy R. (2012). River Linking Project: A Disquieting Judgment. *Economic and Political Weekly* 47(14): 33–40. http://www.jstor.org/stable/23214676.

Kale, V.S. and A. Gupta (2001). *Introduction to Geomorphology*. Kolkata: Orient Longman.

Kale, Vishwas S. (1998). Monsoon Floods in India: A Hydro-Geomorphic Perspective. *Memoirs-Geological Society of India*, 229–259.

Lal, Murari (2008). Implications of Climate Change in South Asia on the Interlinking Project of Indian Rivers. In *Interlinking of Rivers in India: Issues and Concerns*, edited by M Monirul Qader Mirza, Ahsan Uddin Ahmed, and Qazi Kholiquzzaman Ahmad, 1st ed. London: CRC Press. https://doi.org/10.1201/9780203894576.

Ministry of Irrigation (1980). National Perspective for Water Resources Development. Accessed October 19, 2020. http://nwda.gov.in/upload/uploadfiles/files/NATIONAL%20PERSPECTIVE(1).pdf.

Misra, Anil Kumar, Anju Saxena, Manish Yaduvanshi, Ajai Mishra, Yogendra Bhadauriya and Alok Thakur (2007). Proposed River-Linking Project of India: A Boon or Bane to Nature. *Environmental Geology,* 51(8): 1361–1376. https://doi.org/10.1007/s00254-006-0434-7.

Mohile, A.D. (1998). India's Water and its Plausible Balance in Distant Future. Unpublished paper as cited in NCIWRDP (1999a:29).

MOWR (2002). National Water Policy. New Delhi. http://jalshakti-dowr.gov.in/sites/default/files/nwp20025617515534_1.pdf.

Nandargi, S., O. Dhar, M.M. Sheikh, Brenna Enright and Monirul Mirza (2010). Hydrometeorology of Floods and Droughts in South Asia – A Brief Appraisal. *In: Global Environmental Changes in South Asia.*, edited by A.P. Mitra and C. Sharma, 244–257. Dordrecht: Springer, Dordrecht. https://doi.org/10.1007/978-1-4020-9913-7_8.

Nehru, Jawaharlal (1988). *Jawaharlal Nehru on Science and Society: A Collection of His Writings and Speeches*. Edited by Singh Bandev. New Delhi: Nehru Memorial Museum and Library. https://www.indianculture.gov.in/ebooks/jawaharlal-nehru-science-and-society-collection-his-writings-and-speeches.

NWDA (2020a). Feasibility Report of Godavari Inchampalli Krishna Pulichintala Link Project. http://www.nwda.gov.in/content/innerpage/FRofG-Pulichintala.php.

NWDA (2020b). National Perspectives for Water Resources Development : National Water Development Agency. 2020. http://www.nwda.gov.in/content/innerpage/click-more.php.

Platt, Kobi, Sharon Gourdji, Carrie Knowlton and Michael, J. Wiley. (2008). India's Energy Future and Interlinking of Rivers. In *Interlinking of Rivers in India: Issues and Concerns*, edited by M Monirul Qader Mirza, Ahsan Uddin Ahmed, and Qazi Kholiquzzaman Ahmad, 1st ed., 129-139. London: CRC Press. https://doi.org/10.1201/9780203894576.

Radhakrishna, B.P. (2003). Linking of Major Rivers of India—Bane or Boon? *Current Science* 84 (11): 1390–94. http://www.jstor.org/stable/24108378.

Rahman, Munsur, Maruf Dustegir, Rezaul Karim, Anisul Haque, Robert, J. Nicholls, Stephen E. Darby, Hajime Nakagawa, Motahar Hossain, Frances E. Dunn and Marin Akter (2018). Recent Sediment Flux to the Ganges-Brahmaputra-Meghna Delta System. *Science of The Total Environment* 643: 1054–64. https://doi.org/10.1016/j.scitotenv.2018.06.147.

SANDRP (2003). The Mindlessness Called River Linking Proposals. New Delhi. http://wrmin.nic.in/interbasin/default8.htm.

Shah, Tushaar (2010). *Taming the Anarchy: Groundwater Governance in South Asia*. Routledge.

Singh, Dharamveer and Pankaj Singh (2004). Interlinking of Rivers. Practical Lawyer, Eastern Book Company. 2004. https://www.ebc-india.com/lawyer/articles/836.htm.

Singh, Rajendra (2008). The Indigenous Knowledge Systems of Water Management in India. In *Interlinking of Rivers in India: Issues and Concerns*, edited by M. Monirul Qader Mirza, Ahsan Uddin Ahmed, and Q.K. Ahmad, 1st ed., 235. London: CRC Press. https://doi.org/10.1201/9780203894576.

Vombatkere, S.G. (2016). Interlinking Rivers: Cupidity Or Stupidity. The Citizen. 2016. https://www.thecitizen.in/index.php/en/newsdetail/index/4/9255/interlinking-rivers-cupidity-or-stupidity.

Wohl, Ellen (2017). Connectivity in Rivers. *Progress in Physical Geography: Earth and Environment,* 41 (3): 345-362. https://doi.org/10.1177/0309133317714972.

Yadav, Krishna Nand (2007). Interlinking of Rivers in India: Costs and Benefits. In *Interlinking of Rivers in India: Costs and Benefits*, edited by Anil Kumar Thakur and Kumari Pushpa. Deep and Deep Publications.

Chapter 16
Dynamics of Transboundary River Conflicts vis-à-vis Nature Based Negotiated Approach (NBNA) Solution: Case Study River Teesta

Jayanta Basu

INTRODUCTION

As the population across the globe is rising, and in tandem the per capita consumption rate, the pressure on limiting environmental resources is also on the rise; which, in turn, is catering to the enhanced number of environmental conflicts in recent decades including the transboundary ones. According to a global environmental conflicts report, there were nearly 1800 such conflicts in 2018 (Temper et al., 2018). A 2020 study by Scheidel and Martinez-Alier mapped and analyzed 2743 such cases across the world in-between 2011 and March 2019, and found that such conflicts have been increasing with time. The report found that 95% of such cases happened after 1970; while 50% of the cases originated during or after 2008, a clear indicator of the trend (Scheidel et al., 2020).

A more recent mapping of such conflicts done through the dynamic Environmental Justice Atlas process (till January 13, 2021) documented 3342 environmental conflicts, out of which more than one-tenth (342) are located in India; highest for any country in the world. Overall South Asian region shows dominance in this regard having 13.3% of the global conflicts, though it occupies only 3.5% global landmass. Water issues, including those related to trans-boundary rivers, stand high in the list of environmental conflicts (EjAtlas, 2021).

According to EJAtlas, out of 3342 conflicts, water related issues with 693 conflicts (about 21%) is placed only second to land issues. Studies point out that global water withdrawal has nearly septupled over the last century, outpacing population growth by a factor of 1.7 (HDR, 2019); that further vindicates the rise of such conflicts. It is

J. Basu (✉)
Environment Governed Integrated Organisation (EnGIO), Kolkata, WB, India

Department of Environment, Calcutta University, Kolkata, WB, India

The Telegraph, India, Kolkata, WB, India

A. Mukherjee (ed.), *Riverine Systems*, https://doi.org/10.1007/978-3-030-87067-6_16

believed that unless the Nature Based Negotiated Approach (NBNA) is taken, particularly within countries having transboundary river or water conflicts, the situation will only worsen.

WATER CONFLICTS: HYDRO POLITICS OCCUPY THE PIVOTAL POINT

Almost every region of the world has water related conflicts including river related rifts within two or more countries. The rifts mainly occur over access and right to control the water resources and being largely driven by geopolitical interests. This is actually a historical trend as wide range of water conflicts had happened over the years as water is often considered synonymous with the development of any region. An exhaustive database about the water-linked conflicts, the Water Conflict Chronology, has been documented by the Pacific Institute that goes back nearly 6000 years; and clearly highlights the underpinning of politics in the process (Pacific Institute, 2019)

The recent trend is no different with water related conflicts becoming more frequent and formidable as hydro-politics, more specifically river related transboundary rifts, turning increasingly overt and louder; taking a pivotal position in the process. Gradual reduction of resources with increasing impact from climate change has further exaggerated and widened the rifts. In most cases, the upper and lower riparian river conflicts have been found to be dragging over years with no sustainable solutions found in absence of political willingness to settle the same.

The case in point is the Nile basin, which witnesses significant conflict within eleven riparian countries on Nile river water where the negotiations have reached a plateau since 2007 as a result of diverging interests between upstream and downstream countries. **Similarly, Turkey, Syria and Iraq have a running feud over the Euphrates-Tigris** that had been fed by, and also feeds, the inherent political tensions between the countries. Afghanistan's efforts to use Helmand River and the Harirud to support post-conflict reconstruction and development have sent warning signals to Iran, particularly to water security in its eastern and northeastern provinces, which share these rivers. The Mekong basin is also witnessing a mega expansion of dam-building for hydropower generation, especially in China and Laos; which has led to political tensions as lower riparian countries to the dams fear they will be facing several impacts like greater flooding to seasonal lack of water (Reliefweb, 2017). India, Bangladesh, Nepal and Pakistan also share various such transboundary river rifts.

On the other hand, Turkish-Armenian one is a unique and contrasting case study that underlines how political cooperation between two co-riparians can reduce the tension and lead to agreed share of transboundary waters (Altingoz and Ali, 2019).

The rift also operates in intra-country level. Southern states of India have been fighting over the rivers Kaveri (Mantri, 2018), and Krishna (Mohan, 2020) rivers and the rift has not only widened over years but actually got sucked into political quagmire of the region. Competing parties moved to Judiciary for formalizing their rightful claim, as they believe, but judicial directives are unlikely to offer a sustainable way out to solve an issue; which needs socio-political intervention. There is similar discontent within Indian states about sharing Ganges water, and it is claimed that the state like West Bengal, which covers the end part of Ganga in India (called Hoogly river), actually even does not receive 2% of river water generated in Gomukh (the point where Ganges river is born) due to large scale water extraction and utilisation in upper part of Ganga basin that normally leads to significant horizontal disconnect in the main course of river during non-monsoon months (Basu, 2019).

It is felt that unless a holistic river usage policy is framed, considering and integrating social; economical; environmental and legal provisions, irrational and lopsided river use is set to continue being predominantly prodded by narrow political interests. While river politics mainly operates at macro-level, micro-level hydro politics has also been increasing rapidly whether it is about the use of ground water, provision of drinking water or maintenance of waterbodies and wetlands; which, often, feeds to rivers.

In fact, sustenance of natural resources, generally, is hardly considered important by a large section of political leadership, administration and society *per se*; with the financial return, albeit illegal, often proving the driver behind the indifference of a section of decision makers. This, in-turn, leads to large scale organised spoiling of the resources. It is a common practice, especially in Asia, Africa and Latin America, to find politicians and environmental violators working in tandem—both covertly and overtly—to maximise the short-term return of unsustainable exploitation of natural resources following the 'political rent seeking' model as they say in economics (Basu, 2021). Water is no exception.

TRANSBOUNDARY RIVER PARADIGM IN SOUTH ASIA

If one considers South Asia, it is found that most countries in the region and around, like China, India, Pakistan, Bangladesh and Nepal, are at rift over the rivers like Ganga, Brahmaputra, Sindhu or Indus, Teesta and likewise. India and Bangladesh, otherwise geopolitically close, have decade-old disputes over water sharing of Ganga (Fig. 1) and Teesta (Fig. 2), as both countries want to get significant share of dwindling water resources being hampered by climatic change, human intervention and a range of issues. Similarly, river rift adds to already stretched bilateral political relationship of India and Pakistan while Nepal and India also has long term transboundary river issues.

South Asian trans-boundary river issues are linked strongly to regional geopolitics due to three main factors:

Figure 1. Ganga-Brambhaputra-Meghna basin (Source: Dholakia, 2019)

Figure 2. The Teesta channel and some of the studied landmarks along the river(Source: The Third Pole).

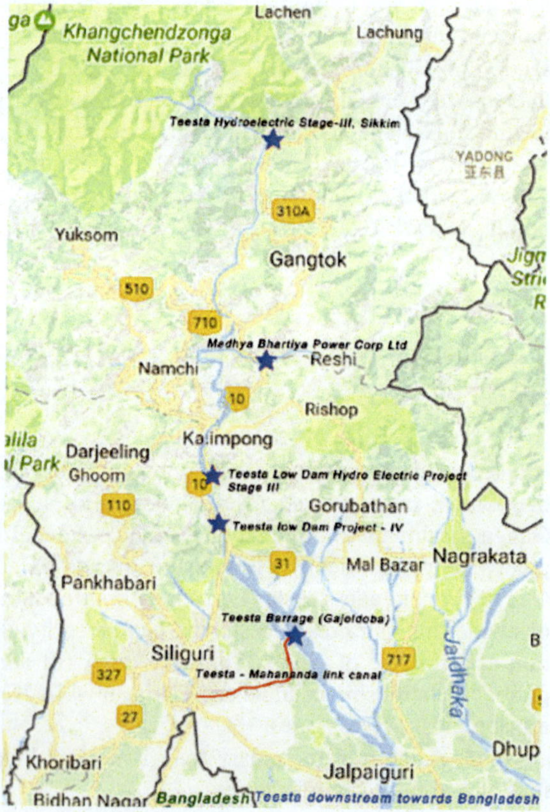

1. The main river systems are generally circum (Himalayan).
2. All countries in the region strongly depend on rivers for agriculture, hydroelectricity and also for a range of reasons.
3. Unequal political power positioning in the region.

India, Pakistan, Bangladesh, Bhutan, Nepal and Afghanistan share twenty major rivers. The Indus basin (consisting of the Indus, Ravi, Beas, Sutlej, Jhelum and Chenab rivers) inter-links India, Pakistan and China, while the Brahmaputra and the Ganges basins inter-link China with India (Fig. 1), Nepal, Bangladesh and Bhutan. The Kosi, Gandaki, and Mahakali rivers join Nepal with India. Major rivers shared between India and Bangladesh include Brahmaputra, Ganges and Teesta. Pakistan and Afghanistan share the Kabul river basin (Dholakia, 2019).

It is seen that the main river systems (Indus, Ganges and Brambhaputra) are connected to Tibet, making China a major player in the process, while there are reasonable basin overlap among the upper riparian and lower riparian countries in the region that enhances the geopolitical angle. The fact that both China and India, the largest countries in the region, has hydropower and river linking aspiration; does not help the situation.

The situation often gets more complicated and terser as the upper and lower riparian water relationship also gets influenced by regional and local political relationships. State of West Bengal has become an important player in the Ganga and Teesta river water sharing between India and Bangladesh, and state political leaderships over the years have played, and are playing, key roles in forging, or not forging, the India – Bangladesh national level agreements on river water sharing (Basu, 2017a).

It is also to be considered that sparring countries often have different, and even contradictory sectoral interests regarding the use of river water, that further magnifies the fissure within countries sharing a common river source. For example, while China tries to milk the hydroelectric potential of river Brahmaputra, India and Bangladesh predominantly use the river water for agriculture (Ghosh, 2020).

INDIA BANGLADESH RIVER DYNAMICS

Within South Asia, India-Bangladesh transboundary river dynamics, and the related political play, needs special mention due to sheer enormity of the split resources. Incidentally, India (almost entirely the state of West Bengal) and Bangladesh shares 54 trans-boundary rivers (Nishat et al., 2014).

The political tension has operated back and forth for about last four decades over the appropriate allocation, access and development of river water; initially with river Ganges and subsequently with river Teesta (Fig. 2). After a longstanding dispute, two countries could ink comprehensive bilateral treaty in 1996 for sharing Ganges water for next 30 years; i.e. 2025 (Haq, 2012). Despite the agreement in place, many in Bangladesh claim that the country has not been getting its fair share of water and criticizes the role of Farakka barrage in the context (Sayanangshu, 2020).

Subsequently, following the Ganges water sharing model, Bangladesh demanded an agreement on Teesta water sharing, claiming that a significant part of the country has been suffering as inadequate non monsoon water flows from Indian state of West Bengal to Bangladesh vis-à-vis undertaking cultivation within the command area of Teesta belonging to Bangladesh part. Despite many rounds of talks and even draft agreements on sharing being arrived at, the official agreement could not be reached yet as Indian state of West Bengal had disagreement on some of the clauses. The endorsement of West Bengal is critical to ink the agreement as water, including river, is on concurrent list of Indian constitution and, hence, union government in India cannot formalize the agreement without the official consent of West Bengal government. The issue has become a major political thorn for Indo-Bangladesh relationship and even several rounds of meetings in the highest administrative levels of the two countries could not resolve the disagreement. (**Majumdar**, 2017).

Since 2015, political posturing and counter posturing has also initiated with river Atreyi, which happens to be coming to India from Bangladesh, with river dependent community in West Dinajpur district in West Bengal complaining that disproportionate volume of non-monsoon flow of the river is being shared to the neighbouring country, leading to significant discontent in Indian side. There have been correspondences between West Bengal government, Union Government of India and Bangladesh government on the agenda; with Inland Waterways Authority of India writing to Inland Water Transport Directorate, West Bengal in March 2020, exploring the possibility of Atrai river dredging within West Bengal as per the request of Bangladesh government (IWAI, 2020).

While the political back and forth continues with select rivers in two countries, the actual scale of the agenda is much bigger and, actually, offers the scope of mutual benefit. A 2014 report of major global environmental organization IUCN (Rivers Beyond Borders: India Bangladesh Trans-boundary River Atlas) shows that trans-boundary rivers in these two countries cover 15322 km; roughly split 60:40 vis-à-vis India and Bangladesh. The report also shows that out of 54 rivers; though 45 have originated in India, 6 actually originated in Bangladesh including important rivers like Punarbhaba, Kulik and Atreyi while another three are born in Tibet and Bhutan (Nishat, et al., 2014). Kalyan Rudra, a river scientist from West Bengal, found in his study that the state receives usually 481949.6 million cubic meter of water in monsoon and 116613.2 million cubic meter of water in non-monsoon months, almost one-fourth compared to monsoon; which vindicates the importance of transboundary rivers in the state of West Bengal (Rudra, 2018).

All these data underline the fact that only upper riparian versus lower riparian country narrative does not hold water in such a dynamic water sharing paradigm as both countries are dependent on one another, albeit in varying scale, for water flowing from one to another; which opens up the argument for holistic intercountry and stakeholder level discussion to arrive at a comprehensive trans-boundary river water usage model. The discussion also needs to consider the current and projected water flowing paradigm in these rivers under stress of climate change and also a range of local or regional factors; to arrive at a pragmatic solution mode.

It is often found that countries, and their political leaderships, fail to delve deep in the huge mutual trans-boundary river resources; and instead concentrate on few rivers like Ganges or Teesta to score political brownie points without properly assessing the ground reality of these rivers.

TEESTA WATER SHARING POLITICS: REALITY VERSUS RHETORIC (CASE STUDY)

Teesta river seems to be a good case study to trace this bigger than upper riparian versus lower riparian straight forward narrative. Over 20 million people are stated to depend on river Teesta. While Bangladesh is keen to have an agreement on Teesta's water sharing for supporting its agriculture, irrigation and livelihoods; and, union government of India is open to such proposal considering the wider geo political implication; Indian state of West Bengal consistently countered such water sharing proposal claiming it would jeopardize the interest of people in northern part of West Bengal. The issue is still hanging with potential of triggering far-reaching political consequences in the region.

Assigned by the global media platform The Third Pole, in 2017 the author of the current paper, an environmentalist and an environment journalist, had undertaken an on-ground investigation from Sikkim (where the river originated) till the entry point of Bangladesh where its journey finally winded in Brambhaputra; and the findings are not only revealing but also contest the popularly believed narrative about the water sharing, or rather the failure of it, between India (West Bengal) and Bangladesh particularly in case of Teesta.

The findings and observations are as follows:

Hardly any water available to split between India and Bangladesh

An internal report prepared by West Bengal government on Teesta river at the peak of its water sharing controversy, accessed by the author, points out as mentioned in The Third Pole that "Two barrages on Teesta, at about 100 km from each other, in India and Bangladesh were planned to cater [to] irrigation in more than 16 lakh [1.6 million] hectares of land together; around 9.2 lakh [920,000] hectares in [the] state of West Bengal in India and 7.5 lakh [750,000] hectares in Bangladesh. According to a rough calculation, such a scale of irrigation for boro crop (dry season paddy) will require around 1600 cumec (cubic metre per second) of water; while through much of the dry period the river hardly has 100 cumec of water, i.e., one-sixteenth of total water requirement in [the] two countries." Interviews with concerned officials confirmed that in recent times, river's peak summer (April and May) water volume has plummeted below 100 cumec (Basu, 2017a).

A committee set up by West Bengal government in 2016-17, under the then principal secretary of the public works department (PWD), suggested that the state should only irrigate 52,000 hectares land with Teesta's water; acknowledging the reduced water volume in Teesta. The referred landmass is only 10% compared to the originally devised plan for irrigation. Sources in West Bengal government claimed that the committee report and suggestion vindicate state's continuing reluctance in parting of Teesta water toBangladesh through a legally binding water sharing agreement; despite continuing pressures from respective union governments of Bangladesh and India.

It may be pointed out in the context that ruling and opposition political parties in West Bengal, are both on the same page over the decision of not parting Teesta water to Bangladesh acknowledging in private the state view that such step would compromise the need of a significant population segment residing in northern part of the state including Siliguri; one of the most important and strategically placed city in state after Kolkata.

Poor Planning: Priorities Changed

India has a barrage on the river at Gojoldoba at Jalpaiguri district of West Bengal, a little upstream of the point where the river enters Bangladesh. The Bangladesh government built a barrage at Doani in Lalmonirhat district, before Teesta rushes to join Brahmaputra.

The report underlined that "while the barrages were planned primarily to provide supplementary irrigation support to aman paddy cultivation during monsoon, subsequently both barrages were used to support boro cultivation during dry season, when there is minimum water in [the] river"; highlighting the role of changed mode of agriculture, and hence changed and increased demand of irrigational water.

The fact that the barrages are not designed for any water reservoir facility, also does not help and, rather, compels major part of the monsoon water to be forced out rather than being held back for dry months.

This changed paradigm of agricultural pattern has also been vindicated in a study carried out in Bangladesh by two agencies - International Food Policy Research Institute (IFPRI) and the Centre for Agri-research and Sustainable Environment & Entrepreneurship Development (CASEED). The study also showed that dry time Boro agriculture has been gradually replacing the monsoon fed Aman agriculture. Even few years back, Aman crop dominated being cultivated on 80% of the land in the area but in recent years the table has turned with Boro crop, highly water intensive, has become the trend. (Basu, 2017a)

Hydroelectricity Factor

Series of large number of hydroelectric projects along Teesta have already been commissioned or in different stages of being commissioned in Indian states of Sikkim and also in West Bengal. Sikkim is upstream to West Bengal. These hydroelectricity projects are stated to be built on run-of-the-river mode, meaning they are not supposed to retain water for a considerable period.

However, the fact on-ground is different. One finds that the riverbed immediately downstream to Teesta Low Dam Project (Phase IV) has become fully dry; in sharp contrast to its immediate upstream where large volume of stagnant water is retained.

The scenario not only gets repeated at Teesta Low Dam Project (Phase III), a 132 MW run-of-the-river hydropower project, but also in case of various projects further upstream in Sikkim. Clearly the volume of the stagnant water being retained gets bigger with the bigger projects; indicating a clear cause-effect relationship between the two.

Discussion with the wide range of stakeholders found that the reason behind the trend is linked to status of India's national electricity grid. The online national power exchange shows that supply remains greater than demand for about four-fifth period of an average day, the dynamics only get reversed during the peak demand period of evening time; around 6 to about 10 pm. This trend pushes most power generation companies to sell their electricity during evening period as they normally get peak rates at that time. The trend, in turn, triggers hydropower projects to predominantly restrict their power generation to evening period; retain water for rest part of the day (20-21 hours) and only release it during the evening hours when the water is required to turn their turbines.

The whole phenomenon, working in series as there are several hydropower projects in the river, triggers a cascading effect throughout the entire river. With power projects and dams holding the water back, hydropower projects in downstream do not get adequate water during the day; only getting it during those 3-4 hours when the water gets released by upstream projects. Naturally the downstream projects often store that to run turbine, and generate power, during peak generation period to come next. The trend impacts the downstream in a critical manner (Basu, 2017b).

This trend often leads to another mismatch between the time when the farmers need the water (during the day) and when it actually gets flown through the river; during very late evening, night and early morning. This lack of coordination not only hits the peasants hard but also often leads to longitudinal cut-off in the river flow over large stretches affecting the river ecosystem in extremely critical manner. This author could confirm the trend through his interview with a senior official of Teesta Low Dam Project (Phase III), who had chosen to remain anonymous; "During impounding, generally 1-2% of the water is released downstream to maintain the river ecosystem". In 2017, the National Green Tribunal (NGT) directed that all the rivers in the country shall maintain a minimum 15 to 20 percent of the average lean season flow of that river, also called the environmental or e-flow (Joshi, 2019).

Officials of privately owned hydropower units like Madhya Bharati Power Corporation (MBPC), which has a power generating capacity of 96 MW, agreed that the practice is in vogue. “During the lean season, all, particularly the larger plants, have to impound the water for power generation”. The MBPC expert alleged, “Sikkim government is just interested to receive about 15-20% proceeds from the plants and hardly monitors [them]. People from the central ministry occasionally come but that is mostly routine in nature.”

In Sikkim only (as on 2017) 29 hydropower projects were planned having capacity from 30 to 1,200 MW each, totalling close to 4,400 MW. “Undertaking so many hydropower projects in Sikkim without any proper comprehensive scientific study is going to impact Teesta’s ecology and ecosystem,” Sonam Wangdi, former Chief Secretary of Sikkim, complained to the author.

The impact of hydropower projects on river flow can be understood from the situation in Rangit river. Based on the water flow measuring point data adjacent to Teesta Bazar, Rangit (one of the major tributaries of the Teesta) has hardly lost any flow during last few years, in telling contrast to the main flow of Teesta. This is not coincidental that while Rangit has one hydropower project upstream; Teesta has many.

The developments trigger the debate that whether such hydropower projects should qualify as run-of-the-river projects, though they were given approvals based on that premise despite the opposition from green lobby.

“The NHPC (National Hydro Power Corporation) has declared that Stage III and IV are ‘low and run of the river dams’. However as both are more than 15 m in height, they cannot be categorised as ‘low’ since the International Commission on Large Dams criteria for a large dam is anything above 15 m in height,” river expert Kalyan Rudra pointed out. Rudra also told this writer in an interview that the entire flow of the Teesta had “lost synchronization due to the various ‘stop and store’ steps applied” (Basu, 2017b).

“These are anything but run of the river projects as they are holding water for considerable periods, often in excess of 10 h at a stretch,” pointed out ecological economist Nilanjan Ghosh, Director of Observer Research Foundation, Kolkata Region. “There seems to be little coordination among the plants and as a result there is hardly any coordinated release, and hence flow,” reiterated the expert (Basu, 2017b).

There are other issues as well linked to long-term holding of the water. “The evaporation and seepage losses of water in other major irrigation projects in West Bengal is very high. There is no reason to believe that the experience of the Teesta Barrage Project (TBP) would be otherwise,” said Rudra. “We have found that in the Ganga, the evaporation may lead to about 15% water loss. It is highly likely with the water being impounded in stagnant pools, the actual evaporation in the Teesta is higher,” added Nilanjan Ghosh who observed that lack of any effective sediment control system further complicates the problem.

Measuring Water Flow During Wrong Time

The timing of measuring water flow in Teesta assumes importance in context to the water retention—release phenomenon being practiced in Teesta hydropower projects.

In the Indian state of West Bengal, Teesta water flow gets measured at three points; most elaborately at Domohani by Central Water Commission (CWC). Interestingly while here, the water flow gets measured in morning hours after 9 am; maximum release of water happens, as explained earlier, either close to midnight after the evening peak generation or early morning after the mini peak. "It's a fact that when CWC measures water at Domohani, the peak flow can be absent as hydropower plants generally release water at night," river expert Kalyan Rudra communicated during the interview.

Apparently, Bangladesh also measures water level of Teesta in their area when the peak flow gets already perished.

The Teesta-Mahananda Link Canal Impact

According to figures, the Teesta water flow started receding significantly since late 1970s. The mean average discharge of water got reduced from 541 cubic metres per second (cumec) during 1979 to 200 cumec in 1999; by a whopping 63%. The minimum discharge (during summer months) got reduced by even 90% occasionally. Figures point out that the reduction in flow initiated in 1979-80, when the minimum discharge became 195 cumec from 361 cumec in previous year (Basu, 2017c).

Downstream to Gajaldoba barrage, the flow got even thinner after the barrage started its full operation in mid-1990s. The Teesta-Mahananda link canal at the barrage, that takes the waters of the Teesta to the Mahananda, contributes. "Around mid-nineties the minimum flow from upstream during the lean months was often 100 to 110 cumec; we normally used to keep 80% and released 20% downstream," said P.K. Basu, a retired irrigation engineer in West Bengal government, to this author in 2017, during an interview (Basu, 2017c).

Scientific researches corroborate the trend, "With reference to the last 15 years, there is declining trend of annual discharge and severe water scarcity can be perceived during non-monsoon (months) due to high demand and declining supply," said geographer Kausik Ghosh in his study.

As per 2010 data, every year West Bengal has been pushing around 10% of Teesta water through the Teesta-Mahananda link canal; which, according to P.K. Basu, could have been more if the irrigation network could be completed in West Bengal. "Though the main canals were completed, but due to the land acquisition policy of the present Trinamul Congress government in West Bengal, hardly any new irrigation network could be created. As a result, though the

Teesta-Mahananda link canal has the capacity to carry around 330 cumec, hardly 190 cumec is taken when there is sufficient water. In the lean season, it must be far less," explained the expert. The laid down policy of the government was not to acquire any land forcefully from people.

"During the driest months of the lean season, West Bengal has to hold back most of the water and channel it through the link canal otherwise not only will agriculture over a vast area be affected, but even Siliguri may be affected as much of its drinking water comes from this canal," reminded Nilanjan Ghosh. Incidentally Siliguri happens not only to be the second largest city in West Bengal but assumes importance due to its strategic positioning being close to various international borders.

This author saw that only 2 out of the 45 gates in Gajaldoba barrage remained open in early June.

Geographer Kaushik Ghosh's study (*Planform Pattern of lower Teesta River after the Gajaldoba Barrage,* published in the Indian Journal of Geography and Environment in 2014) reiterates that mean annual discharge received at the barrage has gone down nearly one-third, from 725 cumec in 1993 to 480 cumec in 2010 leading to the braiding, generation of multiple channels in intertwined manner, in recent years; tell-tale evidences that reducing volume of upstream water further enhancing the sedimentation rate, hence, braiding of the river (Basu, 2017c).

Baishali Mukherjee and Ujwal Deep Saha from geography department of Calcutta University mentioned in their paper (Teesta Barrage Project: A Brief Review of Unattained Goals and Associated Changes, published in the International Journal of Science and Research in 2016) that "water demand in Teesta basin has increased but the available water decreased by 32% from 1990 to 2010, from 69 billion cubic metres to 47 billion cubic metres" (Basu, 2017c).

The combination of reducing water flow along with enhanced water demand has made the situation critical in West Bengal as the river happens to be the major source of water in northern part of the state. "Teesta water, through the Teesta-Mahananda link canal, caters to the paddy and other cultivation, tea gardens and as well as to Siliguri for drinking water," pointed out environmentalist Animesh Bose from Siliguri to this author in an unpublished interview. (Basu, 2017c)

"There has been increasing pressure as the command area for irrigation is consistently increasing but we have inadequate water in link canal," admitted a senior official in the state government during unofficial discussion.

The Climate Change Effect

A 2007-08 study report of the Ministry of Environment, Forests and Climate Change, pointed out that out of the 34 major glaciers contributing water flow in the Teesta, 23 were showing retreat while 8 could be found advancing and rest three glaciers remained unaltered. "While in 1990, 34 glaciers used to cover an area of 305 sq. km, in 2004 the glacial cover on the Teesta basin was reduced by four sq. km," the report added (Basu, 2017c).

According to climatologists, this reduction at source has been contributing over the years to reduce the water volume in Teesta from Sikkim to West Bengal and beyond. The trend is likely to continue in coming years.

NATURE BASED NEGOTIATED APPROACH (NBNA)IS THE KEY

The Teesta study, as elaborated, clearly vindicates the importance of having consolidated dialogue involving all the stakeholders and their linked interests, based on credible nature-based evidences on the agenda; which unfortunately has not been done yet on Teesta, a process may be called Nature Based Negotiated Approach (NBNA).

Overall, if we look at the trans-boundary rivers and rifts within countries based on them, we will find the root cause is over the splitting or sharing of resources, particularly with the resource volume often diminishing for a range of reasons. Such conflicts also percolate down to intra country level. Often political macro level positioning vis-à-vis the conflicts, and the dialogues catapulted by those, tend to ignore on ground nature-based issues involving multiple stakeholders; and thus, fail to lead towards a sustainable discourse.

For a sustainable discourse, the involvement of all major stakeholders (which can be different from case study to case study) like agricultural workers, industries, domestic users, fishermen, transport operators as applicable need to be addressed. Emphasizing on one's stake, however major it seems to be, may lead to inherent long term unsustainability of the discourse.

Analyses show that there are three dimensions of river related conflicts. The first is emergence of direct competition for water as an increasingly scarce resource. Second conflict relates to large scale infrastructure projects, particularly those lead to people displacement and other social and environmental impacts in and around the rivers. The third dimension of water-related conflicts involves disputes over the appropriate levels, roles and access of various stakeholders in river basin management. While the first two, mainly macro-level agendas, get often focused; the third gets overshadowed in multinational process (Hirsch et al., 2005).

All water management techniques have complex and multi-dimensional implications, related to the existing geographical, ecological, socio-political and economic situations. However, these techniques require to be modified, updated and adapted in response to changes in the existing order, or if the primary objectives of adequate and equitable supply and sustainable use of water resources are not achieved. The Nature Based Negotiated Approach (NBNA) to Integrated River Basin Management seems to be the key in addressing transboundary conflicts including those in South Asia.

Integrated River Basin Management (IRBM) is essentially a work concept that aims to conserve and utilize the natural resources within a river basin sustainably, through integrating the needs and skills of various stakeholders like government departments, academics, farmers, and the private sector.

The Negotiated Approach to IRBM is a variant of conventional IRBM, and based on the collective vision that: 'Sustainable and equitable water resources will be enhanced through a negotiated approach that recognizes the river as a unit and embraces local level initiatives, while simultaneously adopting an integrated and ecosystem approach to basin management'. The addition 'negotiated' explicitly indicates that this approach is aimed at creating space for negotiation, including with local stakeholders, on river basin management options.

Negotiation needs to be pursued at and between the local, regional, national, and international governance levels. The 'negotiated approach' should include and responds to local initiatives, and starts from the basis that management policies should build on existing local practices of integrated land and water use. It recognizes the potential of local resources and knowledge to meet the challenges of integrated water management.

It is felt that the proposed NBNA model will be a mix of top-down and bottom-up approaches which should allow local actors to develop river resource usage and broadly basin management strategies specific to their local context, which may then be incorporated and integrated in the larger basin management plan. This allows their knowledge to influence regional and national decisions, ultimately resulting in a truly bottom-up process of policy development and management.

References

Altingoz and Ali (2019). Environmental Cooperation in Conflict Zones – Riparian Infrastructure at the Armenian–Turkish Border. Published in *Journal of Environment & Development*. https://transboundarywaters.science.oregonstate.edu/sites/transboundarywaters.science.oregonstate.edu/files/Publications/%28Altingoz%20%26%20Ali%2C%202019%29%20Environmental%20Cooperation%20in%20Conflict%20Zones-%20Riparian%20Infrastructure%20at%20the%20Armenian%E2%80%93Turkish%20Border.pdf

Basu, J. (2017a). Teesta has one-sixteenth of water needed. Published in The Third Pole, https://www.thethirdpole.net/2017/04/14/teesta-has-one-sixteenth-of-water-needed/

Basu, J. (2017b), Where are the Teesta waters? Published in The Third Pole, https://www.thethirdpole.net/2017/06/19/where-are-the-teesta-waters/

Basu, J (2017c). The canal, the climate and the Teesta. Published in The Third Pole, https://www.thethirdpole.net/2017/06/20/the-canal-the-climate-and-the-teesta/

Basu, J. (2019). Ganga in Bengal has little trace of Gomukh glacier's pristine waters. *The Telegraph*. https://www.telegraphindia.com/west-bengal/ganga-in-bengal-has-little-trace-of-gomukh-glacier-s-pristine-waters-says-waterman-rajendra-singh/cid/1681823

Basu, J. (2021). Geopolitics in Environmental Discourse. *In:* Sikdar, P.K. (ed.). Environmental Management: Issues and Concerns in Developing Countries. Springer (in press).

Dholakia, A. (2019). Transboundary Rivers of South Asia – The Case for Regional Water Management. Published in IMPAKTER https://impakter.com/transboundary-rivers-of-south-asia/#:~:text=Several%20of%20these%20perennial%20South%20Asian%20rivers%20have%20transboundary%20basins%20and%20watercourses.&text=Major%20rivers%20shared%20between%20India,share%20the%20Kabul%20river%20basin

EJATLAS-Global Atlas of Environmental Justice (2021). https://ejatlas.org/
Ghosh, N. (2020). Busting Myths on the Brahmaputra. ORF. https://www.orfonline.org/research/busting-myths-on-the-brahmaputra-59844/
Haq, E. (2012). Ganges Water Sharing. *In:* Islam, Sirajul and Jamal, Ahmed A. (eds.). Banglapedia: National Encyclopedia of Bangladesh (Seconded.). Asiatic Society of Bangladesh http://en.banglapedia.org/index.php?title=Ganges_Water_Sharing
Hirsch, D., Bosch, M.J.V.D.T. and Paranjpye, V. (2005). River Basin Management: A Negotiated Approach. Published by Both ENDS and Gomukh, http://bibalex.org/baifa/Attachment/Documents/161676.pdf
Human Development Report (2019). Beyond income, beyond averages, beyond today: Inequalities in human development in the 21st century. Published by the United Nations Development Programme (UNDP), pg. 190.
Inland Waterways Authority of India: Letter from S.V.K. Reddy, Chief Engineer Technical of IWAI to the Director of Inland Water Transport Directorate, West Bengal, dated March 2, 2020, numbered IWAI/IBP/SCM/2018-2019/Vol-I
Joshi, H. (2019). Wanted: An urgent political saviour for river Yamuna. Published in *Mongabay*, https://india.mongabay.com/2019/10/wanted-an-urgent-political-saviour-for-river-yamuna/
Majumdar, A.D. (2017). Why Mamata Banerjee is opposed to sharing Teesta waters. Published in *Mint*, https://www.livemint.com/Politics/dtlGtxiSUVDJgo7eoBxxDL/Why-Mamata-Banerjee-is-opposed-to-sharing-Teesta-waters.html
Mantri, G. (2018). Explained: What the Cauvery water dispute between Karnataka and TN is all about. Published in *The NEWS Minute* https://www.thenewsminute.com/article/explained-what-cauvery-water-dispute-between-karnataka-and-tn-all-about-76528
Mohan, M.M. (2020). A minor issue, compared to Krishna and Cauvery disputes. Published in *The New Indian Express*, https://www.newindianexpress.com/opinions/2020/jan/13/a-minor-issue-compared-to-krishna-and-cauvery-disputes-2088783.html
Nishat, B., Chakraborty, S.K., Hasan, M.E and Rahman, AJM.Z. (2014). Rivers Beyond Borders: India Bangladesh Trans-boundary River Atlas. Ecosystems for Life: A Bangladesh-India Initiative. IUCN, International Union for Conservation of Nature, Dhaka, Bangladesh, pp. XX + 152.
Pacific Institute Releases Updates to the Water Conflict Chronology (2019). https://www.wateronline.com/doc/pacific-institute-releases-updates-to-the-water-conflict-chronology-0001#:~:text=The%20Water%20Conflict%20Chronology%20is,database%20on%20water%2Drelated%20violence.&text=The%20Chronology%20now%20includes%20over,violent%20events%20in%20recent%20decades.
Reliefweb as Editor's Pick: 10 Violent Water Conflicts, 2017. https://reliefweb.int/report/world/editor-s-pick-10-violent-water-conflicts
Rudra, K. (2018). Rivers of the Ganga-Brahmaputra-Meghna Delta: A Fluvial Account of Bengal. Published by Springer, https://www.springer.com/gp/book/9783319765433
Sayanangshu, M. (2020). The Farakka Fulcrum of Indo-Bangladesh Hydro-diplomacy. Published in ORF, https://www.orfonline.org/research/the-farakka-fulcrum-of-indo-bangladesh-hydro-diplomacy/
Scheidel, A. et al. (2020). Environmental conflicts and defenders: A global overview. Vol. 63. Published by ScienceDirect, https://www.sciencedirect.com/science/article/pii/S0959378020301424#!
Temper, L. et al. (2018). The Global Environmental Justice Atlas (EJAtlas): Ecological distribution conflicts as forces for sustainability. Published by ResearchGate, https://www.researchgate.net/publication/324667848_The_Global_Environmental_Justice_Atlas_EJAtlas_ecological_distribution_conflicts_as_forces_for_sustainability

Part III
Hydro-heritage

Chapter 17
A Deep Ecological Exploration of Indian River Systems: Review of Its Cultural Landscape and Triveni Sangam Case

Joy Sen

INTRODUCTION

Indian river systems (IRS) are a constituent part of India's geographical landscape and her culture that have evolved over thousands of years. Indian philosophy, which is a way of life, is synonymous and synchronous with the flow of her uninterrupted economy, culture and an intrinsic way of life based on predominant patterns of a river-based civilization. In India, the flow of the river has been equated with the flow of overall humanity, nature of life and cycles of livelihood. The flow of the river has also been visualized as inner currents of human evolution based on which an individual aspires for a higher life in harmony with nature. It is evident in the observations forwarded by some great minds.

Nobel laureate Poet Rabindranath Tagore chooses the deeper symbol behind the course of a river; thereby, for him, and for many others thinkers in India, the river becomes symbolic of the cycles of human life, from life to death and beyond and even, a return which is hailed as reincarnation in Indian philosphy. In a mystic poem entitled, 'Sonar Tari' (The Golden Boat), the poet portrays the significance of existence of humanity over time, space and causation and across natural seasons and individual lives (Fig. 1). Though the river is not the direct and the most prominent basis of the story, Tagore has finally related the river to the theme of human life, evolution, and possibilities of salvation.

In India, average individuals within communities and their livelihood are intertwined with the river, as their early childhood shape up on the banks or in the vicinity of a river. And the manifestation of this interrelationship and complementarity has expressed over ages in countless of ways. Stories, narratives, epics have

J. Sen (✉)
Department of Architecture and Regional Planning, Indian Institute of Technology Kharagpur, Kharagpur, WB, India
e-mail: joysen@arp.itkgp.ac.in

A. Mukherjee (ed.), *Riverine Systems*, https://doi.org/10.1007/978-3-030-87067-6_17

Figure 1. The imageries used in 'Sonar Tari'.
Source: https://sandrp.in/2020/01/08/he-spoke-the-language-of-the-rivers-rabindranath-tagore/

emerged from the Indus valley in the west to the Gangetic-Brahmaputra basin in the east to that of Krishna-Cauvery-Godavari plains in the south. The Indian culture in that sense is *Nadi Matrik*, which literally means river-born. As a result, rivers have permeated into the works of most artists, philosophers, musicians, and writers of regions across India, ranging from an ancient Valmiki portraying the Ganga (Ganges) in his Epic Ramayana, to Poet Kalidas delineating the banks of Shipra in the city of Ujjwain in his poems, to the mystic Alvar poets of the South celebrating the flow of river Cauvery as the inner life current heading for spiritual aspiration.

In Indian religion, river has proved to be the dynamic laboratory of the human soul. And the laboratory is the complex ecosystem that has generated a deep ecological ensemble of any Indian river. Of all, the imagery of Ganga, its origin in the snow-capped noble hills and mountains, and the cities that have grown along its backs creating a network of urban system forwards the deep folk embedded within a riparian pattern of Indic civilization. It is evident in the words of Sri Aurobindo (1908):

> This great and ancient nation was once the fountain of human light, the apex of human civilization, the exemplar of courage and humanity, the perfection of good Government and settled society, the mother of all religions, the teacher of all wisdom and philosophy. It has suffered much at the hands of inferior civilizations and more savage people; it has gone down into the shadow of night and tasted often of the bitterness of death. Its pride has been trampled into the dust and its glory has departed. Hunger and misery and despair have become the masters of this fair soil, these noble hills, these ancient rivers, these cities whose life story goes back into prehistoric night. [... But] all our calamities have been but a discipline of suffering, because for the great mission before us prosperity was not sufficient, adversity had also its training; to taste the glory of power and beneficence and joy was not sufficient, the knowledge of weakness and torture and humiliation was also needed.
>
> Bande Mataram (1908) 1.560-561.

An inherent beauty of Indian system of river is that it has always garnered a complementarity between intrinsic and internalized life of contemplation and flow of deep thought, and an externalities of good livelihood livability based on the produce, cottage and village farms and their creative economy based on the river.

Micheal Danino (1990) says:

> For Sri Aurobindo, it can only end if we get rid of a central misconception, a fatal misconception. When we speak of the "laboratory of the soul," of India's wisdom and spirituality, a widespread tendency is to think that all this is fine for those confined to ashrams, or perhaps for old age, but of little practical use to build a nation. Sri Aurobindo frankly disagrees. To him, inner growth can never contradict outer growth, but can alone put it on a sound foundation. Referring to India's extraordinarily creative past, which certainly never neglected material life and achievements.
>
> Sri Aurobindo's View of Indian Culture
>
> http://micheldanino.bharatvani.org/indianculture.html

In the celebrated work, 'The East and the West', Swami Vivekananda (1900) forwards a range of geographical syntax and cultural semantics of Indian culture based on the flow of the rivers. He says:

> Vast and deep rivers—swelling and impetuous—charming pleasure-gardens by the river banks, putting to shame the celestial Nandana-Kânana; amidst these pleasure-gardens rise, towering to the sky, beautiful marble palaces, decorated with the most exquisite workmanship of fine art; on the sides, in front, and behind, clusters of huts, with crumbling mud-walls and dilapidated roofs, the bamboos of which, forming their skeletons, as it were, are exposed to view; moving about here and there emaciated figures of young and old in tattered rags, whose faces bear deep-cut lines of the despair and poverty of hundreds of years; cows, bullocks, buffaloes everywhere—ay, the same melancholy look in their eyes, the same feeble physique; on the wayside refuse and dirt: This is our present-day India!

In the last chapter of the aforesaid work entitled 'Progress of Civilization', he earmarks a deep ecological foundation of India history based on rivers contrasting that of the West, which is based predominantly on the extrinsic features of a maritime, material and imperial expression of civilization. In Indian philosophy, these two tendencies, one deep ecological and the spiritual, and the other, the surface ecological and the material, are termed as the Daivi (divine) and Asuric (material) tendencies of human mind, respectively. He says:

> The whole of tile Asian civilization was first evolved on the plains near large rivers and on fertile soils—on the banks of the Ganga, the Yangtse-Kiang, and the Euphrates. The original foundation of all these civilizations is agriculture, and in all of there the Daivi nature predominates. Most of the European civilization, on the other hand, originated either in hilly countries or on the sea coasts—piracy and robbery form the basis of this civilization; there the Asuri nature is preponderant.....The loom of the fabric of Aryan civilization (in India) is a vast, warm, level country, interspersed with broad, navigable rivers.
>
> https://en.wikisource.org/wiki/The_Complete_Works_of_Swami_Vivekananda/Volume_5/Writings:_Prose_and_Poems/The_East_and_The_West/Progress_of_Civilisation

In the history of human civilization, the sharp contrast between a peace-loving, non-aggressive Indian way of life and living on the one hand, and the more materially aggressive imperial patterns of the Industrial order of Western civilization on the other, is strongly evident. The aggressive and invasive nature of the European powers reigning over colonies in Asia, Africa, and finally, Americas, after the Dark Age and specifically, after the Fall of Granada in 1493, is a living testimony of the broad Asuric nature of the Occident. Of course, there are a few exceptions.

A study of this contrast has taken place in the current academia, and in effect it has led to a turning point in ecosystem studies and sciences in the 20th century world. The ancient ideas of Indian peace and harmony with nature is now known in different names, viz. green peace; sustainability approaches, deep ecology, are just to name a few only. The latter or the Asuric nature is called shallow ecology as compared to the non-aggressive symbiotic nature of deep ecology reflecting ancient Indian values. It is evident in the works of historian Arnold Toynbee:

> It is already becoming clear that a chapter which had a Western beginning will have to have an Indian ending if it is not to end in self-destruction of the human race. At this supremely dangerous moment in human history, the only way of salvation is the ancient Indian way. Here we have the attitude and spirit that can make it possible for the human race to grow together in to a single family.
>
> McNeill, William H. .(1989) Arnold J. Toynbee: A Life. New York: Oxford University Press

DEEP ECOLOGY AND FLOW OF RIVERS

Shallow Ecology is a way of consumption and power processes of a civilization which simply promotes conservations strategies against pollution and the depletion of resources giving highest importance to competition based on excellence in material production by a few as against the rest. On the contrary, Deep Ecology is a movement that promotes a way of life based on 'ecological wisdom' acknowledging the inherent value of all forms of life, and all forms of civilization processes, namely underdeveloped, developing and the developed as different phases of growth and development itself. In 'Deep Ecology', cooperation replaces competition, and a symbiosis of sharing replaces processes that are driven by superiority and aggression.

The Deep Ecology has been a fairly recent movement of the 1970s. Environmentalist and author, Arne Næss had initiated it. There has been others who have joined like author-conservationists Rachel Carson, ecologist and environmentalist David Brower, and system-biologist Paul R. Ehrlich.

The notion of deep ecology, echoing the ancient Indian wisdom of nature-human symbiosis. The symbiosis constitutes of an eco-centric or earth-centered view of planetary consciousness and the existence of interconnected societies in them, rather than an aggressive anthropocentric view that is human centered view where the supremacy over others in terms of greater wealth, higher GDP and material superiority is emphasized. Like the ancient Indian view, the modern seers of deep ecology oppose the rigid anthropocentric narrative that portrays man is an identity separate from nature. Instead man is in charge of nature with full stewardship of nature. On

Figure 2. An ancient perspective of natural sustainability. Death of one state is the resurrection of the next, as portrayed by the water cycle of the river.

the contrary, in deep ecology, the growth and development of an individual is just not a discrete part but an integral part of Mother Nature itself. In Indian ethos, the simile of the river represents this flow (Katz et al., 2000).

Originating from the mountains and heading for the sea, the perennial river represents the perpetual flow of life. The flow is cyclic and therefore a non-linear pattern (Fig. 2). Embedded in natural cycles, rivers continue as clouds higher in the sky, and then they return as rain, from where mountains receive their perennial snow. Finally, they return to the sea, where they again vapourize as clouds. Finally, another cycle takes them back to the mountains where they originate again. Thus, death of one form and regeneration of the other, the basis of life cycle itself, is at the heart of deep ecological foundations of river systems, as envisioned by the Sages (Rishis in Sanskrit) of India.

The condensation-rarefaction cycles constitute the formation of rain clouds to the formation of perennial rivers and their vapourized return to snowline and melting glaciers as rain again, is a significant imagery used in all ancient Indian texts, including the Rig Veda, mandala: 1.164. The imagery also forms an ancient basis of sustainability in other oriental wisdom literature of Taoism and the allegorical cycles of human society in Confucius's 'spring and autumn'.

The rivers of India are seen as carriers of immortalizing rains that comes from the upper worlds, the celestial world as fostered by the sun and its relationship with the orbital and cyclical movement of the earth. The immortalizing rains, as evident in the Rig Veda (5.64.5) are the '*turning of heavens to rain from the luminous worlds, the rains of truth with the yields of bliss*' (Sri Aurobindo, 1956). The rains as portrayed by the Vedas are systems that finally channelize as fertilizing rivers on earth, seen both spiritually and materially. Drawing the imagery from the Rig Veda (5.63.2), it is further said that, "it *is perfectly true that the Vedic Gods are givers of the rain, the abundance (for vristi (in Sanskrit), or rains has both senses) of heaven, sometimes described as the solar waters, 'svarvatr apah', or waters, which carry in them the light of the luminous heaven, the Svara'* (Sri Aurobindo, 1956). Thus, a bridge between the celestial and the material, and the macrocosm and the microcosm is established.

The symbol of these Godheads are initiated by Indra, who destroys darkness in the sky by destroying the covering clouds, earmarked as 'Ahi' or 'Vritta', and by bringing down the fertilizing rains; then *Varuna*, 'who floods the wide earth' (Sri Aurobindo, 1956); and then comes Parjyanas and the Rudras, who restore and fashion the infiltration of rains in the galleries of the earth to charge aquifers and finally, assure the flow of inundating rivers out of the netherworld. Such imageries covering the triad of Indra holding the thunderbolt; Varuna the oceanic expanse; and Rudra-Parjyana, the underworld has parallels in ancient Greek/Orphic mythology in the form of Zeus,Poseidon and Hades or Vulcans.

In the lower end of the rain-cycle, it is 'Prithu' (Rig Veda 4.58.5) that represents the dynamic earth-consciousness, who charges a flow of many rivers fertilizing the mundane earthly plains ('Ptrithvi' in Sanskrit). At the existential plane, the fertilizing river fosters produce and the economy, the livelihood of humanity forwarding the next cycle of change. Rig Veda 4.58.6 adds a double value to the imagery of the lower plain. It says, '*Samyak Sravasti Sarito*', which means that the network of rivers are just not the flows of fertilizing rains that foster, fashions and nourishes the terrestrial level, but they also represent the extent of puissance, the degree of nourishment and the flow of nerve-currents within the individual human, who is the lowest denominator of evolution, 'The Purusha', the epitome of an evolving anthropic principle, within the human form. Thus, '*nadis*' or nerves in Indian Ayurveda (biosciences) are also synonymous with the network of rivers in the internal ecosystem of human being. Therefore, in the Indian system, rivers such as Ganga, Yamuna and Sarasvati are just not external rivers in a geographical and cultural landscape, but they are also the flows of life currents in the human form, the carriers of the system of yoga that lead to a union of the truth within the individual with the vaster truth with the mother nature, cosmos itself. Thus, two images are evident, i.e., spiritual and material. This 'Double intendre' as portrayed by the twin psychic and physical tradition of the Indian wisdom literature is the very basis of the deep ecological foundation of Indian river systems.

Thus, a cyclic pattern of formation of rivers represent a construct of perennial continuity, where there is an involution of higher ecological factors to levels below, i.e., the material and the economic, thereby recharging a return of the lower order back to the higher, the spiritual in the ladder. Thus, involution and evolution characterize a cyclic pattern and eco-balancing integrating the two levels, large and the small. The ancient river systems of India, as portrayed in the wisdom literature of India like the Vedas and the Upanishads, represent this very cycle of continuity, and sustainability. Today, it is the cornerstone of modern environmental and ecological sciences. It is also the cornerstone of the origin of Indian Knowledge systems or IKS, as evident in the Vedas.

The idea and the episteme of Indian Knowledge System has been founded on these foundations that looks for a recovery of an understanding based on both the analytical and intuitive sciences. Normally, analytical sciences pose boundary conditions driven by reductionism. Therefore, exploration of modern sciences is mostly reduced to disjointed analytical blocks that fall short of a holistic eye, and a complete approach. What is needed today is the integral normative-positive paradigm integrating the lower and the higher, or the physical with the psychic.

RIVERS IN INDIAN LANDSCAPE

A later part of the Vedic literature flourished in the Indus Valley. It is termed as the land of Sapta Sindhavah (known as the 'seven rivers'). In the Rig Veda, the 'Saptah Sindhavah' forwards an important role in its hymns, and consequently, it plays an important role in early Indian religion till the genesis of Vedanta or the Upanishads. The sacred landscape of the Vedic lore forwards a vast spatial configuration with an assemblage of natural elements like rivers, mountains and deserts and various summits of the Earth. It is evident in a hymn where rivers continue to play a predominant role reinforcing the landscape (Rig Veda 1.35.8). The Vedic land of Aryavarta (Ancient India) is a land of a multitude of rivers flowing from the snow-capped mountains into the vast sea. The picture represents the highest aspiration of the human soul seeking peace and harmony with nature. It encompasses a variety of regions covering Gandhara to Kurukshetra, from Naimisharanya to Varanasi, and others in the south. Much of this part constitutes the sacred landscape of Epic Mahabharata and the later texts of Upanishads, the Vedanta and its summary, the Gita.

The Yajurveda has a narrative of three Gods connected with the formation and flow of rivers. One is of Indra slaying the Vritra (literally 'the dark clouds or obstacle'), and liberating the fertilizing rains and the rivers. Here, Indra also terminates the Vala cave, releasing the cows from within the human soul. The word 'cow' ('go' in Sanskrit) actually stands for a deeper imagery of light or lighted vision (based on the root word, 'A-bang-manasa-go-charam', which means THAT VAST which is beyond words, mind and vision). It is evident that the two myths are separate, and rivers (representing flow of life) and cows (representing the ray of light) are often associated with a variety of parables in the Rigveda. In Rig Veda 3.33, there is a further description of a 'sangam' or a crossing of two important rivers in relationship with advents by a specific community or a certain tribe. It says, 'there are the two bright mothers in the form of ray-cows and they care for their young offsprings', as represented by two major river streams known as Vipasha and Satadru. Obviously, the description is just not geographical but also metaphorical standing for a certain community in a geographic location in India, but trying to reach the celestial, which is beyond physical limits. Thus, a double movement, viz. physical and psychic is established.

The descriptions of various rivers like the Indus as evident in the Nadistuti sukta, Rig Veda 10.75, and that of Gomati, in northern mid India (close to present Lucknow) in Rig Veda, 8th mandala, contain the imagery of Nadi, or inner neural currents. Nadis are the very foundation of Yoga. Thus, in the ancient Indian tradition both the psychic and the physical representation of the river are to be considered. In the 8th mandala, Gomati is precisely hailed as the glory of the white sun tide, the full blaze of the noon sun called 'Madhyanindan Surya', which becomes the basis of Sukla Yajurveda, the white tradition surpassing the limitations of the Krishna group, or the black. Thus, rivers are also the basis of the iterations of the black and the white symbolizing the mysteries of creation in terms of dissolution and re-emergence of the universe.

An Imagery of Sarasvati

In Rig Veda 6.61.10, one discovers the deeper role of Sarasvati. In this hymn, Srasvati is called 'the Lady with seven sisters and powers'. She is associated with 'saptasvasā', which is a group of eight rivers. Here, the number seven and another one is more important than the individual members. It is comparable to Sapta Rishis or the seven sages and the Astama, which is the 8th stands for the divine word, an ensemble of natural truths.

In the Rig Veda 10.64.8 and in sukta 10.75.1, there are even more significant descriptions. There are three groups of seven rivers and are referred to as three-times seven meandering rivers. It is also evident that the three planes of creation, i.e., the celestial (*Dyu*), the intermediate (*Anatariksan*) and the mundane (Prtitvi) are interlocked with these expressions. In Indian mythology goddess Sarasvati was originally a personification of the river, but later developed an independent identity of a Goddess in Puranic times. In Indian philosophy, the river is considered in a metaphysical formulation. Moreover, the three sacred rivers Ganges, Yamuna and Sarasvati is known as the Triveni Sangam. According to Michael Witzel (1989), the construct of Sarasvati river is derived from the idea of 'Akash-Ganga', the heavenly or celestial river in the Milky Way, which is portrayed as 'a road to immortality and heavenly after-life'. However, in the Rig Veda, there are a range of acceptable and symbolic hints that the river Saraswati is just not the name of a single river, but represents a spiritual recurrence of an inner river venerated in different parts of undivided India, with different geographic and cultural landscapes, but with all having the same spiritual meaning.

There are at least four of many (Fig. 3):

1. The more popular Sarasvati River is connected with the Indus Valley plan and is identified with the ancient Ghaggra-Hakra channel in Western India, connecting the plains of the Sivalik mountains with that of the Indus.
2. A hidden Sarasvati river as the embedded or subterranean river, deeply connected with Ganga flowing northbound making a crescent-like armature of Varanasi. At

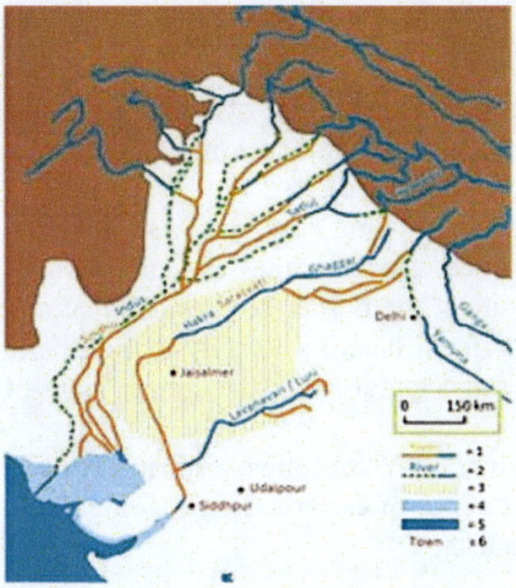

River 1: Sarasvati in Eastern Sivalik-Indus valley, India
https://en.wikipedia.org/wiki/Sarasvati_River

River 2: Saraswati bhumi in Varanasi, Uttar Pradesh, India
https://affordablehousing.live/varanasi-master-plan-2031.html

River 3: Sarasvati in Tribeni-Saptagram, Kolkata, West Bengal
https://en.wikipedia.org/wiki/Tribeni,_Hooghly

River 4: Sarasvati or Padma (River of lotus) in Bangladesh
https://www.researchgate.net/figure/Location-of-Padma-River_fig1_315014183

Figure 3. Four Sarasvati Rivers in undivided India (pre-1947).
River 1: Sarasvati in Eastern Sivalik-Indus valley, India
https://en.wikipedia.org/wiki/Sarasvati_River
River 2: Saraswati bhumi in Varanasi, Uttar Pradesh, India
https://affordablehousing.live/varanasi-master-plan-2031.html
River 4: Sarasvati or Padma (River of lotus) in Bangladesh
https://www.researchgate.net/figure/Location-of-Padma-River_fig1_315014183
River 3: Sarasvati in Tribeni-Saptagram, Kolkata, West Bengal
https://en.wikipedia.org/wiki/Tribeni,_Hooghly

Varanasi, the order of spirituality based on a north-bound or upturned Ganga is equivalent to Yamuna thereby making Varanasi as an ambit of Sarasvati. Thus, Sages say that any pursuit of contemplation ('sadhana' in Sanskrit) at Varanasi is equivalent to a 'Sarasvat-sadhana' (driven by Sarasvati).

3. A third river in south of West Bengal, which is flowing eastbound and linked to an ancient river De-Ganga that had once charged the ancient port of Chandraketugarh, now an archaeological site at the border of West Bengal and Bangladesh. The well-known spot called Tribeni is still a locality in Bansberia Municipality of Hooghly district, called Adi-Saptagram in the Indian state of West Bengal.
4. The fourth river that is known as Padma (river of White Lotus, a spiritual flower associated with the Goddess) after the confluence of eastern Ganga (Ganga) and Yamuna (Brahmaputra) in south of Bangladesh.

Of the four, perhaps the most crucial is the fourth one, which is the confluence of Ganga and Brahmaputra, or Yamuna, which is currently in the divided landscape of Bangladesh, separated from West Bengal. Ganga River is one of the three rivers that form the Ganges Delta. It is the largest delta on earth, which leads to the Bay of Bengal. A part of the larger delta is in Bangladesh, which has been once a part of undivided Indian sub-continent. The tributary of Brahmaputra or the Yamuna River is one of the other three main rivers of Bangladesh. It is the lower stream of the Brahmaputra River, which originates in Tibet as Tsangpo or Sanpo. Then it heads into India and finally enters Bangladesh. Both Ganga (the larger branch compared to Bhagirathi/Hooghly in West Bengal) and Yamuna flow south to form the Padma River, near Goalunda Ghat and then it flows as Meghna (symbolizing the river of clouds) into the Bay of Bengal inundating south-eastern tip of Bangladesh, in the expanse of Chittagong delta in SE Bangladesh.

\Allegory of Triveni Sangam

The Triveni Sangam in a triangulated confluence of three rivers. In the north, it is predominantly Ganga, Yamuna and Sarasvati. On one hand, free or upward (ascending or pointing to Himalayas) triplet is at Prayag in Uttarakhand, North India, on the other, the descending or downward triplet is that of three aforesaid rivers in Bangladesh. In the south, the three rivers, which are Cauvery, Bhavani and Amudha. In South India it is also known as Triveni Sangam. Of the three, the river Amudha is subterranean and finally join the other two from below and then becomes Kooduthurai, where one finds the famous Sangameswarar Temple. Accordingly, Kanchipuram or what is called 'Dakshin-Kasi' or 'Varanasi of the South, had been established by two ancient Sages Agyasta and Scandaswami Murugan (a Rishi who is perhaps an equivalent of the image of Subrahmaniam Swami in the South and Kumar Kartik in the north) in times of remote antiquity.

Network of Nerve Circuits: Plexus centers (chakras) and currents (nadis or rivers)

The allegory of the confluence of the three rivers leads to an esoteric foundation of Yoga. It is in the subtle neural current and their centers called Chakras, which are the nerve centres in human body. There are several nadis (life force channels) and there are various nerve centers for these nadis. There are seven of them, at the coccygeal, sacral, navel, dorsal, cervical, pineal or at basal ganglia (Thalamus), and the final one at fontanelle or crown of upper cerebrum. The location of these Chakras are within the spinal cord that makes the brain float under the medulla oblongata. Of the seven, the sixth one is the *Ajna* or the command zone, formed at the Sinitic base of the eyebrows, where the Yogic features identifies a burning condensed bush ('bindu' in Sanskrit), an eternal light, provided the aspirant is able to reach that level through long-term efforts of deep meditation, sustained moral and humane life and contemplation. It is believed that above the eyebrows in the chakra called *Ajna*, there is the meeting of three nerves, hence called Triveni. It is also because that this point is the convergence point of three basic nadis—the left one is Ida, the right one Pinagla and the center one Sushumna. Figure 4 forwards 3 symbols forwarding the same truth.

The *Ajna* chakra is called Triveni Sangam and the practice of pranayama and meditation focusing on it imparts a burning or calcination of all defects of the lower world. Reaching Ajna or Triveni Sangam, the aspirant or his river of life is purified. It traces back the origin, as Ganga after meeting Yamuna may be traced back to their origin in the perineal snow-capped glaciers of the Himalayas.

Ida and Pingala Nadis are not the gross sympathetic chains. These are the subtle Nadis that carry the Sukshma Prana or condensed life force. In the physical body they tentatively correspond to the right and left sympathetic chains. In the hormonal-physiological sense, Ida and Pingala are comparable to Dura and Pia mater, and are labelled as the lunar and solar nerves. Ida starts from the right and Pingala from the left of the 2nd center, the sacral plexus, the foundation of procreative hormones (*Kama Vija* in Sanksrit). They meet with Sushumna Nadi at the Muladhara Chakra

Figure 4. Depiction of flows or currents in Indian Yogic system (left); in ancient Greek thought (middle); and in modern medical association in the USA (right).
Ref: http://www.energyenhancement.org/Chakras-and-Ida-Pingala-Sushumna-and-the-Caduceus.htm; https://en.wikipedia.org/wiki/Caduceus_as_a_symbol_of_medicine

and make a knot there. This junction of three Nadis at the Muladhara Chakra is known as Mukta Triveni (Kumar, 2013). Ganga, Yamuna and Sarasvati dwell in Pingala, Ida and Sushumna Nadis, respectively. This meeting place is also called Brahma Granthi. They again meet at the Anahata and finally, *Ajna* Chakra, known as the Visgnu, and the Rudra granthis (knots). Like inside, in the external world, there is a Triveni at Prayag where the three rivers Ganga, Yamuna and Sarasvati meet and aspirants seek a deep bath in the conjoint. In the inner sense, the Yogi has a bath in the depths of the nerve currents of his deeper psyche (Eck, 2013).

Caduceus: Three rivers of Medical Corps and World Health Organization

The Caduceus belongs to Roman Mercury, and is derived from the more antique symbol of Hermes in Greece and perhaps, that of Thoth, the God of transmutation and alchemy in Egypt. The symbol is composed of two serpents ('Sarpa' in Sanskrit) who are winding around a central staff that is winged above in the form of a falcon Thunderbird, symbolizing archangels of Semitic religions, or even, images of Ahura mazda (Ashura Mahadeo) in Zoroastrianism, or that of Garuda, a popular winged bird lifting serpents in Indian and Indonesian mythologies. Today it stands as a symbol of medicine. It is also simplified the Rod of Asclepius, especially in the United States, and particularly as a single rod-serpent symbol of the World Health Organization (WHO).

The configuration of two-snake caduceus design has ancient associations with Orphic tales, an intertwining of Zeus as a serpent mating Demeter also as a serpent, and producing young Zeus or Sabazeus, who is also known as Bacchus or Dionysus and its symbolism is also related to that of trade, literature, governance, alchemy and the inner wisdom (Dickson, 2016). But much before the ancient Romans and even the Greeks (about 2500 BCE), there is an older representation that comes from the Syro-Hittites and Mittanis, who worshipped the Vedic pantheon and spoke a language close to Sanskrit. For ages, the circular sign of 'Omkara', looking like the numeric eight or 8, with a moon and dot on top, has been a direct representation of traditional healing and therapeutic treatments in India. Omkara means a mantra, a sacred vibration, the very word of Cosmos, that heals the soul by linking his little nature with the cosmos.

The caduceus was finally accepted by the Medical Department of the United States Army in 1902. It was also accepted in the uniforms of Army medical officers. Despite widespread acceptance of the caduceus as a symbol of health representing the twin serpent, the single serpent driven rod of Asclepius has also been accepted as an emblem of medicine (Oxford, 2001).

DOUBLE INTENDER: THE TWO-WAY FLOW

Thus the allegory of Triveni Sangam takes us to the foundation of complementary between the opposites, the solar and the lunar nerve-forces. There are two opposites: (a) the river in its natural course, in the forward linkage of time, heading from the perennial snow-capped mountains to the ocean, where it is destined to merge with a greater whole; and (b) the river tracing back its origin, a return back to perennial glacier valleys, the frozen form of vapours, after the rain-clouds had shed snow on the highest mountains. The tracing back to origin is a backward linkage in time, if one uses the time-destination arrow as the forward linkage.

Ganga, or the river of *Sangama* ('pleasurable intercourse' in Sanskrit) meaning union, a merger with a greater whole. Yamuna, or the river of *Sanjama* ('ascetic restraint' in Sanskrit) meaning an isolation, a separation from the disorder, and a search for discrete roots. When Ganga meets Yamuna, a third river is formed, which is Sarasvati. The discussion so far has shown that these arguments from the deep semantics of Indian philosophy are metaphorical. Finally, they become significant as the experiences which are based on Gnosis and Realization (yogic experience and wisdom), as encoded in the Indian Vedas and other ancient wisdom literature of India.

To reiterate the argument of the double intender, the paper finally brings in two significant arguments, one from western art and the other from modern science, which resonates with the Indian view point. They are:

- Apollonian-Dionysian conjugate or duality, the two sides of ancient Greek thought and its aftermath in later western Art
- Coat-of-Arms used by physicist Neil Bohr based on the complementarity of opposites of Yin, the lunar female-supple current with Yang, the solar and male-aggressive current. It is an ancient construct hailed by many ancient philosophies in the Asia pacific (China, Korea, Japan and Shamanistic Siberia) (Fig. 5).

The twin couplet of Apollonian and Dionysian duality is a deep Greek concept. It represents opposites, a dichotomy, and a dialectic as a beginning of our understanding of what is good and what is not so good to us. The concept is based on the twin figures of Apollo and Dionysus, which comes from Greek mythology (Fig. 5). Friedrich Nietzsche in his celebrated work, 'The Birth of Tragedy' revived it. It is also used by others like Nobel laureate and writer Herman Hesse to others like Dr. Fritjof Capra, a renowned physicist from the University of Berkeley, USA, in his seminal publication, 'The Tao of Physics'. The concept is also evident in works of many others like poet-narrator Friedrich Holderlin and historian-semantics expert Johann Joachim Winckelmann. The idea is sourced to the ancient Mediterranean world and of late, from later sources like da Vinci and others till renaissance, from where the concept has inspired western philosophy and literature in many ways.

In Greek mythology, both Apollo and Dionysus are seen as sons of a primordial Zeus, but the sources of Dionysius say that he is also known as Sebazeus, who is an

Figure 5. The opposites: Apollonian-Dionysian and Yin-Yang in Neil Bohr's Coat-of-Arms. Source: https://www.pinterest.com/pin/492933121688051351/; https://www.theepochtimes.com/3-concepts-ancient-chinese-science-grasped-modern-physics-just-learning_726622.html

older or an Orphic version of Zeus, who comes from pre-Olympic or Thrace. Relatively, Apollo is a later Olympian God and the champion of Periclean Greece and mainly, Delphi till the time of Aristotle and Alexander.

In western thought, Apollo is the solar god, and his movements stand for reason and order. He has a presence that is allied to straightforward, uncompromising course of direct logic, prudence and purity. On the contrary, the course of Dionysus, who is the god of ecstasy and divine revelry, is a representation of multi-ways and options. Dionysus stands for a realization of primal chaos, that of primordial openness, irrationality and chaos, and appeals more and more to emotional, aesthetic, and instinctive parts of the human mind. The Ancient Greeks actually did not consider the two gods to be opposites, but their original story intertwines a counter polarity, as a juxtaposition of opposites.

In this context the works of American humanities scholar Camille Paglia (1990) are noteworthy. Paglia writes about the Apollonian and Dionysian in her bestseller 'Sexual Personae' 91990). Paglia delineates the concepts comprising a dichotomy, which is also the rudimentary principles of her theory of art and culture.

The presence of the Apollonian shades is that of homogeneous and monotonous light, which is extremely structured and tangible in the world of engineering, analysis and detailing. While the Dionysian way is the movement that is flexible, open and sees the beginning as darker, operating in the formative sheathes of diversity, heterogeneity and variety in free nature. However, there can be an argumentative end to the paper to what extent humanity needs both. It is an important clarification of the pairs that we see as opposites. One is that of *stability and change in political sphere and other that of order and freedom in planning and management. A third may be a pairing of tradition and innovation* in arts and sciences. It is important now to come to that argument and see how both can trigger more holistic ways of life, culture and civilization.

The idea of the duality also appealed physicist Neils Bohr, and more than that Bohr combined it to best explain the dualities of Quantum Physics that defies a straightforward, reductionist approach of Newtonian physics. When Bohr was offered a Danish knighthood, he constructed his own coat of arms and put in it the well-known Taoist Yin-Yang symbol (yin dark, yang light) to best represent his thoughts of a larger all-pervading physics and exceeding the Newtonian world of order and tangibles. He particularly added in Latin that 'opposites are not rivals but complementary', and together they bring the larger picture of reality, which initially looks uncertain and unpredictable.

The symbol of the 'Coat-of-Arms' also appears at the University of Copenhagen under Bohr's bust. But rationality at surface, be it art or physics, are unwilling to accept a complementarity between the sciences and the humanities, or between modern thoughts of strict rationality and order against ancient eastern spiritual semantics of cyclic and long term temporality and natural laissez faire, which mean both disorder and freedom in human evolution and development planning. Bohr's complementarity has attempted to best integrate the pairs *stability and change, or, order and freedom, or, tradition and innovation*, which have otherwise looked opposite to each other, best representing the larger reality, which is just not physical and physics, but also psychic and metaphysical.

CONCLUSIONS

It is evident in the preceding discussions that Indian rivers have played a deeper role in enriching national and cultural landscape. It is further evident that in India, river systems have not only catered to cultural development, but also enriching deep ecosystem and a deeper way of life.

The origin and destination of Indian rivers are seen as a flow, which carries deeper significances, and the imageries are likened to perennial life cycles and goals, which are anthropological at the individual level, cosmological at the larger level, and spiritual as the linkage between the two. What has been true in the outdoors and the external world—the macrocosm, is also a reflection of the changes that are expected in the inner human form, in and around the human soul. River currents, or Nadis have played both a symbolic and a direct neuro-physiological role in galvanizing these changes and formulate a foundation of Indian spirituality, which is Yoga. It is evident from the discussions that this primordial Indian thought had also distant resonances in ancient Greece in some form or the other, and was evident again in the advent of modern science, mainly in quantum and relativistic sciences representing the micro and macro worlds.

Deep ecosystem, as underscored, is therefore a complementarity of the microcosm and the macrocosm, that both ancient Indians and ancient Greeks had realized. In India it has been sustained. In Greece was lost with the dark Ages. However, there has been recovery of late in the modern scientific world.

To sum up, the system of Yoga symbolizes the outer currents of water that are comparable to the inner currents of the human form. The Rig Vedic Rishis label that as *nadis* or neural currents. Thus, the twin explanation, the double intender comprising of both physical and the psychic, represent the ecosystem of India's riparian philosophy. The river currents symbolize the neuro-physiological system of human bio-evolution that garner a higher goal of salvation. From the earliest wisdom literature of India, the Vedas to the recent works in India, rivers are portrayed as the symbol of cycles, of death and resurrection, and reincarnation itself, which have been discussed in the present paper. The place, economy and the deeper folk of India's community systems have evolved along the banks of the rivers of India, which the paper has highlighted.

The present paper has finally depicted a system of three principal rivers namely Ganga, Yamuna and Sarasvati, which collectively reinforces the deep ecosystem, the laboratory of the resurgence of the human soul. It is called the '*Triveni Sangam*', which delineates a higher system of Yoga.

The wisdom behind the flow and network of Indian river systems is therefore a contribution to the holistic ways of fostering human evolution. The holism facilitates an integration, a reconciliation of two extreme ways of life, which are not opposites but complementary. Human society and understanding proceeds only the spirit of integration is facilitated. It is best evident in the words of Schumacher (1993):

> "Societies need stability and change, order and freedom, tradition and innovation, planning and laissez faire. Our health and happiness continuously depend on the simultaneous pursuit of mutually opposite activities or aims' ... "no real understanding is possible without awareness of these pairs of opposites which permeate everything man does."

Human happiness of the highest kind is evident by integrating the pair of opposites which are otherwise complementary and parts of a larger whole. The deep ecological wisdom best configuring a holistic understanding of Indian river systems have always represented this whole.

References

Dickson, Despommier (2016). People, Parasites, and Plowshares: Learning from Our Body's Most Terrifying Invaders, Columbia University Press.

Eck, Diana L. (2013). India: A Sacred Geography.

Katz, E. and Light A. et al. (2000). Beneath the Surface: Critical Essays in the Philosophy of Deep Ecology Cambridge, Mass.: MIT Press.

Kumar, R. Krishna (2013). The Hindu : States/Karnataka : Preparations on for regional version of Kumbh mela. Thehindu.com. Retrieved 15 January 2013.

Michael Witzel (1989). Tracing the Vedic dialects in Dialectes dans les litteratures Indo-Aryennes ed. Caillat, Paris.

Paglia, Camille (1990). Sexual Personae: Art and decadence from Nefertiti to Emily Dickinson. New York: Vintage Book

Chapter 18
Evolution of Water Management Practices in India

Sharad Jain, Aisha Sharma, and P. P. Mujumdar

INTRODUCTION

Since the early times Indians have used various ingenious methods to conserve and use water for various purposes like domestic, irrigation, hydropower, navigation and recreation. This chapter traces the history and evolution of water management since the earliest times in India and briefly covers the history from the Indus Valley (Harappa) Civilisation till recent times. Evolution of water conservation systems under the rule of Chola's, Mughal's, and other empires are also discussed. Steps taken for water resources development (WRD) in the colonial era (under British Raj) and how the advancements were made till the present times are briefly mentioned along with future challenges for water management in India.

The infrastructure for water harvesting has been recognised by many names in different parts of the country, such as khadins, tanks and nadis in Rajasthan; bhandaras, tals and phads in Maharashtra; bundis in Madhya Pradesh and Uttar Pradesh; ahars and pynes in Bihar; kuhls in Himachal Pradesh and Kashmir; ponds in Kandi belt of Jammu region; eris in Tamil Nadu; surangams in Kerala and kattas in Karnataka. Adapting to different changing climate and hydrology over time, these long-established water harvesting and conveyance structures have survived the test of time over the centuries and many of them are still functional.

Different terrains and climatic zones in the country also developed adaptive water management strategies, examples of which are (Sen, NPTEL lecture):

S. Jain
Department of Civil Engineering, Indian Institute of Technology, Roorkee, India

A. Sharma · P. P. Mujumdar (✉)
Interdisciplinary Centre for Water Research, Indian Institute of Science, Bangalore, Karnataka, India
e-mail: pradeep@iisc.ac.in

A. Mukherjee (ed.), *Riverine Systems*, https://doi.org/10.1007/978-3-030-87067-6_18

- *Mountainous regions:* Diversion channels were used for agriculture; rooftop rainwater harvesting and bamboo pipes to carry water for various purposes.
- *Arid and semi-arid regions:* For irrigation purpose, rain-fed and stream-fed storage structures like tanks (e.g. bandharas of Maharashtra) were used. In other areas, rain-fed water harvesting structures capturing runoff and moistening the fertile soil bed were later used for growing crops (e.g. khadins of Jaisalmer, johads of Alwar, ahar of South Bihar). For drinking, groundwater structures to tap the aquifers, like wells and stepwells (e.g. bavdis of Rajasthan, forts of Chittor), rooftop rainwater harvesting (e.g. tankas of Pali), artificial catchments which drain captured rainwater into artificial wells (e.g. kunds of Rajasthan), special structures which do not allow the rainwater to mix with saline water (e.g. virdas of Kutch) and horizontal wells to harvest seepage downhill slopes (e.g. surangams of Kerala) were used.
- *Plains and floodplains:* Inundation channels, constructed on major rivers allowed floodwater to be diverted to agricultural fields (e.g. inundation channel of West Bengal, ahar-pyne of South Bihar) and rainwater stored in agricultural fields surrounded by bunds (e.g. haveli system of Madhya Pradesh) were preferred for irrigation purposes, whereas, dug wells were used to store drinking water.
- *Coastal areas:* Irrigation system was built with a focus to control the ingress of saline river water and was based on regulatory system, especially during high tides (e.g. khazana lands of Goa) and dug wells were used to store drinking water.

The following sections provide details of water management practices during different historical periods.

WATER RESOURCES MANAGEMENT FROM HARAPPAN CIVILIZATION TO GUPTA EMPIRE

This section traces WRD from the beginning of Harappan Civilization (around 3300 B.C.) till the end of the Gupta Empire (500 A.D.). Most of the ancient civilizations, e.g., Indus Valley, Egyptian, Mesopotamian and Chinese, flourished and developed at places where water was readily available like in the vicinity of lakes, rivers and sea (Singh et al., 2020). The Indus Valley (Harappa) Civilization, also known as a *Bronze Age Civilization*, lasted from 3300 B.C. to 1300 B.C. in the North-Western regions of South Asia (present-day Pakistan and North-West India). This civilization was spread over large area in the fertile Indus river basin.

Indus Valley Civilization had an elaborate and efficient sanitary and drainage system which points to meticulous planning, engineering skills and high quality of work. Well-defined drainage and wastewater management systems show that the early Indians possessed advanced knowledge of hydrology and water management. The Harappans, besides excelling in the design of structural and hydraulic infrastructure, were also aware of the seasonal rainfall and flooding of the river which

Figure 1. Ruins of Mohanjo-Daro (left) and The Great Bath (right). (*Source:* https://www.britannica.com/place/Mohenjo-daro)

helped them in planning irrigation system. Singh et al. (2020) have provided a detailed discussion on this topic.

Since the earliest times, water has played a remarkable role in framing the cultural and the religious life of Indian people and the structure and the Great Bath of Mohenjo-Daro supports this fact. The ruins of the bath are shown in Fig. 1, and it is considered as the "earliest public water tank of the ancient world" (Mujumdar and Jain, 2018). Drains were also built to carry used water from the bath (Pandey, 2016).

Another city in the Harappa Civilization, Lothal (Gujarat), had constructed bunds to store rainwater to meet irrigation and drinking needs. Similar water harvesting and conveyance structures were found in northern and western parts of India. Sophisticated sanitisation was practiced and private toilets connected to sewer tanks were also constructed and maintained regularly. The dock-yard in Lothal city, a remarkable lined structure, is another example of excellent hydraulic knowledge of the early Indians (Fig. 2). Channels were provided for inlet and outlet of water. A spill-channel, 7 m wide, connected the dockyard with the nearby Gulf of Cambay (now known as Bhogavo river) (Pandey, 2016).

Provision was made for entry of boats during high tides through the channel and for return from the dock back to the river when the tide was low. Extra discharge was released through a separate spill channel. To regulate the water level, wooden sluice gates were fixed across the spill channel.

Another urban site of the Indus Valley, Dholavira, located in the Rann in Gujarat had an impressive system for harvesting rainwater in several reservoirs (Pandey, 2016). The thick walls in Dholavira confirm that the Harappans were pioneers in coastal disaster management as they were aware of the threats from the tsunamis. Dholavira was located in a drought-prone region and the discoveries of water-harvesting infrastructures reveal that the people were well aware of water scarcity and related issues.

An old water harvesting system was found in Naneghat, in Western Ghats, nearly 130 km from Pune. A large number of rock-cut tanks were constructed along this ancient trade route to provide drinking water to the travellers. Two local seasonal

Figure 2. Dockyard at Lothal (*Source:* Harappa.com).

rivulets, the *Mandsar* (north-west of Dholavira) and the *Manhar* (south-east of Dholavira), collected rainwater in the catchment areas which was diverted to the tanks/reservoirs thereafter. An ingenious system was designed in which the stone bunds were raised across the streams at adequate places which directed the runoff generated to a series of reservoirs. At least 16 reservoirs were constructed to store water within the Dholavira city as reported after archaeological excavations. A network of well-connected storm drains was laid to collect rainwater passing through the citadel. The bathrooms in houses had drains to carry waste water into the city drainage system. Covered manholes were provided at intervals to clean and maintain the drains. Efforts were made to preserve rainwater in the areas where groundwater was brackish and there were no perennial surface water sources (Pandey, 2016). In other Harappan sites, like Kalibangan, Surkotda, Chahnudaro etc., residential units were served by individual wells. A survey by archaeologists reveal that, besides the public wells, there was a well in every third house.

The Vedic period coincided with the later part of Indus Civilization and lasted from about 1500 B.C. to 1200 B.C. (1700 B.C.-1100 B.C. according to some scholars). The references to the water and hydrologic processes in the Vedas signify that during the Vedic period, people understood the concept of hydrology and hydraulics. In Vedas and other ancient texts, one can find numerous references to the continuous movement of water in a cyclic form, similar to those in the modern concept of hydrologic cycle. Many concepts of hydrology are scattered in the Vedic texts in different verses, in the form of hymns and prayers praising the deities (NIH, 2018). Singh et al. (2020) and NIH (2018) have provided a detailed description of this topic. The Vedic literature refers to the significance of preservation of water owing to its medicinal properties (Pandey, 2016). Ground water development and water quality consideration have been mentioned in the *Vrhat Samhita* (550 A.D.).

Water-harvesting infrastructure from the Iron Age were surrounded by megalithic complexes on the riverside and also used as bathing *ghats* (steps) by pilgrims, devotees, and washermen. Water has been associated with the sacred rituals and has been used for drinking, cooking, cleaning and navigation (Morrison, 2015). References to the artesian wells have also been found in the Ramayana (200 B.C.).

An ingenious combined system of irrigation and water management was developed in the Mauryan era, a powerful kingdom which flourished in Magadh between 322 and 185 B.C. The Arthashastra, an ancient manuscript, gives details about water management during the Mauryan period (Cullet and Gupta, 2009). Around 400 B.-C. witnessed one of the driest periods of all times in India which might be the driving force for Kautilya to design and implement better and stringent water management practices to facilitate agriculture in the state (James and James, 2013). The Arthashastra refers to the development in water works, irrigation, navigation and also elaborates on the water tax/water pricing system for the efficient use of water and construction of suitable infrastructure. It specifies that water was the King's property and the users who wish to withdraw water from the irrigation systems installed by the king were levied a water tax (Cullet and Gupta, 2009). Arthashastra elaborates on all the scientific water management practices depicting various aspects of water law. Officers were appointed to monitor the rivers, measure the land and periodically inspect the sluices by which water was let out.

Arthashastra also describes about forecasting of rainfall by observing the position, movement and cloudiness of Jupiter, the rising, setting and movement of Venus and changes in the appearances of the Sun. Kautilya's wisdom is reflected by instructions to install rain gauges to collect the hydrological data and plan agricultural activities accordingly. Arthasastra also refers to check-dams used mostly for agriculture. These dams were later renovated under the reign of King Asoka in the 3rd century B.C. Arthasastra also describes the norms and punishments for the violations against the use of sluice gates of tanks (Karpagaselvi, 2014). Kautilya discovered that water could be harnessed by digging wells with underground feeders. The term '*anudake kupasetu bandhotsan sthapayet*' in Arthasastra refer to the importance and significance of wells for irrigation in the *anudaka* (dry) areas and every ten houses were provided with one tank (James and James, 2013).

The water law, water pricing, construction, knowledge on hydrology and responsibilities of the Irrigation Department prevailing at that time also gives a vivid picture of water management practices in ancient India. An ancient water conservation and management system, the ahar-pyne system is particularly interesting. This was a canal and tank system, where pynes (canals) used to bring water into ahars (tanks) and the ahar recharged the ground water and stored the excessive water, thus allowing the nearby wells to draw good amount of water throughout the year. The ahar-pyne system also diverted some runoff in the catchments thus reducing the potential floods at low lying areas. In recent years, the local bodies are trying to revive this traditional water harvesting system (Koul et al., 2012).

Sudarsan Lake, an ancient artificial lake, was excavated during the rule of Emperor Chandragupta Maurya (3rd century B.C.) in Gujarat by Pusyagpta (Pandey, 2016). Buddhist texts from 3rd century B.C. also mention about irrigation practices and the noted Indian scholar, *Pāṇini,* during the 4th century B.C., referred to the tapping of water from several rivers like, Sindhu, Suvastu, Varnu, Sarayu, Vipas and Chandrabhaga, for irrigation.

The remains of a famous irrigation dam constructed during Chandragupta Maurya's period, which was later repaired by Saka King Rudradaman, is proof of the impressive engineering skills during this period. Another notable structure which depicts an advanced level of hydraulic engineering in those times (1st century B.C.) is the complex of four water tanks at Shringaverapura near Allahabad where a canal was constructed to allow the flow of water from river Ganga to the dam and was more than 250 m long and 38 m wide.

WATER MANAGEMENT AFTER THE GUPTA EMPIRE PERIOD (500 A.D.) TILL 1500 A.D.

The end of Mauryan reign witnessed the emergence of small kingdoms and dynasties in India. Besides the Guptas, other dynasties and rulers in India (Chauhans, Chola, Satavahana, Chutus, Kadamba, Chalukya, Rashtrakuta) contributed to WRD by building tanks, canals, wells and artificial lakes based on the region's topography and climatic patterns. The Chalukyas (6th to 12th century) are remembered for constructing various water harvesting structures like lakes, tanks and canals. A 'golden age' of implementation of high level of scientific and technical expertise during 937 A.D. to 1336 A.D. was witnessed in South India (Ramachandra et al., 2016). The technique of cascading tanks for controlling floods as well as use of water for irrigation was advanced. Structures in the cities of Bagali and Kalyana are the best examples of efficient water harvesting.

Cholas, the longest ruling dynasty of South India (848 to 1279 A.D.), made significant contributions in water management and harvesting systems. Cholas, besides being fierce warriors, paid attention to constructing and maintaining water harvesting structures in their realm. They realised that water conservation and harvesting was a key factor to maintain a long-lasting rule. During the Chola reign, irrigation and water infrastructure had greatly advanced. A Tamil inscription, found in Tirupati, states that people had also planned for emergency arrangement to face scarcity of water. Rules for use of water by the people in proportion of their land holdings were made so that no water is wasted. It was mandatory to mention in the documents about the river, lake or tank to be used for irrigation of the land.

Avvaiyar and Tiruvalluvar, famous poets of that era, have highlighted the importance of conserving water for use by all sections of society. Efficient utilisation of all the resources and excellent water management facilitated the prosperity of Cholas kingdom. The famous early Chola king, Karikala Chola, is credited to have built the world's earliest water-regulating structure at *Kallanai* on Cauvery river around 270 B.C. to promote agriculture. Inscriptions of Tiruppalanam and Tirucchendurai reveal that tax (*Kavirikkarai Viniyokam*) was levied for the maintenance of the banks

Figure 3. Cholagangam lake (Solagangam) by Raja Rajendra Chola (Credit for the figure: Dr. Durgadoss, *Source:* https://www.facebook.com/DrDurgadoss/photos/chola-ganga-ponneri-lake-lakethe-great-lake-lying-about-two-miles-to-the-west-of/1660106954104260/)

of Cauvery river. Cholas also constructed many artificial networks and lakes, as confirmed by inscriptions of those times. Cholas were also experts in engineering designs and execution of the construction work. *Solagangam (Cholagangam)*, a huge tank described in the history as the "**liquid pillar of victory**", was constructed by king Rajendra Chola in the capital city Gangaikonda Solapuram (Fig. 3). This tank was 16 miles long and had sluices and canals to irrigate lands in the neighbouring areas. Other famous lakes constructed in this period are *Madurantakam, Sundra-cholapereri, Kundavai-Pereri* (named after a Chola queen).

A separate board/committee was set-up for maintenance of lakes and members were appointed to help the board. Among the major achievements of king Rajaraja I, *Mayanur* canals, located on the southern bank of Cauvery river at the west of Kulittalai needs to be mentioned. A channel in *Mayanur* was constructed to provide irrigation water to places such as Palaiyur, Analai, Pulivalam and Somarasampettai, on the bank of Trichy. The channel was named after one of the king's titles i.e. *Uyyakkondan*, which exists even today. Similarly, an inscription of Rajaraja I (994 B.C.) found in Tirukkogarnam Siva Temple of *Pudukottai* reveals a construction of a huge lake, *Kavirakulam*, currently known as *Tiruvetpur*.

A large lake was constructed in the times of King Parantaka-1 (907-953 A.D.) of Chola dynasty near Kattumannarkoil in South Arcot district. It was named as *"Veeranam Eri"* (Fig. 4) after the title of king Parantaka-1 and was considered to be the first of the major irrigation works in that period (Rathika, 2016). Villages were provided with exclusive assemblies and tank supervision committees (*eri-variyam*) who maintained the village irrigation works by regulating the water supply and de-silting the channels.

Figure 4. Veeranam tank from Kandakumaran village, Kattumannarkoil Taluk, Tamil Nadu (*Source:* http://www.jothiganesh.com/2014/04/10/cholas-water-management-plan-in-10th-century-is-worth-more-than-200-million-dollar-in-2014/)

Figure 5. Surajkund reservoir (left) and Anangpur dam (right) (*Source*: https://www.inditales.com/surajkund-history-haryana-delhi/ and https://en.wikipedia.org/wiki/Anangpur_Dam#/media/File:Stepped_%22downstream%22_side_of_dam.jpg)

Another large artificial lake was constructed by King Bhoj Parmar, the ruler of Dhar (Malwa), Bhojpur near Bhopal, in the middle of 11th century by creating an embankment across two hills. Water flowed into this lake from at least 365 streams and springs and it originally covered an area of nearly 250 square miles.

The Tomar dynasty, successors of Arjuna of Pandava dynasty, came into power from 736 A.D. to 1180 A.D. Raja Anangpal I of this clan re-established the kingdom in Dhillika (now Delhi) where several major towns were established. They also established the *Surajkund* and *Anangpur* dam (Fig. 5). The reservoir within Lal Kot is known as *Anangtal* (Sharma, 2001).

Water from various Arus (rivers), Vaikkals (Canals) and Kulams (tanks) helped in irrigating the downstream area of Thiruvaiyaru where excellent irrigation system prevailed. An inscription in the *pushpamandapaon*, situated on the banks of Cauvery

Figure 6. The Grand Anicut on river Cauvery *Source:* https://www.thecivilengineer.org/images/Manager/rpaigem/Front_View_Kallanai.jpg)

river in Thiruvaiyaru mentions that the Grand Anicut (Fig. 6), that has an impact on the irrigation system of this region even today, was rebuilt by Achyutappa Nayaka before 1596 A.D.

In Rajasthan, Chand Baori was built by King Chanda during the 8th century. It is one of the deepest and biggest stepwells constructed to solve water scarcity problem in the region. In Karnataka, there are not many big natural lakes. The topography of Telangana and Karnataka favours the construction of storage reservoirs. One of the largest tanks in the region is Sulekera whose circumference is 64 km. Other large tanks are *Ayyankere, Madagakere, Masur-Madagakere, Vyasa-samudra, Ramasagara* and *Moti Talab*. Telangana, where many tanks are present, is known as "The Land of Thousand Tanks". The tanks in Telangana were also developed in series by bunding the same valley at several points (Randhawa, 1980). Major Sankey, one of the first engineers of the Mysore state, undertook systematic repairs of tanks in Bangalore.

In Tamil Nadu, the first tank was excavated by Pallava King Mahendravarman I (during his reign from 600 to 630 C.E.). An inscription of *Nandivarman* states the use of water levers (*jalyantra*). The tank at Uttaramallur, known as *Vayiramega-tataka* was also developed by a Pallava king. King *Dantivikramavarman* and king *Kampavarman* constructed a tank in north Arcot and constructed one *Ukkal*, respectively. During the 9th century *kaveripak* tank, constructed with a bund of about 6.4 km length, stood unrivalled in the district.

From 950 till 1750 A.D., the Chandela and Bundela rulers had ruled over Bundelkhand and had developed a dense network of tanks in their kingdom. The inscriptions from the Bundela times reveal that Bundelkhand had a large number of environment-friendly traditional irrigation methods for storing water. Chandela tanks were constructed to meet water needs of humans and cattle. The flow of rivulets was diverted with the help of massive earthen embankments, of width 60 m or more. Since lime and mortar was used to construct the tanks, they survived even after thousand years (Nair et al., 2014).

During 13th and 14th century, the Gond landlords and villagers had a great impact on the traditional water harvesting structures in Orissa. They had developed an intricate and immensely useful network of *katas* (three-sided reservoirs) with one side kept open to receive runoff. Orissa's *katas, bandhs* and *mundas* have played a major role in mitigating the impact of droughts and famines. In Sambalpur and Balongir districts, these traditional infrastructures proved to be a blessing, during drought years.

Technical advancements in hydraulic systems during the reign of different rulers in Kashmir are described in Kalhan's 12th century text, the '*Rajatarangini*' which mentions about a well-established irrigation system consisting of canals, aqueducts, dykes, barrages, wells, embankments and waterwheels. A vast embankment, the '*Guddasetu*', constructed by king Damodara II and a series of *arghat* (waterwheels), built by King of the Karkota dynasty, Lalitaditya Muktapida, in 8th century are described in this book. The wheels were designed to channelize water from *Vitasta* (Jhelum) river to the villages near *Chakradhara* (Tsakdhar). Suyya, a renowned irrigation engineer is credited for controlling drainage of the Vitasta river by constructing and maintaining a stone dam. To regulate the water flow from the vast Mahapadma lake (or Wular lake), Suyya displaced the confluence of rivers Sindhu and Vitasta, and built stone embankments along river Vitasta. Suyya designed the irrigation system ensuring a fair share of water for everyone (Pandey, 2016).

It is evident from the above discussion that during this period, many kings and rulers constructed tanks and reservoirs, canals, embankments, wells, and other structures to meet the water needs of the population. These developments were in line with the magnitude of demands, technology and knowhow available during that period.

WATER MANAGEMENT DURING THE MEDIEVAL INDIA, THE MUGHAL EMPIRE (1526-1857 A.D.)

This section presents the developments in water management in India during the 14th-18th century. The north Indian subcontinent was dominantly ruled by the Mughal Empire. During this period several powerful Hindu states also emerged, prominently, Vijayanagara, Gajapati, Ahom and Mewar. Several major archaeological monuments and structures were constructed which made their mark in the history. Some of these structures are UNESCO heritage sites today (e.g. Qutub Minar, Rani ki vav, Taj Mahal, etc.). The Mughal and Hindu kings contributed immensely for construction of various hydraulic structures as the population increased. The focus was on canals for water diversion, various attractive waterworks and the development in Ganga-Jamuna basin during this period.

During the medieval period (1300-1700 A.D.), the Vijayanagara empire, in Bellary district (in Karnataka) invested in large water harvesting projects. The

Figure 7. Sulekere (Shantisagara). (*Source:* https://i.pinimg.com/originals/f3/ 97/48/f397488f061 951e3058940361cafc119.jpg)

construction of tanks progressed remarkably during this period. However, in the late 18th century during the British Raj, the tank system degraded and the systems of natural resources management was centralized (Ramachandra et al., 2016). More than 12 anicuts were constructed across the river Tungabhadra in Bellary district which was the major source of water in Vijayanagara (Shivakumar and Cheluvaraju, 2016). Another spectacle from this period, which is functional even today, is *Sulekere*, also famous as *Shantisagara* (Fig. 7), in Karnataka. Construction of tanks was encouraged for agriculture activities and domestic needs. The tanks also acted as an alternative for harnessing water away from the rivers. As the population increased, these structures were accompanied by open wells and spring wells. The commoners started using the groundwater for irrigation and other needs by mechanically lifting the water. Rulers of the empire ensured that certain practices for the maintenance of water structures and irrigation were followed for fair use of water and irrigation.

Datia, which was a part of Orchha state (Madhya Pradesh), was ruled by Maharaja Bir Singh Dev in 1600s. The king had built seven tanks, including, Sita Sagar, Ram Sagar, Karan Sagar, Laxman Tal, Lala ka Tal or Bir Sagar which were connected to each other indicating a fine technical planning in that era. These tanks were developed in a cascading manner—the first tank was located at a high elevation and subsequent tanks were excavated in succession.

There is also mention about a *Haud* (lake) at Palwal in a Persian epigraph which was constructed by *Badr U'ddin Sunqar* (an officer who was incharge of the town in 1211 A.D.) for the benefit of people and cattle. The old *Kolab* reservoir constructed in Sialkot during the period of Delhi Sultanate was a major source for domestic water supply.

Another example which shows the impressive engineering skills of the Mughals are the old water works of Burhanpur town in Khandwa district of Madhya Pradesh which was constructed in 1400s on the banks of Tapti river and was named after

Figure 8. Kundi Bhandara in Burhanpur town. (*Source:* https://i0.wp.com/ranasafvi.com/wp-content/uploads/2019/03/img_5232.jpg?w=720&ssl=1)

Sheikh Burhanuddin. The town's ingenious waterworks allowed provision of freshwater to the population. A groundwater based scheme was conceived by Abdul Rahim Khan in 1615 A.D. Underground network of tunnels and infiltration galleries were discovered by a Persian geologist, Tabkutul Arz, who surveyed the recharge valley between the Satpura ranges in Tapti plains and discovered that these galleries were designed to supply fresh water to the town. The water-supply scheme of Burhanpur constituted of *bhandaras* (storage tanks), as Kundi Bhandara shown in Fig. 8, stored the groundwater captured from underground springs flowing from the Satpura hills towards river Tapti. Groundwater was intercepted at four points (Mulbhandara, Sukhabhandara, Khunibhandara and Chintaharan) which were located to the north-west of Burhanpur town. The subterranean conduits linked by a number of connected wells carried water to a collection chamber called *Jali Karanj*, which supplied water to the town. The underground water-supply system, based on gravity with an adequate gradient depicts the efficiency in design, construction, strategic planning and engineering during the Mughal rule.

The medieval Indo-Persian scriptures mention that the reservoirs were constructed to supply drinking water to the residents of Delhi but according to Batuta (1958), water from the reservoirs was also served for irrigation. Sultan Alauddin Khalji is credited for establishing irrigation system in his empire for the advancements in agriculture. Utilizing big rivers was a significant development that occurred in the irrigation sector. The territories which suffered from the water scarcity were provided with water through the canals. Large artificial canals were constructed in the reign of Sultan Alauddin Khalji near the end of the 13th century.

The nobles of Sultan Iltutmish also promoted construction of lakes, especially in the water scarce areas. An inscription at *Bari Khatu* (District Nagaur, Rajasthan State) reveals that to facilitate the travel and cultivation in the dry desert climate a lake was constructed. The first lake known as *Haud-i-Sultani* (also known as *Hauz-i-Shamsi),* was built by Sultan Shams U'ddin Iltutmish in Delhi near the *Idgah* and outside the Ghazni Gate which was two miles in length and half in breadth. About the same time in the town of Badaun, a lake was constructed. Another very important

Figure 9. Haud-i Khass lake (left) and Haud-i Sultani lake (right). (*Source:* https://upload.wikimedia.org/wikipedia/commons/d/df/Hauz_Khas_Lake.jpg and http://www.spaenvis.nic.in/WriteReadData/links/Picture3-374565400.jpg)

lake *Haud-i Khass* (*Haud-i 'Ala'i* or Ala'i lake) or *Hauz-i-khass* was constructed in the new metropolitan city of Delhi (Siddiqui, 1986). *Haud-i Sultani* and *Haud-i Khass* held an important position in the socio-cultural life of Delhi and became a beautiful recreation centre subsequently (Fig. 9).

Furthermore, the Tughluq period reviews on the advancements which occurred in engineering during the 14th century. Ghiyath-U'd-din Tughluq (1320-24 A.D.) had built aqueducts surrounding the tomb over the lake that was constructed for his own burial.

In two metropolitan cities, Adilabad (near Delhi) and Daulatabad (near old Deogiri in Maharashtra), founded by Sultan Muhammad Bin Tughluq, son and successor of Sultan Ghiyath-U'd-din, beautiful lakes and cisterns were built by him. A lake was built at the palace fortress of Adilabad. Certain administrative charges were collected by *walis* or *muqtas* (governors) of the state while the lakes were used for the public utility in Delhi and Daulatabad. The inscription found at *Bari Khatu* reveals the history of Firuz Saghar (Hind Sagar), constructed by *Mziqtd* (governor) Malik Firuz bin Muhammad.

Beautiful edifices, bridges, aqueducts, lakes, cisterns and irrigation channels, which were also significant from historical perspective, were constructed and developed during the Mughal period. Sultan Firuz Shah built dams in territories that suffered from the scarcity of water and had erected a huge dam to store rain water. He also repaired the steps of the Tomar-era Surajkund reservoir (see earlier section of this article) and the Western Yamuna Canal, which is said to have originally been built by Prithviraj Chauhan.

Another architect, Ali Mardan Khan, had supervised many modifications in the Mughal irrigation system and construction of other structures like Naulakha Garden in Lahore and canals in Delhi and Bansli Madhupur. One of the marvels of his work is a canal which carried water from river Ravi to Lahore, facilitating the construction of Šalamar garden which is a prominent garden of Mughal history. Besides this, he repaired an ancient canal used to bring water to Shahjahanabad, which was initially constructed by Firoz Shah Tughluq. The Shalimar Garden was also constructed by

Ali Mardan Khan in Shahjahanabad on the banks of Yamuna, north of Kashmiri Gate under the reign of Shah Jahan (Rehman, 2019).

Evidences reflect on the availability and use of the mechanical gear devices called "Persian Wheels" used for drawing water from the deep wells in Delhi during early Sultanate period which contributed to the prosperity of the state. This later migrated to South India and Kolar had large number of wells with Persian Wheels (Fig. 10).

Apart from the famous tank system built during the Sultanate period, also known for another very important infrastructure—stepwells or *baolis/bain*. Baolis were the lifeline of water supply during the medieval period in Delhi. These structures were constructed for the water to be used equally. The inscriptions discovered in the stepwells reveal that the old traditions of water harvesting were also practiced in the 15th century and were followed religiously by the regional Sultans and their officers in Gujarat, Malwa, Juanpur, etc. In Rajasthan, the *Raniji Ki Baori* or "The Queen's Stepwell" was constructed in the 17th century by Queen Nathavati of Bundi for religious ceremonies (Fig. 11). Bundi is also referred as the "City of Stepwells" because the city has around 50 stepwells.

Figure 10. A well with Persian Wheel in Kolar district, Karnataka. (*Source:* https://www.indiawaterportal.org/sites/indiawaterportal.org/files/uploads/2008/01/img_0052.JPG)

Figure 11. The remarkable architecture of Raniji ki Baori in Bundi. (*Source:* https://www.tourmyindia.com/images/raniji-ki-baori2.jpg)

During Firuz Shah's rule, the canals, 100-120 miles long, were laid to carry water from the Yamuna and the Ganga rivers. Water flowing through these canals also irrigated the desert and barren lands which had no water before. Some canals were big enough and were even used for navigation. The Sirat-i-Firuz Shahi mentions in detail, about the *Nahr-i Gang* (Ganga Canal). It is believed that during 1365 A.D., the Sultan had constructed a double canal system which drew water from both Yamuna and Sutlej rivers. This laid the foundation of the town *Hisar Firuza* (modern Hisar). The Sutlej canal (*Ulugh Khani*), flowing through towns of Ropar and Sirhind, further merged with the Yamuna canal (*Rajiwah*) which flowed through Karnal town. The outlet for this combined system was located at Hisar Firuza which was a channel draining into the ditch around the city walls. The ditch connected flow to a channel which irrigated the area upto Jhajhar town. Another canal which carried the water from Ghaggar river, was constructed to flow past the Sirsuti fort and irrigated the land up to Harnikhera. *Ju-i-Firozabad*, known as Jamuna canal, drew water from the river Jamuna just below the foothills of Siwalik and flowed upto the capital city of Firozabad. The canal flowed past Shahabad town receiving water from Sirsuti and Salima streams.

The Mughals also carried out some river training works. When they realised that the river Yamuna (Jamuna)[1] was constantly shifting eastwards, emperors Akbar and Shah Jahan tried to embank the river and stop its shift. With the embankments (*bunds*), Akbar's *Band-i-Akbari* and Shah Jahan's *Band-i-Shahjehani*, migration of Jamuna river was largely stopped. The shift was also important to guard the citadels and the newly constructed Red Fort. The shift was also controlled because the significance of such strategic location would have diminished if the region dried up with Yamuna moving away. Such works required hydrologic data and knowledge of hydrologic principles which was gradually being acquired based on their observation of natural processes.

The construction of water infrastructure continued and by 1843, Shahjahanbad city had 607 wells, out of which 52 wells provided sweet water for consumption. This is much away from the reality of today where 80% of the wells are closed because the water is contaminated by the un-monitored sewer systems.

A world-famous monument and one of the marvellous wonders of the world known for its architecture and history is the Taj Mahal. It had sophisticated waterworks to irrigate the garden and an elaborated water channel system was created in the *Khan-i-Alam* complex on the western side of the Taj Mahal. Towards west of the Mahal, after the settlement of silt in the funnel-shaped silt chamber, wide channels carried the water into the huge storage tank in the *Khan-i-Alam*. From these tanks the water was led through the earthenware pipes to various parts of the garden. The conservation works carried out in 1860s revealed the earthen pipes which were embedded in solid masonry about 1.8 m below the ground. The copper vessels attached with the main earthen pipe through copper pipes constituted the fountain system of the central tank. The aqueduct constructed in the Mughal era was operational till 1903.

[1] The words Jamuna and Yamuna are used interchangeably.

WATER MANAGEMENT DURING THE BRITISH PERIOD (1857-1947)

During British rule, India suffered from droughts and famines, especially during 1765-1947. Under the British Raj, irrigation was considered as a major source of revenue. Provision of water to all parts where agriculture was established became a major concern. Moreover, need of water for personal and other commercial purposes was also felt with the growing population. This section briefly discusses the development of irrigation in the colonial times under the British rule. Various structures constructed for irrigation and provision of water, in the driest parts of India are also discussed.

A giant leap in irrigation sector was achieved by the contribution of Sir Arthur Cotton and Major Proby Cautley in 1830s, in southern and northern India, respectively (Naz and Subramanian, 2010). In India, before the British rule, there was no control on water resources in such a strong centralised manner. Rather, control on water resources was exerted by different categories of groups within local societies and commoners wherein the distribution of water was carried out according to the political level of such different groups within the local communities. In southern deltas, Sir Arthur Cotton had initiated the structural renovation of the Grand Anicut situated on the Cauvery river. Reconstruction started in 1838 and a phase of enormous activity related to the construction of canals under the colonial rule began leading to a paradigm shift in irrigation. In the Upper Ganga areas where rainfall was low, a large number of new canals were developed which resulted in vast interconnected canal network and became a trending mode of progress.

Irrigation being a major source of revenue for the British, focus was on 'productive' irrigation works of canals so that it could lead to revenues equal to the interest on their capital cost. The first Famine Commission was appointed in 1878, as its need was realised after the widespread suffering caused by successive famines. In 1880, when Indian Famine Commission announced curtailing famine relief expenditure as the indirect returns of irrigation works, it forced the British Raj to align with the 'protective' irrigation works. Constructions of large-scale storage dams and canals in Bombay Deccan were a few examples of such works. A highly centralized irrigation management system evolved with a hierarchal bureaucratic structure extending even to Britain (Naz and Subramanian, 2010). The Punjab 'canal colonies' are one of the finest examples of the colonial irrigation policy. The lands were cultivated using these interlinked irrigation canals. In 1880s, the construction of a series of interlinked irrigation canals in Western Punjab by the colonial engineers agriculturally colonized and developed about 14 million acres of arid land.

The disastrous Great Famine during 1837-1838 in the north-western parts of India made the British administration realise that it was important to protect the fertile belt of Ganga-Yamuna from recurring drought. It compelled them to take follow up actions. Some other significant works were the construction and alignment of Sutlej and Chenab canals (in the Punjab), the lower Ganga and Betwa canals (in the United Provinces), and canal between Cuttack and Hooghly for the purpose of irrigation and navigation. The then newly created Famine Relief and Insurance Fund financed some 'protective' works such as Betwa Canal project in Bundelkhand which was one of the first 'protective' work financed. The first of its kind, Ganga canal, was completed in 1854. It was an impressive monumental work by Sir Proby Cautley and at that time, it was the largest canal in the world. Some of the impressive features of the lower Ganga canal (constructed later) were its suitable location and construction of Narora weir. It was the first large masonry structure developed in the alluvial plains of Ganga (Sharma and Singh, 2011).

James Thomason, Lieutenant Governor of the north-western provinces directed the construction of Thomason College of Civil Engineering (1854) to produce "well educated artifices" for repair, operation and completion of Ganga canal system and for the other similar works as indicated by Proby Cautley. This college was elevated as the University of Roorkee in 1949 and later as the Indian Institute of Technology, Roorkee in 2001.

Another famine relief work, the Nira left bank canal came into existence in the Deccan region of Bombay Presidency during 1876-1885 and was initiated with the concept of '*protective irrigation*'. 'Block system' based on the traditional crop rotation system, the bhandara or phad (used in Maharashtra) was started in Nira canal. The canal projects were highly centralized and bureaucratically controlled and successfully served the interests of users. This system practiced in northern district of Bombay Presidency was suitable according to the circumstances and thus the 'block system' and was successful for the cultivation of sugarcane along with other food crops in rotation.

The first Indian Irrigation Commission, constituted in 1901, gave comprehensive recommendations for the upgradation and construction of irrigation projects. The Commission's focus was on providing irrigation to prevent or mitigate 'the horrors and cost of famine'. Pragmatism of the Commission in its development proposal of irrigation prospects recommended both productive and protective programmes in irrigation works. Following their recommendations, many projects were constructed. Some *productive* works like 'Triple Canal Project' in the Sind and *protective* works such as Tribeni Canal in Bihar, Ken Canal in Bundelkhand in 1906, and Tendula Canal in Central Provinces are a few significant projects. Great irrigation projects e.g. Sukkur Barrage Project (in Sind) and Sarda Canals (in Central Province) were constructed. Some of the other noteworthy projects undertaken were the Sutlej Valley and Cauvery-Mettur Projects for irrigating southern districts of Punjab and some districts in Bikaner and Bhawalpur and improving the water supply in the existing irrigation system of the Cauvery delta, respectively.

Construction of new canals continued in Punjab, extending to Sindh in 1920s and 1930s. For further development in the water and irrigation sector, upper river valleys of South India and Deccan plateau were selected for large and medium canals schemes. The upland areas were identified for small surface water systems including tanks and mechanized groundwater schemes by the British engineers. After the construction of Betwa canal in north India, a need for storage networks was strongly felt. The Dhukwan Weir, Pahuj and Garhaman canals in 1901 and Dhasan canal in 1910 were constructed for irrigation. High cost of construction of canals, uneven topography and non-uniform rainfall conditions lead to an uncertainty of economic returns. Therefore, it did not result in vast upstream canal development in the south and west of India. In those times, revenue generation was the main aim of the government, which led to construction of many canals during British rule. This changed concept of irrigation of using vast, unused and unpopulated areas for irrigation by diverting and using large untapped rivers of the continent. This practice had significant impacts on basin hydrology.

The colonial government promoted canal construction, but well construction was encouraged in Gujarat through tax exemption as it was believed that wells were most suitable for irrigation in Gujarat. Thus, in Gujarat, there was no major project on canal development at that time and wells became the most important source of irrigation. Water was lifted from the wells through bullock-powered lift during the 1900s. Contrary to this, use of mechanized tube wells was promoted in the province of Uttar Pradesh and Punjab by the Agricultural Department.

During the early 1930s, a detailed survey was conducted to locate suitable tubewell sites in India. By 1936, a total of 743 tubewells were already installed in various parts of India and by 1939 the number of tubewells increased to 1474 (Sharma and Singh, 2011). By 1943, minor irrigation works became significant and "Grow More Food" drive was initiated. The net irrigated area in the Indian sub-continent during 1944-1947 was about 28.2 million hectares. This was largest in any country in the world at that time and only $1/4^{th}$ of the Indian cultivated area.

In British era, WRD was initially seen as a means to fight famines. Later, it was also seen to help in generation of additional revenues. The hallmark of this period was small dams/diversions and canals although wells were also constructed in many places. Colonial masters in India advocated that the full potential of rivers can be harnessed for the improvement of human welfare by the state with the help of improved technologies. The colonial government took a significant decision to create the Provincial Agriculture Departments to provide professional expertise on the issues faced by water and agricultural sector. This paradigm shift and ideology progressed until the end of British era. Post-colonial independent India saw an exponential growth in irrigation bureaucracy. Much of this growth was necessary as a number of WRD projects were to be constructed but this also saw that the society was gradually alienated, an aberration which is being gradually addressed now. A critical contribution of growth in irrigated area and thereby growth in productivity was that the policy makers were convinced that irrigation was necessary for food security. Therefore, immediately after independence, water sector received full attention of the government of India.

POST-INDEPENDENCE TO 2000 AD

Independence brought the partition of India resulting in loss of large productive irrigated lands in Indus basin and majority of the irrigation networks built by the British, went to Pakistan (Naz and Subramanian, 2010).

This section covers various initiatives of the Government of India post-independence till the end of the century. With a focus on institution building and capacity development, a number of national commissions, research and higher education institutions were created. These included the Planning Commission, Central Water Commission (created before independence), National Institute of Hydrology (NIH), Water and Land Management Institutes (WALMIs), Central Soil and Materials Research Station (CSMRS), National Water Development Agency (NWDA), Indian Institutes of Technology (IITs), National Institutes of Technology (NITs) etc. which were established at various times.

During the early post-independence years, large dams were seen as 'temples of modern India' and many new projects were launched. During this period, the water storage capacity increased from about 15 BCM to more than 200 BCM by constructing over 4000 dams across the country. Large projects like Bhakra-Nangal, the Damodar Valley and Hirakud were built post-independence. Other important projects post-independence included Matatila dam, Ramganga dam, Sarda Sahayak dam, Saryu canal, Gandak canal, Madhya Ganga canal, Eastern Ganga canal, Tehri dam project, and rehabilitation of vulnerable reach of upper Ganga canal (Sharma and Singh, 2011). The Indira Gandhi Canal, previously known as Rajasthan canal, which is also the longest canal of India was also constructed. Progresses in hydrology and WRD accelerated with the advent of space technology. The agricultural engineering also gained a new impetus (Irrigation Commission, 1972).

There is a significant variation in water availability in different parts of the country. For example, water resource potential of Ganga-Brahmaputra-Meghna system, home to nearly 100 million people (Nicholls et al., 2018), is very high contributing to about 60% of India's water resources, while Rajasthan, with 68 million population has only 1% of the country's water resources, inclusive of groundwater, rainfall and surface flows. It is estimated that 70% of the geographical area and 76% of population will be facing threats in terms of health and economy with availability of water less than 1000 cum per capita per year, which is identified as scarcity level (Planning Commission, 2002).

Post-independence, a few traditional practices still continue according to the demand and climate changes in different parts of India. For example, in Madhya Pradesh, the upper part of the Narmada Valley, a unique cultivation system, Haveli, based on water harvesting and runoff farming was applied and is still in practice. It was developed in the fertile plains towards the north-west of Jabalpur city, from where the system went across the Narmada into the Narsinghpur district in Madhya Pradesh. Rainwater was held in embanked fields, which were enclosed on four sides and water was let out as soon as the land was dry and the fields were sown, but unfortunately today this tradition is at stake due to introduction of sprinkler irrigation (Nair et al., 2014). In Nagaland, north-eastern India, an ingenious system known as Zabo (impounding the runoff), which combines water conservation with forestry, agriculture and animal care is still known to be practised (Fig. 12).

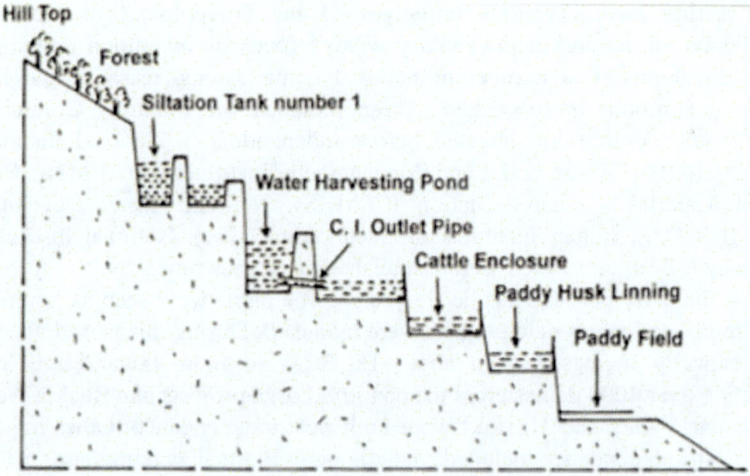

Figure 12. Zabo farming system of Nagaland (*Source:* Singh and Gupta, 2002).

Figure 13. Teesta-V hydropower station, in Sikkim. (*Source:* https://www.hydropower.org/news/indian-hydropower-project-an-example-of-good-practice-in-sustainability)

During the past 20 odd years, some long-pending major projects have been commissioned, notably, Tehri dam and the Sardar Sarovar Dam. The Teesta-V project (Fig. 13), a 510 mega watt (MW) hydroelectric power and flood control project was commissioned in 2008.

A few soil and water conservation projects started in 1970s/80s, viz., Sukhomajri, Ralegaon Siddhi and Pani Panchayat, were very successful. *Bhungroo* is a water management system in Gujarat to store rain water underground which is later used for irrigation during summers (NITI Aayog, 2017). *Sujalam Sufalam Yojana* is another water conservation scheme by the Gujarat government which focuses on increasing rainwater storage during monsoons. The programme promoted the participative approach and focused on desilting of water bodies in the state (CWMI, 2019). *Mission Narmada* was initiated in Rajasthan where the use of microirrigation technology like sprinkler and drip irrigation was made mandatory. Participatory Irrigation Management (PIM) was encouraged and enforced that led to the formation of 2236 Water Users Associations (WUAs). The *Neeru-Chettu* programme initiated by Andhra Pradesh government is to make the state a drought-proof region and reduce economic stress by following better water conservation and management practices (CWMI, 2019). Madhya Pradesh under the MGNREGA programme started *Kapil Dhara Yojana* to create irrigation facilities for small and marginal farmers, and promotes the development of dug wells, farm ponds, check dams etc. '*Paani Bachao, Paisa Kamao*' scheme of the Government of Punjab aims to give monetary benefits to farmers for conserving and using electricity and water efficiently in agriculture. In addition to participatory water management and awareness programmes that promote conjunctive use of water, various states like Madhya Pradesh, Punjab and Maharashtra, have initiated incentive schemes to farm owners for the construction of irrigation structures on private lands.

The annual average rainfall in Ladakh is about 100 mm and people are dependent on glacier meltwater for their water requirements. However, meltwater is not sufficient for irrigation and domestic requirements. To overcome this problem, an innovative concept of artificial glaciers (Fig. 14) was developed by Mr. Norphel, an engineer from Ladakh, popularly known as the 'ice man'. Canals were constructed to divert the glacial melt water to a small catchment where artificial glaciers were made to save frozen water. This gave a secondary source of water as the glaciers melt in summer and water is carried to villages to serve domestic and irrigation needs.

The steps taken by *Jal Jeevan Mission*, encourage use of technologies like IoT based sensors for an improved water management. The idea of 'water as a service' at village level management is encouraged in this mission as it also gives a transparent system to levy water service charges by Paani Samiti (sub-committees of Gram Panchayats) (Jal Shakti, 2020).

Figure 14. Ice stupa (left) and artificial icefall glacier (right) in Ladakh. (*Source:* https://www.colorado.edu/tibethimalayainitiative/2018/03/11/ladakhs-artificial-glaciers-ice-stupas-and-other-attempts-survive-warming-planet)

One of the major challenges India is facing in the water management sector today is the growing competition between demand sectors. At national level the irrigation sector consumes most of the water withdrawn from sources (above 85%). Due to excessive use and intensive irrigation, depletion of water tables, drying of aquifers, groundwater pollution, water logging and salinity are being faced by the farmers and the rural society of India. In many parts of the country groundwater extraction exceeds annual recharge, resulting in depletion of groundwater beyond sustainable limits.

PERSPECTIVES FOR FUTURE DEVELOPMENT

In the previous sections, we have given a historical snapshot of development of water resources in India during various periods. It is clear from the description that decisions were taken at various times keeping in view the source of water, requirements/demands of the population of the regions, technology available, and perspective of the decision makers.

Challenges in management of water resources in India in the future will arise from a host of factors. Population of India is likely to continue to rise till about 2060 and may stabilize thereafter. In tandem, demands for water, food and energy will also rise before they stabilize. Climate change is likely to bring additional uncertainties including those due to changing precipitation patterns. Due to this and rising temperatures, snow/glacier melt timings and magnitude and the river flows are likely to change significantly in terms of magnitude and variability. Further, in recent past, groundwater has played a major role in water and food security but water tables have substantially fallen in many places in the country. Quality of water in rivers has deteriorated over the past four decades or so. More worrying is deteriorating quality of groundwater over larger areas and emergence of toxic pollutants such as fluoride and arsenic in many parts of the country. Increasing awareness among citizens of the deteriorating status of environment and possible impacts of developmental projects on ecology and environment also play a major role in planning and execution of new projects.

To summarize, future is likely to bring new and stiffer challenges and the decision-making space is likely to become more constrained. It is likely that very few new surface water projects will be taken up in future. Conservation and use of flood water is very important for India but this will be difficult without requisite surface water projects. Also, the cost of the projects is rising rapidly and many of them may become financially unviable by following the current norms in near future. In the backdrop of increasing variability and spatial and temporal mismatch between demands and availabilities of water, it will become increasingly important to focus on increasing water use efficiency, use of recycled water for all secondary uses and, in general, explore soft options rather than structural measures. Extensive use of technology at all scales would become essential to address the increasing water stresses in the country. A detailed review of existing and emerging technologies in this context is provided in Mujumdar and Tiwari (2019).

In the past decade or so, provision of environmental flows has become an important concern in India. With some defined quantity of water flowing in the rivers, gradually the health of rivers will improve and this would help in rejuvenation of the environment. Allocation of water for environmental needs will reduce the allocation to other sectors, thereby increasing the sectoral competition.

A review of status of water resources of India shows that to meet the various demands for water in a sustainable manner in future, the regulatory framework—making appropriate rules and their strict implementation—needs to be strengthened. Strict regulation of groundwater usage is necessary to check rapid fall in water tables. Most problems of water quality and environmental degradation have arisen either because the enabling rules were not framed or not enforced. However, this will be a tough challenge.

In near and far future, climate change will occupy the central position in planning, design, and operation of hydro-infrastructure. It is comforting that research is being undertaken in India to understand, predict and manage the impacts of climate change. In addition, research is also underway to develop new and improved understanding of hydrological processes, pollution of water bodies and emerging contaminants. The outcomes will be definitely beneficial for hydrological sciences and for water sector in India. Non-conventional as well as traditional water conservation practices for water utilisation such as artificial groundwater recharge, wetland management, sea water desalination, rainwater harvesting by all means, etc. need to be practiced to further increase the water resource potential.

For the successful implementation of decision making on development aspects, correct data and information are necessary tools. As the information on water resource development is scattered, agencies involved in the implementation of a hydro-project find it difficult to make plans for optimal water resources development and management. High intensity monsoons but lesser number of rainy days are foreseen and although the monsoon rainfall over India is likely to increase, a decrease in annual precipitation in some regions is also expected. Studies reveal that shorter intense floods and dry summer months will also likely to be more frequent. As these extreme events become more frequent, access to clean water availability becomes arduous as bacterial contamination, cyclones, floods and unmaintained sewage system makes the limited resources more contagious. Water guzzler crops in places where climate and soils are not suitable, are pushing the water tables lower besides impacting groundwater quality. Future climate change is expected to escalate this problem as higher temperatures would increase evaporation and further lower the water table. For sustainable solution to these problems water experts will have to closely work with agronomists and social scientists.

The studies reveal that the increasing temperatures in India will lead to increase in potential evapotranspiration, glacial melt will increase for some years and summer monsoon precipitation may increase throughout. This is believed to be more marked in the north-east as there may be a decrease in winter precipitation leading to an increase in rainfall variability further making the date of onset of summer monsoon more variable (NWM, 2008). It is evident that the demand for finite water resources and the fight for it will continue to increase. Frequency of extreme events, uncertainty in water availability and rapid return flows of water to the atmosphere are likely to increase in future, posing a greater challenge for water management.

SUMMARY

Based on extensive survey and review of literature including internet, this chapter describes major activities and milestones in WRD in India in different periods over the past 5000 odd years. This chapter is not meant to be a complete chronicle of such activities. The purpose behind describing these activities is to provide an understanding of how WRD has evolved in India in response to the needs at that time and the technology and other resources that were at the disposal of people. Browsing through the material, it is evident that in earlier times, demands were low and the tools were primitive. With time, demands increased and better options became available. Indian engineers have constructed large and marvellous projects in challenging geographical settings which have stood the test of time. In recent times, Indian professionals are deploying emerging technologies for observation of hydrologic data, processing and using them for decision making.

References

Batuta, I. (1958). The Travels of Ibn Battuta, AD 1325-1354 (Vol. 141). Ashgate Publishing, Ltd.

Cullet, P. and Gupta, J. (2009). Evolution of Water Law and Policy in India 10.2 The Pre-colonial History of Water Law. http://www.ielrc.org/content/a0901.pdf%0Awww.ielrc.org

CWMI (2019). Composite Water Management Index. Ministry of Jal Shakti and Ministry of Rural Development, New Delhi.

Irrigation Commission (1972). Report of the irrigation commission, 1972. Vol. I. Ministry of Irrigation and Power, New Delhi.

Jal Shakti (2020). Jal Shakti - Jan Shakti, Making Water Everyone's Business. Ministry of Jal Shakti, Government of India. May, 2020.

James, J. and James, J.E. (2013). Culture and Heritage for Environment Management. Proceedings for Kerala Environment Congress 2013. KSCSTE, Kerala.

Karpagaselvi, S. (2014). The State of Irrigation in Colonial Tamilnadu, 1800-1850. Thesis submitted to Manonmaniam Sundaranar University.

Koul, D.N., Singh, S., Neelam, G. and Shukla, G. (2012). Traditional water management systems—an overview of Ahar-pyne system in South Bihar plains of India and need for its revival. *Indian Journal of Traditional Knowledge*, 11(2): 266-272.

Morrison, D.K. (2015). Archaeologies of flow: Water and the landscapes of Southern India past, present and future. *Journal of Field Archaeology*, 40(5): 560-580. DOI: https://doi.org/10.1179/2042458215Y.0000000033.

Mujumdar, P.P. and Jain, S. (2018). Hydrology in Ancient India: Some Fascinating Facets. 20th EGU General Assembly, EGU2018, 20(2011), 8690. https://doi.org/10.1080/02626667.2011.587425.NIH

Mujumdar, P.P. and Tiwari, V.M. (eds.) (2019). Water Futures of India: Status of Science and Technology, INSA and IISc Press, ISBN: 978-81-939482-0-0.

Nair S.S., Singh, S. and Gupta, K.A. (2014). Traditional Water Management Systems for Drought Mitigation in India.

Naz, F. and Subramanian, S.V. (2010). Water Management across Space and Time in India . ZEF Working Paper Series 61, University of Bonn, 61.

Nicholls, R.J., Hutton, C.W., Adger, W.N., Hanson, S.E., Rahman, M.M. and Salehin, M. (2018). Integrative Analysis for the Ganges-Brahmaputra-Meghna Delta, Bangladesh BT - Ecosystem Services for Well-Being in Deltas: Integrated Assessment for Policy Analysis (R.J. Nicholls, C.W. Hutton, W.N. Adger, S.E. Hanson, M.M. Rahman and M. Salehin (eds.); pp. 71–90). Springer International Publishing. https://doi.org/10.1007/978-3-319-71093-8_4

NIH (2018). Hydrology in Ancient India. Jal Vigyan Bhawan. Second Edition. Roorkee, Uttrakhand.

NITI Aayog (2017). Selected best management practices in water management.

NWM (2008). Comprehensive Mission Document Comprehensive Mission Document of National Water Mission National Water Mission. Volume II, New Delhi.

Pandey, A. (2016). Society and Environment in Ancient India (Study of Hydrology). *International Journal of Humanities and Social Science Invention*, 5(2): 26-31.

Planning Commission (2002). National Human Develoment Report, 2001. Oxford University Press, USA.

Rathika, C.R. (2016). Irrigation system in Thanjavur district under the Cholas. *International Journal of Interdisciplinary Research in Arts and Humanities*, 1(1): 198-200.

Rehman, A. (2019). Garden of Nobility: Placing Ali Mardan Khan's Baradari at Peshawar in the Context of Mughal Architecture. *South Asian Studies*, 35(1): 129-144. https://doi.org/10.1080/02666030.2019.1605577

Ramachandra, T.V., Vinay, S., Mahapatra, D.M., Varghese, S. and Aithal, B.H. (2016). Water Situation in Bengaluru. ENVIS Technical Report, Environmental Information System, 114 (September).

Randhawa, M.S. (1980). A history of agriculture in India. Volume 1. Beginning to 12th century. ICAR, New Delhi.

Sen, D. NPTEL Lecture on Water Resources Engineering. https://nptel.ac.in/courses/105/105/105105110/

Siddiqui, H.I. (1986). Water works and irrigation system in India during pre-mughal times. *Journal of the Economic and Social History of the Orient*. 29(1): 52-77.

Sharma, Y.D. (2001). Delhi and its Neighbourhood. Surjakund and Anagpur Dam. New Delhi: Archaeological Survey of India. Archived from the original on 31 August 2005. Retrieved 5 September 2009.

Sharma, N. and Singh, R. (2011). History of Irrigation in Uttar Pradesh. Indian National Committee on Irrigation & Drainage, New Delhi. https://www.researchgate.net/publication/275222005_History_of_Irrigation_in_Uttar_Pradesh

Shivakumar, M.G. and Cheluvaraju, R. (2016). Ancient Irrigation Systems of Karnataka. https://www.icid.org/pd3_pap_Shivakumar_Ancient.pdf

Singh, P.K., Dey, P., Jain, S.K. and Mujumdar, P. (2020). Hydrology and Water Resources Management in Ancient India. Hydrology and Earth System Sciences Discussions, 1–20. https://doi.org/10.5194/hess-2020-213

Singh, R. and Gupta, R. (2002). Traditional land and water management systems of north-east hill region. *Indian Journal of Traditional Knowledge (IJTK)*, 01: 32-39.

Chapter 19
Implications of Geodynamics on Extinction of Vedic River Sarasvati

Abhishek Kaushik, P. K. Champati Ray, Suresh Kannaujiya, and Goutham Krishna Teja Gunda

INTRODUCTION

River migration and disappearance of channels in Himalayan front primarily involve geodynamic processes originated from the collision of Indian and Eurasian plates. The geodynamic framework plays an important role to understand the cause of disappearance of once existed river Sarasvati during Neolithic and Harappan civilization in the northwest India (Wheeler, 1953; Parpola, 1994; Kalyanaraman, 1997). The first historical record of the lost river Sarasvati was in the oldest Hindu scripture Rigveda where many hymns mentioned its presence and indicate its geographical extent by placing it between Sutlej and Yamuna river (Bhardwaj, 1999; Danino, 2010). In the Last Glacial Maximum (LGM) period, the extent of glaciers was 10 km farther than the present-day margin (Owen et al., 2002). Tamsa river now called Tons and many other streams were emerging from glaciers and flowing vigorously in the ending period of LGM when the temperature rose up (Eugster et al., 2016; Sharma et al., 2006). In 1874, British geologist, Charles Frederick (Oldham, 1874) prepared a report of existence and disappearance of a big river of Vedic time. Later, Richard Dixon Oldham worked on similar paleo-channels in Haryana-Punjab plain. His findings and details helped in building a framework for further researchers for a clear understanding of disappeared river Sarasvati. A few decades back, satellite images and ground surveys indicate that two major rivers, Sutlej and Yamuna once contributed to Sarasvati and the dry bed of Ghaggar is an impetuous bed of erstwhile Sarasvati (Pal et al., 1980; Chopra, 1990; Valdiya, 1996, 2016). Ancient river systems show their remnants in the form of geomorphic features such as buried channels or paleo-channels, which have been explored by number of earlier workers (Bridge, 1985; Valdiya, 2017). Bhadra and Sharma (Bhadra et al., 2009; Bhadra and

A. Kaushik · P. K. Champati Ray (✉) · S. Kannaujiya · G. K. T. Gunda
Indian Institute of Remote Sensing, Dehradun, India
e-mail: champati_ray@iirs.gov.in

A. Mukherjee (ed.), *Riverine Systems*, https://doi.org/10.1007/978-3-030-87067-6_19

Sharma, 2011) delineated a number of such paleo-channels and validated with hydrological and archaeological evidences. Resistivity survey and sedimentological analysis carried out on deposits of old channels in the northwestern part of Haryana, Rajasthan and Punjab also validate the presence of paleo-channels and their disposition in the past and present (Sinha et al., 2013). Evidence of archaeological sites along the old abandoned courses and extreme migration of streams as main contributing factors of decline of Harappan civilization also supports the evidence of paleo river system (Dixit et al., 2014; Raikes, 1964; Sharma, 1949; Staubwasser et al., 2003; Stein, 1942; Valdiya, 1980).

ROLE OF GEODYNAMICS ON MIGRATION OF RIVER SARASVATI

The presence of ancient Sarasvati has been proved with sufficient geological, geomorphological, geophysical, sedimentological and archaeological evidences starting from foothills of Himalaya through plains of Punjab, Haryana and Rajasthan upto its confluence in Arabian Sea (Rann of Kachchh, Dholavira). However, the most intriguing fact is the role of geodynamics at the frontal Himalaya, which seems to play a significant role ever since the formation of entire mountain range, particularly the outer Himalaya or the Siwalik range. This has been revealed by excellent data products now available in the form of satellite images and aerial photographs taken in the last century. Satellite images and aerial photos although provide quantitative measures of river migration, largely these have been used for qualitative description on river migration and regional tectonics. Recent data sets available from Global Navigational Satellite System (GNSS) provide unprecedented details of subtle crustal deformation that is observed at frontal Himalaya and have potential for reinterpretation of tectonic features and their evolution. This phenomenon is worth exploring for understanding the major river dynamics at the frontal Himalaya. In this endeavour to explore the role of geodynamics and its varied role in shaping northeastern part of the Indo-Gangetic plain, evidences from satellite images on river migration, paleo drainage detection, sub-surface probing by geophysical techniques, present day crustal deformation using GNSS observations and geological and tectonic analysis carried out by various earlier workers have been used to provide an insight into the disappearance of Sarasvati at least at the frontal Himalaya. As articulated and constrained by various earlier workers, the role of climate is neither negated nor overemphasised to overrule the imprint of geodynamics, which was the main reason for leaving many loose ends while interpreting the relevant data both from geological as well as anthropological domain by the previous workers.

The Himalaya has started building up when India and Eurasia plate first collided around 60 million years ago (Hu et al., 2016). Continuous stresses between continental plates generated a number of successive fault zones (Molnar, 2015; Valdiya, 2015). Active faulting in Himalaya has caused the formation of imbricate thrust zone resulting in the uplift of Himalayan Frontal Zone (Nakata, 1989; Yeats and Thakur, 2008). The Himalayan Frontal Thrust (HFT) was demarcated as the high potential zone for seismic activity. Paleoseismic research and scarps of 15-38 m height reported in HFT region suggested revival of HFT several times (Malik et al., 2008; Seeber & Gornitz, 1983; Valdiya, 1980; Wesnousky et al., 1999). Active tectonics study in recent decades found new faults like Jainti Devi Fault near Chandigarh and Trilokpur Fault near Nahan, that have offset the streams by 780 m and 1.5 km, respectively (Chaudhri, 2012). Other paleoseismic evidences support occurrences of big earthquakes in a few thousands of years i.e. 5800 years back (Philip et al., 2012).

The area attains special significance because of Indo-Eurasian plate collision and its influence on streams draining the adjoining hills and foreland basin. As the area just south of Himalayan foothills experiences much pronounced tectonic activity due to continuous rising of Himalaya. Many strike-slip faults in the realm of HFT divide the area into number of blocks that have also experienced relative movements reflected by strike-slip faults (Tapponnier and Molnar, 1976). Markanda and Bata basin isolated by a divide called Uttamwala Katasen high, a kind of fan deposit by river Jalmuseka Khala. This channel blockage by fan and upliftment has migrated the river Jalmuseka Khala in southeast direction from its earlier northwest direction (Virdi et al., 2006).

The present study aims at understanding the morphotectonic evidences to unravel the mystery of disappearance of the Sarasvati river at frontal Himalaya. The basic objective of the research involved the delineation of present and old drainages in frontal Himalaya in the context of geodynamics of the region. Further, resistivity survey aided in locating water saturated zones and geometries of channel for validation of the paleo-channels. Although the study appears to be singular in nature but we have carried out a lot of investigation in the entire frontal area of H.P. and Uttarakhand Himalaya, where we operate 16 GNSS stations, 5 seismic stations and in the recent past at several sites satellite imaging, drone surveys, geophysical surveys and trenching have been carried out to identify active faults (Kannaujiya et al., 2020). Results presented here are only with respect to drainage dynamics and disappearance of Sarasvati river at the frontal Himalaya. The study area includes the region between Sutlej and Yamuna rivers flowing through Haryana and Punjab (Fig. 1) which is surrounded by Siwalik hills in north and alluvial plains in the south. Many streams have drained the area in the past and some of them are still present that includes perennial rivers like Sutlej and Yamuna and seasonal streams like Ghaggar , Markanda, Somb and Boli.

DATA SETS AND METHODOLOGY

Remotely sensed satellite data sets such as Landsat 3 MSS, Landsat 5 TM, Landsat 8 OLI, LISS-IV and ALOS PALSAR DEM were used in this study. Topographic maps of scale 1:50000 from Survey of India was also utilized to digitize the present

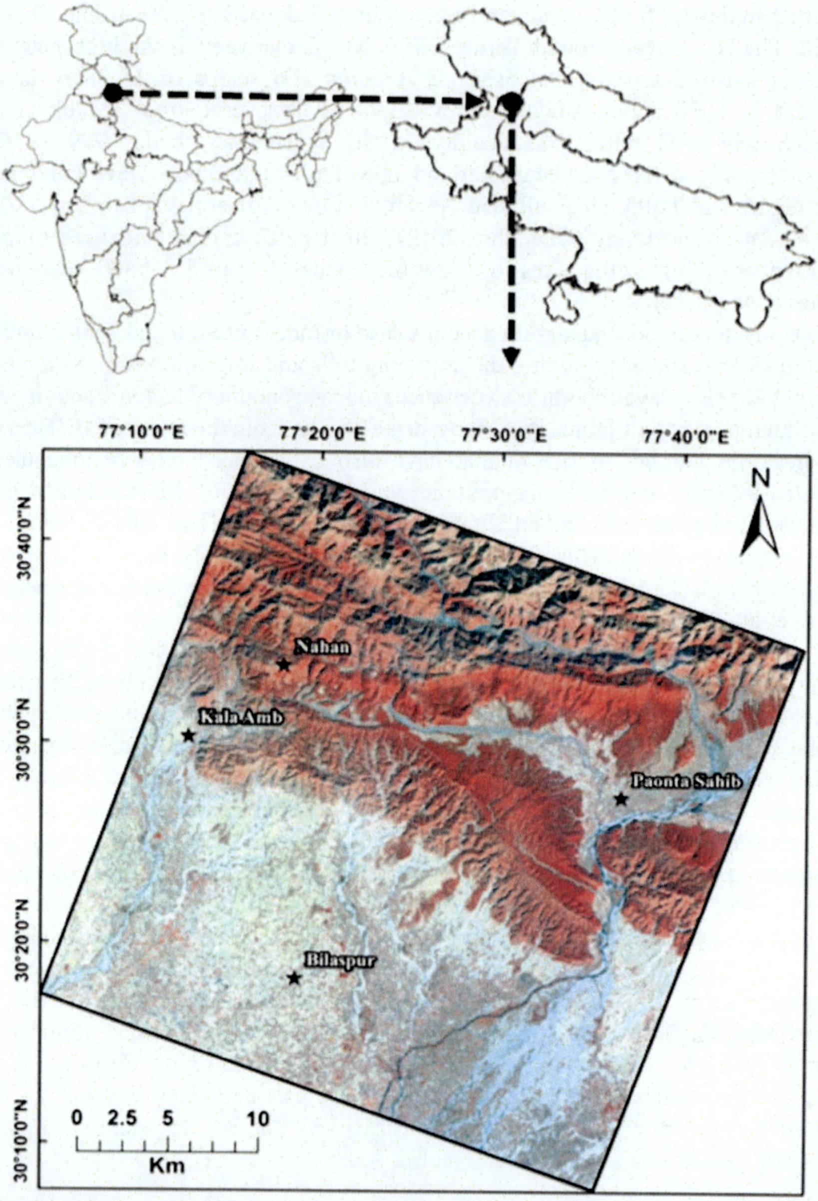

Figure 1. Study area in northern India at the foothills of Himalaya and Indo-Gangetic Plain shown on Landsat-3 image.

drainages. Landsat 3 MSS, Landsat 5 TM, Landsat 8 OLI were downloaded from Earth Explorer (EE), United States Geological Survey (USGS). The LISS-IV (IRS-P6) was made available from National Remote Sensing Centre (NRSC),

Hyderabad. ALOS PALSAR DEM downloaded from the website of Alaska Satellite Facility (ASF), NASA, was used for digital topographic details of area.

Established channels of drainages as a reference has been mapped from topographical maps produced from aerial photographs with proper field validation by Survey of India. Geographically referenced Landsat 3, Landsat 5, Landsat 8 and LISS-IV images were visually interpreted for analysis of changing courses of present rivers. Atmospheric corrections were applied to Bands 4, 5, 7 of Landsat 5 TM (1989) images and then stacked into a single multispectral file for further analysis. The band combination of 4, 5 and 7 shows minimum atmospheric effect and was mainly used for soil moisture and texture analysis. Image enhancement technique using Principal Component Analysis was carried out on the stacked image bands resulting in three principal component bands. Then colour composite images were obtained through resulted principal components and used for further mapping of paleo-channels. It is important to mention that due to similar materials in the piedmont region, the spectral and spatial signature of paleo-channels show subtle variation that limits interpretation. Secondly, due to cultivation and urbanization, the original course of abandoned streams are obliterated and difficult to map. It is quite likely that due to good moisture availability, the paleo-channels have been extensively cultivated thus obscuring spatial and spectral details for their delineations. Therefore, image enhancement and temporal satellite data analysis was necessary to map paleo-channels which was also confirmed by other evidences and field investigation. In our case, surface expressions of paleo-channels were verified using spatial profiling of ALOS PALSAR (12.5m) DEM followed by field investigation and geophysical investigation at selected places.

Paleo-channels mapped on Landsat 5 TM imagery were validated using Electrical Resistivity Tomography (ERT) survey. A 2-D electrical resistivity survey was carried out with 40 electrodes deployed at 5 m separation along a straight line with multi-core cable attached to electronic switching units and laptop. Different settings such as the method of survey and other specifications (current type) were applied through Geo Test software. Then the system generated the pseudo section of apparent resistivity values. Bad points were removed from the pseudo-section editing manually. The data sets were inverted to determine a two-dimensional (2-D) resistivity model for further analysis and interpretation.

RESULTS AND DISCUSSION

Drainage Migration

Drainages present in frontal Himalaya are continuously experiencing the tectonic activities as a result of which shifting and migration has been noticed in Markanda and Somb river at six prominent locations as shown on Landsat 3 image of 1980 and LISS-IV image of 2016. At those sites (Fig. 2, A to F), considerable eastward migration has been observed within a period of 36 years, which indicates Markanda,

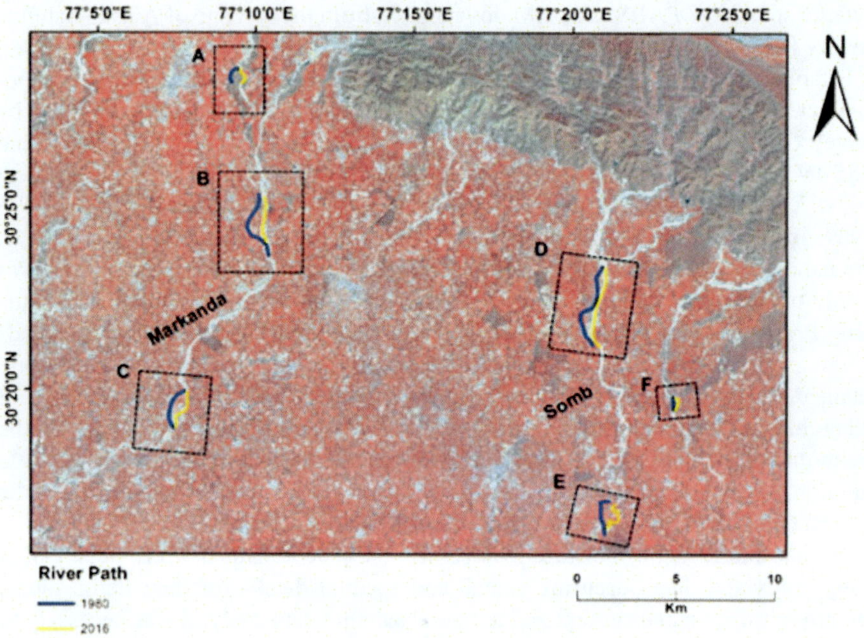

Figure 2. Drainage shift as observed along Markanda and Somb rivers at frontal Himalaya on LISS IV image (2016).

Somb and other tributaries draining this area have the tendency to shift in eastward direction.

Locations A, B and C show the derangement of drainage by shifting almost 500 to 800 m in eastward direction. This shifting in river course changes are attributed to the tectonic regime at frontal Himalaya. Active tectonics can be responsible for the changes in the river like avulsion and shifting. Close inspection at location D, E, F shows considerable migration in the range of 200 to 500 m towards east in last 36 years.

Landsat 5 TM (1989) and Landsat 8 OLI (2018) reveals significant migration of Yamuna river, while entering the Gangetic plains after its descent from mountains passing through Paonta Fault during last 3 decades (Fig. 3). At three locations A, B, C, river conspicuously changes its course ranging from 1000 to 2000 m towards east. Archaeological data and political history also show that capital Delhi has been constructed and relocated many times and there is substantial movement towards east following the same trend of Yamuna river migration (Singh, 2006). Some of the sources of the drinking water of Delhi flow through the paleo-channels of Yamuna, which of late ceased to be a main source of drinking water due to both shifting in location and pollution. The most conspicuous eastward shifting of Yamuna is corroborated by recent mapping of a major paleo-channel zone between south of Delhi and Mathura as published in a committee report on Paleo-channels in India by Ministry of Water Resource, Govt. of India (MWRRDGR, 2018).

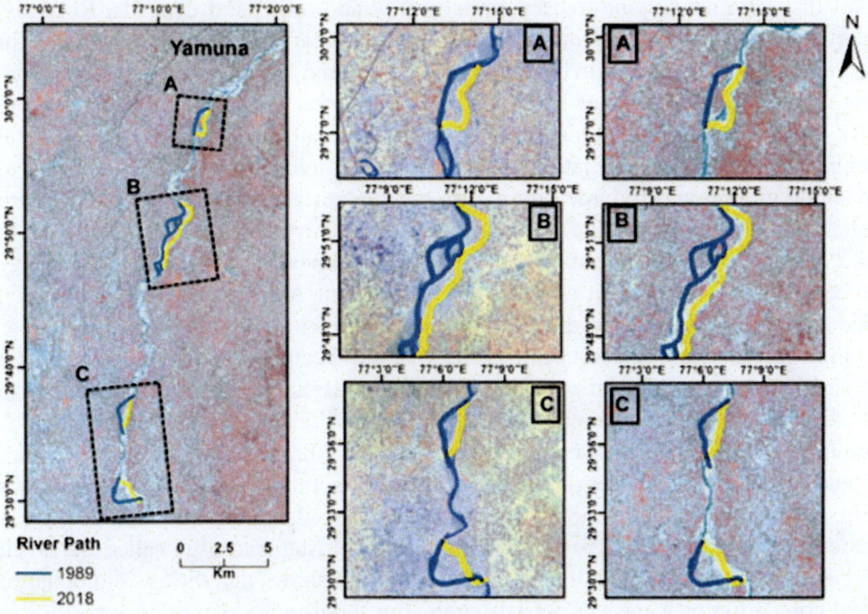

Figure 3. Drainage migration of Yamuna river during 1989 and 2018.

Paleo-channel Identification and Mapping

Paleo-channels are typical fluvial geomorphic features in a location representing past drainage system which were flowing either as ephemeral or perennial streams in the past and now stands abandoned and buried or lost or shifted due to tectonic, geomorphologic, anthropogenic process/activities, as well as climatic changes. These features are synonymous with paleo-rivers, lost river, buried river, buried channel, or buried valley. Paleo-channels vary greatly in age starting from recent historical past to several thousands of years and beyond. It consists of deposits of unconsolidated sediments or semi-consolidated sedimentary rocks deposited in ancient, currently inactive river and stream channel systems. As once river had flown through, these possess good water bearing or holding capacity represented by highly porous gravel, coarse-medium sand (in the alluvial fan area), medium fine sand, and fine silt (in the river channel zones). It may contain deposits which are different from the overbank deposits of currently active river channels, because the river bed was filled with sedimentary deposits which are unrelated to the normal bed load of the current drainage system due to change in the catchment area or the sediment source region. In the present context, it is not only important to evaluate the paleo-channels that is remnant of the river Sarasvati, but also its numerous tributaries, which are small and may be seasonal like many such streams present today. Another challenge is related to frequent migration of such streams in the foothill region due to tectonics. Thirdly, due to climate change and prevailing of dry period, a lot of such channels have been buried by wind-borne sand deposits. Recently, many such areas due to favourable water bearing characteristics, are preferred for agriculture which results in levelling of the ground thus obscuring the spatial/ morphological signatures of such features. Therefore, their identification is often tricky and challenging requiring both optical satellite data analysis in temporal domain to exploit the variation in moisture and microwave remote sensing to penetrate ground and provide information on sub-surface strata, particularly in dry condition. In case of satellite data, it is advised to carry out a temporal analysis to take advantage of variable moisture content in delineating paleo-channels. Image processing techniques such as Principal Component Analysis and IHS (Intensity, Hue and Saturation) transformation are also expected to facilitate delineation.

In the present case, paleo-channels show subtle spectral variation due to variation of soil texture and moisture content in the Himalayan frontal region which was highlighted by Principal Component Analysis using Landsat 5 TM (1989) image. A number of older channels were mapped between Markanda and Somb river where tributaries of Sarasvati emerging from the Siwalik mountain range might have flown (Fig. 4). These paleo-channels subsequently moved southward and now pass through some major localities like Bilaspur and Mustafabad now called Sarasvati Nagar and Thanesar. Locality named after Sarasvati shows that the age old belief of local population on Sarasvati persists even after thousands years of its extinction.

Important older channels initiating from Adh Badri in the emerging Siwalik hills and meeting the Ghaggar river at Shutrana are shown as S1, S2 and S3 (Fig. 4).

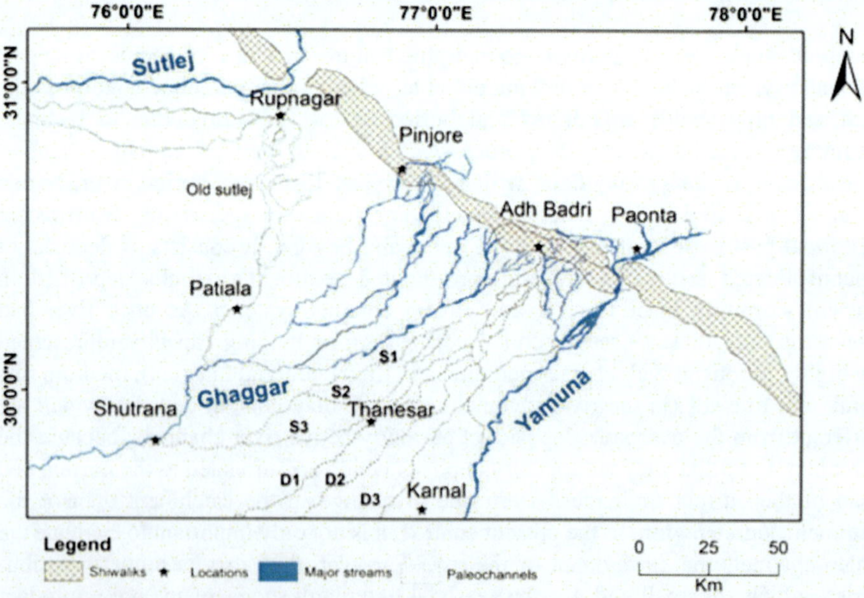

Figure 4. Present and past drainage systems between Sutlej and Yamuna rivers in frontal Himalaya.

These channels show an interesting migration pattern resulting in a number of streams that finally gave rise to more paleo-channels. Significant shifting in old channels of Sarasvati has caused the nourishment of channel D1, D2 and D3 that are associated with Drishadwati river (presently Chautang river). Yamuna river flows towards southwest and southeast and finally meet river Ganga at Allahabad. Old channels of Sutlej river were passing through Rupnagar, Patiala and finally were contributing to the Ghaggar course. Two older channels (remnants of migrating Sutlej river towards west) also traced running parallel to the present-day Sutlej river which takes a conspicuous westward turn after passing through Rupnagar.

Topography Profiling and Electrical Resistivity Tomography (ERT) Survey for Paleo-channel Characterization

In order to assess the topographical condition across several paleo-channels, the spatial topographical profiles are drawn across paleo-channels using ALOS PALSAR DEM at several locations (Fig. 5). Topographic information extracted through most of the profiles depicts depression along the paleo-channels. Spatial profiles from A to F are shown in Fig. 6. Profile A and B show depressions of 8 m and 4 m in 500 m wide channel, respectively. Profile C represents 2 m deep and 250 m wide channel. Similarly, all other profiles drawn across paleo-channels show the topographic low. At some locations, the area does not give the exact surface expression of paleochannel due to agricultural activities which has changed the local slope of channels. All of these profiles match with the general trend of paleo-channels in the region.

Two-dimensional resistivity survey was carried out for validation of paleo-channels at five different locations. Profiles were obtained by two different resistivity data acquisition methods, viz. Wenner and Schlumberger. The Wenner array has strongest signal strength and, therefore, is more suitable for noisy areas. It gives high vertical resolution which is better for resolving more number of horizontal units. Schlumberger array has less signal strength compared to Wenner but has more than pole-dipole and dipole-dipole array methods. Secondly, it is easier to carry out, less time consuming and provides greater depth of investigation.

Paleo-channels were detected on satellite images followed by field verification and ERT investigation at Dehdra Village (30°42'8.07"N 76°26' 54.70"E) using Wenner and Schlumberger methods which shows shallow water table and moisture saturated zone.

The ERT profile represents two different layers of resistivity values ranging from 10 to 30 Ωm that demarcates different materials present across the profile (Fig. 7). Very low homogeneous value of resistivity in left side signifies the presence of the water-saturated zone in the paleo-channel which is approximately 140 m wide and

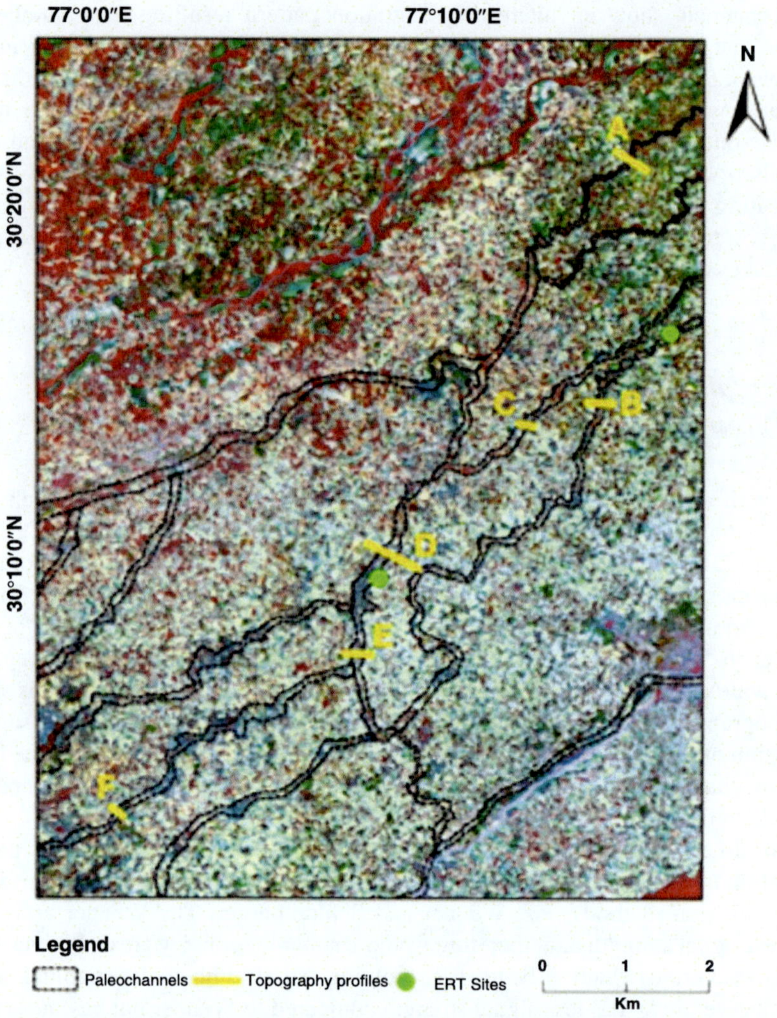

Figure 5. Spatial profiles and resistivity investigation sites mapped on paleo-channels.

7 m deep. It is important to note that some of the channels receive water based on their connectivity with streams in the upstream direction. In the present case, point B located towards centre of the channel, have good moisture zone. Many such sites were visited in the foothill region and some of the shallow wells were investigated and basic sedimentation supporting the evidence of a river flowing in the region has been observed at Mughalwali village (30°23'14.37"N, 77°20'26.36"E) which shows the shallow groundwater table (2-3 m) in the paleo-channel (Fig. 8).

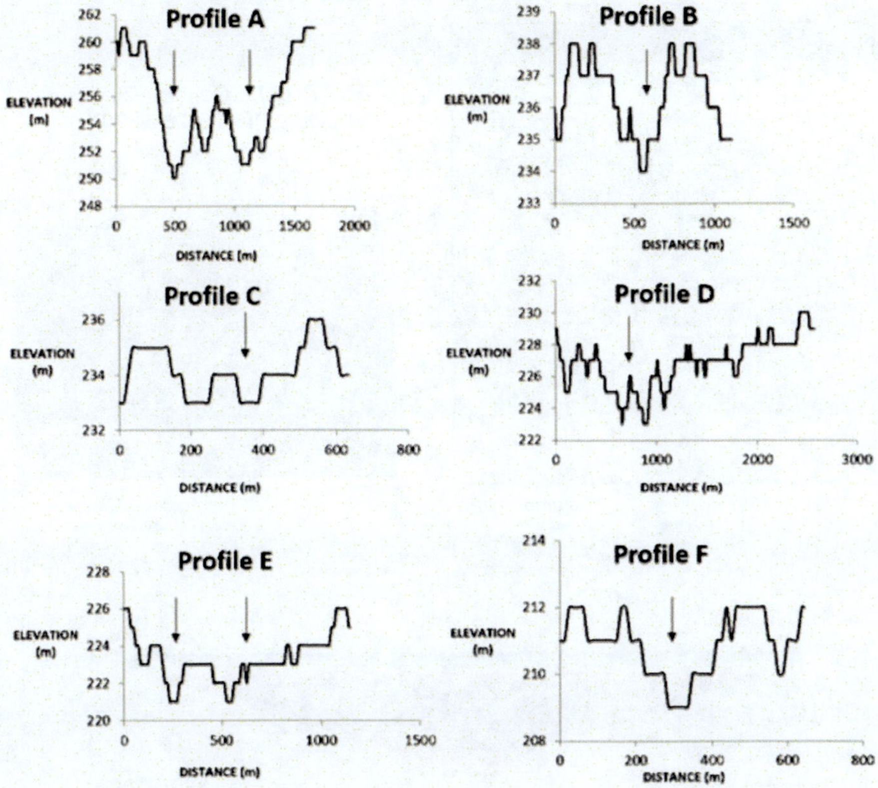

Figure 6. Spatial profile drawn across paleo-channels reveal topographic depressions, which suggests flow along paleo-channels.

Paleo-drainage Modifications under Tectonic Influence

Image interpretation followed by geomorphic analysis of paleo-channels mapped on Landsat 5 TM (Fig. 9) depicts a unique trend of movement of the paleo-drainages in the form of knee bends across channel length or major flow direction i.e. NE to SW. Such bends are mostly parallel to main Himalayan range and are reasonably consistent across major streams and presence of which suggests the influence of tectonic deformation at the active plate boundary. Similar spectacular bends showing diversion of Solani river towards east is attributed to presence of fault in the piedmont region (Thakur, 2004). Many drainages in the piedmont region show eastward deviation or migration. These knee bends when join together across adjoining streams from NW to SE, a linear feature emerges parallel to Himalayan range and interestingly, in the present area three such parallel linear features could be interpreted (Fig. 9).

These linear features are interpreted as piedmont faults that occur in an echelon or imbricate manner in front of the rising Himalaya. Considering their consistency and

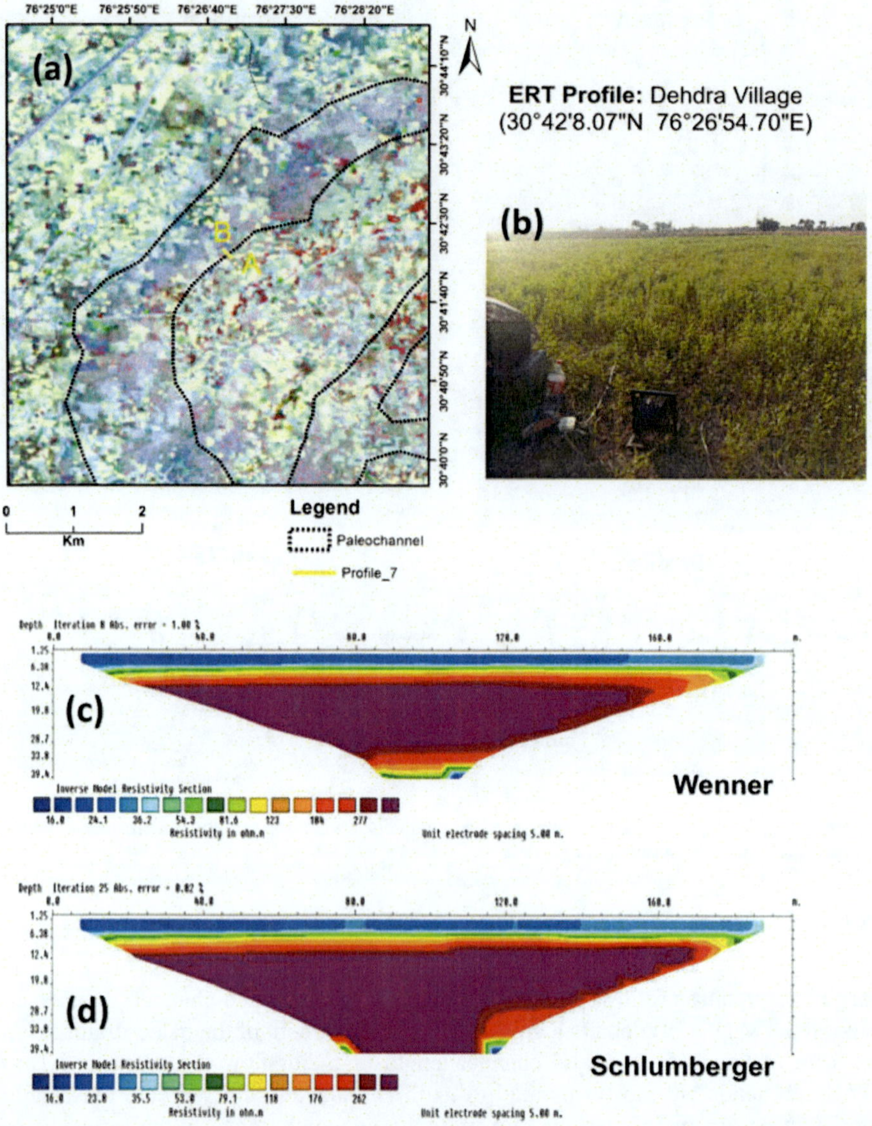

Figure 7. Paleo-channel detection on satellite image followed by field verification and ERT investigation at Dehdra Village (30°42'8.07"N, 76°26'54.70"E) shows resistivity profiles by Wenner and Schlumberger methods.

repetivity in a close proximity, these are considered as a part of a major fault system named as Himalayan Piedmont Fault System (HPFS). Although the fault system occurs in the Indo-Gangetic Plains, as its origin is attributed to Himalayan orogeny, it is suggested to name it as HPFS for the first time in this paper. Different segments of this mega system has been mapped by earlier workers (Yeats and Thakur, 2008;

Figure 8. At a another paleo-channel location near Mughalwali village (30°23’ 14.37”N, 77°20’26.36”E), water found at depth of 2-3 m which shows shallow groundwater table in the paleo-channel.

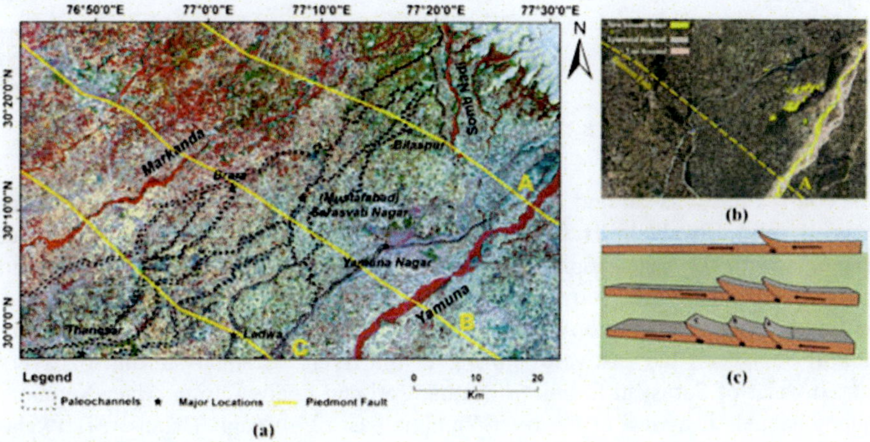

Figure 9. Paleo-drainage modification by imbricate structure of Himalayan Piedmont Fault System (HPFS) at the frontal Himalaya.

Thakur et al., 2020) and different local names have been applied. As many splays and imbricates have been observed and many are yet to be discovered, one collective terminology is applied to represent all such faults, supported by one unified theory. Many warpings and minor foldings in the near vicinity can also be attributed to kinematics of this mega system. It is very interesting to note that the same mechanism resulting in imbricate structure is also noticed in areas between HFT and MBT in Doon valley as revealed in trench excavations carried out to resolve the fault geometry.

These faults in the piedmont region are responsible for changes in the present and past drainage system. At these fault intersection, the river channels take prominent sharp turns in eastward direction. These faults in the piedmont realm are extended imbricate of Himalayan Frontal Thrust and are similar to low dipping thrust fault causing the upliftment of piedmont region. At some places in the piedmont zone, subtle uplift has caused reversal of drainages i.e. streams have flown in northward direction against the regional slope of south and southeast. These faults formed in order of A, B and C (Fig. 9), i.e. the fault closer to HFT is the oldest followed by successive younger ones towards south and southeast direction. The streams of first and second order respond directly to the inferred imbricate fault zone and move in a curvilinear form while high order streams remain unaltered due to high energy which is a function of slope and discharge. In this study, three prominent traces have been mapped and the 1st fault A correspond to just south of areas that show significant water logging due to presence of seasonal water in the recent past due to marginal uplift of the downstream area. The presence of seasonal water has been confirmed by satellite data analysed for 33 years (1985 to 2018) (Source: Global Surface Water Explorer). This fault has also caused disorientation of river Yamuna in upstream and downstream areas adjacent to the fault intersection. The third fault corresponds to the already reported Gula-Ladwa high (topographical) mainly due to uplift of southern areas (Thussu, 1999).

Sarasvati River Course and Tectonics

The courses of Sarasvati and Drishadwati rivers between Sutlej and Yamuna were mapped from Siwalik hills upto Suratgarh in Rajasthan by integrating present drainages with the paleo-drainage of the rivers. Fig. 10 shows that old Sutlej and Sarasvati channels were contributing to Ghaggar-Hakra in earlier times. Continuous shifting has occurred due to northward drift of Indian plate and series of imbricate faulting enhanced the eastward shifting in the rivers. Gradual shifting of rivers caused water of Sarasvati to link to Drishadwati and finally into the river Yamuna. Paleo-channels just west of the river Yamuna passing through Thanesar, Barwala, Nohar were contributing to Ghaggar river at Suratgarh. The Sutlej moved westward and Yamuna eastward along a major water divide that separates the two drainage basins even today. This major drainage reorganization that took place in the past has also resulted in reduction of surface water flow in plains of Haryana. Although the chronology of river reorganization continues to be a major scientific debate, recent studies by Singh (2017) have confirmed with the help of extensive chronological data to postulate that the withdrawal of the paleo-Sutlej was complete a little after 8 ka. Thus, it is conluded that the tributaries of Vedic river Sarasvati have been partly captured by Yamuna, which now flows a part of its discharge and finally at Prayagraj, its confluence with Ganges brings the water from three major Himalayan river systems, Ganga, Yamuna and Sarasvati at Triveni, the holi confluence, well documented in historic and pre-historic period religious texts of India (Oldham, 1874).

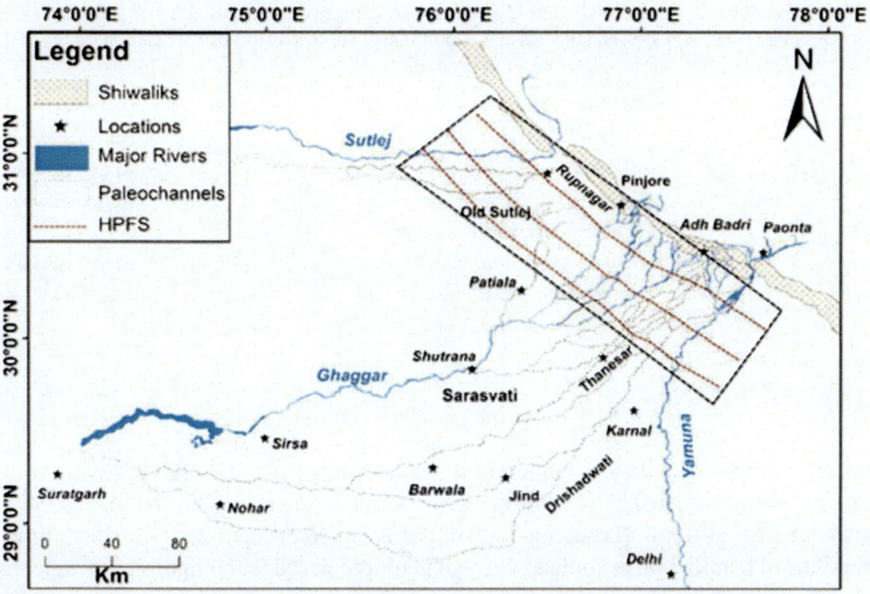

Figure 10. Course of Sarasvati river and other streams upto Suratgarh and its association with imbricate fault structure as interpreted from satellite based observation.

Global Navigation Satellite System (GNSS) Observations

GNSS data sets from several stations closer to the study area have been analysed to assess the crustal deformation in recent times. Preliminary analysis of International GNSS service stations (IGS) on Indian and adjoining plates had indicated about the rotational component of the Indian plate motion with respect to Eurasia plate. Hence, in the Himalayan frontal part several stations were setup and additional data was also acquired using campaign mode survey for about 5 days continuously using dual frequency GNSS receivers. The same survey was repeated at an interval of 6 months and at least 3 sets of data were used for campaign mode station and continuous data was used from GNSS/Continuously Operating Reference Stations (CORS).

The rotational rate at triangular zones of three non-collinear GNSS sites were calculated. The GNSS sites are distributed from Indo-Gangetic Plain to Southern Lesser Himalaya. The rotation rates are calculated at centroid of 6 triangular zones of sites as listed in Table 1. The rate of vertical axis rotation ranges between 2.15° and 6.06°/Ma towards anticlockwise direction. The calculated rotation rate is consistent with the previously estimated by Ponraj et al. (2010) and Dumka et al. (2018) and the clockwise rotation of Himalayan meta-sedimentary wedge system in fixed Indian plate is consistent with the counterclockwise rotation rate of rigid Indian plate as observed. This rotational component has a very strong bearing on eastward migration of streams and provides very crucial information in support of our hypothesis on stream migration and capture as described in the previous sections.

Table 1. Rotation rates obtained from GNSS stations using triangulation network located in the Indo-Gangetic plain and Lesser Himalaya. All points just lie to east of area as shown in Fig. 10. Longitude and latitude refer to centroid of the triangle

	GPS sites	*Direction*
1	Mohand-Dehradun-Haridwar	Anticlockwise
2	Roorkee-Buggawala-Haridwar	Anticlockwise
3	Roorkee -Haridwar-Buggawala	Anticlockwise
4	Roorkee-Mohand-Haridwar	Anticlockwise
5	Saharanpur-Rajpur-Haridwar	Anticlockwise
6	Roorkee-Rajpur-Haridwar	Anticlockwise

CONCLUSION

Starting from the Himalayan mountain range, a number of tributaries were contributing to the Ghaggar-Hakra river system which earlier constituted the ancient Sarasvati river system. These channels of Sarasvati river in frontal Himalaya have experienced continuous tectonic activity in geologic past due to ongoing movements of the Indian and the Eurasian Plates. Investigations in eastern interfluve of the Sutlej and the Yamuna reveal the eastward shift of rivers on LANDSAT 3 (1980) and IRS LISS IV (2016). Paleo-channels identified from satellite images were confirmed by geophysical investigations using Electrical Resistivity Tomography (ERT). Topographic and morphometric analysis using ALOS PALSAR DEM further aided to decipher the influence of tectonic activity on surface manifestation of paleo-channels. A series of imbricate faults at almost parallel to the Main Frontal Thrust (MFT) or Himalayan Frontal Thrust (HFT) have been interpreted as the main cause and mechanism of active faulting in frontal Himalaya that has significantly influenced the migration and amalgamation of streams. A unique trend of eastward stream migration with knee bends as observed on satellite images is attributed to the rotational component marked by anticlockwise movement (corroborated by GNSS observation) of under thrusting Indian plate. The series of imbricate faulting coupled with anticlockwise motion of the region has resulted in migration of the streams towards Yamuna which finally captured those including Drishadwati originally flowing towards west contributing to Saraswati system. Therefore, geodynamics and tectonics play a crucial role in understanding river migration and water resource availability in many parts of thickly populated Indo-Gangetic alluvial plain.

ACKNOWLEDGEMENTS Authors would like to acknowledge the support from IIRS and Indian Space Research Organisation (ISRO) to the project on "Geodynamics and seismicity of northwest Himalaya" under Earth Observation Application Mission (EOAM) which resulted in the present study.

References

Bhadra, B.K., Gupta, A.K. and Sharma, J.R. (2009). Sarasvatinadi in Haryana and its linkage with the Vedic Sarasvati River—integrated study based on satellite images and ground based information. *Journal of the Geological Society of India*, 73(2): 273-288.

Bhadra, B.K. and Sharma, J.R. (2011). Satellite images as scientific tool for Sarasvati palaeochannel and its archaeological affinity in NW India. *In:* Proceedings of Int. Seminar on How Deep are the Roots of Indian Civilization by Draupadi Trust and Ministry of Culture, New Delhi, 30-49.

Bhardwaj, O.P. (1999). The Vedic Sarasvati. Vedic Sarasvati - Evolutionary History of a Lost River of Northwestern India, Radhadrishna B.P., Merh S.S. (Eds.), Geological Society of India, Bangalore, *Memoir* 42, 15-24.

Bridge, J.S. (1985). Paleochannel patterns inferred from alluvial deposits—A critical evaluation. *Journal of Sedimentary Research*, 55(4): 579-589.

Chaudhri, A.R. (2012). Tectonic morphometric studies as a tool for terrain characterization in the Himalayan foothill region—A case study. *Journal of the Geological Society of India*, 79(2): 210-218.

Chopra, S. (1990). A geological-cum-geomorphological framework of Haryana and adjoining areas for land use appraisal using LANDSAT imagery. *Journal of the Indian Society of Remote Sensing*, 18(1-2): 15-22.

Danino, M. (2010). The lost river: On the trail of the Sarasvatī. Penguin Books India.

Dixit, Y., Hodell, D.A. and Petrie, C.A. (2014). Abrupt weakening of the summer monsoon in northwest India~ 4100 yr ago. *Geology*, 42(4): 339-342.

Dumka, R.K., Kotlia, B.S., Kothyari, G. Ch., Paikrey, J. and Dimri, S. (2018). Detection of high and moderate crustal strain zones in Uttarakhand Himalaya, India. *Acta Geodaetica et Geophysica*, 53(3): 503-521.

Eugster, P., Scherler, D., Thiede, R.C., Codilean, A.T. and Strecker, M.R. (2016). Rapid Last Glacial Maximum deglaciation in the Indian Himalaya coeval with mid latitude glaciers: New insights from 10Be-dating of ice-polished bedrock surfaces in the Chandra valley, NW Himalaya. *Geophysical Research Letters*, 43(4): 1589-1597.

Hu, X., Garzanti, E., Wang, J. Huang, W., An, W. and Webb, A. (2016). The timing of India-Asia collision onset—Facts, theories, controversies, Earth-Science Reviews 160: 264-299

Kannaujiya, S., Philip, G., Champati Ray, P.K. and Pal, S.K. (2020). Integrated geophysical techniques for subsurface imaging of active deformation across the Himalayan Frontal Thrust in Singhauli, Kala Amb, India. *Quaternary International*. https://doi.org/10.1016/j.quaint.2020.05.003.

Kalyanaraman, S. (1997). Sarasvati River (circa 3000 to 1500 B.C.). Sarasvati Sindhu Research Centre. Chennai, India.

Malik, J.N., Nakata, T., Philip, G., Suresh, N. and Virdi, N.S. (2008). Active fault and paleoseismic investigation: Evidence of a historic earthquake along Chandigarh Fault in the Frontal Himalayan Zone, NW India. *Himalayan Geology*, 29(2): 109-117.

Molnar, P. (2015). Plate tectonics: A very short introduction (Vol. 425). Oxford University Press, USA.

MWRRDGR (2018). Review of the available information on paleo-channels in India, Ministry of Water Resources, River Development and Ganga Rejuvenation, Govt. of India, New Delhi.

Nakata, T. (1989). Active faults of the Himalaya of India and Nepal. Geological Society of America, Special Paper, 232: 243-264.

Oldham, C.F. (1874). Notes on the lost river of the Indian desert. Calcutta Review 59, 1-27.

Owen, L.A., Finkel, R.C. and Caffee, M.W. (2002). A note on the extent of glaciation throughout the Himalaya during the global Last Glacial Maximum. *Quaternary Science Reviews*, 21(1-3): 147-157.

Pal, Y., Sahai, B., Sood, R.K. and Agrawal, D.P. (1980). Remote Sensing of the 'lost' river Sarasvati river, Proceedings of Indian Academy of Sciences, 89(3): 317-331.

Parpola, A.H.S. (1994). Deciphering the Indus script. Cambridge University Press.

Philip, G., Bhakuni, S.S. and Suresh, N. (2012). Late Pleistocene and Holocene large magnitude earthquakes along Himalayan Frontal Thrust in the central seismic gap in NW Himalaya, Kala Amb, India. *Tectonophysics*, 580: 162-177.

Raikes, R.L. (1964). The End of the Ancient Cities of the Indus. *American Anthropologist*, 66(2): 284-92.

Ponraj, M., Miura, S., Reddy, C.D., Prajapati, S.K., Amirtharaj, S. and Mahajan, S.H. (2010). Estimation of strain distribution using GPS measurements in the Kumaun region of Lesser Himalaya. *Journal of Asian Earth Sciences*, 39(6): 658–667.

Seeber, L. and Gornitz, V. (1983). River profiles along the Himalayan arc as indicators of active tectonics. *Tectonophysics*, 92(4): 335-367.

Sharma, B.R. (1949). The Vedic Sarasvati. Cal. *Review* 112(1), 53–62.

Sharma, J.R., Gupta, A.K. and Bhadra, B.K. (2006). Course of Vedic River Sarasvati as deciphered from latest satellite data. *Puratattva*, 36: 187-195.

Singh, U. (Ed.) (2006). Delhi: Ancient History. Berghahn Books.

Singh, A., Thomsen, K.J., Sinha, R., Buylaert, J.P., Carter, A., Mark, D.F. and Paul, D. (2017). Counter-intuitive influence of Himalayan river morphodynamics on Indus Civilisation urban settlements, *Nature communications*, 8(1): 1617. https://doi.org/10.1038/s41467-017-01643-9.

Sinha, R., Yadav, G.S., Gupta, S., Singh, A. and Lahiri, S.K. (2013). Geo-electric resistivity evidence for subsurface palaeochannel systems adjacent to Harappan sites in northwest India. *Quaternary International*, 308: 66-75.

Staubwasser, M., Sirocko, F., Grootes, P.M. and Segl, M. (2003). Climate change at the 4.2 ka BP termination of the Indus valley civilization and Holocene south Asian monsoon variability, *Geophysical Research Letters*, 30(8).

Stein, A. (1942). A survey of ancient sites along the "Lost" Sarasvati River. *The Geographical Journal*, 99(4): 173-182.

Tapponnier, P. and Molnar, P. (1976). Slip-line field theory and large-scale continental tectonics. *Nature*, 264(5584): 319-324.

Thakur, V.C. (2004). Active tectonics of Himalayan frontal thrust and seismic hazard to Ganga Plain. *Current Science*, 86(11): 1554-1560.

Thakur, V.C., Joshi, M. and Jayangondaperumal, R. (2020). Active Tectonics of Himalayan Frontal Fault Zone in the Sub-Himalaya, *In:* Gupta, N., Tandon S.K. (Eds.), Geodynamics of the Indian Plate, Springer Geology, https://doi.org/10.1007/978-3-030-15989-4_12

Thussu, J.L. (1999). Role of tectonics in drainage migration in Punjab-Haryana Plains in recent times, *In:* Vedic Sarasvati-Evolutionary history of a lost river of northern India, Radhakrishna, B.P. and Merh, S.S. (Eds.), Geological Society of India, Memoir 42: 205-217.

Valdiya, K.S. (1980). The two intracrustal boundary thrusts of the Himalaya. *Tectonophysics*, 66(4): 323-348.

Valdiya, K.S. (2015). The making of India: Geodynamic Evolution. Springer. ISBN 3319250299.

Valdiya, K.S. (2016). Palaeochannels of northwest India: Review and Assessment. Central Groundwater Board, New Delhi.

Valdiya, K.S. (1996). River Piracy Sarasvati that Disappeared. *Resonance*, 1(05), 19-28.

Valdiya, K.S. (2017). Prehistoric River Sarasvati, Western India. Springer International, ISBN 978-3-319-44223-5.

Virdi, N.S., Philip, G. and Bhattacharya, S. (2006). Neotectonic activity in the Markanda and Bata river basins, Himachal Pradesh, NW Himalaya: A morphotectonic approach. *International Journal of Remote Sensing*, 27(10): 2093-2099.

Wesnousky, S.G., Kumar, S., Mohindra, R. and Thakur, V.C. (1999). Uplift and convergence along the Himalayan Frontal Thrust of India. *Tectonics*, 18(6): 967-976.

Wheeler, M. (1953). The Cambridge History of India: Supplementary Volume: The Indus Civilization (Vol. 3). University Press.

Yeats, R.S. and Thakur V.C. (2008). Active Faulting South of the Himalayan Front: Establishing a New Plate Boundary, *Tectonophysics* 453(1-4): 63-73.

Chapter 20
Evaluating Groundwater Sources and Dynamics along the Paleochannel Network of Northwest India Using Isotope Tracers

Tirumalesh Keesari, Annadasankar Roy, Uday Kumar Sinha, Diksha Pant, Hemant Mohokar, Gyanendra Rai, and S. K. Jain

INTRODUCTION

The word paleochannel is formed from two words, viz. paleo and channel, meaning an old channel. Paleochannels hold unconsolidated or semi-consolidated sediments deposited by ancient, currently inactive river/stream channel systems. Paleochannels often have good groundwater potential because of *channel-lag* deposits, which are coarse and generally hold good quality groundwater. Depending upon the nature and shape, the paleochannels are grouped into three categories, viz. (a) older river channels that were re-buried due to sedimentation, (b) remnant scars of shifting rivers and (c) the streams or rivers which were flowing either ephemeral or perennial in the past but now are lost either due to tectonic activities or changes in climate or geomorphology. In order to undertake any groundwater development activity through paleochannels rejuvenation, it is important to evaluate (i) path and geometry, (ii) source and extent of recharge, (iii) quality of water in the paleochannels vis-a-vis the surrounding areas and (iv) age of groundwater. In Indian context, the studies on paleochannels have been mostly dedicated to the tracing of the history of the ancient rivers with respect to their geology and to assess the influence on local hydrogeology (Anon, 1874; Yashpal et al., 1980). During late 18th century, researchers identified a few major paleochannels in Punjab plains along the alluvial tracts of Gagghar River basin (Oldham, 1893). Later, many studies were carried out in this region (NW India) to further understand their occurrence and significance by employing multiple techniques such as aerial photography, remote sensing, geophysical studies and hydrogeological studies (Ghose et al., 1979; Kar and Ghose,

T. Keesari (✉) · A. Roy · U. K. Sinha · D. Pant · H. Mohokar
Isotope Hydrology Section (IRAD), BARC, Trombay, Mumbai, India
e-mail: tirumal@barc.gov.in

G. Rai · S. K. Jain
Central Ground Water Board, MOWR, Government of India, Chandigarh, India

A. Mukherjee (ed.), *Riverine Systems*, https://doi.org/10.1007/978-3-030-87067-6_20

1984). Sinha (1996) used both satellite images and field investigations to infer the migration trends and paleo-hydrogeologic conditions of the Quaternary sedimentation in North Bihar. During 21st century detailed studies were undertaken by many researchers (Saini et al., 2009), which helped in mapping the buried channels in alluvial formations of northwestern India and reconstructed the history of the past fluvial activity (Bhadra et al., 2009). These researchers used sub-surface litho-facies data from well-logs and also Optically Stimulated Luminescence (OSL) technique to arrive at the conclusions. Studies by Valdiya et al. (2002, 2013 and 2016) inferred that there was a large river fed by glacier existed in the past. In the case of Northern Plains of India, the studies have clearly indicated the potential prospects of paleochannels in recharging groundwater in Yamuna flood plains (Rao et al., 2009) and Ganga plains (Samadder et al., 2011). These studies infer that paleochannels located in the Western Ganga plain are composed of coarse grained sandy material and are well interconnected with the adjacent shallow aquifers. The hydraulic characteristics of the paleochannels indicate that these aquifers hold good promise for the artificial recharge to groundwater. Presence of buried paleochannels in the Ghaggar plains and their upstream connectivity was established using geophysical investigation (Sinha et al., 2013). Few more paleochannels were identified in the Ghaggar River basin by Singh et al. (2016). Satellite imagery of the western parts of the Jaisalmer district (western Rajasthan) indicated the buried course of a river orientated in the NE-SW direction (Bakliwal and Grover, 1988), which was confirmed by the fresh quality of groundwater at a depth of 30 m along this buried channel. Researchers have also used IRS LISS-III imagery based on the tonal difference and inferred paleochannels that have rich groundwater potential in the coastal parts of Tamil Nadu, India (Suganthi et al., 2013).

Environmental isotopes have been applied as potential tools to understand the groundwater source and its dynamics (Keesari et al., 2010, 2014; Gupta and Deshpande, 2004). Deuterium and oxygen-18 have been extensively used in studies like groundwater recharge, source, origin and movement of groundwater contamination and groundwater salinisation (Clark and Fritz, 1997). Isotope studies have been very helpful in determining the origin and age of groundwater along the buried river channels (Rao and Kulkarni, 1997). Isotopes have been employed in understanding the role of paleochannels on local groundwater hydrology and also to assess the climate of recharge (Navada, 1999). Isotope investigation on River Saraswati by Rao (2003) identified groundwater of up to 22,000 years BP using ^{14}C dating method in Jaisalmer, Rajasthan. Nair et al. (1999) carried out isotope study to investigate the source and residence time of groundwater along the paleochannels in Jaisalmer and Ganganagar districts of Rajasthan. Studies in the arid regions of the western Rajasthan have indicated that past episodic rain may have recharged the shallow aquifer but in recent times there is no recharge. Reconstruction of the past climate of this region suggests that cooler and pluvial conditions were favourable for the recharge of these aquifers in Holocene period. Turner et al. (1995) have used environmental tracers ^{36}Cl,^{14}C, δ^2H and $\delta^{18}O$ in addition to chloride mass balance and hydrogeological information to evaluate the groundwater recharge-discharge process in hypersaline paleochannels of the Eastern Goldfields of Western Australia.

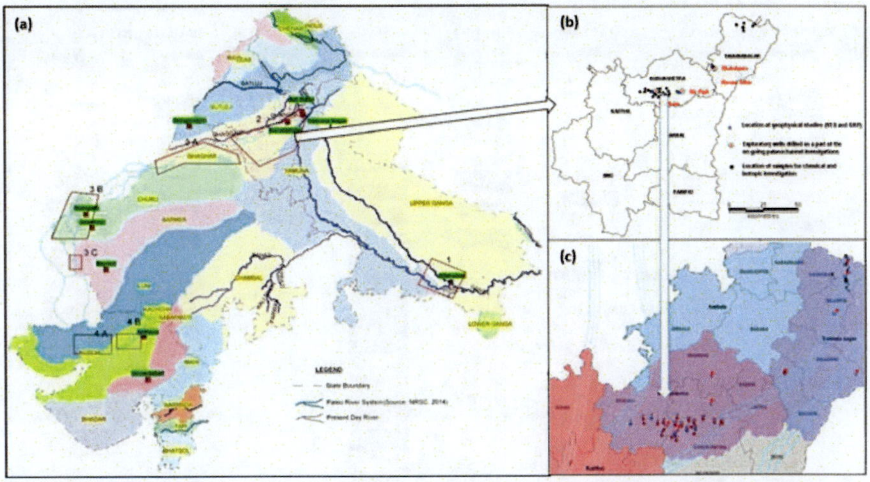

Figure 1. (a) Paleochannel network NW India (boxes indicate CGWB studied sites), (b) Paleochannel tract in Haryana (this study) and (c) sampling points for isotopes.

The study area belongs to the northwest part of India comprising alluvial aquifers, which are under tremendous stress due to quantity and quality issues. In order to overcome the issue of prevalent dryness over large parts of this region artificial recharge was planned through the paleochannels. These paleochannels with an aggregate length of 2200 km have been identified and comprehensively investigated using conventional and high-end techniques (CGWB 2014, 2016). The studies sites are shown by boxes in Fig. 1(a). In order to assess the sustainability of these paleochannels and the possibility of replenishing them through artificial means, it is important to obtain information on the surrounding groundwater and its recharge and dynamics. In this study, an integrated approach consisting of groundwater age with stable isotopic characterisation and interpretation of chemical data was adopted to discern the vital information on the source and dynamics of groundwater. This information is crucial for planning sustainable groundwater development and management in regions where the water needs are increasing exponentially due to irrigation and domestic requirements.

STUDY AREA

The study area falls on the upstream of the Paleochannel (Saraswati Nadi), which extends from the foot of Siwalik (Haryana) to Rann of Kachchh (Gujarat). The location of the study area is shown in Fig. 1(b). Geophysical survey, exploratory drilling, grain size distribution, lithological modelling and aquifer hydraulic parameters were studied by CGWB (2016). The main districts covered in this project are Yamunanagar, Kurukshetra and Kaithal. This region forms a part of mainly Ghaggar River basin and in few parts it belongs to Yamuna basin. Though multi-aquifer

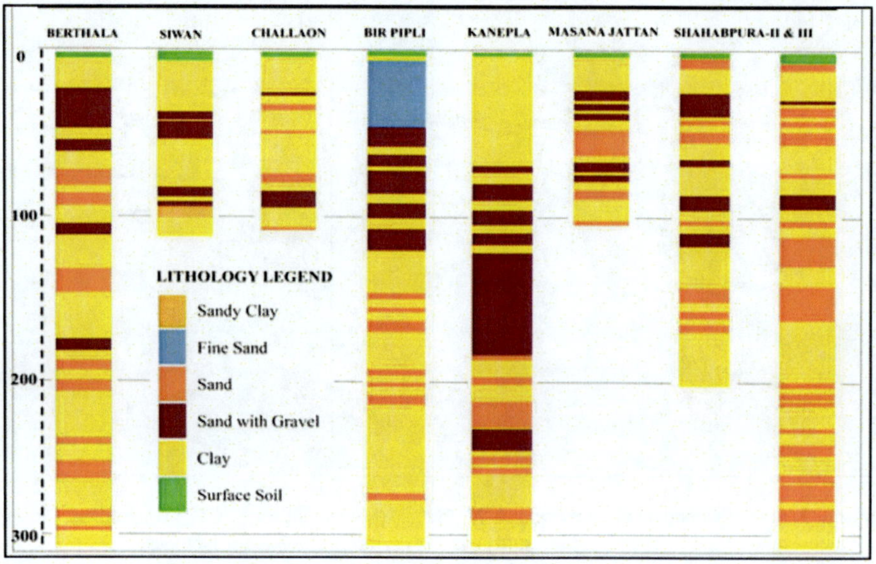

Figure 2. Lithologs of exploratory wells drilled along the paleochannels.

system to a single aquifer system of alluvial sediments is commonly envisaged in this region, there are mainly three aquifers systems delineated based on the subsurface stratigraphy. Difference in topographic elevation is 274 to 241 m above mean sea level. Groundwater at shallow depth occurs under unconfined and semi-confined conditions while it is under confined condition in deeper aquifers. Depth to water level ranges from 20.18 to 32.64 m bgl during premonsoon period and 21.80 to 34.41 m bgl during postmonsoon period. The annual rainfall is about 582 mm, which is mostly received from June to October.

A typical borehole cross-section is shown in Fig. 2. The details of the aquifer delineation based on lithological analysis are as follows:

(i) Yamunanagar district: Three Aquifer System
(Aq –I: 5-186 m, Aq-II: 123-324 m, Aq-III: 245-386 m)
(ii) Kurukshetra District: Three Aquifer System
(Aq –I: 6 to 146m, Aq-II: 172-231: Aq –III: 236-274 m)
(iii) Kaithal District: Single Aquifer System
(up to 300 m)

SAMPLING AND MEASUREMENT

A reconnaissance sampling programme was carried out during the postmonsoon period of 2014 where four samples were collected from deep wells of CGWB monitoring sites. Another sampling programme was carried out during postmonsoon of 2015 from two selected sites, Pipli (Kurukshetra district) and Mugalwali (Yamuna Nagar district). Groundwater was collected from three zones in Pipli (149-153 m bgl, 196-200 m bgl, 270-277 m bgl) and two zones in Mugalwali (6 m bgl and 33 m bgl). Based on the information gathered from these two reconnaissance surveys, two comprehensive sampling programmes were carried out for a complete set of isotopes (^{2}H, ^{18}O, ^{13}C, ^{3}H and ^{14}C) during postmonsoon of 2015 (October 2015) and premonsoon of 2016 (June 2016). The sample location map for environmental isotopes is given in Fig. 1(c). The samples were collected from the custom-built piezometers after pumping out the stagnant water.

The water samples were filtered using 0.45 μm pore size membrane filters and stored in polyethylene bottles. Prior to sampling the samples bottles were washed with concentrated HNO_3 and then rinsed with distilled water. For cation measurements, the collected samples were acidified with concentrated ultrapure HNO_3 to bring down the pH below 2. The physicochemical parameters, viz. temperature (°C), electrical conductivity (EC in μS/cm), pH, alkalinity (in terms of HCO_3^- in ppm) were measured in situ in the field using a multiparameter kit (HI 9828). Alkalinity was measured in field by titration of 10 ml of water sample with 0.02 N H_2SO_4, the mixed indicator (bromocresol green—methyl red) was used for indicating the end point of the titration. Cations (Li^+, Na^+, K^+, Mg^{2+} and Ca^{2+}) and anions (Cl^-, SO_4^{2-}, NO_3^- and F^-) were analyzed in DIONEX 500 ion chromatography instrument with an electrochemical detector (ED 40) in conductivity mode. To check the accuracy of the data charge balance error was calculated for each sample using Eq. (1), and was found to be within the permissible limit (Hounslow, 1995).

$$\mathrm{CBE} = \frac{\mathrm{meq(cations)} - \mathrm{meq(anions)}}{\mathrm{meq(cations)} + \mathrm{meq(anions)}} \times 100 \tag{1}$$

Environmental stable isotopes (^{2}H and ^{18}O) were measured from water samples collected in 50 ml airtight high density polyethylene bottles and using continuous flow Isotope Ratio Mass Spectrometer (IRMS, Isoprime 100). For measuring $\boldsymbol{\delta}^{2}$H, 1 ml of the water was allowed to equilibrate with H_2 gas in presence of the Pt-coated Hokko beads catalyst at 50°C for 90 min and the resultant gas was introduced into the IRMS (Ohsumi and Fujino, 1986). For $\boldsymbol{\delta}^{18}$O measurement, the water sample was equilibrated with CO_2 gas at 50°C for 8 hours and the equilibrated gas was introduced into the IRMS (Epstein and Mayeda, 1953; Horita et al., 1989). The δ-notation is used to report the isotope results, which is expressed in units of parts per thousand. The δ values are calculated using Eq. (2) (Coplen, 1996):

$$\delta(\%) = \left(\frac{R_x}{R_s} - 1\right) \times 100 \tag{2}$$

where R is the ratio of heavy to light isotope (e.g. ^{2}H/^{1}H or ^{18}O/^{16}O) and R_x and R_s are the ratios in the sample and standard, respectively. The measured values are normalized on VSMOW/SLAP scale as per the methodology established by Nelson

(2000). The precision of measurement for δ^2H is ± 0.5 ‰ and for $\delta^{18}O$ ± 0.1 ‰ (2σ).

The tritium content of the water samples was measured by a liquid scintillation counter (Perkin Elmer Quantulus—Model No.1220) after electrolytic enrichment (Nair, 1983). Tritium concentration is expressed in tritium unit (TU), where 1 TU corresponds to one tritium atom per 10^{18} protium atoms. The precision of the measurement is ± 0.5 TU (2σ). Carbon-14 measurement involves precipitation of total dissolved inorganic carbon (TDIC) in water samples in the form of $BaCO_3$ precipitate, conversion of $BaCO_3$ precipitate to carbon dioxide (CO_2), absorption of CO_2 into carbasorb containing scintillator mixture and finally counted in ultra-low level liquid scintillation counter for 1000 min (Nair, 1983; Keesari et al., 2019). ^{14}C activity is expressed in percent modern carbon (pMC), where 100 pMC corresponds to an activity of 13.56 dpm per g of carbon.

RESULTS AND DISCUSSION

Hydrochemical Inferences

All the samples indicate fresh quality of water with all the measured chemical parameters within the drinking water limits. From the chemical data it is found that in general the cation dominance follows: $Na^+ = Ca^{2+} > Mg^{2+}$ while anion dominance is in the order $HCO_3^- > Cl^- > SO_4^{2-}$. A trilinear Piper's plot was constructed to understand the geochemical nature of groundwater in different aquifers (Fig. 3) where it can be observed that most of the groundwaters are represented by Ca-Na-HCO_3, Ca-Mg-HCO_3, Na-HCO_3, Mg-Ca-HCO_3, Mg-Na-HCO_3 and Na-Ca-HCO_3 types. Based on the hydrogeological zonation, the groundwater samples are divided into three groups (shallow, intermediate and deep) and their relative position in the diamond field is shown in Piper's plot (Fig. 3). It can be clearly seen that shallow and intermediate zones have more or less same chemical signature and fall in one cluster while the deep zone samples show spread in the diamond field and represent different water types. The shallow and intermediate zone samples are mostly Ca-Na-HCO_3, Ca-Mg-HCO_3, Mg-Ca-HCO_3 and Mg-Na-HCO_3 types while deep zone samples are Na-HCO_3 and Na-Ca-HCO_3 types. One intermediate (zone II) sample shows a similar water type to that of the deep zone (zone III), which could be due to interconnection between aquifers at that location.

From the hydrochemical behaviour it can be inferred that shallow groundwater (zone I) are Ca-type water, intermediate groundwater (zone II) are Mg-type water and deep groundwater (zone III) are Na-Cl type. This observation indicates that both zones I and II waters are dominantly recharging waters while groundwater in zone III is not connected with the above two aquifers and the low EC with a different geochemical facies indicate ion exchange as the governing mechanism rather than dissolution of aquifer material.

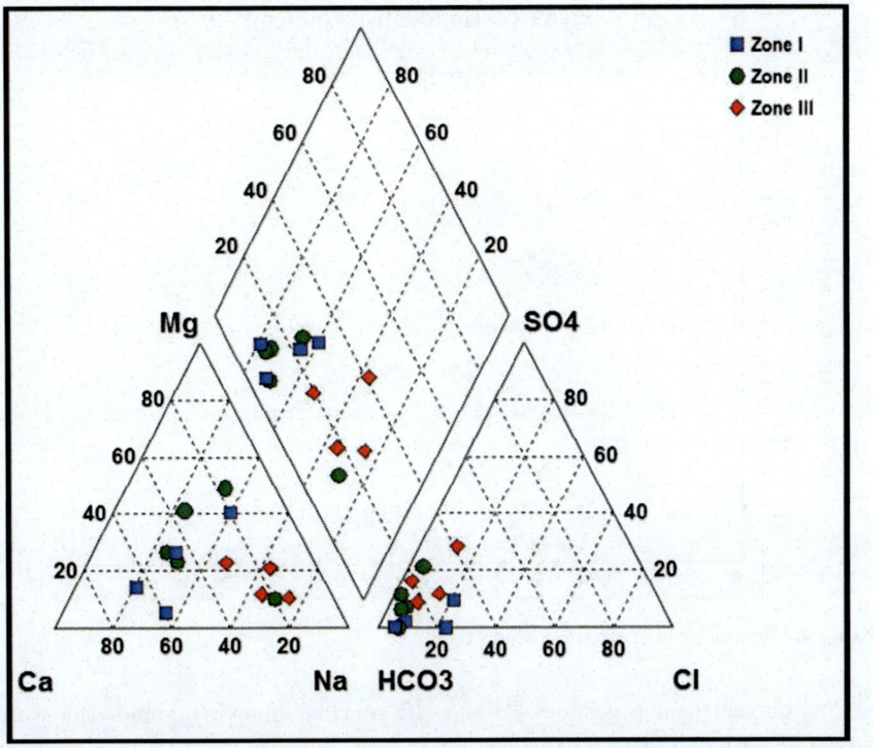

Figure 3. Trilinear Piper's plot depicting the hydrochemical facies variation in different zones of the alluvial formations.

Vertical Trends

This variation in hydrochemistry along the depth is further evaluated by well depth versus EC plot (Fig. 4) which shows that the electrical conductivity of the samples decreases with the depth of the sample in both pre- and postmonsoon seasons. It is observed that premonsoon samples show greater variability in EC compared to postmonsoon samples. Depth variations were also investigated for $\delta^{18}O$, ^{3}H and ^{14}C (Fig. 5). Unlike the EC profile, there is no clear trend in isotopic composition ($\delta^{18}O$) of the groundwater with depth, however, there are subtle differences in the isotopic values of the groundwater with depth (Fig. 5 (i)). On the whole, the samples show a wide range of isotopic signatures from −7.5 to −5.0‰ with a few samples especially from zone III showing $\delta^{18}O$ close to −9.5‰. Groundwater from zones I and II shows a larger variation in isotopic content while zone III falls in two groups: −7 to −6 ‰ and also at −9.1‰. The stable isotope composition of zone III samples shows a narrow range of isotope values compared to zone I and zone II. A few samples from the zones II and III show highly depleted isotopic composition (−9.1‰), indicating a different source of recharge compared to other samples.

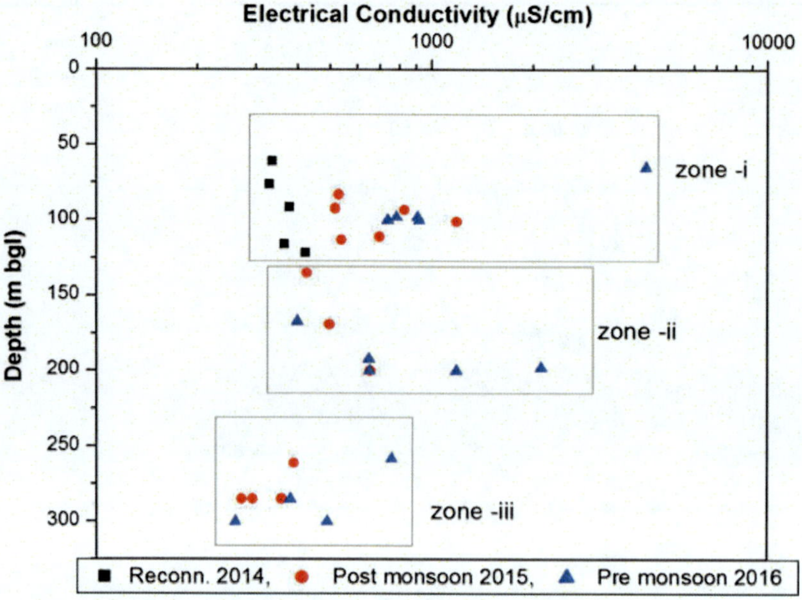

Figure 4. Vertical trend of electrical conductivity.

Environmental radio isotopes ^{3}H and ^{14}C concentrations of groundwater were plotted against well depth to evaluate the overall dynamics of the groundwater in different zones. From the depth-wise variation of tritium (Fig. 5 (ii)), it can be inferred that deep zones are relatively older and less dynamic compared to shallow zones. The tritium content in groundwater shows a wide range of values from 0.4 to 9.5 TU. There is a decrease in overall tritium concentration as well as its spread with the depth, like in the case of electrical conductivity. In general, in shallow groundwater samples of zone I, the tritium content is high (2.7 to 9.5 TU) indicating their equilibrium with present day precipitation. The high value of 9.5 TU is observed in a shallow zone well (65 m bgl) which also shows high electrical conductivity, signifying anthropogenic contamination. Modern nature of groundwater is also seen in intermediate and deep zones. However, few locations from the intermediate and deep groundwaters (zone III) show negligible tritium content, ^{3}H: 0.3 to 1.2 TU. These groundwaters are very old and are being recharged by a distant source.

In the case of ^{14}C variation with respect to depth, it can be observed that groundwater upto 200 m bgl showed ^{14}C in the range of 50 to 108 pMC, while below 200 m bgl, it is mostly < 30 pMC (Fig. 5 (iii)). A few exceptions are noted which can be attributed to vertical mixing with the zone I and II groundwater.

In order to examine the vertical distribution of stable isotope signatures along the groundwater flow, the study area is broadly divided into three divisions: north, central and south. The δ^{18}O variation with depth in these three divisions is shown in plots (Fig. 6, i-iii). The north division shows two groups of samples belonging to shallow (zone I) and intermediate aquifers (zone II) as shown in Fig. 6 i. Mostly

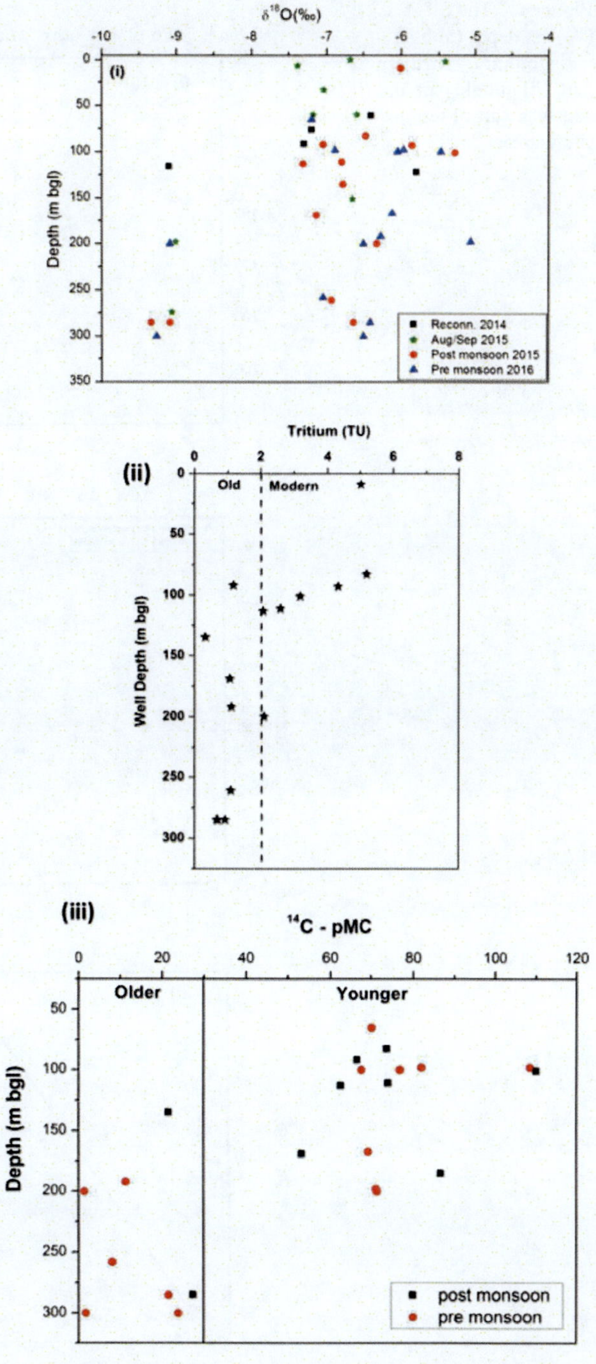

Figure 5. Vertical variation of: (i) $\delta^{18}O$ (‰), (ii) tritium (TU) and (iii) ^{14}C (pMC).

Figure 6. The $\delta^{18}O$ (‰) vs. depth (m bgl) variation in: (i) northern part, (ii) middle part and (iii) southern part of the study area.

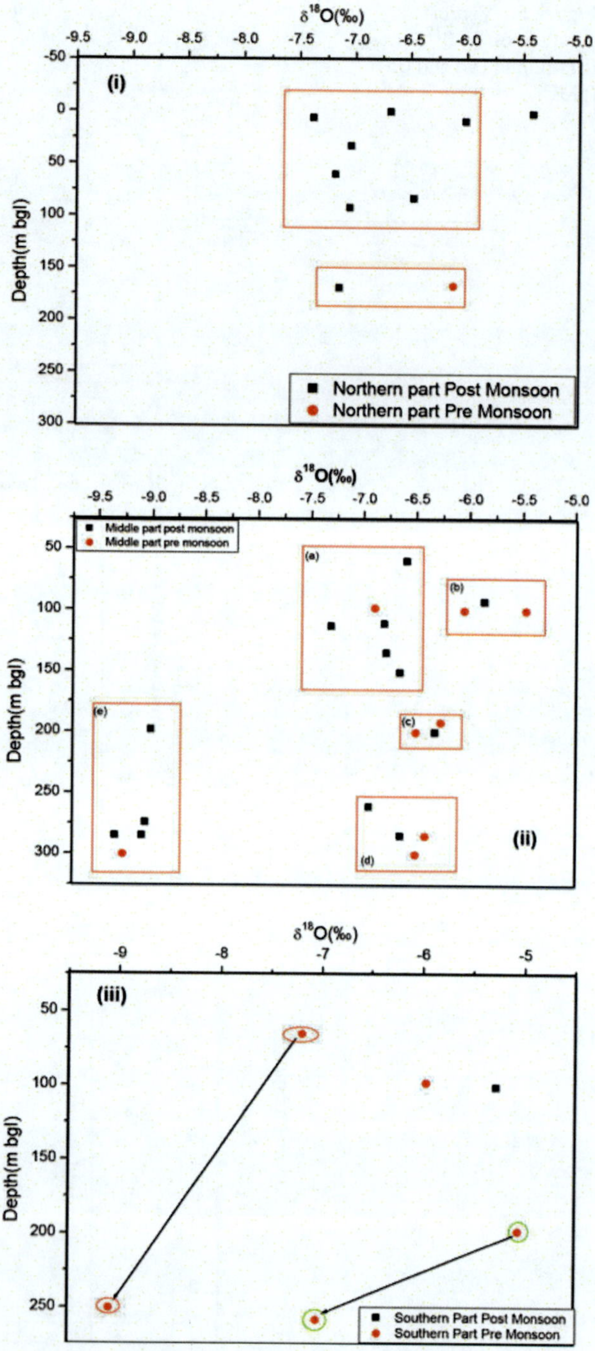

wells are confined to shallow depths up to 110 m bgl and both these groups show more or less similar variation of $\delta^{18}O$ values (−7.5 to −6.25‰). This observation suggests that the groundwaters of north division are interconnected or have the same source water isotopic signature.

A complex scenario was noticed from the stable isotopic signature of groundwaters in the middle division. About five groups were observed as shown in Fig. 6 ii. The shallow zone seems to be influenced by local precipitation (group (a)) as well as surface water sources (group (b) in Fig. 6 ii). The highest enrichment in groundwater was found to be $\delta^{18}O$: −5.5‰. Zone II and zone III samples showed isotope signature in a very narrow range $\delta^{18}O$: −7.5 to −6.5 ‰ (group (c) and (d) in Fig. 6 ii). This infers either the zones are interconnected with zone I or are derived from other recharge source but with similar isotopic signature. A remarkably different group (e) of samples is identified at depths 200 m bgl and below 250 m bgl as shown by Fig. 6 ii. The sample locations belong to Pipli and Shahpur villages. The isotopic signature of these groundwaters is $\delta^{18}O$; −9‰ in both zones II and III. This might be due to interconnection between intermediate and deep zones at Pipli site. The depleted signature of groundwaters of this group clearly indicates that recharge mechanism is different from other zones and possibly from higher elevations.

Very few samples were collected in the south division of the study area. On the whole, it can be observed that with depth the isotopic composition depletes at Kaithal and Khanpur (Fig. 6 iii). The lithology of this part of the study area suggests a single aquifer system, however, isotopically it can be seen that groundwater characteristics are not similar. Hence, the recharge sources are different for different zones and there is no interconnection between different zones.

Temporal Variation

In order to estimate the impact of monsoon on the groundwaters from different zones, bar plot of EC, ^{18}O and ^{3}H for common samples were plotted (Fig. 7). Most of these common samples represent intermediate and deep zones (zones II and III). From the plot (Fig. 7 (i)), it can be inferred that the EC of the groundwater is mostly less during premonsoon compared to postmonsoon. This is in contrary to the generally observed behaviour that due to rainwater recharge the EC of the groundwater decreases during postmonsoon.

A similar observation is noticed in case of tritium variation (Fig. 7 (ii)). A decrease in tritium content is noticed in most of the groundwaters during postmonsoon compared to premonsoon; however most of the samples indicate modern values. This decrease can be attributed to delayed recharge to the aquifer. Since these alluvial formations consist of alternating clay and sand beds, the vertical percolation of the rainwater might be slow depending on the thickness of the clay beds. The stable isotope variation with season was found to be insignificant (Fig. 7 (iii)).

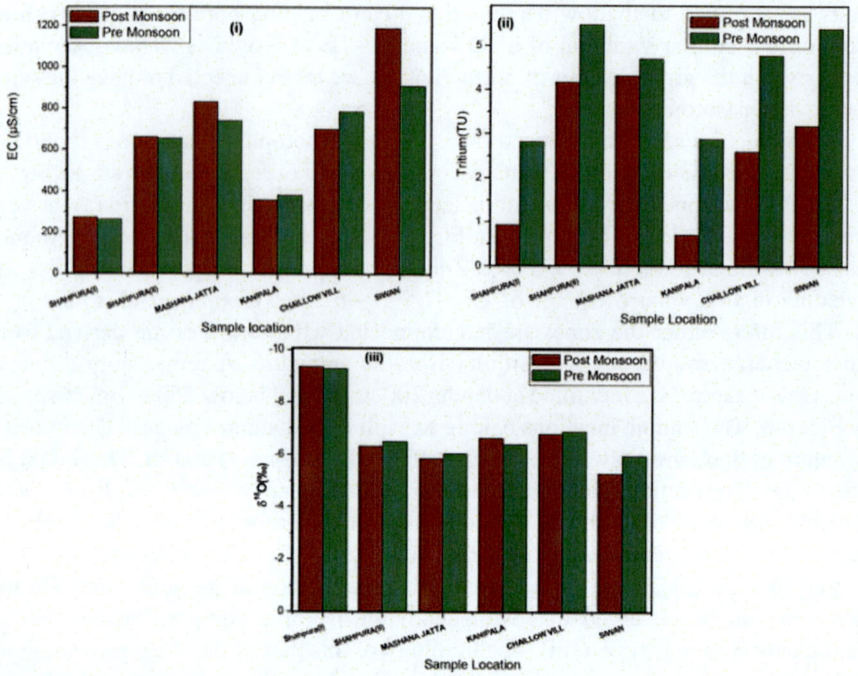

Figure 7. Temporal variation of: (i) EC (μS/cm), (ii) Tritium (TU), (iii) $\delta^{18}O$ (‰) in post-and premonsoon samples.

Environmental Stable Isotopes

The δ^2H vs. $\delta^{18}O$ plot of water samples is presented in Fig. 8. The groundwater samples fall in two broad clusters—depleted and enriched. In order to evaluate the probable source of groundwater, local meteoric water lines for Western Himalaya, New Delhi, Indian Meteoric Water Line and Global Meteoric Water Line were also drawn in the plot. The equations for these meteoric water lines are:

GMWL – $\delta D = 8.17*\delta^{18}O + 11.27$ (Rozanski et al., 1993)
LMWL (New Delhi) – $\delta D = 7.20*\delta^{18}O + 2.70$ (Gupta and Deshpande, 2005)
Western Himalaya – $\delta D = 7.95*\delta^{18}O + 11.51$ (Nachiappan and Kumar, 2002)
IMWL – $\delta D = 7.93*\delta^{18}O + 9.94$ (Kumar et al., 2010)

Additionally, three locations representing higher (Dabrani), middle (Uttarkashi) and lower elevations (Roorkee), where long term isotope data in precipitation is available, are also shown (Fig. 8) for comparison. The data is taken from Kumar et al. (2010).

In the enriched cluster, groundwaters can be separated into three groups. Group (a) samples fall on the Western Himalayan line. These samples represent zone I in the northern part, zone II in central part and zone III in southern part. The rainwater

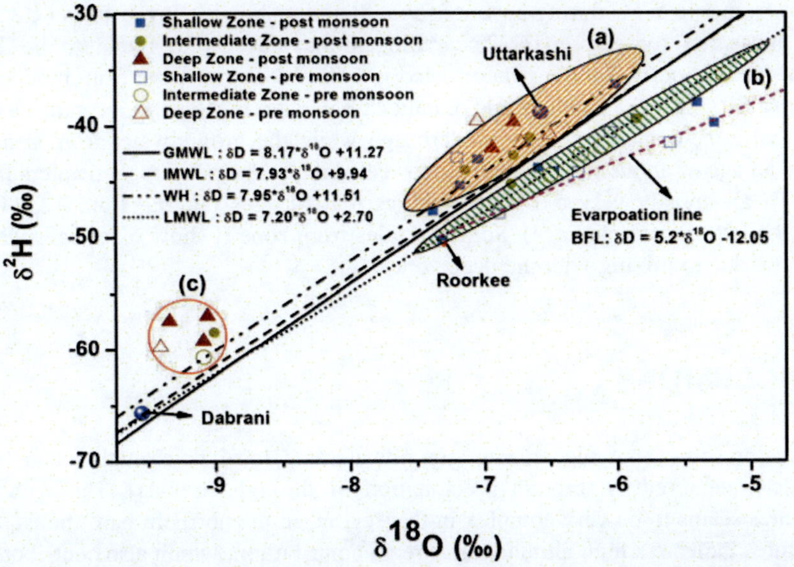

Figure 8. The $\delta^{18}O$ (‰) vs. $\delta^{2}H$ (‰) plot of the water samples collected from the study area.

composition representing middle elevation (Uttarkashi) also falls in this group. This clearly indicates the possibility of groundwater recharge by the western Himalayan precipitation. In group (b) samples from zone I and II fall on the local meteoric water line of New Delhi. Most of the samples belong to the central and south parts of the study area. These samples represent recharge from local precipitation. A few samples from zones I and II show similar isotope values indicating the possibility of interconnection at few places where clay zones are not prominent. The rainwater composition representing lower elevation (Roorkee) also falls in this group. This further proves that the recharge is mostly through local precipitation. None of the samples from zone III falls in this group indicating that the contribution from local precipitation to zone III is very remote. A few samples from zone I fall on a line below LMWL with a slope of 5.2 indicating evaporation effect (group (c)). These samples receive a part of recharge from the evaporated surface water bodies (Figs. 5 and 6). Samples falling in depleted cluster show a very narrow range of isotope values around −9.1‰ (group (c)). The rainwater composition representing higher elevation (Dabrani) falls close to this group. These samples are mostly derived from the precipitation occurring at very high elevations.

Radioisotopes

Environmental tritium and carbon-14 isotopes were measured in water samples of all the zones. The tritium values range from 3 to 7 TU during premonsoon while it ranges from 1 to 5 TU during postmonsoon. Samples showing low tritium (< 2 TU) were measured for carbon-14. The radiocarbon values were found to be 2 to 108 pMC. The corresponding uncorrected groundwater ages range from modern to 35,000 a BP. From the depth profile of radiocarbon data (Fig. 5 (iii)), it can be clearly seen that zone I samples show modern age while the groundwater from zone II shows an age of about 10,000 a BP (uncorrected age). The deep groundwaters from zone III are mostly very old water with carbon-14 ages ranging from 12,500 to 35,000 a BP (uncorrected age). Some samples from zone II show older ages which might be due to mixing with the deep zone.

CONCLUSIONS

The major sources of recharge to groundwater are local precipitation and high altitude waters (either through precipitation or through streams). The recharge dynamics seems to be very complex in this region. In the northern part, the isotope signatures indicate a high altitude rain/stream contribution, which also contribute to zone II of the central part and zone III of southern part. Zones I and II of the central and southern parts show two major sources of recharge, viz., local precipitation and evaporated surface water, which could be open water body or irrigation return flow. Hydrochemical and isotope data clearly indicate the presence of multi-aquifers in this region. Based on the isotope and EC variations presence of three aquifers can be inferred in the central region of the study area.

The hydrochemical and $\delta^{18}O$ vertical trends indicate interconnections among the three identified aquifers are not very common. However, some interconnections between zone I and II in central part of the study area can be feasible. The groundwater of zones I and II at locations Mashana Jattan and Shahpura show similar isotope and hydrochemical signatures. Similarly, groundwater in zones II and III near KhanpurJattan and Kanipala also show similar values of isotope and hydrochemical parameters. In the southern part, groundwater from different zones indicate different signatures of hydrochemistry and isotopes, indicating less chances of interconnection between the zones.

The environmental tritium data indicates that groundwater is both modern and old depending on the zone of occurrence. The tritium data indicate that groundwater collected during premonsoon is more recent compared to the postmonsoon season, which may be due to delayed recharge. A few groundwater samples from zones II and III show negligible tritium (< 1 TU) hinting the possibility of old groundwater, or slow and distant recharge to groundwater. The groundwater age was quantified using uncorrected C-14 data. The uncorrected ^{14}C ages of groundwater corroborate findings from environmental tritium data. The groundwater age ranges from modern to about 35,000 a BP. The deep groundwater shows long residence time of 12,000 to 35,000 a BP (uncorrected). Some samples of zone II also indicate very old groundwater, which could be mixing with zone III, which needs further verification. A few of zone II and III samples show very young groundwater (modern nature). This could be due to mixing with zone II samples with zone I. Vertical mixing is also inferred in the case of a few samples based on their stable isotope signatures.

FUTURE STUDIES

In order to implement groundwater recharge through paleochannels, it would be beneficial to undertake a comprehensive understanding of the local groundwater system with regard to its recharge, water quality and dynamics. A multi-disciplinary approach consisting of geophysical surveys, borehole data, chemical quality of groundwater and isotopic composition would form a good data base for preparing integrated maps. These maps can be helpful in deciding the efficacy of the recharge structures on groundwater recharge, which in turn, will assist in prioritizing the zones where artificial recharge to groundwater can be taken up. Especially in northwest India where groundwater level depletion is highest in the world, the paleochannels can form a good conduit to replenish the fast depleting groundwater resources. Paleochannel hydraulic connectivity with the local aquifers is one of the major challenges, which can be addressed using isotope tracers. Once the hydraulic connections are established with the use of abovementioned said integrated maps, artificial structures such as tanks, trenches, pits and wells can be constructed to augment the groundwater recharge. Another means of replenishing groundwater levels would be to allow the floodwater/overflowing water in the canals to spread through these identified zones to induce faster infiltration.

ACKNOWLEDGEMENT Authors thank Dr. P.K. Pujari, Director, Radiochemistry and Isotope Group (BARC) and Dr. H.J. Pant, Head, Isotope and Radiation Application Division (BARC) for their constant support and guidance throughout the course of this project. Authors also thank the scientists from Central Ground Water Board, Chandigarh, for their help during field work.

References

Anon (1874). Notes on the lost river of the Indian Desert, *The Calcutta Review*, 59: 1-29.

Bakliwal P.C. and Grover, A.K. (1988). Signature and migration of Sarasvati river in Thar Desert, western India, *Rec. Geol. Surv.* India Rec., 116(3–8): 77-86.

Bhadra, B.K., Gupta, A.K. and Sharma, J.R. (2009). Saraswati Nadi in Haryana and its linkage with the Vedic Saraswati — integrated study based on satellite images and ground-based information, *Jour. Geol. Soc. India*, 73: 273-288.

CGWB (2014). A Review of Studies on Vedic River Saraswati, Central Ground Water Board, Ministry of Water Resources, River Development and Ganga Rejuvenation, Govt. of India.

CGWB (2016). Palaeochannel of North West India: Review and Assessment, Central Ground Water Board, Ministry of Water Resources, River Development and Ganga Rejuvenation, Govt. of India.

Clark, I.D. and Fritz, P. (1997). Environmental isotopes in hydrogeology, New York, Lewis Publishers.

Coplen, T.B. (1996). New guidelines for reporting stable hydrogen, carbon and oxygen isotope-ratio data, *Geochim. et Cosmochim. Acta*, 60: 3359-3360.

Epstein, S. and Mayeda, T. (1953). Variation of ^{18}O contents of water from natural sources, *Geochim. Cosmochim. Acta*, 4: 213-224.

Ghose, B., Kar, A. and Husain, Z. (1979). The lost courses of the Saraswati River in the great Indian desert: new evidence from LANDSAT Imagery, *The Geographical Journal*, London, 45(3): 446-451.

Gupta, S.K. and Deshpande, R.D. (2005). Groundwater isotopic investigations in India: what has been learned? *CurrSci.* 89: 825-835.

Gupta, S.K. and Deshpande, R.D. (2004). Water for India in 2050: first-order assessment of available options, *Curr. Sci.*, 86: 1216-1224.

Horita, J., Ueda, A., Mizukami, K. and Takatori, I. (1989). Automatic δD and $\delta^{18}O$ analyses of multi- water samples using H_2- and CO_2- water equilibration methods with a common equilibration set-up. *Int. J. Radiat. Appl. Instrum.*, Part A, *Appl. Radiat. Isot.* 40(9): 801-805.

Hounslow, W.A. (1995). Water quality data-analysis and interpretation, New York: Lewis, 73.

Kar, A. and Ghose, B. (1984). The Drishadvati River System of India: An assessment and new findings, *The Geological Journal*, 150(2): 221-229.

Keesari, T., Kulkarni, U.P., Deodhar, A., Ramanjaneyulu, P.S., Sanjukta, A.K. and Saravana, U.K. (2014). Geochemical characterization of groundwater from an arid region in India, Environmental Earth Sciences, 71(11): 4869-4888.

Keesari, T., Shivanna, K., Sriraman, A.K. and Tyagi, A.K. (2010). Assessment of quality and geochemical processes occurring in groundwaters near central air conditioning plant site in Trombay, Maharashtra, India, *Environmental Monitoring and Assessment*, 163: 171-184 doi: https://doi.org/10.1007/s10661-009-0825-9.

Keesari, T., Chatterjee, S., Pant, D., Singh, M., Sakhare, V., Sinha, U.K., Mohokar, H.V., Jaryal, A., Roy, A. and Maitra, A. (2019). Dating of hot springs at Attri, Tarabalo and Athmalik sites in Odisha, India using radiocarbon technique, *Journal of Radioanalytical and Nuclear Chemistry*, doi:https://doi.org/10.1007/s10967-019-06867-1.

Kumar, B., Rai, S.P., Sarvana, U.K., Verma, S.K., Garg, P., Vijaya, S.V.K., Jaiswal, R., Purendra, B.K., Kumar, S.R. and Pande, N.G. (2010). Isotopic characteristics of Indian precipitation, Water Resour. Res 46: W12548 doi:https://doi.org/10.1029/2009WR008532.

Nachiappan, R.P. and Kumar, B. (2002). Estimation of subsurface components in the water balance of Lake Nainital (Kumaun Himalaya, India) using environmental isotopes, *Hydrol. Sci. J.*, 47: 41-54 doi:https://doi.org/10.1080/02626660209493021.

Nair, A.R. (1983). Possibilities of liquid scintillation counting for tritium and radio carbon measurements in natural water, *In:* proceedings of the Workshop on Isotope Hydrology, Mumbai, 41-56.

Nair, A.R., Navada, S.V. and Rao, S.M. (1999). Isotope study to investigate the origin and age of groundwater along palaeochannels in Jaisalmer and Ganganagar districts of Rajasthan, *Memoir Geol. Soc.* India, 42: 315-319.

Navada, S.V. (1999). Groundwater problems studies in the Thar desert, India using isotope techniques, Water and Environment News, Quarterly No. 7: 10-12

Nelson, T.S. (2000). A simple, practical methodology for routine VSMOW/SLAP normalization of water samples by continuous flow methods. *Rapid Communications in Mass Spectrometry*, 14: 1044-1046.

Ohsumi, T. and Fujino, H. (1986). Isotopic exchange technique for preparation of hydrogen gas in mass spectrometric D/H analysis of natural waters. *Analytical Science*, 2: 489-490.

Oldham, C.F. (1893). The Saraswati and the lost river of the Indian desert, *Jour. Royal Asiatic Soc.* London, 34: 49-76.

Rao, A.L.K., Ali, S.R., Rahman, Z.U., Siddiqui, R., Priyadarshi, H., Sadique, M. and Islam, Z.U.L. (2009). Hydro-geomorphological studies for groundwater prospects using IRS-1D, LISS III Image, in parts of Agra district along the Yamuna River, U.P., India, *Jour. Environ. Res. & Develop.*, 3(4): 1204-1210.

Rao, S.M. (2003). Use of isotopes in search of Lost River, *J. Radioanal. and Nucl. Chem.* 257(1): 5-9.

Rao, S.M. and Kulkarni, K.M. (1997). Isotope hydrology studies on water resources in western Rajasthan, *Current Science*, 72: 55-61.

Rozanski, K., Aragus-Araguas, L. and Gonfiantini, R. (1993). Isotopic patterns in modern global precipitation. *In:* Swart, P.K., Lohmann, K.C., McKenzie, J. and Savin, S. (eds) Climate change in continental isotopic records. American Geophysical Union, Monograph 78. AGU, Washington, DC.

Saini, H.S., Tandon, S.K., Mujtaba, S.A.T., Pant, N.C. and Khorana, R.K. (2009). Reconstruction of buried-floodplain systems of the northwestern Harayana Plain and their relation to the 'Vedic' Saraswati, *Current Science*, 97: 1634-1643.

Samadder, R.K., Kumar, S. and Gupta, R.P. (2011). Palaeochannels and their potential for artificial groundwater recharge in the western Ganga plains, *Journal of Hydrology* 400: 154-164.

Singh, A., Paul, D., Sinha, R., Thomsen, K.J. and Gupta, S. (2016). Geochemistry of buried river sediments from Ghaggar Plains, NW India: Multi-proxy records of variations in provenance, palaeoclimate, and palaeovegetation patterns in the late Quaternary. Palaeogeography, Palaeoclimatology, *Palaeoecology*, 449: 85-100.

Sinha, R., Yadav, G., Gupta, S., Singh, A. and Lahiri, S. (2013). Geo-electric resistivity evidence for subsurface palaeochannel systems adjacent to Harappan sites in northwest India. *Quaternary International* 308: 66-75.

Sinha, R. (1996). Palaeohydrology of Quaternary river systems of North Bihar plains, India. Surface Water Hydrology, V.P. Singh and B. Kumar (Eds), Kluwer Academic Publishers.

Suganthi, S., Elango, L. and Subramanian, S.K. (2013). Groundwater potential zonation by remote sensing and GIS techniques and its relation to the groundwater level in the coastal part of the Arani and Koratalai river basin, southern India, *Earth Sci. Res. Jour.*, 17(2): 87-95.

Turner, J.V., Michael, R.R., Fifield, L.K. and Allan, G.L. (1995). Chlorine-36 in hypersaline palaeochannel ground-waters of Western Australia, Application of Tracers in Arid Zane Hydrology (Proceedings of the Vienna Symposium, August 1994). IAHS Pub. no. 232: 15-33.

Valdiya, K.S. (2013). The River Saraswati was a Himalayan-born River, *Current Science*, 104(1): 42-54.

Valdiya, K.S. (2002). Saraswati: The River That Disappeared, Universities Press, Hyderabad, 116p.

Valdiya, K.S. (2016). Prehistoric River Saraswati: Geological Appraisal and Social Aspects, Springer, 136p.

Yashpal, S.B., Sood, R.K. and Agrawal, D.P. (1980). Remote sensing of the lost Saraswati river, Proc. Indian Acad. Sci. (Earth & Planet. Sci.) 89: 317-337.

Chapter 21
Delineation of Two Ancient Rivers (Chandrabhaga in Konark and Saradha in Puri, Odisha): A Multi-disciplinary Approach

Subhamoy Jana, William Kumar Mohanty, Saibal Gupta, Priyadarshi Patnaik, and Chirashree Srabani Rath

INTRODUCTION

The east coast of peninsular India (which encompasses numerous world heritage sites extends from Tamil Nadu in the south to West Bengal in the north) is the intervening land between the Bay of Bengal and Eastern Ghats. It was formed after separation of India from Antarctica-Australia in the Mesozoic time (Ramana et al., 1994). A number of river systems such as Mahanadi, Krishna, Godavari and Kaveri flow through the Peninsular India. The Mahanadi River system comprises many existing as well as abandoned channels which traverse the Mahanadi delta region (Mahalik et al., 1996). Some of these abandoned channels and their names are mentioned in the ancient texts. Two such major ancient channels are Chandrabhaga and Saradha which are associated with two ancient temples, namely, Konark Sun Temple in Konark and Jagannath Temple in Puri, respectively.

Chandrabhaga

The mythical river called *Chandrabhaga* is believed to be present around 2 km from the 13th century Sun Temple of Konark, a UNESCO world heritage site in the state of Odisha, India. However, presently no trace of any water body is found in the

S. Jana · W. K. Mohanty (✉) · S. Gupta
Department of Geology & Geophysics, Indian Institute of Technology Kharagpur, Kharagpur, India
e-mail: wkmohanty@gg.iitkgp.ac.in

P. Patnaik · C. S. Rath
Department of Humanities & Social Sciences, Indian Institute of Technology Kharagpur, Kharagpur, India

A. Mukherjee (ed.), *Riverine Systems*, https://doi.org/10.1007/978-3-030-87067-6_21

vicinity of the temple, although associated myths strongly suggest the presence of water nearby. Rath et al. (2015) reported the existence of water bodies proximal to the temple in the past based on the preserved palm-leaf drawings, sketches and rare old photographs. At present, a swampy area which is believed to be a remnant of the ancient river (known as Chandrabhaga Lake), is used for the ritual holy bath during the festival of *Magha Saptami*.

Saradha

Puri town on the coast of Odisha is location of a number of ancient temples from 12th century CE, comprising the region of Mahanadi delta, about 60 km south of Bhubaneswar. Two historic landmarks of the Puri town, the Jagannath and Gundicha temples, are connected by the Grand Road or *Badadanda*. It is believed that an ancient river Saradha once flowed across the present Grand Road (or *Badadanda*) and bisected the town. Many ancient texts like *Skanda Purana*, *Brahma Purana* and *Kapila Samhita*, mention King Indradyumna of Vaishnava tradition and the Puri Jagannath temple. According to legend the king found a temple on the blue coloured mountain top in Puri also known as *Neelachala/Neelashaila* (Tagare, 1994; Parameswaranand, 2001; Mishra, 2005). He was guided by the Lord in his dream to collect a wooden log on the seashore and subsequent to this he established the wooden idols of Lord Jagannath, Balabhadra and Devi Subhadra in the temple at Puri. Many *Puranic* texts and temple chronicle like *Madala Panji* also mention ancient river 'Saradha' in the vicinity of the temple (Mohanty, 1969). According to Kulke and Tripathi (1987), another ancient text *Katakarajavamshabali* (a compiled work from old historical records, prepared and preserved in the temple archive since the early part of 19th century CE) refers existence of the same ancient river in-between the Jagannath and Gundicha temples, dividing the Grand Road. The text also mentions six chariots, out of which three used to carry the idols up to one side of the river, and the remaining three on the other bank waited to carry the idols to the temple during *Rathayatra* (or car festival) from the other bank of the river up to Gundhicha temple. There is also another myth that Lord Jagannath appeared and advised Saradha Devi, wife of Puri Gajapati Narasingha Dev, in her dream to have a single journey for making the *Rathayatra* smoother by closing the river with sand (Nayak, 1998). The river bed as well as the name of *Saradha Bali* which represents sacredness and purity of holy sand is mentioned by the poets of 16th-18th century CE. Still, in the vicinity of the Gundicha temple, the name *Saradha Bali* exists in the mind of the devout and thus the place attains its sacredness for pilgrimage in Puri (Mohanty, 1978).

Location and Geology

The region that comprises Puri and Konark lies within the Mahanadi delta, geologically a part of Paleo to Mesozoic era Gondwana basin that is at present covered with recent alluvium.

In Puri region, the area is taken from latitude 19.80°N to 19.83°N, and longitude 85.77°E to 85.86°E while in Konark region, the study area is from latitude 19.85°N to 20.08°N and longitude 86.08°E to 86.25°E (Fig. 1). In the northern part of Puri, river Bhargabi flows towards west, while river Dhaudia branches off from Bhargabi and flows towards south encircling Puri town. The southern part comprises heart of Puri town which includes Jagannath and Gundicha temples connected by the Grand Road (Jana et al., 2018). In Konark region, the study area is traversed by Kadua and Prachi rivers in the eastern side of Konark Sun Temple. The geology of the coastal Odisha is already described in Jana et al. (2016) and Jana et al. (2018). In brief, the region comprises alluvium underlain by Gondwana sediments. On the western side the Eastern Ghats rocks like khondalites, charnockites and migmatitic are exposed (Behera et al., 2004). The subsurface structures include alternate depressions and ridges which are traversed by various sets of lineaments (Behera et al., 2004; Nayak et al., 2006). These lineaments control various distributaries of the Mahanadi river.

ANCIENT LITERATURE STUDY

The available literature and folklore like *Samba Purana, Madala Panji* (1078-1150 CE), the *Kapila Samhita* (14th century CE), *Baya Chakada* (12th Century CE) and *Sarala Mahabharat* (15th century CE) study reveals possibility of the existence of these two rivers in ancient times (Rath et al., 2015; Jana et al., 2016; Jana et al., 2018). Thus, in spite of the absence any physical evidence of these rivers, these ancient literatures point to the existence of these two rivers in the mind of devout. Often such myths and traditions of beliefs are grounded in truth. This encourages the question if these ancient rivers (Chandrabhaga and Saradha) really existed in the past, or are they mythical and representative of some other symbolic meaning. Hence, objective of this study is to determine if any paleo-channel (related to the name Chandrabhaga in Konark and Saradha in Puri) really existed in the vicinity of the present Konark Sun temple (here Chandrabhaga river) or somewhere cutting across the Grand Road between the Jagannath and Gundicha temples (Saradha river).

Myths Related to these Ancient Rivers

There are several myths and popular tales in ancient texts which suggest the presence of these ancient rivers. One such myth related to river Chandrabhaga is the story of

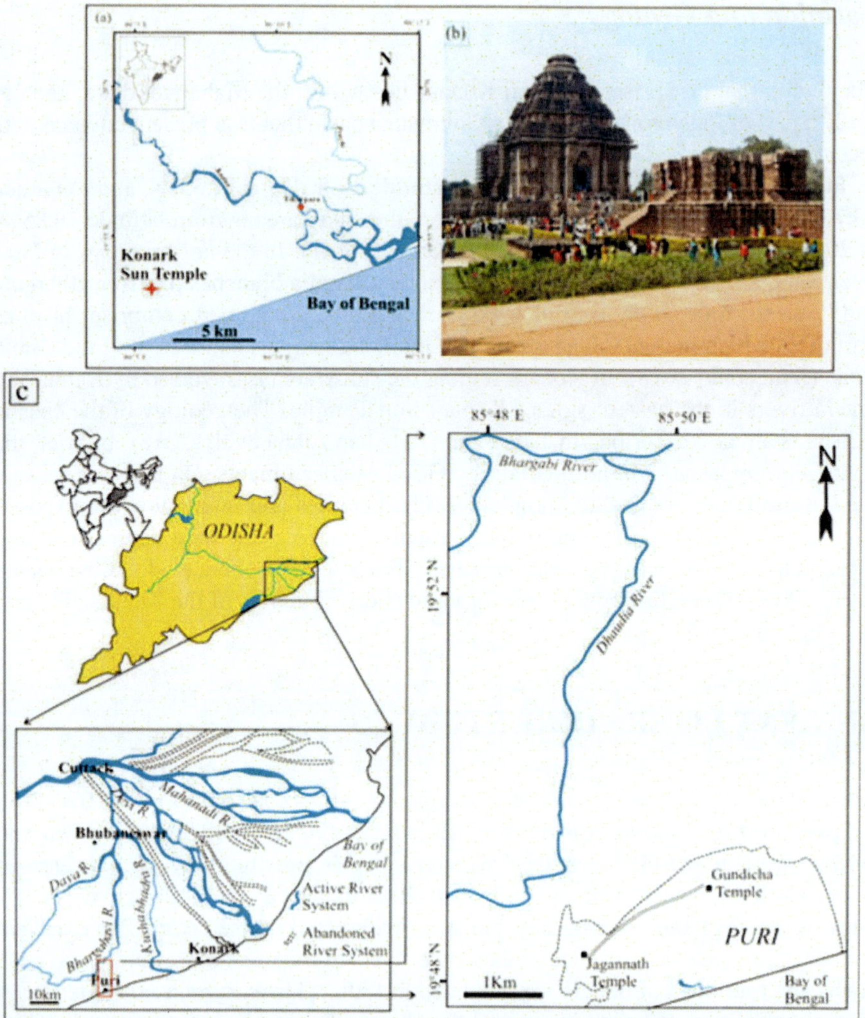

Figure 1. (a) Location map of the Konark Region (retraced from satellite images) shows Konark Sun Temple in the Odisha coast. (b) Photograph of present-day Konark Sun Temple taken during field trip (modified after Jana et al., 2016); (c) Location map of the Puri town in Mahanadi delta region (modified after Mahalik et al., 1996; Jana et al., 2018).

Dharmapada, the son of Bishu Maharana who was the chief architect of the Konark Sun Temple. In order to save his father and twelve hundred craftsmen, Dharmapada sacrificed his life by jumping from the top of the temple to river Chandrabhaga (Rath et al., 2015). This suggests there was water body in the proximity of the temple. Another more ancient myth narrates the story of Samba, the son of Lord Krishna. In *Samba Purana*, it is mentioned that Samba was cursed with leprosy by Lord Krishna and in order to cure the ailment he worshipped the Sun God on the bank of river Chandrabhaga after taking holy bath in the river as per the advice of sage Narada. After meditation with devotion for long twelve years, Samba constructed the temple on the bank of river Chandrabhaga (Rath et al., 2015). Similarly, there are various myths as well as reference in Odishan *Bhakti* poetry, related to the river Saradha in Puri.

Old Paintings and Photographs

The paintings made by James Fergusson (1837 CE) and rare photographs by William Henry Cornish (1890 CE) from British Gallery also depict water body in the vicinity of the temple (Rath et al., 2015; Jana et al., 2016). This suggests the ancient river Chandrabhaga existed proximal to the Sun Temple.

SCIENTIFIC DELINEATION OF TWO ANCIENT RIVERS

In the previous section, it has been mentioned that the available literature and folklore give the evidence of the possible existence of these two rivers in ancient times. This encourages to verify the existence of the two rivers based on scientific evidence. Hence, the current work, drawing from mythical and historical evidences for identification of approximate locations and antiquity of the streams, is based on an integrated study of ancient literature, historical evidences, satellite imagery study as well as near surface geophysics such as GPR.

Satellite Imagery Study

Satellite imagery reveals the presence of vegetation band which is one of the evidence of paleochannel trail of these two rivers (Figs. 2 and 3). In the False Color Composite (FCC) which is the combination of band 4, 3 and 2 of Landsat 7 ETM+ reveals vegetation as red, water bodies as dark blue or black and shallow water bodies or wet land as greenish blue. In ETM +Pan Mosaic, the paleochannel is more clearly visible in the vicinity of the Konark Temple (Fig. 2). Here the vegetation appears green embedded with blue color water patches. On the other hand, the vegetation index or Normal Difference Vegetation Index (NDVI) as well as water index (Modified Normal Difference Water Index or MNDWI) shows bright tone (indicates vegetation and water patches) in-between the two temples in Puri region (Jana et al., 2018). The embedded water bodies within the vegetation band are

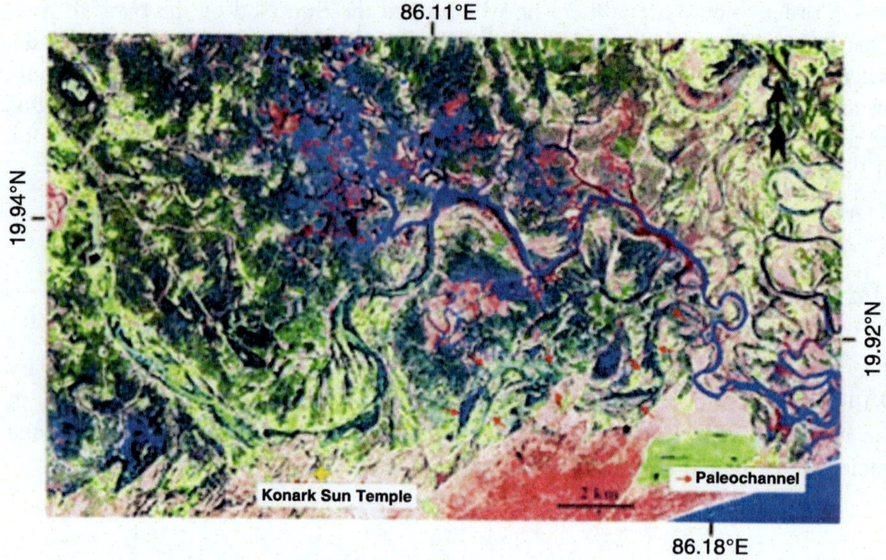

Figure 2. The ETM+Pan mosaic image shows paleo-channel trail marked by red arrows. Green coloured paleochannel along with its sinusoidal bend comprises blue patches of water bodies suggesting the presence of vegetation band embedded with remnant water bodies (Jana et al., 2016).

identified by the help of Linear Spectral Unmixing (LSU) method which quantifies the relative percentage of water components (Jana et al., 2018).

The spatial profiles (plotting of elevation against distance) of both the rivers show the depressed topographic outline which may represent the remnants of river valley.

GPR Survey

The presence of the paleochannel of these two rivers identified in the satellite imagery are evidenced in the field through Ground Penetrating Radar (GPR) survey along the selected profiles (Fig. 4).

As paleochannel is the remnant river valley filled with loose sediments, hence, there is a di-electric difference between the valley wall and the filled up sediments. GPR sends the electromagnetic waves which reflect back due to this di-electric difference and as a result various litho structures are seen in the subsurface imagery (Lunt et al., 2005; Conyers, 2012; Jana et al., 2016; Jana et al., 2018). Based on this principle, GPR survey was carried out in the Konark and Puri regions in the vicinity of Konark Sun Temple and Jagannath Temple, respectively. In Konark region GPR profile shows the subsurface river valley in the north east of the Sun Temple. On the other hand the subsurface river valley is seen along the Grand Road in eastern side of the Jagannath Temple.

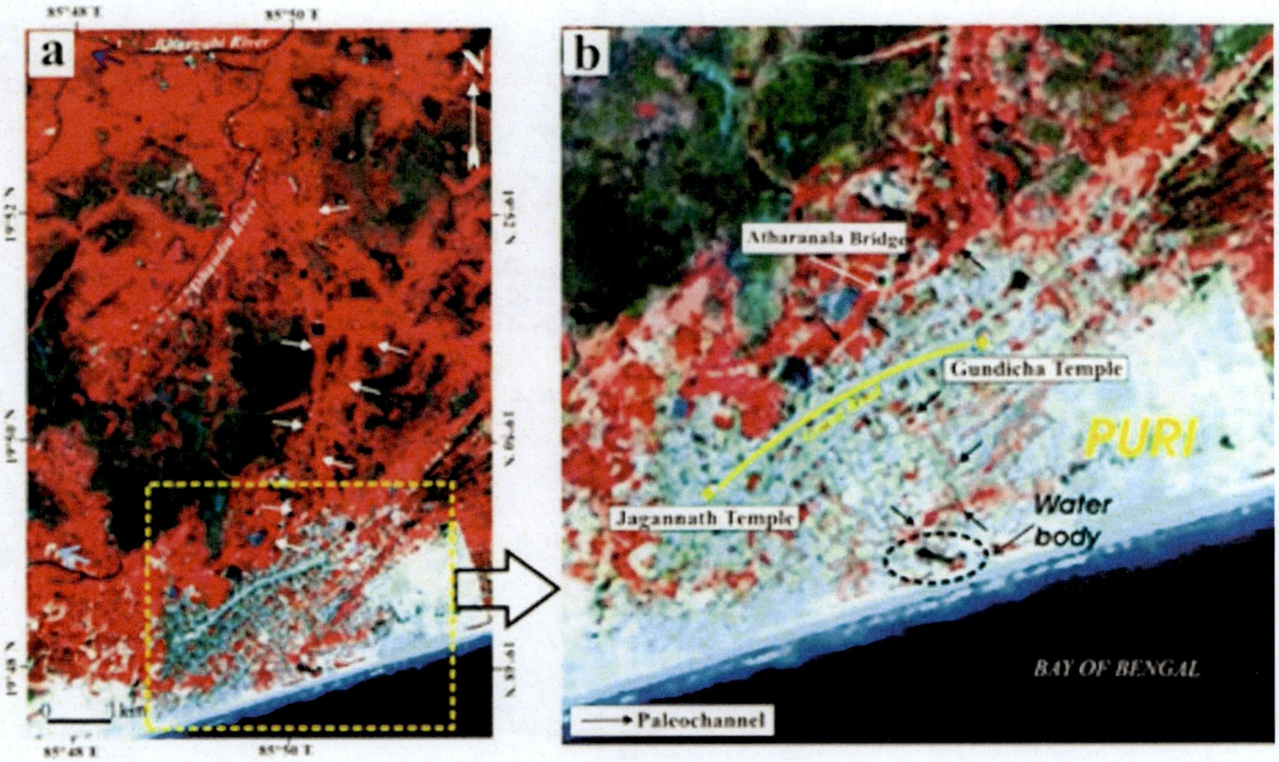

Figure 3. (a) FCC image of the Puri region shows red colour vegetation band as one of the evidence of paleochannel (white arrow) branching off from the Dhaudia River; (b) Curvilinear vegetation band in south of Grand Road within heart of Puri town is connected with remnant water body near sea beach (circle). Vegetation band and remnant water body suggest the trail of paleochannel (modified after Jana et al., 2018).

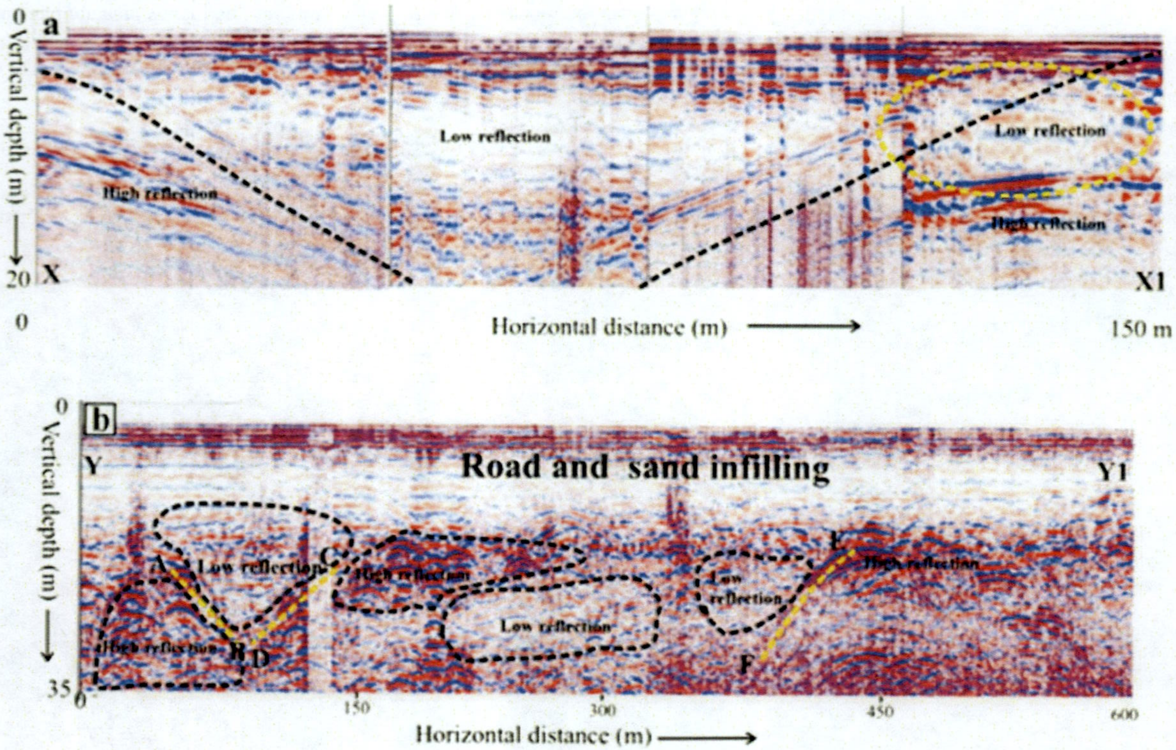

Figure 4. (a) Subsurface river valley along with shifting signs (yellow dashed circle) found in the north east of the Konark Temple. Here high reflection represents wet clay while low suggests loose sediments or sand; (b) Similar type of subsurface river valley is also seen along the Grand Road in the east of Jagannath Temple.

Field Validation

Field survey, carried out in the suspected paleochannel zones of Puri and Konark region revealed the presence of wet land covered with aquatic weeds in the vicinity of these ancient temples in Puri and Konark (Jana et al., 2016; Jana et al., 2018). In Tikarpada on the Konark Kakatpur road, the connection of the ancient river Chandrabhaga with the existing Kadua River proves the existence of this ancient river (see field locations A and B, Fig. 5). Similarly, the large water bodies which are suspected to be remnant of ancient Saradha river are located on north and south of the Grand Road in Puri. The water body in south of the Grand Road near Puri sea beach is found to be debouching into the sea and remnant river mouth is still existing in the sea beach (Fig. 6).

EVIDENCE OF EXISTENCE OF TWO ANCIENT RIVERS

Ancient literature and old folklore mention the names of these two rivers. However, scientific evidences reveal the existence of paleochannels which can be located on the places of these ancient rivers mentioned in the old text. Some of these evidences are sinusoidal river bend; variation in soil colour and moisture content; thick vegetation band as observed from the high value of NDVI. Greenish blue coloured water patches within the red colored vegetation band in FCC image of both Konark and Puri region (Figs. 3 and 5); low land surface temperature zone which is represented by low value in the Thermal Infra-Red (TIR) band. Other evidences are remnant river valley observed in the spatial profile; V-shaped subsurface river valleys which are seen from the GPR image in both Puri and Konark region. The river shifting marks are found in both GPR profiles and are represented by low reflection loose sediments (Fig. 4). LSU reveals the presence of embedded water bodies in the vegetation band in the northern part as well as within the Puri town (Jana et al., 2018). Thus, several evidences collected in both Konark and Puri area to delineate the paleochannel can be merged together on the GIS platform to view the highest probable zone of paleochannel along with paleochannel trend. Such type of data integration was conducted in Konark area and is described in the following section.

Layer Integration for Chandrabhaga River

In the earlier studies, data integration was conducted in various fields such as mineralogy, seismology, hydrology and even in archaeology also to get the probable zonation map (Mohanty et al., 2007; Mohanty and Walling, 2007, 2008; Rajani and

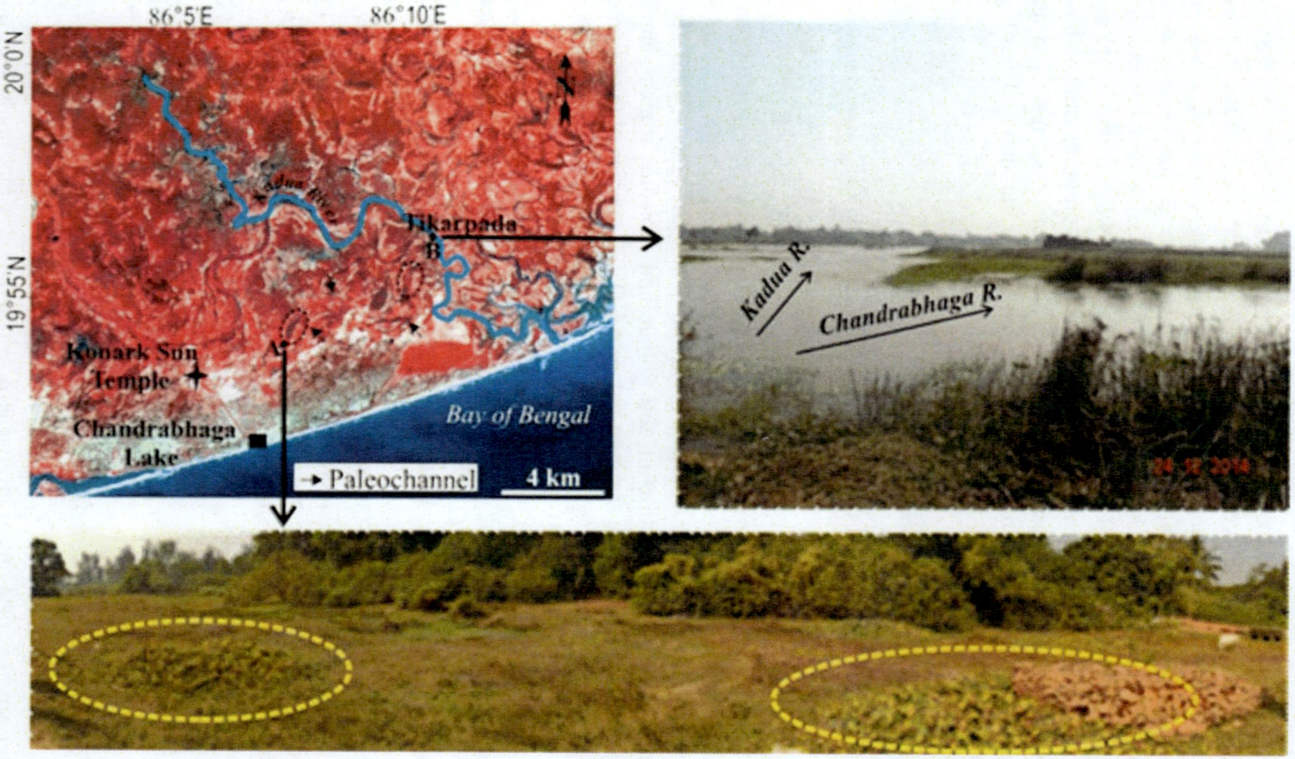

Figure 5. FCC map of the Konark region shows sinusoidal paleochannel trail (marked by arrow) along with bluish colour water patches (black circle). Two field locations such as location A remanant of past river as swampy land covered with aquatic weeds (yellow circles). Location B in Tikarpada shows the existing connection of Chandrabhaga and kadua Rivers.

Figure 6. Results of field validation. (**a**) Large water body near Atharanala covered with water hyacinth plants; (**b**) Remnant river mouth near Puri sea-beach that finally opens into the sea (taken from Jana et al., 2018).

Rajawat, 2011). Here different weights against each of the evidence are calculated based on the Saaty's Analytical Hierarchy Process (AHP) (Saaty, 1980; Jana et al., 2016) Similarly, different evidence of paleochannel (each represent as thematic map) like thermal, elevation, vegetation, gravity anomaly, geology/lithology and tectonic structures found in Konark area are integrated to get final probable map of paleochannel (Jana et al., 2016). Relative importance of the six evidences and details of the calculation of their respective weights are described in Jana et al. (2016). However, the reasons behind taking each of the evidences as thematic layer is described below.

Thermal: As paleochannel is characterized by high moisture content zone than the adjacent region, so the land surface temperature on that region will be relatively low than adjoining area and thus thermal parameter is the most important evidence in detecting the paleochannel. Here low temperature zone is given more priority than the high temperature zone.

Elevation: The paleochannel is an old river valley which is characterized by low elevated river bed and high elevated river banks. Here low elevated zone which may represent the depressed zone of the river valley is given more priority.

Vegetation: Vegetation is an important factor in delineating paleochannel zone. It grows along the channel due to availability of water and organic nutrients. In the study area we got sinusoidal vegetation band that can be interpreted as natural growth along the channel. The vegetation is marked by the vegetation index or Normal Difference Vegetation Index (NDVI) in the satellite imagery. The high NDVI value which represent dense vegetation is given more priority than the low NDVI value.

Gravity Anomaly: The study region shows alternate ridges and depression in the subsurface structures. Depression zone is represented by low gravity anomaly possibly due to deposition of low density sediments. These sediments can be carried down by the rivers and deposited in the depression zone. Thus, low gravity anomaly can be another evidence in delineating the sedimentary deposit and the presence of ancient river.

Geology/Litho structures: The region comprises alluvium which is one of the characteristic deposit of the rivers flowing through this region. Thus, alluvium is prioritized more than other litho structures which are exposed in different places.

Tectonic structures: The region is traversed by alternate subsurface ridges and depressions which trend along NE-SW. Our delineated channel also trends along that direction. As depression zones are relatively weak zones, so rivers will get easy passage to flow and thus have the tendency to flow through these weak zones. That is why depression zones are given more priority than the ridge areas. However, these subsurface structures, especially depression zones, are overlain by thick sediments, so there still remains an uncertainty whether these depression zones at such depth are responsible for giving easy passage to the channels. Hence, these tectonic structures including ridges as well as depressions are taken as the less important evidence and given low weightage value (Jana et al., 2016).

Mathematically all the integrated evidences or thematic maps along with their weights can be shown as (Jana et al., 2016):

$$
\begin{aligned}
&[\text{Thermal}]*0.2857 + [\text{Elevation}]*0.2381 + [\text{Vegetation}]*0.1905 \\
&+[\text{Anomaly}]*0.1429 + [\text{Geology}]*0.0952 + [\text{Tectonic}]*0.0476 = \\
&\text{Weighted Sum/Final Output Map}
\end{aligned}
$$

Here weights which are obtained from Analytic Hierarchy Process (AHP) are multiplied with each of the thematic map and are integrated all together to get the output map. This output map represents the probable paleochannel zones where 1 represents the highest probability and 0.115 represents the lowest probability of the presence of paleochannels. This highest probability of paleochannel zone is matched well with the Google Earth image and the obtained paleochannel trail is found to extend towards east and west on the north of the Sun Temple (Jana et al., 2016). However, this obtained probable paleochannel zone does not comprise the existing Chandrabhaga Lake in the south of the temple. Nevertheless, the possibility of the original flow path from the identified paleochannel, into Bay of Bengal through the present site of Chandrabhaga Lake cannot be completely overruled (see question-marked dashed line, Fig. 7(a)). The presence of sand dunes in the intervening region between the identified paleochannel and the lake suggest that probably small yet unidentified connecting channels are almost completely covered with these sand dunes. Therefore, the possibility of a smaller and as yet unidentified channel passing through the lake from the main river cannot be discarded entirely. It is worth mentioning that near the coast, rivers change course frequently; this possibility, therefore, cannot be completely negated on the basis of the present evidence. Similarly, in Puri region, based on the observations and evidences, it can be stated that delineated paleochannel was the branch of Dhaudia river which itself is the branch of Bhargabi river. The delineated branch of Dhaudia river flowed towards south through the Puri town to debouch into the Bay of Bengal and the location of this delineated channel match well with the location of the ancient Saradha mentioned in old texts (Fig. 7(b)). This study is an illustration of a multidisciplinary and convergent approach to scientifically validating mythical, textual and oral references to the existence of physical structures and formations which currently are not visible. It is expected that such studies would hold significance in terms of connecting disciplines and revalidating history through scientific studies.

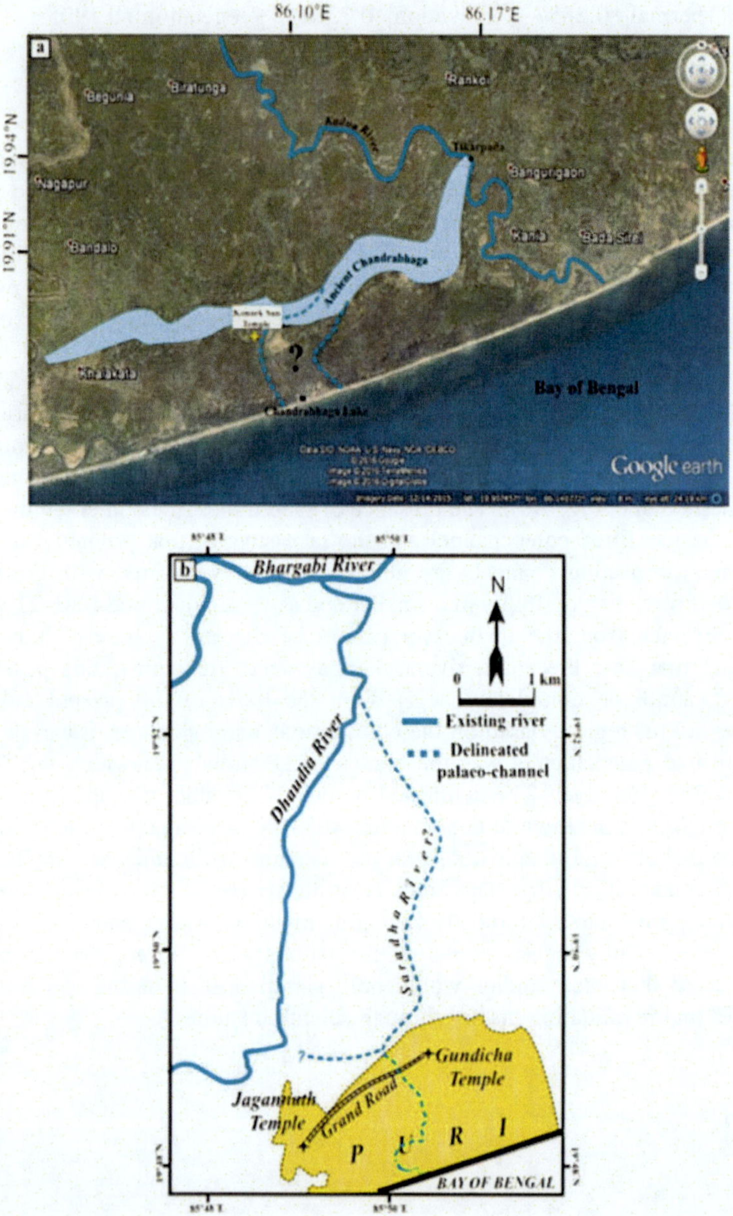

Figure 7. (a) The probable trace of Chandrabhaga paleochannel (blue area), trends almost parallel to the coast as obtained from high probability zone. Paleochannel (blue dashed line and question mark) show a possible, but as yet unconfirmed flow path through the present site of Chandrabhaga Lake (from Jana et al., 2016); (b) The delineated palaeochannel of the Saradha river (blue dashed line), branching off from the Dhaudia River and flowing towards south through Puri town. Delineated palaeochannel (Saradha River) divides the Grand Road which connects the Jagannath and Gundicha Temples (from Jana et al., 2018).

References

Behera, L., Sain, K. and Reddy, P.R. (2004). Evidence of underplating from seismic and gravity studies in the Mahanadi delta of eastern India and its tectonic significance. *Journal of Geophysical Research*, 109: 1-25.

Conyers, L.B. (2012). Interpreting Ground-penetrating Radar for Archaeology, California, Left Coast Press, Inc.

Jana, S., Mohanty, W.K., Gupta, S., Rath, C.S., Behera, R.R. and Patnaik, P. (2016). Multi-pronged search for paleo-channels near Konark temple, Odisha – implications for the mythical river Chandrabhaga. *Current Science*, 111: 1387-1393.

Jana, S., Mohanty, W.K., Gupta, S., Rath, C.S. and Patnaik (2018). P. Palaeo-Channel Bisecting Puri Town, Odisha: Vestige of the Lost River 'Saradha'?; *Curr. Sci.* 115(2): 300-309.

Kulke, H. and Tripathi, G.C. (Allahabad, 1987). Katakarajavamshabali: A Traditional History of Orissa, p. 78.

Lunt, I.A., Hubbard, S.S. and Rubin, Y. (2005). Soil moisture content estimation using ground-penetrating radar reflection data, *Journal of Hydrology,* 307: 254-269.

Mahalik, N.K., Das, C. and Maejima, W. (1996). Geomorphology and Evolution of the Mahanadi Delta, India. *Journal of Geosciences*, 39: 111-122.

Mishra, P. Kapila (2005). Samhita: Text with English Translation and Critical Study New Delhi, New Bharatiya Book Corporation, p. 316-319.

Mohanty, A. (1969) Madala Panji. Third Ed., Prachi Publication p. 26.

Mohanty, B. (ed.). (1978). Nilachakre Ho Dekha Uduchi Bana, Prachina Odia Kabyadarshana (Odisha Sahitya Akademi), Bhubaneswar, p. 61

Mohanty, W.K. and Walling, Y.M. (2007). Seismic Hazard in megacity Kolkata, India. *Natural Hazards*, DOI:https://doi.org/10.1007/s11069

Mohanty, W.K. and Walling, Y.M. (2008). First order Microzonation of Haldia, Bengal Basin (India) using a GIS platform, *Pure and Applied Geophysics,* 165(7):1325-1350.

Mohanty, W.K., Walling, Y.M., Nath, S.K. and Pal, Indraji (2007). First Order Seismic Microzonation of Delhi, India Using Geographic Information System (GIS), *Natural Hazards,* 40: 245-260.

Nayak, A. (1998). Bhaktara Jagannatha. S.B. Publications, Binod Bihari, Cuttack, p. 119-120.

Nayak, G.K., Rao, C.R. and Rambabu, H.V. (2006). Aeromagnetic evidence for the arcuate shape of Mahanadi Delta, India. *Earth Planets Space*, 58: 1093-1098.

Parameswaranand, S. (2001). Encyclopaedic Dictionary of Puranas, New Delhi, Swarup & Sons., 1: 320-321.

Rajani, M.B. and Rajawat, A.S. (2011). Potential of satellite based sensors for studying distribution of archaeological sites along paleochannels: Harappan sites a case study. *Jour. of Archaeological Science*, 38(9): 2010-2016.

Ramana, M.V., Nair, R.R., Sarma, K.V.L.N.S, Ramprasad, T., Krishna, K.S., Subrahmanyam, V., D'Cruz, M., Subrahmanyam, C., Paul, J., Subrahmanyam, A.S. and Chandra Sekhar, D.V. (1994). Mesozoic anomalies in the Bay of Bengal. *Earth and Planetary Science Letters*, 121: 469-475.

Rath, C.S., Behera, R.R., Jana, S., Patnaik, P. and Mohanty, W.K. (2015). Exploring the Lost River (s) at Konarka: A Multi-Disciplinary Approach. *Odisha Review*, 72: 92-96.

Tagare, G.V. (1994). The Skanda Purana, Motilal Banarasidas Publisher, Part-5, p. 31-112.

Saaty, T.L. (1980). The Analytical Hierarchy Process, New York, McGraw-Hill.

Chapter 22
Changing River Courses in Bengal (1780 to 2020)

Kalyan Rudra

The undivided Bengal is a linguistic region which incidentally synchronises with the Ganga-Brahmaputra-Meghna (GBM) Delta covering an area of more than 2×10^5 km^2. The region was politically divided in 1947; while the eastern part went under territorial jurisdiction of erstwhile East Pakistan (now Bangladesh), the western counterpart emerged as a State of India and was named 'West Bengal'. The vast alluvial plain with numerous wetlands, patches of relatively high lateritic tracts, intricate drainage network and many islands being separated by labyrinth of creeks flowing to the sea have given Bengal an identity making it a distinct geomorphic unit (Rudra, 2018). Since the country was divided based on religion, the international boundary was drawn across 54 rivers which continue to flow downstream from West Bengal to Bangladesh. The bright sunlight, fertile alluvial tract and huge water resources have had made Bengal agriculturally productive. Bernier, a French traveller who came to India in 1656 and 1668, described Bengal as more productive than Egypt (Smith, 1916). Abul Fazal, a member of court of emperor Akbar (1556 to 1605) saw floodplain of Bengal producing three crops in a year and noted that a single grain of paddy could produce two-three seers (1 seer= 0.933kg) of rice (Gladwin, 1800). This prosperity of Bengal attracted many merchants from European countries since the early 16th century and was described as 'Paradise of India' (Campos, 1919). The riverine trade and agrarian economy of Bengal flourished due to its river system and has been proverbially described as "Nadimatrik" (meaning land nourished by rivers) in local dialect.

The rivers in this newly built delta are very dynamic and have changed their courses dramatically during preceding centuries. The writing of a dossier on changing river courses of Bengal is difficult as it encompasses many allied subjects like Geology, Geomorphology, Archaeology, History, Palaeonotology, Cartography and

K. Rudra (✉)
West Bengal Pollution Control Board, Department of Environment, Govt. of West Bengal, Kolkata, India

A. Mukherjee (ed.), *Riverine Systems*, https://doi.org/10.1007/978-3-030-87067-6_22

Remote Sensing; Even travelogues, especially those by the foreign merchants, may throw light on the history of rivers. The first scientifically correct survey of Bengal was conducted during 1764 to 1777 by James Rennell, the first Surveyor General of East India Company and published as "A Bengal Atlas" (Rennell, 1781). The province of Bengal in the second half of the 18th century included parts of present Bihar, Jharkhand, Odisha and Assam. All maps published by foreign travellers prior to Rennell were sketches without any correct scale and co-ordinates. The maps by Rennell and his associates were so accurate that those can be geo-referenced and compared with modern maps and images. This was the reason for accepting 1780 as the base year to initiate the present discussion. But comparison of this base map with the satellite image of 2020 does not reveal the changes which took place in-between. The most authentic maps published after 1780 were topographical sheets published by the Survey of India in the second decades of the 20th century. The survey continued till 1970s in phases. Many of these maps are not available in public domain due to obvious restriction imposed by Government of India. Still some old maps were procured and digitised to create an interim layer. Though published in different years, these maps still help to appreciate the dynamics of rivers.

The river research either in India or in Bangladesh had never been given the heed it demands. The flowing rivers have been looked upon as the stocks of water being exploited ruthlessly even making the downstream stretches dry. The ecological services rendered by rivers were grossly overlooked by the planners. This seems to be a colonial legacy. We embanked the rivers for flood control and hardly allowed them to oscillate within its meander belt nor we like them to overtop its bank to deposit fertile silt on the floodplain. The British left India in 1947 but old philosophy of command and control over the nature continues. The issue of holistic eco-hydrology never grasped the mind of planners (Rudra, 2015). The rivers in the soft terrain of Bengal are inherently dynamic and change their courses either by avulsion or by lateral oscillation of meander bends. While most cases of avulsion of rivers had been governed by neo-tectonics of Bengal basin, the lateral swing of river is the tendency to achieve dynamic equilibrium coping with fluctuating discharge. Both the events took place in the monsoon months when discharges in the rivers gradually exceeded a threshold limit. It is observed that wavelength and amplitude of a meandering river changes with the increasing volume of water. The large rivers like Ganga, Jamuna (Brahmaputra), Meghna, Teesta, Jaldhaka, Torsa, Damodar achieve bankful stage during the monsoon and shrink in width during lean months when they flow through the narrow channels within their wide valleys.

The rivers of Bengal can be divided into five groups, viz. (a) rivers of North Bengal, namely Mahananda, Teesta, Jaldhaka and Torsa, flowing through Barind tract between the Ganga and the Jamuna; (b) Ganga and its distributaries; (c) western tributaries to the Bhagirathi-Hugli river; (d) Jamuna-Meghna system and (e) the tidal creeks of coastal tract. All these rivers have changed their courses appreciably in many ways since the late 18th century. One needs to have in-depth understanding of the Holocene evolution of the Bengal basin to appreciate the dynamics of rivers.

BASIN FILL HISTORY

The area encompassed by the Chhotagpur plateau in the west, the Rajmahal-Meghalaya hills in the north, Manipur-Tripura-Chittagong mountains in the east and the Bay of Bengal in the south is known as the 'Bengal Basin'. The underlying structure of the basin was gradually known to the scientists since 1950s when exploration in search of petroleum started. Subsequently, researches conducted by Morgan and McIntire (1959), Sengupta (1966), Allison et al. (2003) added further knowledge.

The origin of the Bengal basin dates back to 145 million years BP, when a broken part of the Gondwanaland confronted with the SIBUMASU (Siam, Burma, Malaysia and Sumatra) and the Asian block. Thus, the basin was formed due to a soft collision (59-44 million years BP). The underlying basement of the basin is tilted towards east and central part in Bangladesh has nearly 22 km. thick Cretaceous-Holocene sedimentary successions (Alam et al. 2003). A hinge zone having a width of 25 km is aligned in north-east direction from the subsurface of Kolkata to Mymansingh and divides continental and oceanic parts of the Indian plate. The dip of the basement changes (from 2-3° to 6-12°) along the hinge which also differentiate the stable shelf province of north-west from the central deep basin province (Uddin and Lundberg, 2004). Geologist have identified a series of *en-echelon* faults along the western edge of the Bengal basin and a series of compression-faults in Indo-Burma mountain ranges.

The land building in the delta was initiated by the western tributaries to the Bhagirathi-Hugli river along the eastern front of the Chhotanagpur plateau. The palaeo/sub-deltas built by those tributaries (Ajoy, Mayurakshee, Damodar, Rupnarayan etc.), have formed the Rarh plains along the oldest strandline, which extended from Digha to Nabadwip (Agarwal and Mitra, 1991). The land building in Doors and Tarai region started 20 million years BP in Miocence period (Banerjee and Sen, 1987). The post-Pleistocene melting of the Himalayan glaciers and consequent increasing sediment load in rivers had been the cardinal factors inducing early Holocene delta development. Since 12,000 years BP, substantial increase of snow-melt water and sediment load in the Himalayan rivers led to the formation of two mega-fan, viz. the Barind tract and Madhupurgarh. At the same time, the south-west monsoon established over Indian Subcontinent and the sediment load in the GBM system increased to 2.5 billion tonnes per annum (compared to present 1 billion tonnes per year). The sea level rise and accelerated deposition helped the growth of many Holocene deltas including the Ganga-Brahmaputra-Meghna delta (Stanley and Warne, 1994).

Many scholars opined that the deposition of sediment layers facilitating delta building has had been governed by fluvial process (Oldham, 1870; Majumder, 1942; Bagchi, 1944). But it should be described as fluvio-marine processes as the ingress of silt-laden water from the sea into the estuaries plays most dominant role. It is observed in the coastal tract of Bengal that silt-laden water spills over the inter-tidal space twice in 24 hours leaving behind layers of sediment on the floodplain. As the coast receded southwards during last 20 million years, the processes of accretion also shifted and it is now active in the inter-tidal space in the coastal front (Rudra, 2018).

DYNAMIC RIVERS

The Ganga and Brahmaputra (known as Jamuna in Bangladesh) enter Bengal through the 200 km. wide passage known as 'Rajmahal-Meghalaya gap'. The Surma and Kusiyara, two headstream of Meghna approaches Sylhet plain from the north-east hilly tracts. The three mighty rivers have a combined catchment of more than 17×10^6 km^2 and carry 1375 BCM water annually. These three rivers with their numerous tributaries and distributaries have created a complex drainage network. The soft and flat alluvial plains of Bengal render the rivers opportunity of free meander swing and avulsion. The cardinal factor governing the changes in the courses of the Ganga, the Brahmaputa and the Teesta is believed to be the subsidence of land in central Bengal. While Hirst (1915) presumed existence of a subsiding trough between Jalpaiguri and Barisal and attributed it to the changing courses of the Teesta, the Brahmaputa and the Ganga, Morgan and McIntire (1959) identified a sinking trough along the course of Jamuna between Dhubri and Goalundo. This trough is possibly governed by the existence of a deep-seated fracture, or by the excessive load of overlying sediment. The two mighty fans, Barind and Madhupurgarh lying along both sides of Jamuna are treated as areas of compensatory upheaval or terraces (Rizvi, 1957). The subsidence and upheaval of alluvial plain alters the topographic slope and may compel the rivers to change their courses and this happened in Bengal during proceeding few centuries. Thus, dynamics of the rivers had been governed by the neo-tectonic as well as the stratigraphy of the Bengal Basin. The peak flow in rivers during the August and September also play important role.

Changing Courses of the Ganga

The Ganga enters the territorial jurisdiction of West Bengal after dashing the Rajmahal hill and flows for about 210 km. before it crosses the Indo-Bangladesh border near Jalangi town and flows further 135 km. upto Goalundo where it meet the Jamuna. The combined river continues to flow south-east for 115 km to meet the Meghna at Chandpur and finally discharges into the Bay of Bengal through a mighty estuary. This 460 km stretch of the Ganga-Padma between Rajmahal and Chandpur has changed appreciably. One significant change is the decay of left-bank distributaries, namely Kalidri, Chhoto Bhagirathi and Pagla in Malda which have been beheaded and gone dry. The Boral which took off from the Ganga at Sarda in Bangladesh and flowed eastward through Chalan bill receiving Karatoya at Sazadpur and rejoined again the Ganga at Jafferganj is now moribund. An anastomosing channel taking off at Pabna no longer exists. The Dhaleswari which had been an anabranching channel of the Ganga has turned into a branch of Jamuna since 1830. The Rajmahal, Gaur, Pandua and Tanda had been four capital towns of Bengal during 1203 to 1608 AD. Rajmahal had also been administrative headquarters of Bengal from 1639 to 1660 AD in its second term. The capital towns shifted to cope with the changing course of the Ganga and its distributaries. The navigation route to Gaur and Tanda was through the Chhoto Bhagirathi which had been a navigable channel but now remains fordable (Fig. 1). But when Rennell surveyed Bengal during 1764-1777, the Ganga had migrated far away from Gaur. Rennell (1793) in his memoir noted "*No part of the site of ancient Gour is nearer to the present bank of*

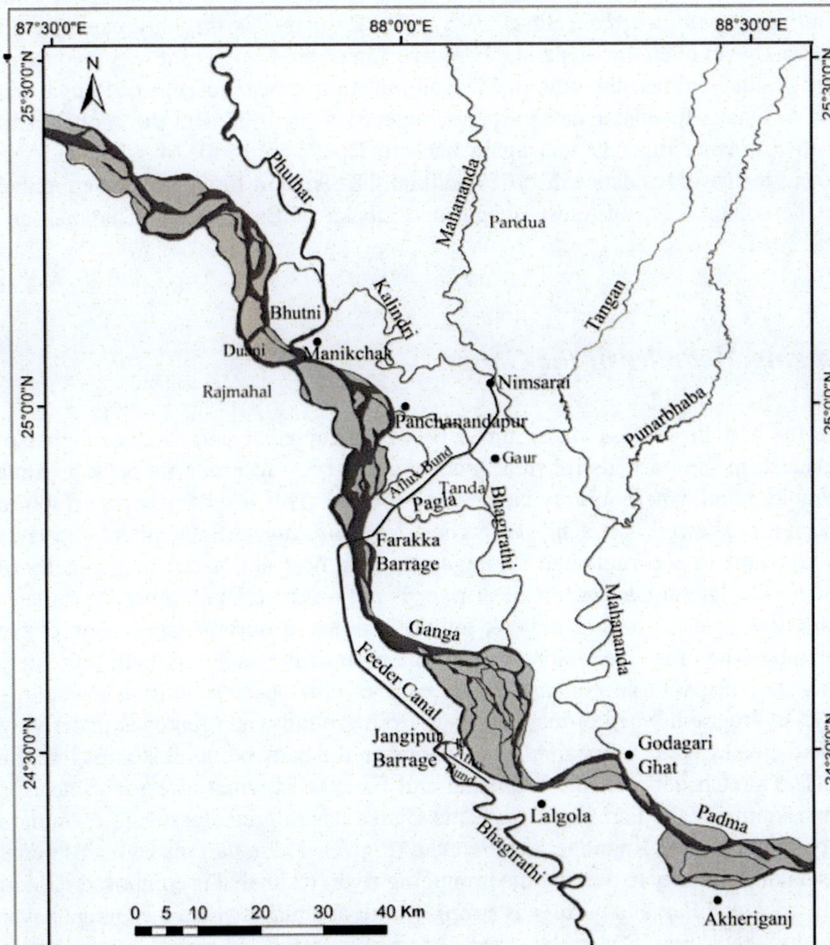

Figure 1. The Ganga and old capitals of Bengal.

Ganges, ..., are now 12 miles from it". The navigation to and from Gaur was favoured by its location on interfluves delineated by Kalindri in the north, Mahananda in the east and Chhoto Bhagirathi along south-west. Both the Kalindri and Chhoto Bhagirathi carried substantial flow when Gaur was the capital of Bengal. Sarkar (1973) opined that bulk of water from the Ganga flowed through Kalidri in the 13th century and noted "*Leaving the hills of Rajmahal the Ganges seemed to have passed northwards through modern Kalindri, and then southwards into the lower course of the Mahananda, east of the ruins of ancient Gaur*", i.e. the portion of the Mahananda below its confluence with the Kalindri was part of the Ganga. The Ganga flowed through these two channels in the past and nourished three capital towns—Gaur, Tanda and Pandua. The decline of Gaur was attributed to the "*severe earthquake in 1505 AD and shortly after it, the Ganges left its old course past Gour and retreated southwards*" (Hirst, 1915). The Rajmahal is the only capital town which had never been threatened by bank erosion of the Ganga as it was favourably located on the hard basaltic outcrop. The administrative headquarters of Bengal was shifted to Dhaka for better navigational connectivity in 1608. But the capital came back to Rajmahal in 1639 and again back to Dhaka in 1660. Murshidabad was chosen as the royal residence in 1702 and finally Kolkata in 1757. Kolkata continues to be the capital town of West Bengal and Dhaka of Bangladesh (Chapman and Rudra, 2015).

Changing Meander of the Ganga

While the Ganga avulsed many times between Rajmahal and Godagari ghat of Bangladesh in the past, its inherent tendency has been to erode its bank at some vulnerable points where mighty bends have formed. This discharge hydrograph of the Ganga is skewed and achieves highest level in August-September. The river erodes its bank to accommodate the huge monsoon flow and alters the geometry of meander. The lateral encroachment of river is not unidirectional rather it oscillates within wide meander-belt. The bank failure is initiated during rising stage of the hydrograph when the river starts to scour the base of the shelving bank ultimately causing its collapse. The collapse of the bank is also observed during the falling stage of hydrograph when effluent seepage from ground water removes underlying sand and creates void. A linear crack develops on the bank before it slumps into the river. The stretch between the Rajmahal and Farakka is worst affected. Since the commissioning of the Farakka barrage the Ganga has migrated eastward forming a mighty bend between Rajmahal and Farakka (Fig. 2). These two places act as nodal points having stability to prevent the impinging river. Rajmahal is composed of hard basaltic rock and Farakka barrage is strongly built at a place where the Ganga never changed its bank-line. But in the stretch lying in-between, the river has swallowed more than 200 km^2 fertile land and many villages displacing more than 0.10 million people (Rudra, 2010). The erosion-victims settled either on the newly emerged *char* or *shoal* along opposite bank or migrated inland to find new shelters. The barrage

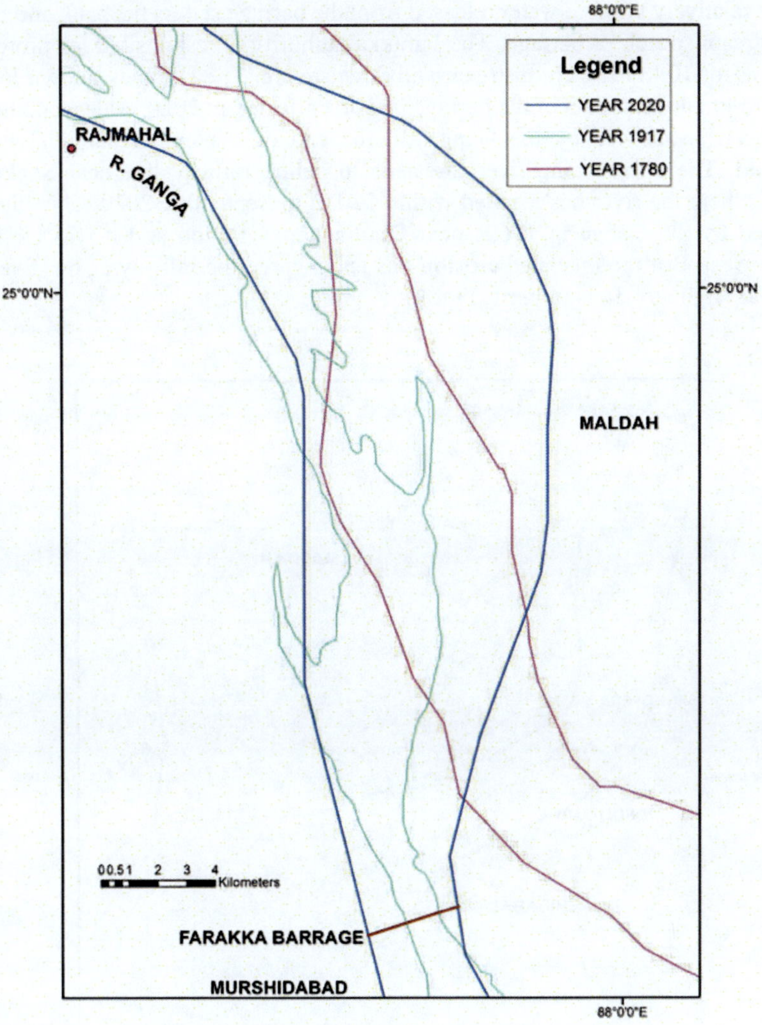

Figure 2. The Ganga between Rajmahal and Farakka.

impounds more than 87 million m^3 water and accelerates sedimentation reducing the cross-sectional area and water-holding capacity. The Ganga has so altered its course that its flow is no longer co-axial to the barrage. In the 80 km stretch between the Rajmahal and Farakka barrage, the water flows through three other channels barring the main artery. These channels are separated by char-land lying in-between and again united in the Farakka pond. The oblique flow leads to swelling of water along southern bank and each bay of the barrage having discharging capacity of 500 cusec cannot render passage to the huge volume exceeding threshold limit. The impounded water strikes left bank causing bank failure.

The relatively silt-free water released from the barrage scours the bank and bed in downstream stretch of barrage. The Farakka authority is responsible for protecting the bank for 10 km stretch upstream and downstream. The Dhulian, a town located about 15 km downstream of the barrage is worst affected in 2020. At least 43 houses have been swallowed by the Ganga till the end of September and 263 houses damaged. The encroaching river threatens to delink railway line near Sankopara station where the river has reached within 220 m in September 2020. This line was engulfed by the Ganga in 1960s near Dhulianganga station and it was realigned further west. But the unabated erosion has endangered the railway again. Even the National Highway 34 is in peril (Fig. 3).

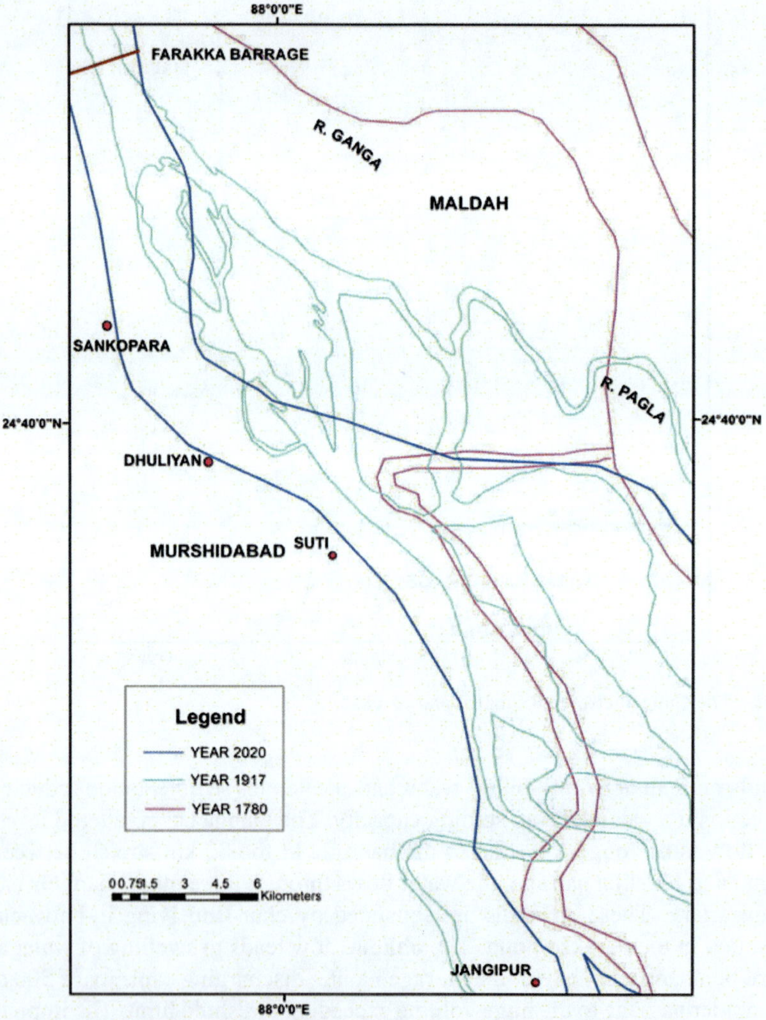

Figure 3. The Ganga between Farakka and Jangipur.

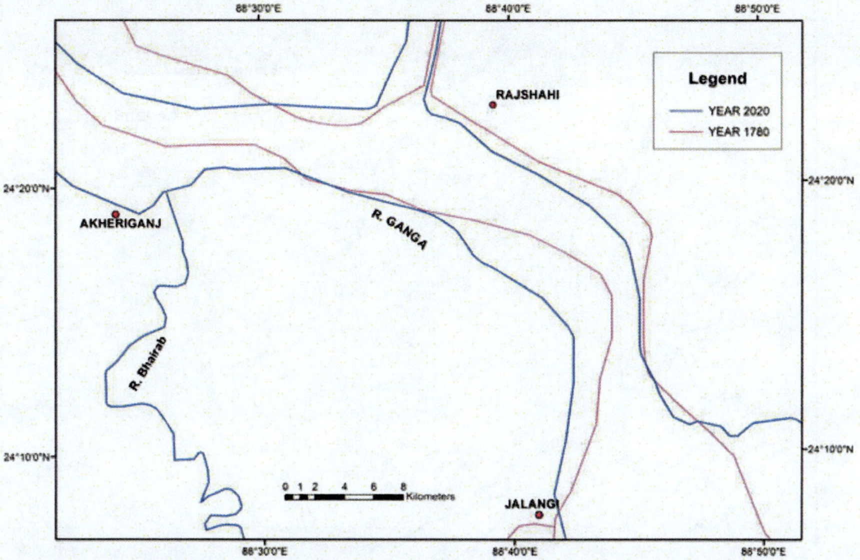

Figure 4. The Ganga between Akherigang and Jalangi.

There are four vulnerable points in Murshidabad where the Ganga has formed large bends are Suti, Mithipur, Akherigang and Jalangi (Fig. 4). These sites were severely affected between 1980 and 2000. The Ganga followed Indo-Bangladesh border in 1947 but continuously migrated towards Indian territory at three vulnerable points except at Suti where the river migrated towards Bangladesh. The possession of *char* emerged along the opposite bank had been an issue between security forces of both the countries. The southward encroachment of the Ganga in the Murshidabad district was not unilateral and it again migrated towards north (i.e. Bangladesh). In fact the river oscillates laterally within a stretch having a width of approximately 10 km. There are some nodal points where the river remained more or less stable. The population displacement, loss of agricultural land created a class of neo-refugees and the smuggling emerged as an organised business along the Indo-Bangladesh border (Rudra, 2006). The erosion and consequent homelessness has been an age-old problem but large-scale human migration from Bangladesh (erstwhile East Pakistan) since 1947 changed the geography of the river-front of West Bengal (Colebrooke 1801, Sherwill 1858, Rudra 2018). The situation is no exception in Bangladesh. In its lower 135 km stretch in Bangladesh between Jalangi and Goalundo, the Ganga or Padma is so aggressive that it is known as *kirtinasha* or destroyer of creation. The most drastic change in river courses happened after 1830 when the Ganga altered its outlet towards the Bay of Bengal through Arialkhan and opened a new route widening a narrow channel which met Meghna opposite to Chandpur. This happened after unification of the Ganga and Jamuna at Goaludo. The combined flow widened the channel and opened a more convenient outlet through the Meghna estuary. The other dramatic change which took place simultaneously was the

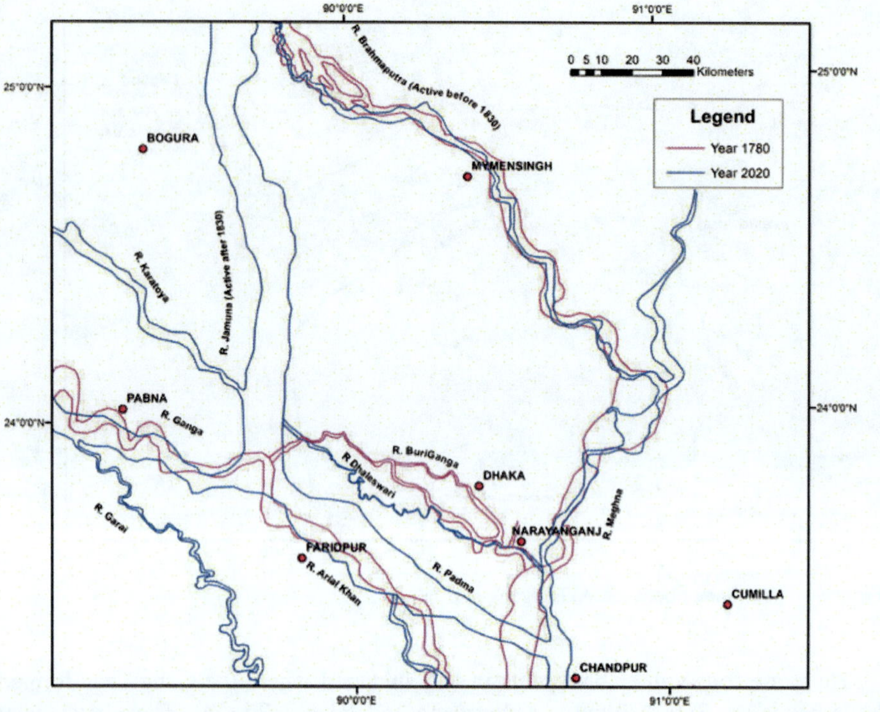

Figure 5. Changing courses of the Ganga and Jamuna after 1830.

disconnection of the Dhaleswari from the Ganga. It was a distributary of the Ganga connecting Meghna. But while changing its course, the Jamuna appropriated Dhaleswari as a link channel towards the Meghna (Fig. 5).

EASTERLY FLIGHT OF WATER

The Holocene eastward tilt of the delta empowered the Ganga to deepen and widen its valley. The incision of bed and eastward flight of flow through the Padma sealed the scope of outflow of water from the Ganga into the formerly active distributaries, namely, the Bhagirathi, Bhairab, Jalangi and Mathabhanga (Rudra, 2018; Fig. 6). The diminishing upstream water in the Bhagirathi caused decay of its distributaries, namely, the Saraswati, Bidyadhari, Jamuna and Adiganga which had been referred in mediaeval Bengali literatures. Those channels were found sluggish in the second-half of the 18th century (Rudra, 2020).

It is opined by many scholars that the Bhagirathi-Hugli river had been main outlet of the Ganga water and gradual eastward flight of the water towards Bangladesh got a momentum since the 12th century AD (Oldham, 1870; Hirst, 1916; Majumder,

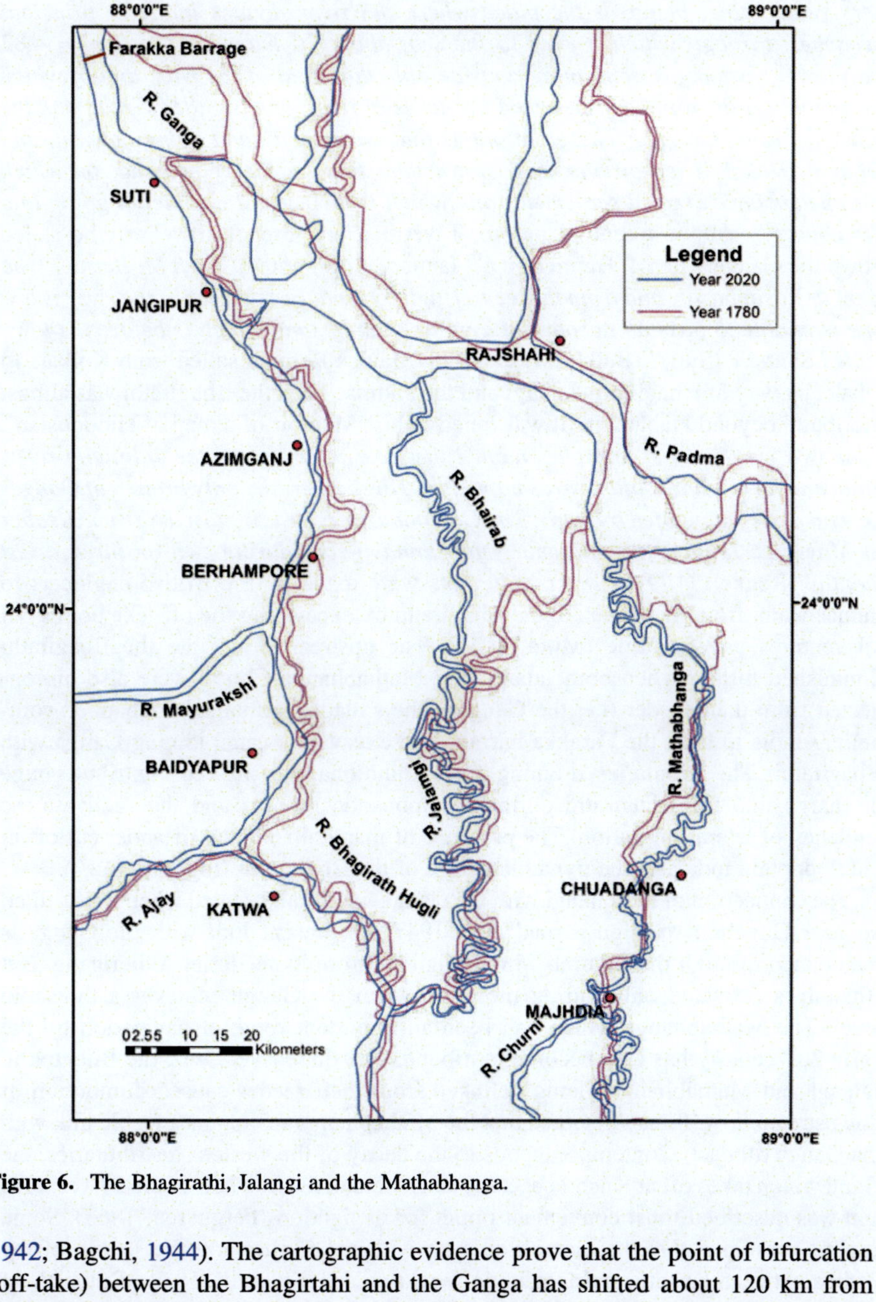

Figure 6. The Bhagirathi, Jalangi and the Mathabhanga.

1942; Bagchi, 1944). The cartographic evidence prove that the point of bifurcation (off-take) between the Bhagirtahi and the Ganga has shifted about 120 km from Rajmahal to Mithipur (in Murshidabad district). J.B.B. d'Anville (1752) while describing the geography of Rajmahal noted, "*Its situation was on the Ganges is very remarkable, being at the place where the river divides into two principal branches, through which it runs to the sea, about 70 leagues* (1 league = 4.82 km) *lower, forming a delta more considerable than that of the Nile, and of which Raji-mohol is the top.... Of these two branches of river, one is called Great, and the other the Little Ganges: The great one is that on the left hand going down it, and leads to Daka: It is however less known than that on the right hand, on which Europeans have erected settlement; and which is their usual channel to go up into the country*". About a century before, Tavernier, and French traveller who sailed down the Ganga with M. Bernier on 6th January, 1666 noted, "*I left M. Bernier who went to Kasimbazar, and from thence to Hugli by land, because when the river is low one is unable to pass on account of a great bank of sand which is before the town called Soutque* (Suti)" (Ball, 1889). In 1764, James Rennell sailed from Kolkata to Dhaka by way of Bhagirathi-Jalangi and the Padma. Then the Bhagirathi was almost moribund beyond Nabadwip. It was noted in his 'Memoir of a map of Hindoostan', "*The two westernmost branches named the Cossimbazar and the Jallinghy river, unite and form what is afterwards named the Hugli river, the only branch of Ganges i.e. commonly navigated by ships. The Cossimbazar river is almost dry from October to May and Jallinghy is in some years unnavigable during two or three driest months*" Rennell (1793). Several attempts were made by the British engineers to induce water from the Ganga to the Bhagirathi by excavating the off-take points but all ventures proved futile (Mitra, 1953). The upstream flow into the Bhagirathi diminished further when both Jalangi and Mathabhanga-Churni were also disconnected from their feeder (i.e. the Ganga). The situation remained unchanged compelling India to built the Farakka barrage and excavate a canal joining Ganga with Bhagirathi. The tributaries draining from Chhotonagpur plateau contribute huge discharge into the Bhagirathi during the monsoon months and thus activate the tendency of lateral migration. The presence of many abandoned meander cut-off in the floodplain indicates the dynamic nature of the Bhagirathi (Fig. 7(A to F)).

The connection of the Jalangi with the Ganga was finally sealed when it was filled up near Gopalpur to align a road after 1947. Its present link with the Ganga is maintained through the Bhairab which originate from its feeder at Akherigang. But Bhairab is activated only during monsoon when the Ganga achieves a threshold level. The Mathabhanga which had been an important route of navigation till the early 20th century has also become moribund. The three rivers, viz. the Bhagirathi, Jalangi and Mathabhanga, being delinked from their feeder caused diminution of downstream flow. It may be noted that the Mathabhanga maintains a feeble link with the Ganga till date. Contemporaneous to the decay of the western distributaries, the Garai which takes off at Kushtia became active and in the first half of the 19th century and was described most convenient outlet for navigation (Fergussen, 1863). Some experts presumed that the Garai would be widened further and the stretch of the Padma between Kushtia and Goalundo would go dry (Sherwill, 1858). But such an idea subsequently proved futile.

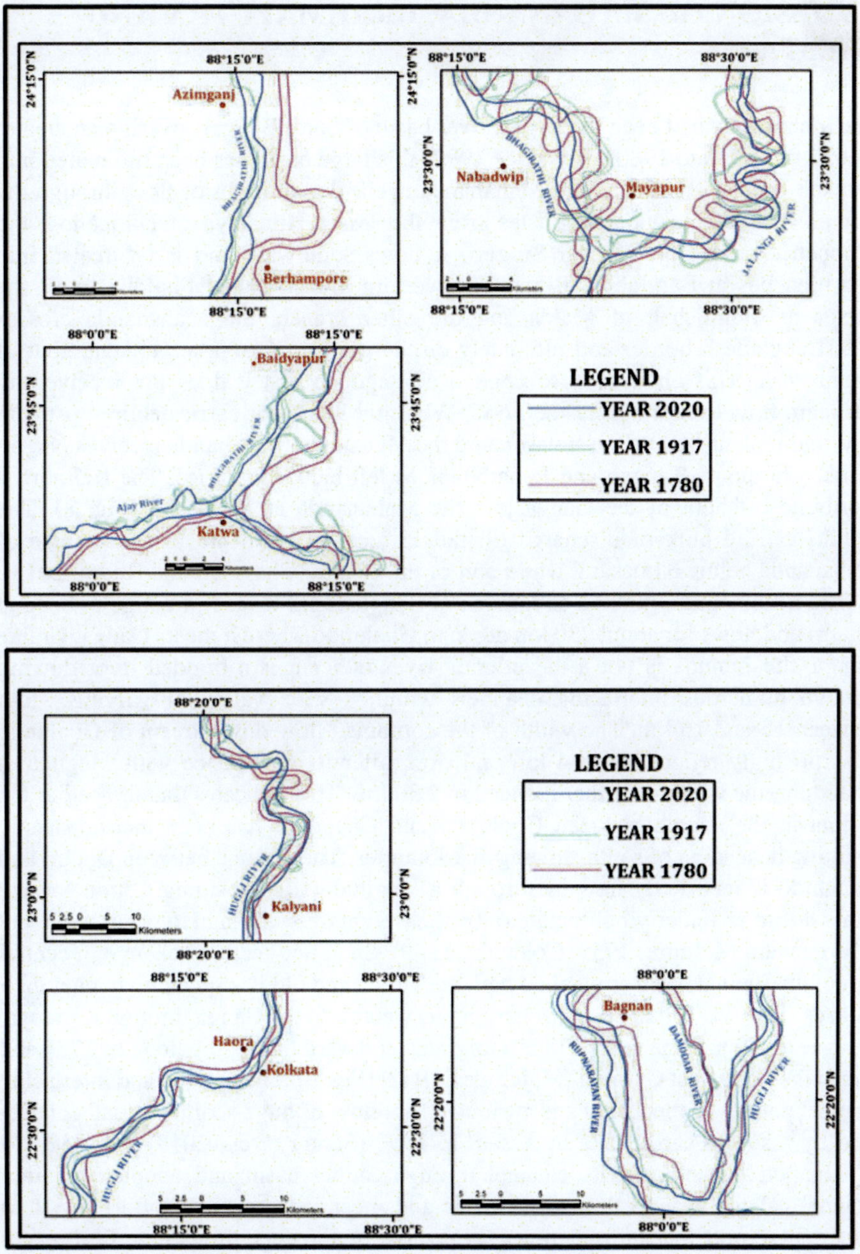

Figure 7. The Changing courses of the Bhagirathi (7A to 7F).

AVULSION OF MAHANANDA, BRAHMAPUTRA AND TEESTA

The Mahananda had been the largest river basin of north Bengal covering an area of 27,653 km^2 till late 1980s. But since 1990s this large basin has been bifurcated into two sub-basins—Phulohar and Mahananda itself, due to flight of flow through the former. The Mahananda originates from the lesser Himalaya of Darjeeling and debouches on the plains near Siliguri. It flows southward and is bifurcated into two branches at Bagdob of Bihar. The western branch called Phulohar meet the Ganga at Manikchak of Malda and the other branch, the Mahananda crosses Indo-Bangladesh border and ultimately discharges into Ganga at Godagarighat of Bangladesh. This channel is so choked at Bagdoab that it does not receive any upstream flow except during the flood. While the Phulohar carries entire upstream flow and replenished by the Balason and the Mechi, the Mahananda receives Nagor, Kulik, Chiramati, Tangan and Punarbhaba as left bank tributaries. The Kalindri, a moribund offshoot of the Ganga join the Mahananda at Nimasarai (Fig. 8). The Kalidri carried bulk of discharge till the 13th century and the present course of Mahananda below Nimasarai was a part of the Ganga (Chapman and Rudra, 2015).

The Brahmaputra (known as Jamuna in Bangladesh), takes right-angular bend at Dhubri and flows for about 230 km down to Goalundo where it meets Ganga. Unlike Ganga, the Jamuna is not a meandering river rather it is a braided river flowing through numerous interlacing channels within a wide valley of varying width between 13 and 16 km. The width of the combined flow downstream of Goalundo is surprisingly reduced to 3-4 km and even after its confluence with Meghna at Chandpur, the width is further reduced to 2.30 km. This indicates that the bed of the Padma in the stretch between Goalundo and Chandpur has been incised deeply resulting beheading of south-flowing distributaries. The Jamuna between Dhubri and Goalundo is reportedly flowing through a longitudinally subsiding trough formed either due to an underlying fracture or by the weight of overlying Holocene sediment (Morgan and McIntire, 1959, Coleman, 1969). Such neo-tectonic change in central part of the Bengal basin compelled both the Teesta and old Brahmaputra to alter their courses. The Teesta had been discharging its water into the Ganga through Karatoya, Atrayee and Punarbhaba till 1786. It migrated eastward during the flood of 1787 and joined the Jamuna at Chimari (Majumder, 1942) (Fig. 9). The braided and interlacing channel pattern of the Jamuna is attributed to inflow of huge sediment load contributed by a severe earthquake in Assam in 1950 (Sarkar et al., 2013). In 1830, the Brahmaputra abandoned its channel through Sylhet basin and adopted a minor channel called Janai to flow southwards and meet the Ganga at Jaffargang. It is important to note that Jaffergang was wiped out by the huge flow of the Brahmaputra. Since then it is known as Jamuna in Bangladesh but its old outlet through Mymansingh continues to be known as old Brahmaputra. The southward migration of Jamuna and its joining with the Ganga brought about three important changes. First, the Dhaleswari, which had been a distributary of the Ganga was converted into an offshoot of Jamuna and consequently Buriganga also lost its connectivity with the

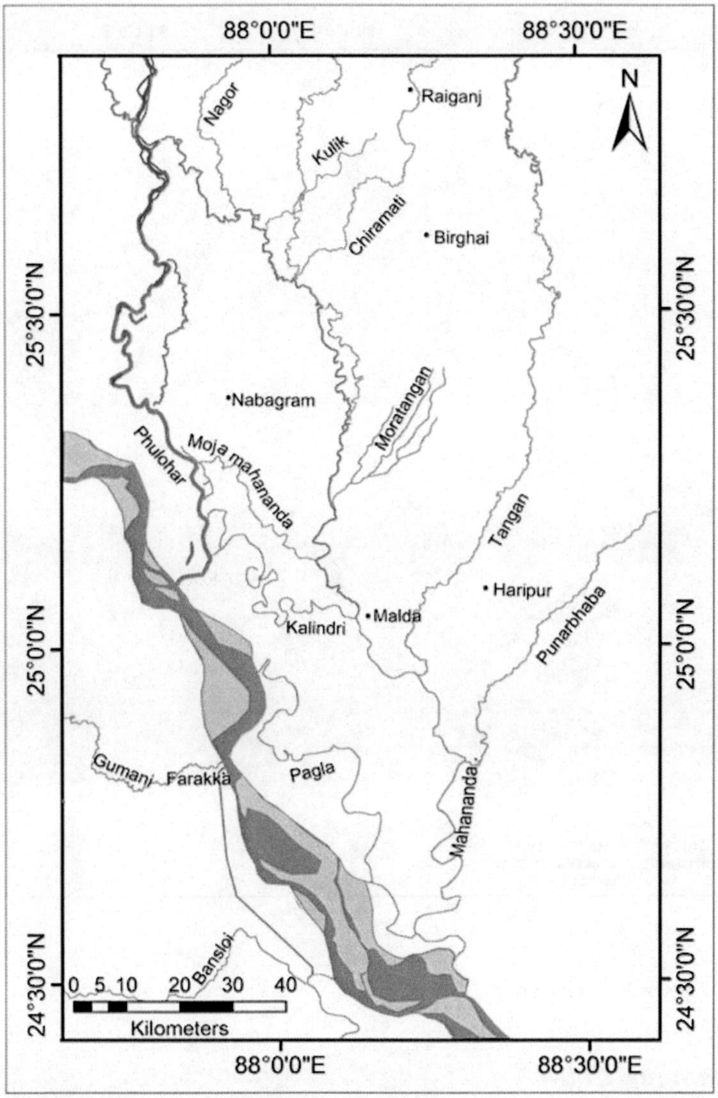

Figure 8. Bifurcation of the Mahananda and the Phulohar.

Ganga. Secondly, the combined flow of Ganga and Jamuna widened a narrow channel then taking off at Rajnagore and joined the Meghna opposite to Chandpur (Fig. 5). Thirdly, the Ganga abandoned its former passage towards the sea through Arialkhan which became moribund. It is important to note that the Ganga and Brahmaputra-Meghna discharged into the Bay of Bengal separately till 1830; the united outlet of the Ganga-Jamuna and Meghna is now named after the latter.

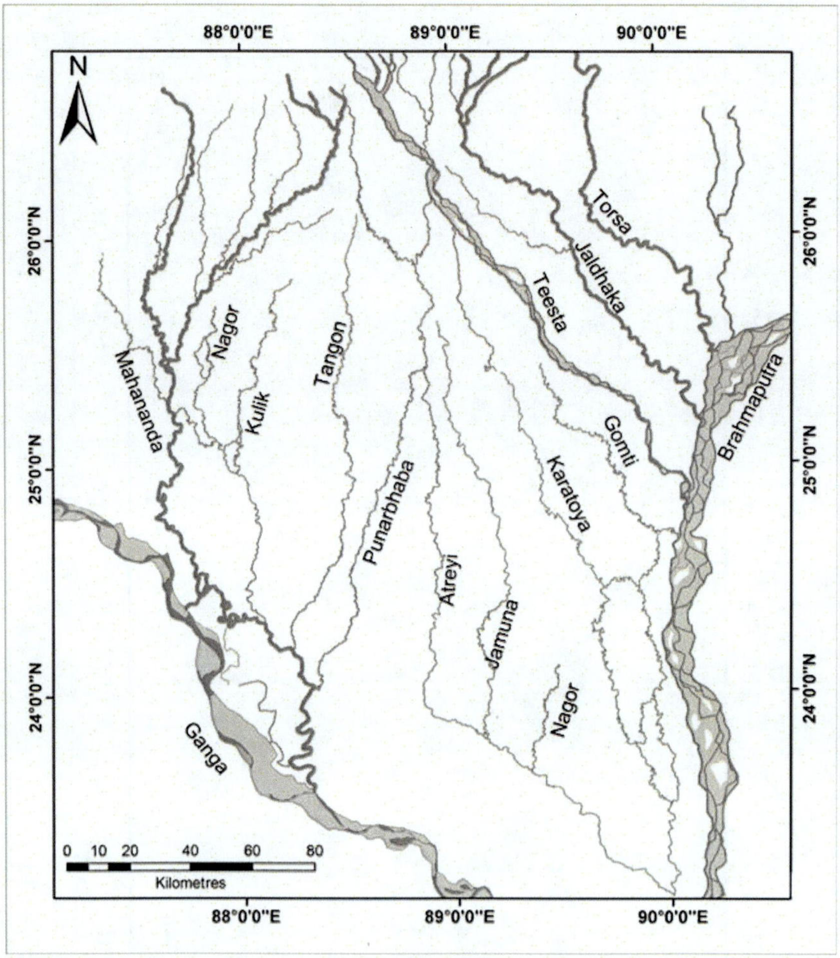

Figure 9. Eastward migration of the Teesta.

The Changing Coast

The coastal tract of Bengal has changed dramatically since 1780 and the configuration of the littoral tract continues to change unabated till date. The most striking change is noticed along the Hugli and Meghna estuary. While the coastline has retrograded along the Indian side, the Meghna estuary in Bangladesh has grown southward. The dense mangroves of the Sundarbans lying in-between acted as barrier against storm surge and tidal invasion keeping the coastline more or less stable. Comparison of multi-dated maps and image clearly shows this change. Report of the IPCC (2019) noted that the sea-level went up at the rate 3.6 mm/year during 2006-2015 due to combined impact of melting of glaciers and thermal

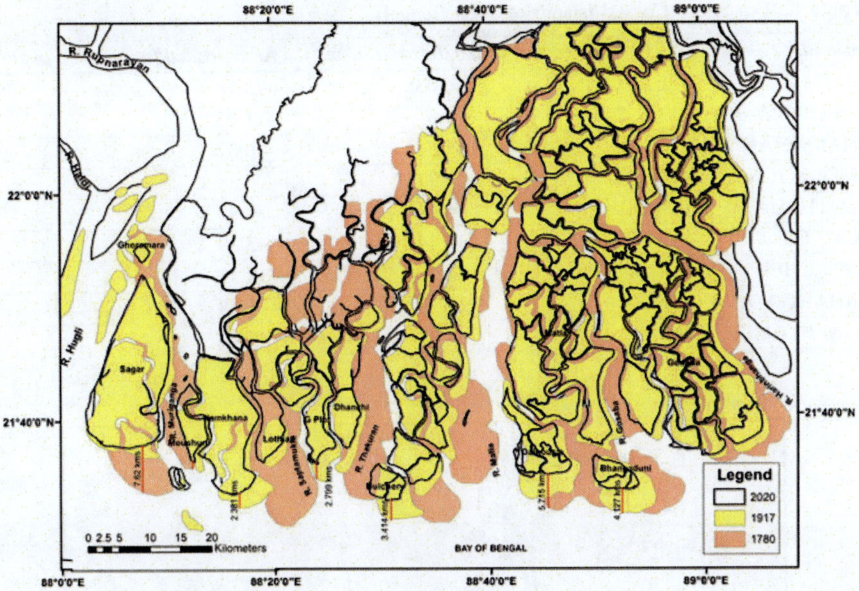

Figure 10. The Changing Hugli estuary and islands (Kaberi).

expansion of the sea. This alarmingly high rate is 2.6 times higher than the average rate of 1.4 mm/year for the period commencing from 1901 to 1990. The situation in Bengal coast is further aggravated due to subsidence of land at the average rate of 2.90 mm/year (Brown and Nicholls, 2015). Thus, sea-level of Bay of Bengal is changing at the rate +6.5 mm/year and this seems to be highest since 1850. The increasing sea-surface-temperature favours the formation cyclones and storm surge over the Bay of Bengal. The two recent disastrous cyclone, viz. Bulbul (09.11.19) and Amphan (20.5.2020), have changed the coastal configuration at Sagar, Mousuni and G-plot swallowing chunks of land.

The sea has encroached about 8 km northwards along the Hugli estuary since 1780 (Fig. 10). The Sagar Island which had been a cluster of six islands is now coalesced to one. It is estimated that eight sea-front islands of West Bengal have lost 201 km^2 land since 1917 and only Lothian has gained 9.83 km^2 (Table 1). The sea has encroached drastically along Saptamukhi, the Thakuran and Matla estuaries. The islands of Namkhana, Mousuni, G-plot and Bulcheri have been reduced in size. Meghna estuary, on the contrary, has grown fast and deposition of huge sediment load has negated the impact of sea-level rise pushing the sea far southwards (Fig. 11). The estuarine islands of Hatiya, Lalmohan, Rangabali and Kuakata in Bangladesh have been enlarged and the coastline has shifted southwards more than 33 km from Noakhali since 1780. The Meghna estuary, being the principal conduit of sediment flow into the sea, is obviously the most fast changing littoral tract of Bengal. The retrogradation and progradation of sea in Bengal coast between 1780 and 2020 is shown in Figs. 10 and 11.

Table 1. Erosion of Coastal Islands of West Bengal

Islands	*1917 (area in km^2)*	*2020 (area in km^2)*	*Area eroded/(accredited in km^2)*
SAGAR	261.622	236.038	–25.58
GHORAMARA	15.52	4.5	–11.02
NAMKHANA	157.39	146.52	–10.87
MOUSHUNI	33.29	27.91	–5.38
LOTHIAN	26.61	36.44	9.83
G-PLOT	84.63	41.44	–43.19
BULCHERY	41.49	22.96	–18.53
BHANGADUNI	67.44	25.76	–41.68
DALHOUSIE	106.43	61.28	–45.15

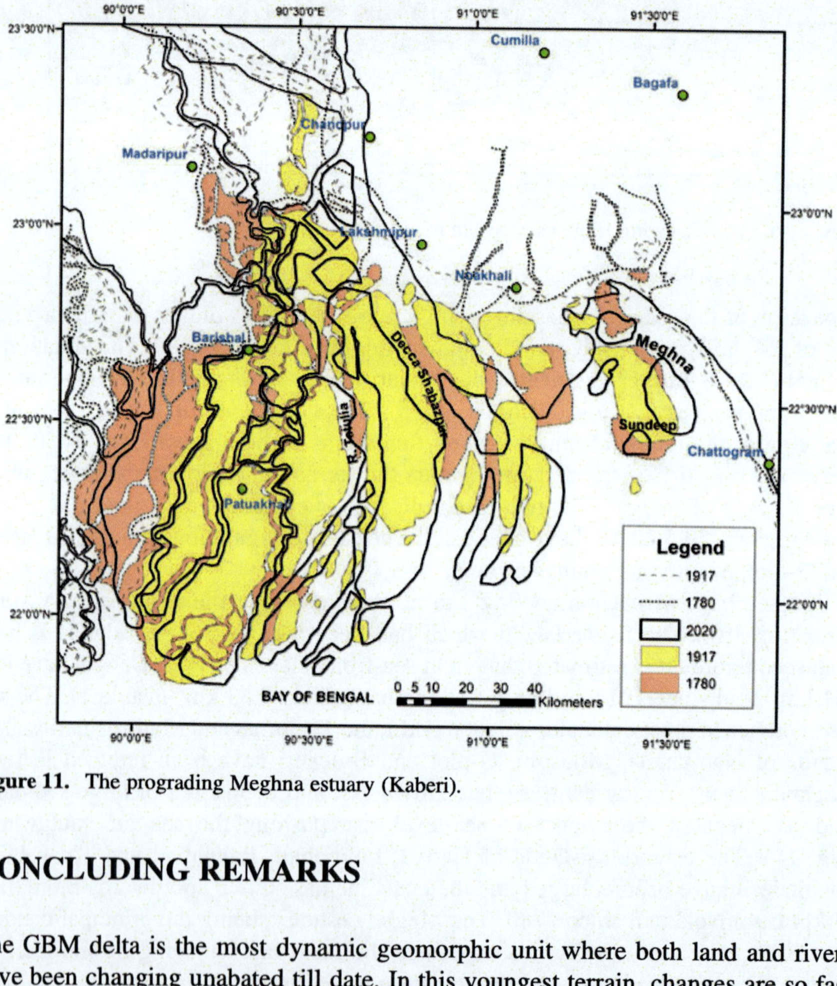

Figure 11. The prograding Meghna estuary (Kaberi).

CONCLUDING REMARKS

The GBM delta is the most dynamic geomorphic unit where both land and rivers have been changing unabated till date. In this youngest terrain, changes are so fast that it became difficult for the human society to cope with, especially when the rivers erode their banks and displace the population. It is important to identify the widths of meander belts of the Ganga, the Jamuna and Meghna, and also the inter-tidal spaces in the littoral tracts. The human settlement should not be allowed in these tracts which should be declared as the 'space for rivers'. While Indo-Bangladesh river valley commission is mainly interested in sharing of trans-boundary water, but many important issues are grossly neglected. The management of rivers in both the countries is governed by a reductionist idea of embankment building and plan to divert riverine flow to satisfy the agricultural, industrial and other human demands. We need to appreciate holistic eco-hydrology of the Bengal basin and ecosystem services rendered by the rivers. Since a common fluvial system drains both the countries, a consensus on the sustainable river management needs to be developed between India and Bangladesh.

ACKNOWLEDGEMENT The author is grateful to Dr. Kaberi Samanta, Research Assistant, West Bengal Pollution Control Board, for cartographic assistance.

References

Anville, d' (1752). Geographical Illustration of the Map of India: London. p. 29.

Agarwal, R.P. and Mitra, D.S. (1991). 'Paleogeographic reconstruction of Bengal delta during Quaternary period', in Vaidyanadhan, R. ed. *Quarternary Deltas of India, Memior of the Geological Society of India*, 22: 13-24.

Alam, M, Alam, M.M., Curray, J.R., Chowdhury, M.L.R. and Gani, M.R. (2003). 'An overview of the sedimentary geology of the Bengal Basin in the regional tectonic framework and basin fill history', *Sedimentary Geology,* 155, 179-208.

Allison, M.A., Khan, S.R., Goodbred Jr., S.L. and Kuhel, S.A. (2003). Stratigraphic Evolution of the late Holocene Ganges-Brahmaputra lower delta plain. Sedimentary Geology, 155: 317-342.

Bagchi, K. (1944). *The Ganges Delta*, University of Calcutta, Calcutta.

Ball, V. (1889). Travels in India by Jean Baptista Tavernier, Macmillan and Company, London. 1, pp. 125-126.

Banerjee, Manju and Sen, Prasanta K. (1987). Palaeobiology in understanding the change of sea level and coast line in Bengal basin. Indian Journal of Earth Sciences, 14(3-4): 307-320.

Brown, S. and Nicholls, R.J. (2015). Subsidence and human influences in mega deltas: The case of the Ganges–Brahmaputra–Meghna. Science of the Total Environment 527-528: 362-374.

Campos, J.J.A. (1919). History of Portuguese in Bengal. Butterworth & Co. India Ltd. Calcutta.

Chapman, G.P. and Rudra, K. (2015). Time Streams/History and Rivers in Bengal. Centre for Archaeological Studies & Training, Eastern India, Kolkata.

Colebrooke, R.H. (1801). On the courses of the Ganges through Bengal. Asiatic Researches, Asiatic Society, Kolkata. 7: 1-31.

Coleman, J.M. (1969). 'The Brahmaputra river, channel processes and sedimentation', *Sedimentary Geology*, 3(2&3): 123-239.

Fergussen, J. (1863). Recent changes in the delta of the Ganges. Quarterly Journal of Geological Society (of London) 19: 321-354.

Gladwin, Francis (1800). Ayeen Akbery or the Institute of the Emperor of Akbar; Vol. II. J. Swan & Co. London.

Hirst, F.C. (1915). Report on the Nadia Rivers, Reprinted by Gazetteer Department, Government of West Bengal, Kolkata. pp. 3-61.

IPCC (2019). Summary for Policymakers. *In:* IPCC Special Report on the Ocean and Cryosphere in a Changing Climate.

Majumder, S.C. (1942). Rivers of Bengal Delta. Reprinted by Gazetteer Department, Government of West Bengal, Kolkata. pp. 7-102.

Mitra, A. (1953). History of the mouth of the Bhagirathi River 1781-1925, Selection of Records of the Government of Bengal relating to Nadia Rivers (from 1848-1926), Reprinted in District Census Handbooks/Murshidabad 1961, Appendix vi, pp. 111-133.

Morgan, J.P. and McIntire, W.G. (1959). Quaternary geology of the Bengal basin, East Pakistan and India, Bulletin, Geological Society of America 70: 319-342.

Oldham (1870). President's address. *Proceedings, Asiatic Society of Bengal for February*, 1870, Calcutta.

Rennell, J. (1781). A Bengal atlas, UK. p. 60.

Rennell, J. (1793). Memoir of a Map of Hindoostan or the Mugul Empire, Reprinted by Editions Indian in 1976, Kolkata. pp. 148-150.

Rizvi, A.I.H. (1957). Pleistocene Terraces of the Ganges Valley. Oriental Geographer, I(I): 1-18.

Rudra, Kalyan (2006). 'Shifting of the Ganga and Land Erosion in West Bengal/A Socio-ecological viewpoint', CDEP Occasional Paper 8. IIM Kolkata.

Rudra, Kalyan (2010). Dynamics of the Ganga in West Bengal, India (1764–2007): Implications for science-Policy Interaction, Quaternary International (2010), 227/2: 161-169.

Rudra, Kalyan (2015). Rivers of West Bengal/Dying, Living, In Living Rivers/Dying Rivers; edited by Ramaswamy R. Iyer. Oxford, New Delhi; 188-204.

Rudra, Kalyan (2018): Rivers of the Ganga-Brahmaputra-Meghna Delta/A Fluvial Account of Bengal. Springer, Switzerland.

Rudra, Kalyan (2020). *Dui Banglar Nadikatha* (in Bengali). Sahitya Samsad, Kolkata

Sarkar, J.N. (1973). The History of Bengal/Muslim Period (1200-1757), Academica Asiatica, Patna. pp. 318-319.

Sarkar, M.H., Akter, J. and Rahman, M. Md. (2013). Century-Scale Dynamics of the Bengal Delta and Future Development. Fourth International Conference on Water and Flood Management; pp. 91-104.

Sengupta, S. (1966). Geological and geophysical studies in western part of Bengal basin, India, *Bulletin of American Association of Petroleum Geologists*, 50(5): 1001-1017.

Sherwill (1858). Selections from the Records of the Bengal Government/Reports on the Rivers of Bengal; G.A. Savielle, Calcutta Printing and Publishing Company.

Smith, V.A. (1916). Travels in Mughal Empire/AD 1656-1668 by Francois Bernier; OUP, London.

Stanley, D.J. and Warne, A.G. (1994). Worldwide initiation of Holocene marine deltas by deceleration of sea-level rise, Science, 265 (5189): 228-231.

Uddin, A. and Lundberg, N. (2004). Miocene sedimentation and subsidence during continent-continent collision, Bengal basin, Bangladesh. Sedimentary Geology 164: 131-146.

Chapter 23
Investigating the Evolution of Architectural Elements and Patterns of New-urbanism along Ganga: A Case of Varanasi

Vidhu Bansal, Sunny Bansal, and Joy Sen

INTRODUCTION

"The city illuminates truth and reveals reality. It does not bring new wonders into the scope of vision, but enable one to see what is already here. Where this eternal light intersects earth, it is known as Kashi."
– Diana L. Eck.

Varanasi is long known as the salient feature of India which exists on the banks of Ganga. Varanasi and Ganga are the conjointed twins in the historical legacy of India. Varanasi is, and will always be, Ganga's Kashi, the city of light. Varanasi without Ganga is an improbable idea and non-existent in its very essence. Lives of people are bound with threads of the divine flow as it meanders its course reflecting the changing hour to the rhythm of the sun. Morning walk along the *ghats* (stepped banks), daily dip in the holy river, morning worship at the ghats, all initiate the day for the old *benarasis* (People living in Varanasi for many generations). The diurnal rhythm gets encased with the hypnotic chants emanating from the temples which dot the sacred landscape. Ganga flows over 2,525 km of length in India from north to south and east. It is in Varanasi only that its course changes and it starts flowing in a south to north direction (Singh, 2004). The prominent crescent shape and the majestic palatial buildings lining onto its banks gives it a surreal outlook (Figs. 1 and 2). The sacredness of Ganga has permeated in the whole city but on its banks it is an entirely different energy vibrating into every grain. In all there are 96 sacred spots along the banks of Ganga in Varanasi which are spread over 84 ghats. These are

V. Bansal (✉) · S. Bansal
Indian Institute of Technology Kharagpur, Kharagpur, West Bengal, India

J. Sen
Department of Architecture and Regional Planning, Joint Faculty, Ranbir and Chitra Gupta School of Infrastructure Design and Management, Indian Institute of Technology Kharagpur, Kharagpur, West Bengal, India

A. Mukherjee (ed.), *Riverine Systems*, https://doi.org/10.1007/978-3-030-87067-6_23

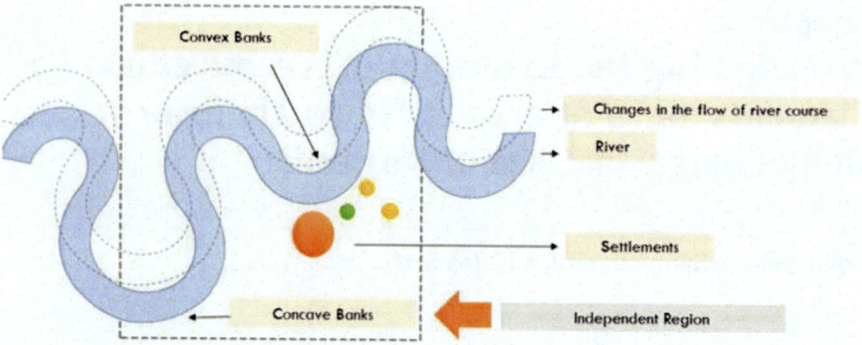

Figure 1. Changing course of Ganga

Figure 2. Varanasi Riverfront

called as *jal-teerths* or sacred water spots. Many of the ghats have religious significance associated with them which is again related to the festivals or other significant dates (*tithi*) of the year.

ROLE OF GANGA RIVER IN ATTRACTING VARIOUS KINGDOMS ACROSS INDIA

The Varanasi has documented history which dates back to 11th-12th century. The archaeological excavations done under the guidance of renowned archaeologist Dr. Vidula Jaiswal (Professor, BHU) have proved the fact that the city which is now Rajghat area (northern part of Varanasi) probably represents the ancient Kashi of 12th-11th century BCE (Shah, 2019). The city regions in ancient times were classified as *Janapadas* and *Mahajanpadas*. Kashi was among the 16 *Mahajanpadas* as described in the ancient texts (Shah, 2019). The strategic trade route and the location of Varanasi, flourishing learning centers, and religious importance popularized it as an attractive hub. Kings from other states started sending their representatives to claim a piece of this dynamic city. The different *havelis* (palatial houses) that dot the banks of river ganga in Varanasi are a consequence of it. Different kingdoms started developing different part of the *ghats*. Nuclei became the haveli or palace for the aristocrats. The neighbourhoods were also developed surrounding these palatial structures which housed the people working for these aristocrats. They mostly belonged to different regions hence with this, in-migration they brought different traditions and architectural styles. The Maharashtrian architectural style at *Bhonsle ghat* to Darbhanga Palace at *Darbhanga ghat*, the Aurangzebi Mosque at *Panchganga Ghat* to *Gauri Kedareshwar* temple at *Kedar Ghat*. Piece-meal contributions have given Varanasi a very diverse architectural heritage to cherish and hence evolved the unique architectural style of Varanasi.

ARCHITECTURAL STYLES OF THE GHATS

For this particular study following *ghats* would be examined namely *Adi Keshav Ghat* and *Panchganga Ghat* in the northern portions of the city, *Dashashwamedh Ghat* and *Kedar Ghat* in the central portions of the city and *Assi Ghat* as the southernmost tip of the city. *The zones taken under this study are from southern, central, and northern parts of the ghats' armature so as to understand how the physical setting is changing at an intra-city level.* These zones have evolved by virtue of multiple layers of cultural amalgamation embedded into them. Though these layers may be interpreted in manifold ways, the present study picks up a specific standpoint of organic urban design and attributes it to a series of acts of public participation, implanted over hundreds of years. *These zones with the virtue of time are giving certain urban patterns which are long term identifiers for this city. They are functional but may be lost under the layers of urbanization.*

Adi Keshav Ghat

The ghat's signature element is the Adi Keshav temple which is one of the most revered Vishnu Temples in the city (Fig. 3). It is disconnected from the other ghats which are majorly connected and form a promenade. Major land in this zone is under the ownershio of Jiddu Krishnamoorthy Trust. Various colleges and schools under the trust are situated in this zone. The Varuna river flows adjacent to this zone. Rajghat is the neighbouring ghat to the Adi Keshav Ghat. The Rajghat forms the part of Ancient Kashi.

Panchganga Ghat

The Panchganga ghat (Fig. 4) zone is among one of the northernmost *ghats* of the city. The *ghats* are very steep in this area which are evident by the overwhelming staircase which can be found in this part of the city. The rise of the slope can also be observed in the neighbourhoods as one walks along the interior lanes. The urban fabric of neighbourhood adjacent to this ghat is denser than the other neighbourhoods of the area. The serpent like streets are organically laid out. They are devoid of signs which easily makes the city illegible to the person who is not familiar to the system of Varanasi. The 21st century navigators failed in this part of the city and became dependent on the information provided by the locals to understand the spatial structure of the area. There are many niches and corners in this zone which hide abundant spaces, a sudden shock to the constrained streets. Most of these spaces are either religious in nature or privately owned. The most familiar axis is called as Thatheri bazaar street which merges into Chaukhamba road. The area is bereft of any green spaces and is mostly residential.

Sherwali kothi, a 200 plus year old palatial mansion is situated along the Thatheri bazaar street as one approaches from the main road. This mansion is privately owned and has an impressive open area which houses a temple. As one moves along Sri Gopal Mandir which is located off the main axis houses a big residential complex and a temple inside. The backyards also contain good amount of green areas. The photography is not permitted in this complex. Sri Tailang Swami temple is another religious point which houses a huge Shivling which is revered by the native citizens. At the end of this trail is the Aurangzebi mosque or Alamgiri mosque which is an architectural delight. The tall minarets, carved jalis and intricate work on the outer

Figure 3. Adi Keshav Ghat and glimpses of the site.

Figure 4. Panchganga Ghat and glimpses of the site.

façade adorn this mosque. The history is controversial for this structure as is for many other structures in the city. The site was historically a Vishnu temple which was demolished during the reign of Mughal emperor Aurangzeb. The mosque was established in its place and the Vishnu temple is now opposite to this site. Many people still refer to it as 'Bindu Madhava ka Dharara', the ancient name for the Vishnu Temple.

Dhashashwamedh Ghat

Dhashashwamedh ghat (Fig. 5) is one of the most prominent ghats famous for its evening Aarti (Chanting and ritualistic worship offered to River Ganga). The road leading to Dashashwamedh ghat from the main crossroad also known as Godowliya Chowk, is quite broad and is lined with shops selling fabric and dress materials, food outlets, worship items, trinkets, and other materials. The Bengali tola area starts from this zone with majority of people from Bengal residing in the city. The main street mostly has the markets dealing with wholesale goods while interiors are mostly residential zones.

Kedar Ghat

Kedar ghat zone can be characterised by a space amidst the flow of migration. The mohalla of *Bengali tola* flows into it. The nature of spatial organisation used to be

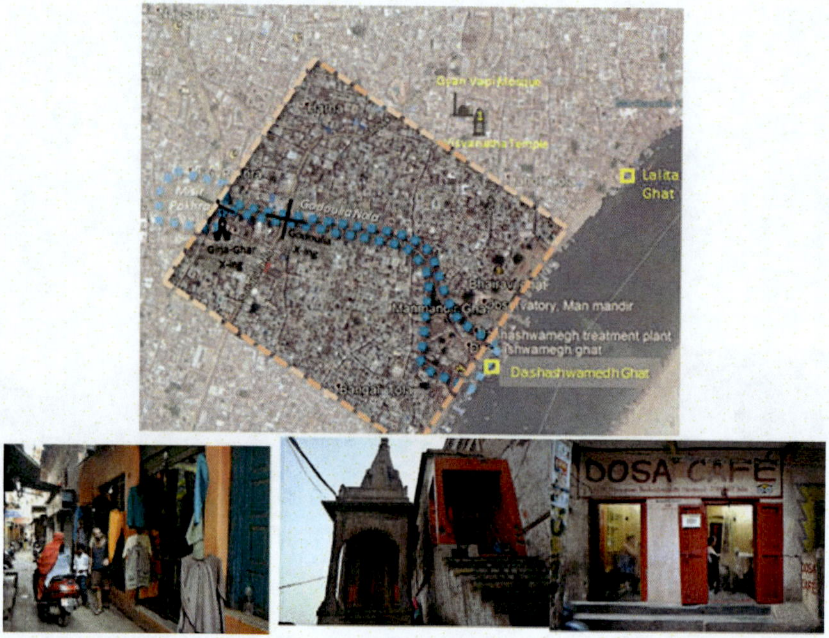

Figure 5. Dashashwamedh Ghat and the glimpses of the site.

highly influenced by the majority population that was *Bengali*. But slowly the growing popularity of Gauri Kedareshwar temple lead to a continuous shift in the socioeconomic dynamics of the zone. The *kachauri* shops got changed into idli business (comparison between a shift from northern cuisine to a souther cuisine). Slowly the Bengali population departed from this part of the city to pave way for the pilgrims from the south. Many of the lands were donated in charity to various NGOs, which has changed them into ashrams for the pilgrims and young scholars (Figs. 5 and 6).

Assi Ghat

Assi ghat is one of the new ghats of the city making it a local favourite among the youths. The urban fabric is moderately dense in this part of the city as compared to the northern parts. Dedicated parking lots, wider streets and accessibility, cafes of various cuisines and frequent cultural events helped into achieving the status of one of the most popular ghats overturning the status of Dashashwamedh Ghat. The Tulsi Manas Haveli of late poet Shri Tulsi Das is found in this zone. The zone also has some kunds of religious significance like Pushkar Kund, Lolark Kund, Durga Kund, Kurukshetra and others (Fig. 7).

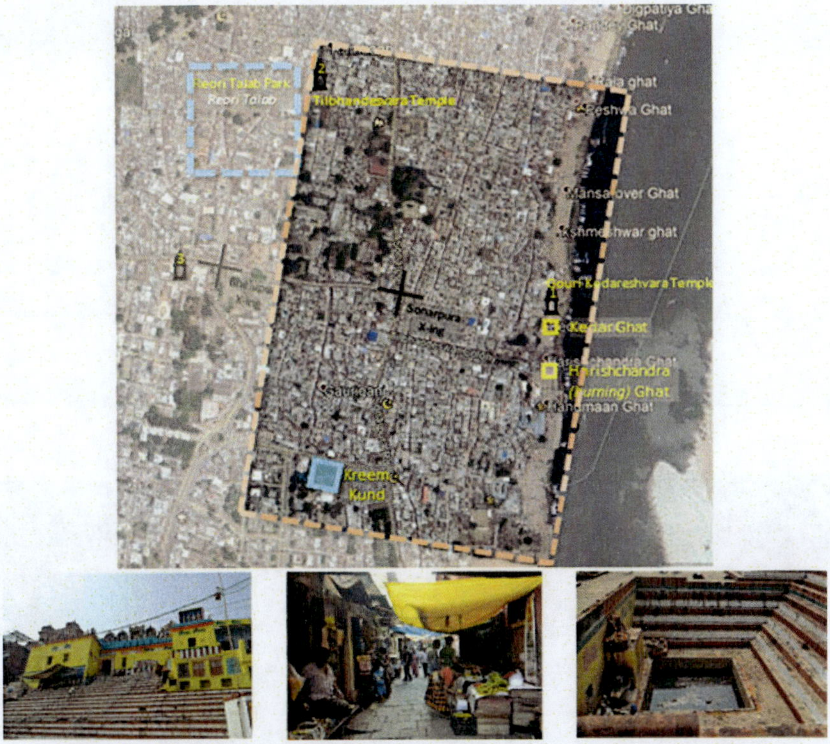

Figure 6. Kedar ghat and glimpses of the site.

KUNDS AND GANGA: SMALLER BODIES AS ENACTMENT OF LARGER BODY

In ancient times all the water bodies of Varanasi be it large or small were interlinked with each other. The lanes which we see today were drained from their original form to make way for the movement of the citizens. These used to be small rivulets. The kunds and river was connected forming a seamless interface of drainage system which helped in keeping the floods of Indo-Gangetic plane at bay (Cunha, 2018). These were drained and clogged development authorities and hence the whole system choked. Now there is no interconnection left. Most of the kunds vanished into waves of urbanization while others got stagnated. Initially, these tanks were supposed to collect the drainage and direct it either through the Varuna river or through the Godaulia *nala* into the river Ganges. However, with the densification of the city, most of these ponds and *nalas* have been filled up (Jalais, 2008).

These kunds had almost equal importance as the sacred river since the water of kunds was the same as that of Ganga. They formed the central focus of any worship held in the community. But a gradual disconnection from ancient rituals and

Figure 7. Assi Ghat and the glimpses of the site.

changing religious, cultural and economic dynamics, draining of kunds during the time of British colonization, and rapid urbanization have led to a change in the status quo of these entities. The community participation in older times ensured a certain level of maintenance for such spaces. But they are gradually fading from the cognitive maps of citizen's mind. The situation tends to get worsened if immediate actions are not taken. The new generation is generally not aware about the ancient traditions and hence a sense of connectedness gets lost as temporal scale moves on. Diligent efforts are required on the parts of authority and citizens to preserve this aspect of the Varnasi's heritage. Pushkar Kund and Kurukshetra Kund are found in the Assi Ghat zone which are live examples of how a paradign shift in cultural aspects can lead to dilapidation of heritage. Matsyodhari talab (northern part of the city) is another example of ill maintained water bodies. There are examples of other kunds in this Assi ghat zone like Durga Kund and Lolark Kund which are in a comparatively better shape. Another example could be of the Beniya Park which lies in the northern part of the city. It used to be kund in earlier times but was drained to form a park altering the drainage capacity.

CITY AND ELEMENTS OF NEW URBANISM

Built heritage, be it individual building or whole neighbourhood, make a very essential part of the inheritance. As observed in various examples around the world, cities of the past had a unique charm inherent to them, with each being a distinctive identity in itself. This study tries to break down the language of the built environment of Varanasi into its components to form a vocabulary of new-urbanism elements which can be interpreted for various urban levels and can be defined for Indian context. New Urbanism is a planning and development approach based on the principles of how cities and towns had been built for the last several centuries: walkable blocks and streets, housing and shopping in close proximity, and accessible public spaces (Fig. 8). In other words: New Urbanism focuses on human-scaled urban design (CNU, 2020). The study takes its basis from Christopher Alexander's pattern language. Pattern language when first conceived was based on generative schemes that existed in traditional cultures. When the generative scheme is carried out, the results are always different, because it always generates a structure that starts with the existing context and creates things which relate directly and specifically to that context (Alexander, 1999). Varanasi has been continuously habituated city since past 3000 years and more. The probability of finding element which can be related to new urbanism is very high. Following are few examples which translate the Christopher Alexander's Pattern Language (Alexander et al., 1977) into Indian context specifically Varanasi (Fig. 8). How shopping street of Christopher Alexander is also valid case as a shopping street in Varanasi (Fig. 9), how promenades in the context of Christopher Alexander are also valid as connected ghats in the context of Varanasi, how identifiable neighbourhoods in context of pattern language are also valid as *Mohallas* in Varanasi and many others (Fig. 10). The elements find a new meaning in Varanasi as the Ganga gets entwined with the life of the citizens and govern their daily lives.

EVOLUTION OF RELATIONSHIP OF GANGA WITH THE CITIZENS

River Ganga is termed as the holiest river among all the rivers in India. Its significance transpires mainly from spiritual and religious mythology and faith. It also offers many socio-cultural and economic prospects, especially to the city of Varanasi. Holiness of these socio-cultural events and activities is as deep-rooted as the River Ganga itself. The River is mother for the people of Varanasi, who descended from the heaven to purify and nourish their lives. This diverse and incredible relationship of River Ganga with the people, culture, and life of Varanasi is unique and well-established. The River is an important part of the life of citizens in the form of various rituals, ceremonies, and festivals practiced around it. The following section discusses these rituals in detail.

Figure 8. Christopher Alexander's Pattern vs on site new-urbanism elements: shopping street.

Rituals Related to the River Ganga

To understand the evolution of relationship of River Ganga with the citizens of Varanasi, it is important to understand the relevance of the River in their life. According to Hinduism, water is one of the five elements of which human body is made. In every day rituals, the essence and auspiciousness of the holy *Gangajal*-the sacred water is unmatched. The rituals and ceremonies on the banks of the holy River attract thousands and millions from the city as well as the country. Most of these rituals date back to the Vedic era while some rituals have evolved over time. Table 1 gives some key rituals practiced along the river.

These rituals are more or less unchanged over the years. However, their prevalence and participation has grown drastically in recent times. This can be attributed to the growing population, improved accessibility and connectivity, and enhanced affordability of the country and its citizens. All this has resulted in extra pressure on the river Ganga, the city of Varanasi, and its infrastructure. City of Varanasi has definitely kept all the customs and traditions alive but their understanding and real meaning is being largely lost. Ignorance regarding the spiritual aspects is a major reason for this diminishing value. This may also be accredited to the general wandering away of the people from the time-worn traditions and beliefs.

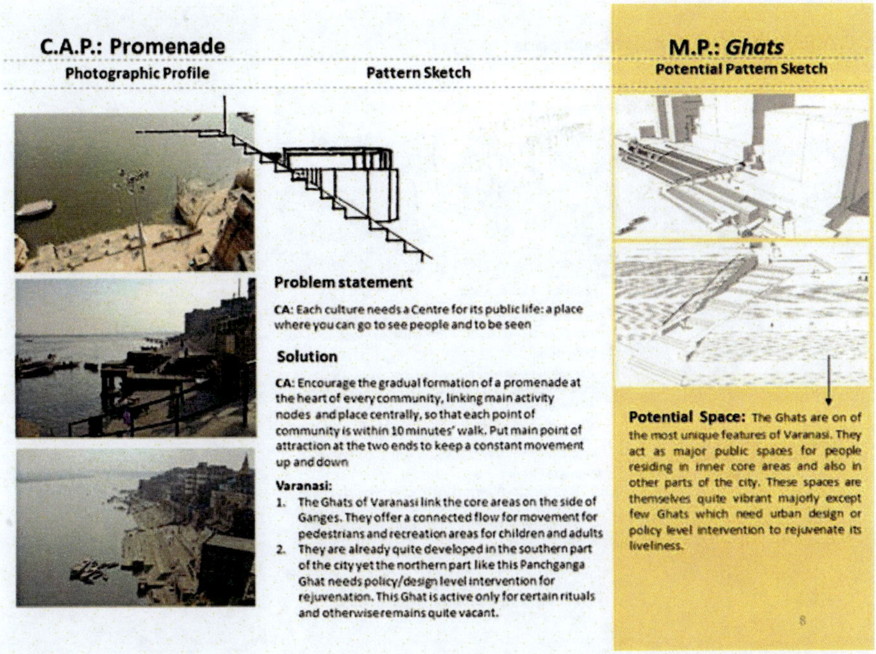

Figure 9. Christopher Alexander's Pattern vs on site new-urbanism elements: Promenades and Ghats.

Changing Perspectives

Drastic changes have been observed in the perception of the locals, visitors, and pilgrims regarding the rituals and its impact on the River. With time, people have become quite insensitive and careless towards the health and condition of the River (Basak et al., 2015). This lack of awareness might be accredited to the belief (or myth) that the river has infinite strength to assimilate whatever is being thrown into it. This could be due to the growing lack of concern by the people towards the spiritual and ecological aspects of the river. For people who are performing rituals, these offerings are expected to be absorbed by the river as it is meant to wash away all the sins (Sinha, 2019). The pollution caused by such activities may not always be chemical or biological but it is definitely physical and impairs its aesthetic value. The quality of water has become unacceptable for potable or farming purposes and it is becoming worse by the day. The discharge of untreated sewage and industrial waste into the river is also a major concern which is not being addressed (Kala, 2018). Only one-third of the total sewage produced is being treated before discharge (Kumar et al., 2012).

One major reason for the current state of the river is the exponential increase in the number of pilgrims. The massive growth of visitors was not matched with proportionate increase in the required infrastructure facilities. Thus, it eventually

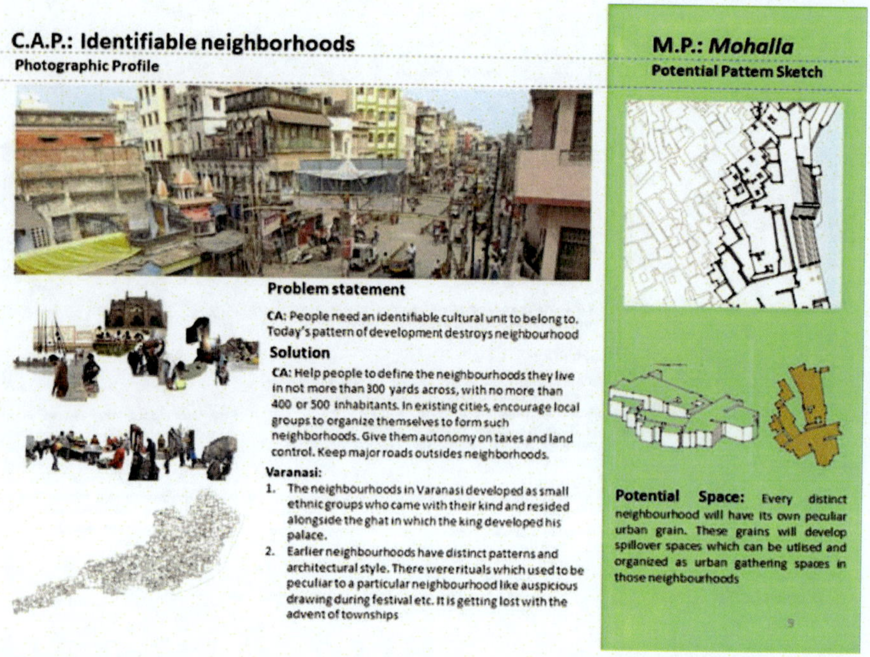

Figure 10. Christopher Alexander's Pattern vs on site new-urbanism elements: Identifiable neighbourhoods vs Mohallas.

led to high concentration of activities in small areas of significance creating issues like traffic congestion, air, water, and noise pollution, crowding and stampede, etc. Moreover, the increase in the number of visitors has also provoked an unregulated and unplanned growth of *ashrams*, hotels, *dharamshalas* in the city which further aggravates the issues related to town planning, transportation, land use, safety, and infrastructure shortage. Increased construction activity in and around the region has also resulted in illegal sand mining on the river banks. Due to this the river morphology has been disturbed and the local flora and fauna is severely effected (Das and Tamminga, 2012).

Numerous plans have been attempted in the past four decades to conserve and beautify the *ghats* by focusing on the cultural and environmental aspects and modern needs (Singh, 2018a, b). However, the deep sacred ecology of the river and the *ghats* is not given its due consideration. Till date, only municipal waste and industrial discharge is being blamed for the worsening condition of the river. The contribution of the people and the rituals is not being considered much. It is interesting to note that almost 60% of people who represent various *ashrams*, local groups, and religious institutions do not recognize the rituals as a source of pollution (Behera

Table 1. Rituals along the River Ganga (Kala, 2018; Behera, et al., 2013; Jayapal, 1996; Singh, 2018a, b)

Ritual/ Ceremony	*Description*	*Activity w.r.t. Ganga River*
Mundana ceremony or a child's first haircut	It is tonsuring a child before his/her first birthday or after his/her third birthday. It is believed that the hair from birth links to some unwanted qualities of earlier life	Ceremony takes place on the bank of River and the hair shaved off are offered to the River symbolically
Antim Samskara or death ceremony	In Hinduism, the mortal remains of a person are cremated in flames. This is believed to be the most hygienic method of disposal of dead body and an attempt to break-off the connection between the body and the soul, thus, facilitating the further journey of the soul	River banks are preferred location for cremation and Manikarnika Ghat in Varanasi is believed to be the best place for that
Asthi Visarjan or immersion of ashes	Cremation of the mortal remains is followed by the immersion of the ashes into a river or waterbody	It is believed that the immersion of ashes into Ganga means that the soul will attain salvation or *moksha*
Jal Samadhi	A *sadhu* or sage does not require cremation after death as his/her soul is believed to be liberated already	His/her mortal remains are released into a river, especially Ganga
Shraaddha	A ritual performed by individuals to pay homage to their ancestors or *Pitṛs*, especially to the dead parents. It is performed on special days or death anniversaries as per Hindu calendar	It can be performed at home but it is more auspicious to perform it on the banks of the river Ganga
Aarti	It is a routine ritual of worship performed once or more than once every day. Here, sacred fire is offered as reverence to the deity along with incense, flowers, praising-songs, music, etc.	Regular aarti is performed at several *ghats* in the honor of the River during early morning and evening hours
Snana or bathing	Significance of taking dips in water at holy places and on special occasions is believed to get rid of sins (or *karmic* debt) of the current and previous births	Varanasi and its *ghats* are famous for such dips throughout the year
Tarpana or offering	As per the Vedic texts, this refers to an offering (water, flowers, food, etc.) made to divine entities like Gods, Sun, Moon, ancestors, etc.	This ritual is preferably performed on the banks of the River especially in morning

et al., 2013). On the other hand, majority of tourists perceive these rituals as polluting activities and are open to modifications or alternatives. Similarly, a consensus on the need to regulate the footfall of visitors is seen among all the stakeholders involved. It is agreed that mere awareness will not serve the purpose but strict administrative and legal measures are required for it (Zhuang et al., 2019).

CONCLUSION

According to Hinduism, Varanasi is the place that liberates the soul from the body to *moksha* (salvation). But this divine place is incomplete without the river Ganga which is the fountain head of spiritual, emotional, physical, ecological, and economic hope and prosperity for the people of Varanasi. With the development of *ghats*, the river has been an integral part of Varanasi's planning, architectural layout, and design (Singh, 2018a, b). Slowly, the daily lives and the rituals which were followed in older times are getting lost. Lack of knowledge regarding cultural heritage and the dismal state of the river due to urban sewage, industrial waste, and activities and rituals around it has contributed to it. This cultural and spiritual heritage of India needs to be protected.

River Ganga, being a manifestation and soul of the Indian civilization, needs an incessant reflection from the citizens for their own growth and identity (Kala, 2018). Seeing its socio-cultural value and the status of 'National River of India', there is a need to plan effective policies to restore the river to its old glory. The previous attempts to save the river and its ecosystem failed miserably. The main issue in such policies is a casual approach of incorporating one-shoe-fits-all approach which gravely impacts the development of such sensitive cities. The need of the hour is to explore whether this river is to be treated as a mere geographical/hydrological feature or an entity of spiritual and religious importance. This will in-turn govern daily lives and places which people are inhabiting. The 'Namami Gange programme', 'Swacch Bharat Abhiyan', HRIDAY are few of the programmes which are trying to revive the city but still there is a long way to go. The human scale of the city needs to be brought back and assimilated as it was once in the ancient times. Also, it is required to gauge it rigorously according to needs of modern paced life. Only then an appropriate perspective could be derived to tackle the complex issues in and around the holy river Ganga.

References

Alexander, C. (1999). *The Origins of Pattern Theory: The Future of the Theory, and the Generation of a Living World.* [Online] Available at: doi:https://doi.org/10.1109/52.795104. [Accessed July 2017].

Alexander, C., Alexander, P.D.A.C., Ishikawa, S., Silverstein, M., Jacobson, M., Fiksdahl-King, I. and Shlomo, A. (1977). Center for Environmental Structure. *In:* A Pattern Language: Towns, Buildings, Construction. Center for Environmental Structure Berkeley, Calif: Center for Environmental Structure series.

Basak, S., Sarkar, A., Ghosh, R. and Chaudhuri, A. (2015). Ghats of Varanasi: An Emerging Centre of Pollution. *IOSR Journal of Humanities and Social Science*, 20(9): 19-24.

Behera, B. et al. (2013). *Cultural-Religious Aspects of Ganga Basin.* s.l.: s.n.

CNU, 2020. *CNU: Congress for New Urbanism.* [Online] Available at: https://www.cnu.org/resources/what-newurbanism#:~:text=New%20Urbanism%20is%20a%2 planning,on%20human%2Dscaled%20urban%20design. [Accessed August 2020].

Cunha, D.D. (2018). *Water Urbanism Studio-Varanasi (Joint studio by IIT Kharagpur and Columbia University)* [Interview] (January 2018).

Das, P. and Tamminga, K.R. (2012). The Ganges and the GAP: An Assessment of Efforts to Clean a Sacred River. *Sustainability*, 4: 1647-1668.

Jalais, S. (2008). *The Riverfront of Benares: Between 'Sacred' Waters and Sewage Water*. France, HAL Archives.

Jayapal, P. (1996). *Varanasi: City of the Living, Dying and Dead*. ICWA Letters (PJ-10) South Asia, April, pp. 1-15.

Kala, C.P. (2018). Cultural Significance and Current Conservation Practices of the Ganga's Ecosystem and Environment. *Applied Ecology and Environmental Sciences*, 6(4): 128-136.

Kumar, S. et al. (2012). Pollution of Ganga River Due to Urbanization of Varanasi – Adverse Conditions Faced by the Slum Population. *Environment and Urbanization Asia*, 3(2): 343-352.

Shah, A. (2019). *The Rise of Kashi* (800-600 BCE). [Online] Available at: https://www.livehistoryindia.com/cover-story/2019/10/29/indias-first-cities-kingdoms

Singh, R.P. (2004). The Ganga Riverfront in Varanasi: A Heritage Zone in Contestation. *In:* Context: Built, Living and Natural. s.l.:DRONAH.

Singh, R.P. (2018a). The Ganga Ghats, Varanasi (Kashi): The Riverfront Landscapes. *In:* R.P. Singh (ed.), Banaras (Varanasi): Cosmic Order, Sacred City, Hindu Traditions. Varanasi: Varanasi-Studies Foundation, pp. 65-102.

Singh, R.P. (2018b). Urbanisation in Varanasi and interfacing Historic Urban Landscapes. Chennai, C.P.R. Institute of Indological Research.

Sinha, A. (2019). Ghats on the Ganga in Varanasi, India: A Sustainable Model for Waste Management. *Landscape Journal*, 37(2): 65-78.

Zhuang, X., Lin, L. and Li, J. (2019). Puri vs. Varanasi destinations: Local residents' perceptions, overall community satisfaction and support for tourism development. *Journal of the Asia Pacific Economy*, 24(1): 127-142.

Index

A. Mukherjee (ed.), *Riverine Systems*, https://doi.org/10.1007/978-3-030-87067-6

H

I